10

Springer Series in Chemical Physics

Edited by Fritz Peter Schäfer

Springer Series in Chemical Physics

Editors: V. I. Goldanskii R. Gomer F. P. Schäfer J. P. Toennies

Volume 1 **Atomic Spectra and Radiative Transitions**
By I. I. Sobelman

Volume 2 **Surface Crystallography by LEED** Theory, Computation and Structural Results By M. A. Van Hove, S. Y. Tong

Volume 3 **Advances in Laser Chemistry** Editor: A. H. Zewail

Volume 4 **Picosecond Phenomena**
Editors: C. V. Shank, E. P. Ippen, S. L. Shapiro

Volume 5 **Laser Spectroscopy** Basis Concepts and Instrumentation
By W. Demtröder

Volume 6 **Laser-Induced Processes in Molecules** Physics and Chemistry
Editors: K. L. Kompa, S. D. Smith

Volume 7 **Excitation of Atoms and Broadening of Spectral Lines**
By I. I. Sobelman, L. A. Vainshtein, E. A. Yukov

Volume 8 **Spin Exchange** Principles and Applications in Chemistry and Biology
By Yu. N. Molin, K. M. Salikhov, K. I. Zamaraev

Volume 9 **Secondary Ion Mass Spectrometry SIMS II**
Editors: A. Benninghoven, C. A. Evans, Jr., R. A. Powell, R. Shimizu, H. A. Storms

Volume 10 **Lasers and Chemical Change**
By A. Ben-Shaul, Y. Haas, K. L. Kompa, R. D. Levine

Volume 11 **Liquid Crystals of One-and Two-Dimensional Order**
Editors: W. Helfrich, G. Heppke

Volume 12 **Gasdynamic Lasers** By S. A. Losev

Volume 13 **Atomic Many-Body Theory** By I. Lindgren, J. Morrison

Volume 14 **Picosecond Phenomena II**
Editors: R. Hochstrasser, W. Kaiser, C. V. Shank

Volume 15 **Vibrational Spectroscopy of Adsorbates** Editor: R. F. Willis

Volume 16 **Spectroscopy of Molecular Excitons**
By V. L. Broude, E. I. Rashba, E. F. Sheka

Volume 17 **Inelastic Ion-Surface Collisions**
Editors: W. Heiland, E. Taglauer

Volume 18 **Modelling of Chemical Reaction Systems**
Editors: K. H. Ebert et al.

A. Ben-Shaul Y. Haas K.L. Kompa
R.D. Levine

Lasers and Chemical Change

With 245 Figures

Springer-Verlag Berlin Heidelberg New York 1981

PHYS
Sep/ae

Professor Dr. Avinoam Ben-Shaul
Professor Dr. Yehuda Haas
Professor Dr. Raphael D. Levine

Department of Physical Chemistry, The Hebrew University of Jerusalem,
91904 Jerusalem, Israel

Professor Dr. Karl Ludwig Kompa

Max-Planck-Institut für Laserforschung,
D-8046 Garching, Fed. Rep. of Germany

Series Editors

Professor Vitalii I. Goldanskii

Institute of Chemical Physics
Academy of Sciences
Vorobyevskoye Chaussee 2-b
Moscow V-334, USSR

Professor Robert Gomer

The James Franck Institute
The University of Chicago
5640 Ellis Avenue
Chicago, IL 60637, USA

Professor Dr. Fritz Peter Schäfer

Max-Planck-Institut für
Biophysikalische Chemie
D-3400 Göttingen-Nikolausberg
Fed. Rep. of Germany

Professor Dr. J. Peter Toennies

Max-Planck-Institut für Strömungsforschung
Böttingerstraße 6–8
D-3400 Göttingen
Fed. Rep. of Germany

ISBN 3-540-10379-1 Springer-Verlag Berlin Heidelberg New York
ISBN 0-387-10379-1 Springer-Verlag New York Heidelberg Berlin

Library of Congress Cataloging in Publication Data. Main entry under title: Lasers and chemical change. (Springer series in chemical physics ; 10). Bibliography: p. Includes index. 1. Lasers in chemistry. 2. Photochemistry. 3. Chemical lasers. 4. Molecular structure. 5. Molecular spectra. I. Ben-Shaul, A. II. Series. QD63.L3L38 541.3'5 80-22277

Offset printing: Beltz Offsetdruck, 6944 Hemsbach/Bergstr. Bookbinding: J. Schäffer OHG, Grünstadt.
2153/3130-543210

6/15/81/(SD) 8B

הַזֹּרְעִים בְּדִמְעָה בְּרִנָּה יִקְצֹרוּ:

תהלים ,קכו:ה

Those who sow in tears shall reap in joy

(Psalms, 126 : 5)

Preface

Lasers and chemical change is the study of radiation and molecules in disequilibrium. The distinguishing feature of such systems is the extreme departure from thermal equilibrium: the radiation is usually confined to a narrow frequency range, is well collimated, and is far brighter than black body radiation; the chemical composition and also the distribution of molecules over their different energy states are often markedly displaced from that expected at equilibrium. Such systems can be used as a source of laser radiation and, reversedly, lasers can rapidly and selectively displace molecular systems from equilibrium. The subsequent evolution of the initially prepared state can then be monitored - again using lasers.

One purpose of this book is to introduce the concepts required to discuss systems of radiation and molecules in disequilibrium. These include the physics of (laser) radiation and of radiation-matter interaction and molecular structure and spectroscopy. Excellent textbooks of these topics are available and our survey (in Chap.3) is only intended to accent the essential points, with special reference to atomic and molecular radiation physics. Considerably more attention is given to the topic of disequilibrium in chemical systems (Chap.2). In particular we consider both inter- and intramolecular dynamics with special reference to energy requirements and energy disposal in chemical reactions and to what goes on in between - intramolecular energy migration. Disequilibrium in macroscopic systems and their temporal evolution is then discussed in terms of the underlying molecular events. The discussion throughout is in terms of a thermodynamiclike formulation motivated by information theoretic considerations and is illustrated by examples drawn from current studies. The principles of the design of such experiments, the experimental setups as used in practice, and the nature and interpretation of typical results are discussed in detail in Chaps.4 and 5.

Lasers and chemical change is primarily the study of the phenomena of interconversion of radiant and chemical energy. Exoergic chemical reactions can be employed to generate laser radiation and lasers can be used to induce and to interrogate chemical reactions. The first chapter is an introduction of these two broad classes of phenomena (which are then treated in more detail in Chaps.4 and 5). It also serves to motivate the need for the theoretical concepts introduced in Chaps.2 and 3. The first chapter is self-contained, but many details are glossed over. The basic phenomena and the essential interpretation can however be found there.

Large sections of Chaps.4 and 5 are devoted to case studies. Attention is given to the experimental arrangements with special reference to the more commonly used techniques. Chapters 4 and 5 describe the practice of chemical lasers and laser-induced chemistry, including the blending of experimental studies and theoretical interpretation.

The material in Chap.1 is suitable for inclusion in advanced undergraduate physical chemistry courses. The other chapters are for the graduate level. It is however our intention and hope that they will also prove useful to the specialist. The selection of topics and their relative emphasis reflect not only our judgment but also our expertise and research interests, the overall progress in the field, and the availability of other sources. In particular, laser-induced processes in condensed phases (solutions, matrices, mixed crystals) and in biological systems are not covered. We have not tried to be exhaustive and the list of references is by no means complete and is only meant to offer an access to the literature. We have tried to offer an integrated picture, to emphasize chemical reactions and chemical lasers, to draw attention to the complexities and the promises associated with the use of larger molecules, and to stress the theme of disequilibrium on both the microscopic and macroscopic levels.

The rapid progress in this field has been made possible by the elegant and probing experiments and the incisive interpretation carried out by many of our colleagues. Inspection of the contents of this volume demonstrates the magnitude of our debt to them. Our own work has been supported by the U.S. Air Force Office of Scientific Research, the U.S.-Israel Binational Science Foundation, the Israel Academy of Sciences and Humanities, the "Bundesministerium für Forschung und Technologie", and the "Max Planck Gesellschaft zur Förderung der Wissenschaften e.V.". We also wish to thank Dr. M.J. Berry for his involvement during the initial stages of this project and Dr. H.K.V. Lotsch of Springer-Verlag for his constant advice and for not giving up hope. Last but not least, we are very grateful to our families for their continuous encouragement and support.

Jerusalem, Israel, and
Garching, Fed. Rep. of Germany
January, 1981

A. Ben-Shaul
Y. Haas
K.L. Kompa
R.D. Levine

Contents

1. *Lasers and Chemical Change* 1

1.1 Light Amplification and Population Inversion in Chemical Processes 3
1.2 Molecular Rate Processes 13
1.3 Photoselective Chemistry 21
1.3.1 The Discrete Spectrum 22
1.3.2 The Quasicontinuum 29
1.3.3 Radiationless Transitions 37
1.3.4 Dissociative Continuum 40
1.3.5 Ionization 42
1.4 The Road Ahead 43

2. *Disequilibrium* 45

2.1 Specificity and Selectivity of Chemical Reactions 45
2.1.1 Overview: Microscopic Disequilibrium 45
2.1.2 The Detailed Rate Constant 49
2.1.3 Detailed Balance 51
2.1.4 Energy Disposal and Energy Consumption 53
2.1.5 The Reaction Probability Matrix 55
2.1.6 Measures of Specificity and Selectivity 58
2.1.7 The Maximum Entropy Formalism 62
2.2 Surprisal Analysis 65
2.2.1 The Prior Distribution 65
2.2.2 The Surprisal 68
2.2.3 Vibrational Surprisal 70
2.2.4 The Rotational State Distribution 74
2.2.5 Electronic Excitation 79

2.2.6 Polyatomic Molecules 80

2.2.7 Surprisal Analysis and Collision Dynamics 83

2.2.8 On the Role of Reagent Translation 87

2.3 Molecular Reaction Dynamics 89

2.3.1 Computational Studies 89

2.3.2 Potential Energy Surface(s) 92

2.3.3 Bond-Tightening Models 94

2.3.4 Kinematic Models of Collision Dynamics 97

2.3.5 Unimolecular Processes - The RRK Approach 99

2.3.6 Unimolecular Processes - Selectivity and Specificity 106

2.3.7 Preparing the Initial State 109

2.4 State-to-State Processes 113

2.4.1 The Prior Detailed Rate Constant 113

2.4.2 The Exponential Gap Representation 119

2.4.3 Reactive Collisions 122

2.4.4 The Adiabaticity Parameter 124

2.4.5 Polyatomic Molecules 129

2.4.6 Temperature Dependence 131

2.4.7 Electronic Energy Transfer 134

2.4.8 A Laser Bridge for the Exponential Gap 139

2.4.9 Intramolecular Electronic to Vibrational Energy Tansfer: Radiationless Transitions 141

2.5 Macroscopic Disequilibrium 144

2.5.1 The Master Equation: Relaxation of Harmonic Oscillators ... 144

2.5.2 Rotational Relaxation 150

2.5.3 Separation of Time Scales 152

2.5.4 Vibrational Anharmonicity and V-V Up-Pumping 154

2.5.5 Intermode V-V Transfer 159

2.5.6 From Macroscopic Relaxation to Microscopic Information 163

2.5.7 Thermodynamics of Molecular Disequilibrium 168

2.5.8 Laser Thermodynamics 172

Appendices

2.A. The Prior Distribution 177

2.B. Practical Surprisal Analysis 185

2.C. Statistical Models, Prior Distributions, and Collision Dynamics 187

2.D. Derivation of the Treanor Distribution 190

3. *Photons, Molecules, and Lasers* 192

3.1 Interaction of Molecules with Radiation 192
3.1.1 The Golden Rule 193
3.1.2 The Line Shape Function 201
3.1.3 Coherent Interaction 205
3.2 Essential Physics of Lasers 209
3.2.1 The Gain Coefficient 209
3.2.2 Laser Oscillators 211
3.2.3 Laser Radiation and Modes 216
3.2.4 The Laser Rate Equations 225
3.3 Survey of Atomic and Molecular Spectroscopy 232
3.3.1 Atomic Spectra 232
3.3.2 Molecular Spectra 235
3.3.3 Electronic Spectra of Diatomic Molecules 238
3.3.4 Infrared Spectra of Diatomic Molecules 242
3.3.5 Energy Levels of Polyatomic Molecules 246
3.3.6 Infrared Spectra of Polyatomic Molecules 257
3.3.7 The Near Ultraviolet Spectrum of Carbonyl Compounds 261
3.4 Laser Sources 266
3.4.1 Laser Specifications 267
3.4.2 Exciplex Lasers 272
3.4.3 Dye Lasers 278
3.4.4 CO_2 Lasers 284

4. *Chemical Lasers* 289

4.1 Survey of Chemical Lasers 289
4.2 Lasing Conditions in Chemical Lasers 295
4.3 Operation 304
4.3.1 Flash Photolysis: The Iodine Laser as a Model Case 305
4.3.2 Hydrogen Halide Chemical Lasers 314
4.3.3 The Chemical CO Laser 324
4.4 Chemical Laser Kinetics 326
4.4.1 The Rate Equations 327
4.4.2 Rotational Equilibrium 334

4.4.3 Rotational Nonequilibrium 337
4.4.4 cw Chemical Lasers 340
4.5 Some Applications of Chemical Lasers 347
4.5.1 Total Rate Constants, Kinetic Isotope Effects 348
4.5.2 Vibrational Population Ratios from Threshold Time Measurements 352
4.5.3 Gain Probing 354
4.5.4 Energy Transfer Measurements: An Example 357
4.5.5 An Industrial Diagnostic Application 358

5. *Laser Chemistry* 360

5.1 The Laser Evolution 361
5.2 Bimolecular Reactions 363
5.2.1 Molecular Beam Studies 365
5.2.2 Reactions in the Bulk 371
5.3 Electronic Excitation of Polyatomic Molecules 380
5.3.1 Direct Photodissociation: The A State of ICN 380
5.3.2 Photopredissociation: Formaldehyde 385
5.3.3 Excitation of Bound Electronic States: Biacetyl and Glyoxal 391
5.4 Multiphoton Activation and Fragmentation 398
5.4.1 The Nature of Multiphoton Excitation 401
5.4.2 The Rate Equation Approach 410
5.4.3 Nondissociative Reactions Induced by Multiphoton Absorption 427
5.4.4 Multiphoton Ionization (MPI) 430
5.5 The "Compleat" Laser Chemist 434
5.5.1 Preparing the Sample 435
5.5.2 Excitation and Probing Techniques 438
5.5.3 Laser-Oriented Absorption Measurements 446
5.6 From the Laboratory to Large-Scale Laser Chemistry 452
5.6.1 Practical Photoselective Chemistry 454
5.6.2 State Selective Chemistry 457
5.7 Synergism 459

References 461
Author Index 477
Subject Index 485

1. Lasers and Chemical Change

The dictionary defines laser as a device for the amplification of light. Clearly there is more to it since the acronym itself stands for light amplification by stimulated emission of radiation. Moreover there must be some source for the enhanced light energy coming out. In a chemical laser this energy is provided by a chemical reaction. Chemical lasers combine the physical mechanism of lasing with the versatility and scope of chemical kinetics and spectroscopy. They are the present-day analogues of the electrochemical cells and history again repeated. In addition to their many practical applications chemical lasers have also led to considerable progress in our understanding of the fundamentals of chemical reactions and of bulk systems in disequilibrium.

Electrochemical cells can be used to generate energy or, reversedly, chemical reactions can occur upon supply of electrical energy. Here too, laser radiation can be used to drive chemical reactions. Photochemists have long been using visible or UV light to promote reactions which do not take place upon heating [1.1,2]. Upon absorption of such light molecules are promoted to excited electronic states. Chemical reactions mediated by such different potential energy surfaces can proceed along pathways not easily traversed on the ground potential energy surface [1.3,4] (which is typically the only one accessible to thermal reactants). Laser pumping differs from conventional photochemistry in several essential ways, made possible by the high power and the sharply defined frequency available from laser sources. First and foremost, laser chemistry involves not only selectivity due to excitation to different potential energy surfaces but also selectivity on a given potential energy surface. It was long realized that chemical reactivity is dependent on the symmetry and other constraints implied by the electronic state [1.3,4]. The new feature is selectivity, i.e., the presence of constraints on a given potential energy surface. Total available energy, even when restricted to one potential energy surface, may not be the only factor determining reactivity. It may also depend on the distribution of the energy among the different degrees of motion of the nuclei [1.5,6]. Chemical lasers are possible because of this selectivity and, in turn, lasers can be employed to pump molecules to very specific initial states. Their high power, monochromaticity, and short pulse durations enable one to prepare a high concentration of molecules in a sharply defined initial state (corresponding to a particular division of the total energy among the electronic and nuclear degrees of freedom). This is the prerequisite for achieving state-selective chemistry. The initially prepared state is not necessarily stationary, and may evolve even in the absence of collisions. Given sufficient energy it may eventually dissociate. In practice, the creation of nonstationary states is the rule rather than the exception in the case of polyatomic molecules. The subsequent redistribution of energy in the pumped molecules is a major issue in laser chemistry studies. At higher pressures collisions may engage the excited molecules in bimolecular processes. The selective and rapid preparation of specific initial states is not the only application of lasers in chemical dynamics research. Lasers are also extensively used to monitor the time evolution and energy distribution of chemical processes.

In this chapter we consider the essential concepts of chemistry and physics necessary for the understanding of the operation principles of a chemical laser or of a laser-pumped molecular process. What is light amplification by stimulated emission and how does a chemical reaction sustain such a process? Why are chemical lasers so intimately related to bulk systems in disequilibrium and how do they depend in their operation (and, in turn, provide information) on the energy disposal and energy requirements of chemical reactions? How can gaseous systems be displaced from molecular equilibrium, what relaxation pathways restore equilibrium, and on what time scales? What is selectivity in a chemical reaction and how do we know that there are constraints? How can polyatomic molecules be pumped up selectively by lasers and why does intramolecular energy migration affect the prospects for laser-induced chemistry? How do we describe rate processes in systems in molecular disequilibrium and what can we learn from equilibrium considerations? Of course, the answers provided in this chapter will be sketchy, at best, and the detailed story will only unfold in the subsequent chapters. What we propose to find out in this chapter is what basic phenomena of chemical interest are possible for a system of molecules and radiation in disequilibrium and what is it that we need to know to describe them.

In this chapter there will occur terms not yet fully discussed, to which the readers should give their colloquial meaning. Subsequent chapters will provide more detail and, in particular, will also consider the experimental arrangements which made the experiments possible. Only then can we turn to such questions as what can be measured, what we would like to measure, and what have we learned thus far about the generation of laser radiation, its acquisition by molecules, and its disposal by diverse molecular processes.

1.1 Light Amplification and Population Inversion in Chemical Processes

The amplification of light upon passage through matter is not the normal state of affairs. Consider a dilute atomic or molecular gas, at or near thermal equilibrium. A beam of light of a suitable frequency will be absorbed, i.e., the intensity of the beam will decrease along its path through the gas. It is however sometimes possible to bring (to "pump") the system to a nonequilibrium situation where the intensity of the incident light beam is enhanced as it propagates (the additional energy of the beam being provided by the pumping process). In chemical lasers the required nonequilibrium situation is often maintained by a specific disposal of the energy of an exoergic chemical reaction. As an example, Fig.1.1 contrasts the distribution of vibrational energy in DF molecules in a gas at thermal equilibrium with the distribution in newly formed DF molecules produced in the

$$F + D_2 \rightarrow D + DF \tag{1.1.1}$$

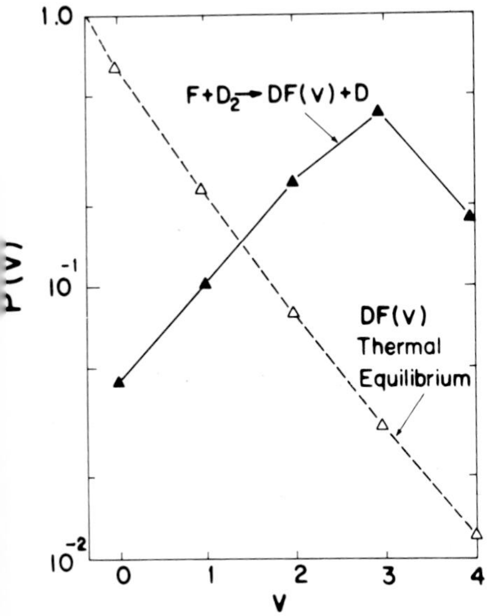

Fig. 1.1. Distribution of vibrational states in DF under equilibrium and non-equilibrium conditions in the gase phase. (△) DF in equilibrium at 4000 K. Even at this high temperature the population of the higher vibrational states is negligible. (▲) DF as produced in the $F + D_2$ reaction (at 300 K) before any subsequent vibrational relaxation. ((▲) adapted from [1.7])

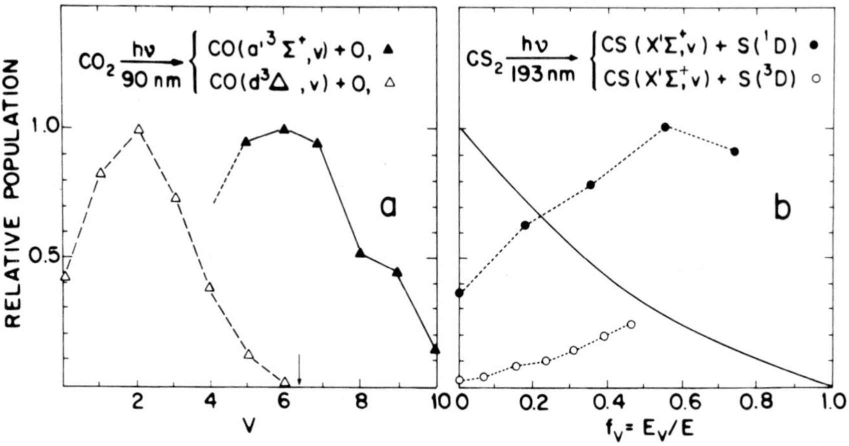

Fig. 1.2a,b. Specific energy disposal in photodissociation. (a) The high-energy photolysis produces CO in two different excited electronic states. The maximal energy available for vibrational excitation is indicated (by an arrow) for one case. It is given by $h\nu$-D-E* where D is the bond dissociation energy of CO_2 to produce electronic ground state CO and E* is the electronic excitation of the CO. A large fraction of the available energy is channelled into CO vibration. The balance is partitioned among CO rotation and the relative translation of CO and O. (Experimental data from [1.8]). (b) Vibrational distribution of the CS fragment produced in laser photodissociation of CS_2, plotted vs $f_V = E_V/E$, the fraction of the available energy in the vibration. Here $E = h\nu$-D-E*, where E* is the electronic excitation of the S atom. The solid curve describes the distribution if E is equipartitioned among the vibration, rotation and relative translation of the products. For high vibrational states the signal (of laser-induced fluorescence, Sect.5.5.2) is too weak to be detected (see also Fig.2.22). (Adapted from [1.9])

reaction. The two distributions are seen to be qualitatively different. Similarly, collisions of excited molecules can lead to a specific channelling of the available energy. Thus, the vibrationally excited DF molecules [produced, say, by reaction (1.1.1)] will, upon collision with CO_2, transfer their energy preferentially to one vibrational mode, namely to the asymmetric stretching vibration of the CO_2. Energy-rich molecules, produced by light absorption or otherwise, will also often show nonequilibrium partitioning of the excess energy in the nascent products. Figure 1.2 shows the energy disposal for two photodissociation processes,

$$CO_2 \xrightarrow{h\nu} \begin{cases} O + CO(a'\,^3\Sigma^+, v) \\ O + CO(d^3\Delta, v) \end{cases} \qquad (1.1.2)$$

and

$$CS_2 \xrightarrow{h\nu} \begin{cases} CS(v) + S(^1D) \\ CS(v) + S(^3P) \end{cases} , \qquad (1.1.3)$$

where in (1.1.2) the CO molecule is formed in two different excited electronic states and in (1.1.3) the S atom is formed in two different electronic states.

The CO_2 photodissociation was induced by flashlamp excitation and the products' vibrational distributions were determined by monitoring the fluorescence to the ground electronic state. The short lifetimes of electronically excited species allows their detection prior to collisional deactivation. In (1.1.3) the CS molecules are formed in the ground electronic state and measuring their vibrational distribution by fluorescence is harder due to the long radiative lifetimes of vibrational transitions in the same electronic manifold. Therefore the CS vibrational distributions were determined by an alternative method, laser-induced fluorescence to be described in detail in Chap.5. In this particular experiment the excitation source was also a laser (an excimer laser, Sect.3.4).

The energy available for vibrational excitation of the photodissociation products is $h\nu$-D-E*, where $h\nu$ is the energy of the exciting photon, D is the bond dissociation energy, and E* is the electronic excitation of the products. The energy not present as vibrational excitation appears in the rotation of the molecule (CO,CS) or in the relative translation of the atom and the diatomic molecule. If the available energy was equipartitioned among the vibration, rotation and translation, the vibrational distribution would appear as in the solid line in Fig.1.2b (further discussed in Sect.2.2). The extreme deviation of the actual distribution from this behavior reflects the very specific energy disposal in the dissociation process.

Chemical lasers are feasible due to the specific release of the available chemical energy in the nascent products. How specific need this release be; what is the essential characteristic of the nonequilibrium situation where a light beam will be amplified upon passage through the system?

Absorption of light of frequency ν corresponds to excitation of molecules from some initial internal state i to a final internal state f at an energy $h\nu$ higher (h is Planck's constant). At thermal equilibrium the populations (number of molecules per unit volume) of the two states are related by

$$(N_f/N_i)_{eq} = \exp(-h\nu/kT) \quad , \qquad (1.1.4)$$

where k is Boltzmann's constant. The lower energy states are more heavily populated and light will be absorbed. The very act of light absorption increases

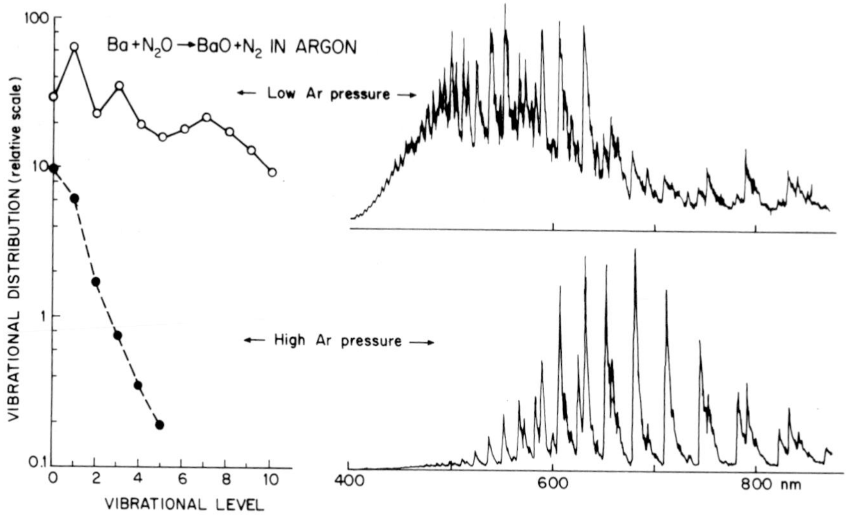

Fig. 1.3. (*Right*) The spectra of the emitted visible light (the "flame") when Ba atoms, in an inert carrier gas, react with N_2O. At the higher carrier gas pressure most of the emission is from the process $BaO(A^1\Sigma) \rightarrow BaO(X^1\Sigma) + h\nu$. (*Left*) The vibrational distribution of the $BaO(A^1\Sigma)$ molecules, as determined by their emission, at a low and a high carrier gas pressure. The important role of collisions (with the carrier gas molecules, which are more frequent at higher pressure) is considered in Sect.1.2. (Adapted from [1.10])

the population of the higher state f and displaces the system from equilibrium. One mechanism for the return to equilibrium which is always available is the emission of light of frequency ν during a transition from f back to i[1]. At most, all the absorbed photons will be emitted and no amplification will result.

Consider now the case when we pump molecules into the state f by some subsidiary mechanism, say a chemical reaction with a specific energy release such that f is one of the preferentially populated states. We could recover some of the pumping energy through the spontaneous emission of photons from the state f, i.e., through the so-called "chemiluminescence". Figure 1.3 shows the spectrum of the emitted (visible) light from the reaction of $Ba + N_2O$ which produces electronically excited BaO. Also shown is the vibrational distribution of the BaO molecules in the (electronically excited) $A^1\Sigma$ state.

1 There usually are additional pathways. Molecules may lose excess energy by collisions, or through photochemical processes, or by emission (fluorescence) at a lower frequency from the state f to some intermediate state, which is higher in energy content than the state i.

Having displaced the equilibrium so that one (or more) excited state(s) is preferentially populated we need to determine the rate of depopulation by photon emission.

The radiative transition $f \to i$ can take place in two ways. One is a spontaneous (or "unimolecular") photon emission with a rate $A_{fi}N_f$. Here A_{fi} is a "first-order" rate constant such that the lifetime of the state f, due to spontaneous emission, is $1/A_{fi}$. There is also a "second-order" mechanism for photon emission whose rate is proportional to both the density[2] ρ of the radiation present and to the concentration (i.e., number density) of the state f. The rate of this so-called "stimulated emission" is then $B_{fi}\rho N_f$. Here B_{fi} is a second-order rate constant. The combined rate of emission is $k_{fi}N_f$,

$$k_{fi} = A_{fi} + B_{fi}\rho \quad . \tag{1.1.5}$$

Of course, when photons (at the frequency ν) are present they will be partially absorbed, due to $i \to f$ transitions, at the rate $k_{if}N_i$,

$$k_{if} = B_{if}\rho \quad . \tag{1.1.6}$$

Absorption, like stimulated emission, is a second-order process whose rate is proportional to the density of the photons. The two rate coefficients are known as the Einstein A and B coefficients, following his original [1.11] derivation of (1.1.14) below.

When an external light beam is incident on the system, those photons emitted through stimulated emission have the same phase as the external radiation. These photons therefore reinforce the stimulating beam; the radiation emitted via stimulation is coherent with the beam. The spontaneously emitted photons are not coherent and hence will not contribute to the intensity of the stimulating radiation; the spontaneous radiation will be emitted into all spatial directions as it need not be collimated in the direction of the external beam.

In order to achieve amplification of the external light beam it is thus not sufficient to achieve a net excess of emission over absorption. The mere introduction of some external pumping mechanism which raises the concentration N_f of state f to such a level that

$$k_{fi}N_f > k_{if}N_i \tag{1.1.7}$$

---

2 $\rho = \rho(\nu)$ is the radiation energy per unit volume and unit frequency ($erg \cdot s \cdot cm^{-3}$), i.e., $\rho(\nu)d\nu$ is the energy per unit volume due to photons in the frequency interval $\nu - \nu + d\nu$.

is not enough. Condition (1.1.7) is the requirement of net emission of photons. To achieve amplification we need that the rate of stimulated emission will exceed the rate of absorption, $\rho B_{fi}N_f > k_{if}N_i$, or

$$B_{fi}N_f > B_{if}N_i \quad . \tag{1.1.8}$$

Since A_{fi} is positive, the amplification condition (1.1.8) is more stringent than the luminescence condition (1.1.7), which can be written as

$$[B_{fi}+(A_{fi}/\rho)]N_f > B_{if}N_i \quad . \tag{1.1.9}$$

In order to obtain a quantitative comparison and to express (1.1.8) as an explicit condition on the populations, we need to determine the ratios $B_{fi}\rho/A_{fi}$ and B_{fi}/B_{if}. We shall do so by a method which will have many other applications and which is a particular example of the principle of detailed balance.

In a real system there may be many mechanisms for the transfer of systems from state i to state f and vice versa. The absorption and emission of photons of frequency ν is just one of those possible mechanisms. At equilibrium the overall rate of transfer from i to f is necessarily equal to the reverse rate. The principle of detailed balance states that at equilibrium not only are the overall forward and reverse rates equal but the same applies to each particular mechanism. At equilibrium the rate of $i \to f$ transitions by photon absorption equals the rate of $f \to i$ transitions by photon emission, irrespective of any other processes which can take place.

Consider now our system at equilibrium with radiation of density ρ at the temperature T. Then, by detailed balance $k_{fi}N_f = k_{if}N_i$ or

$$(k_{if}/k_{fi}) = (N_f/N_i)_{eq} = \exp(-h\nu/kT) \quad . \tag{1.1.10}$$

Substituting into (1.1.6) and (1.1.5)

$$B_{if}\rho(\nu) = \exp(-h\nu/kT)[A_{fi}+B_{fi}\rho(\nu)] \quad . \tag{1.1.11}$$

Solving for the equilibrium density $\rho(\nu)$ of the radiation, we obtain

$$\rho(\nu) = A_{fi}/[B_{if}\exp(h\nu/kT)-B_{fi}] \quad . \tag{1.1.12}$$

The equilibrium density of radiation has been determined separately by Max Planck to be

$$\rho(\nu) = (8\pi h\nu^3/c^3)/[\exp(h\nu/kT)-1] \quad . \tag{1.1.13}$$

Comparing (1.1.12) and (1.1.13)

$$B_{if} = B_{fi} = (c^3/8\pi h\nu^3)A_{fi} \quad . \tag{1.1.14}$$

These results, first derived in essentially this way by Einstein [1.11], relate the three rates. They show that under equilibrium conditions stimulated emission is of negligible importance compared to spontaneous emission,

$$[B_{fi}\rho(\nu)/A_{fi}]_{eq} = c^3\rho(\nu)/8\pi h\nu^3 = 1/[\exp(h\nu/kT)-1] \quad . \tag{1.1.15}$$

Only at very low frequencies $h\nu \sim kT$ do the two emission processes have comparable rates. At even lower frequencies, $h\nu << kT$, the ratio can be approximated by $kT/h\nu$. Under ordinary conditions, the lifetime for photon emission will be essentially that of spontaneous emission $\tau = (1/A_{fi}) \propto \nu^3$. The lifetime for spontaneous emission will be much shorter (typically 10^{-8} s) for transitions in the visible region (i.e., changes in the electronic state) as compared to transitions in the infrared (as slow as 10^{-1} s for the far IR) which are typically associated with vibrotational and pure rotational changes. Explicity, quantum mechanical analysis provides that

$$A_{fi} = (64\pi^4/3hc^3)|\underline{\mu}_{fi}|^2\nu^3 \quad , \tag{1.1.16}$$

where μ_{fi} is the "transition dipole moment" for the $i \to f$ transition[3,4]. More-

---

3 In practical units $A_{fi} \simeq 10^{-38}\nu^3\mu^2$ s^{-1} with ν in s^{-1} and $\mu_{fi} = |\underline{\mu}_{fi}|$ in Debyes (1 Debye = 10^{-18} esu·cm). For $\mu_{fi} \simeq 0.1$D and $\nu \simeq 10^{14}$ s^{-1} (corresponding approximately to the $v = 1 \to 0$ transition in HF) $\tau = A^{-1} \simeq 10^{-2}$ s. Note the difference between the transition dipole μ_{if} ($f \neq i$) and the permanent dipole moment μ_{ii} (for a molecule in state i); e.g., the vibrational transition dipole in HF is $\mu_{10} \simeq 9.85 \cdot 10^{-2}$ Debye whereas $\mu_{ii} \simeq 1.82$ Debye.

4 The magnitude of the transition dipole moment determines the rate of spontaneous emission, stimulated emission, and absorption; cf. (1.1.14) and (1.1.16). Equivalently, $\mu_{if} = |\underline{\mu}_{if}|$ measures the response of the molecular system to the radiation field (see Sect.3.1 for more details). Transitions with negligible μ_{if} (e.g., the "overtone", $v = 0 \to 2$, transition of HF, $\mu_{02} \simeq 10^{-2}$ Debye) are termed (nearly) forbidden while those with large μ_{if} (e.g., the $v = 0 \to 1$, transition, $\mu_{01} \simeq 0.1$ Debye) are termed (strongly) allowed. Based on the classical description of the interaction between atoms and radiation, where the electron is regarded as an oscillating dipole, the distinction between allowed and forbidden transitions is sometimes expressed in terms of the "oscillator strength" of the transition [1.12,13]. $f_{fi} \simeq (8\pi^2 m\nu/3he^2)\mu^2_{fi}$ where m and e are the electron mass and charge, respectively. Strongly allowed (forbidden) transitions are said to carry (not to carry) oscillator strength.

over, since $B_{fi} = B_{if}$ we can never hope to achieve amplification [i.e., (1.1.8)] merely by using an external light beam with a high (nonequilibrium) photon density at the frequency ν. No matter how high $\rho(\nu)$ is, the most that we can achieve is $N_f = N_i$ [when $B_{fi} >> (A_{fi}/\rho)$; cf. (1.1.9)].

The condition for amplification (1.1.8), combined with the relation (1.1.14) between the Einstein coefficients, can be now expressed as the condition

$$N_f > N_i \tag{1.1.17}$$

on the populations of the two states. Amplification requires that the population of the upper state exceeds that of the lower state. This represents such an extreme deviation from the common, equilibrium, situation that it is characterized by a special term "population inversion". The normal order of occupancy is reversed. By way of comparison, the condition for net emission of light (1.1.9) (due to auxiliary pumping, chemical or otherwise) can be expressed, using (1.1.14), as

$$N_f/N_i > (N_f/N_i)_{eq} = \exp(-h\nu/kT) \quad . \tag{1.1.18}$$

For most transitions, (1.1.17) is much harder to realize than (1.1.18), unless ν is quite low or T is high. Thus, while any preferential pumping will suffice to disturb the equilibrium in the direction required by (1.1.18), amplification (i.e., population inversion) imposes special conditions on the pumping process. It needs to maintain a strict excess ($N_f > N_i$) of systems in the upper state f.

Thus far we have assumed that there is just a single quantum state f at the energy $h\nu$ above the state i. If there are several, say g_m, such states of the same energy we shall refer to the "level" m with a degeneracy g_m. In this case (1.1.14) becomes

$$g_m B_{mn} = g_n B_{nm} \tag{1.1.19}$$

and the condition for amplification now reads

$$(N_m/g_m) > (N_n/g_n) \quad . \tag{1.1.20}$$

By way of contrast, at equilibrium

$$(N_m/g_m)_{eq} = (N_n/g_n)_{eq} \exp[-(E_m - E_n)/kT] \quad . \tag{1.1.21}$$

All lasers operate on the basis of some pumping mechanism which can bring the system to a state where the amplification condition (1.1.20) obtains. In Chap.4 we shall be concerned with exoergic ("energy releasing") chemical processes which are sufficiently specific to create a population inversion. Section 4.1 lists many of the different possible processes which have been used to pump chemical lasers [1.14]. They include the following examples: a) photodissociation

$$RI \xrightarrow{h\nu} R + I(5^2P_{1/2}) \quad , \quad (R=CH_3, CF_3, C_3F_7 \dots) \tag{1.1.22}$$

where lasing occurs at the iodine $5^2P_{1/2} \rightarrow 5^2P_{3/2}$ electronic transition (Sects. 3.3.3 and 4.3.1). This was the pumping mechanism of the very first chemical laser [1.15]. The energy delivered by the laser is here provided by the ultraviolet light used to photodissociate RI. A similar source applies in b) photoelimination [1.16],

$$CH_2 = CHCl \xrightarrow{h\nu} CH \equiv CH + HCl^{\dagger} \quad . \tag{1.1.23}$$

Here, the HCl diatomic product carries a fraction of the excess ($h\nu$-D, where D is the minimal energy for dissociation) energy as vibrotational excitation. Note that such energy-rich polyatomics can also be produced by c) chemical activation [1.17],

$$\begin{aligned} &CH_3 + CF_3 \rightarrow CH_3CF_3^{\dagger} \\ &CH_3CF_3^{\dagger} \rightarrow CH_2 = CF_2 + HF^{\dagger} \quad . \end{aligned} \tag{1.1.24}$$

Here some of the energy of the newly formed C-C bond (~100 kcal/mole) appears as vibrotational excitation of HF. The most common type of chemical lasers obtains when d) a weaker chemical bond is replaced by a stronger one as in (1.1.1) [1.14], or Fig.1.4)

$$Cl + HI \rightarrow I + HCl^{\dagger} \quad . \tag{1.1.25}$$

In all four cases the lasing transitions are at frequencies in the infrared. This of course will always be the case when the transitions occur between different vibrotational levels (of the same electronic state). To achieve chemical laser action in the visible or UV regions of the spectrum it is necessary to preferentially produce electronically excited reaction products.

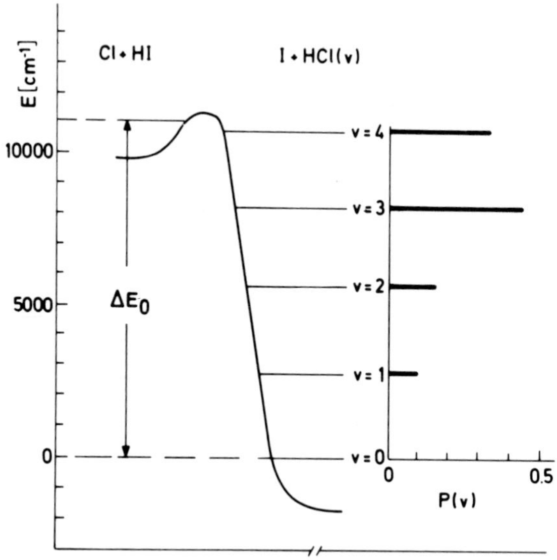

Fig.1.4. Schematic illustration of the energetics and vibrational energy disposal of the Cl + HI → I + HCl reaction. ΔE_0 is the difference between the ground vibrotational states of HI and HCl (the "zero-point to zero-point" exoergicity). Shown also are the energies of the lower vibrational states of HCl and their nascent populations. A significant fraction of the available exoergicity is channelled into vibrational excitation of HCl. (Experimental data from [1.18])

The observation that for elementary chemical reactions specific energy disposal is often the rule is a fairly recent one. It reflects our current ability to analyze the nascent collision products, immediately after they emerge from their encounter. When chemical reactions are studied in the bulk, the nascent products undergo secondary collisions with the other gas molecules. The excess internal energy is redistributed by these energy-transfer collisions until it is equilibrated between all the molecules of the gas. The reaction exoergicity will thus ultimately appear as heat. Under ordinary conditions this collisional degradation of the initially specific energy disposal is the rule. It is the major mechanism for the return to equilibrium, successfully competing with the relaxation via photon emission. Knowledge of the initial energy partitioning and the magnitude of the spontaneous and stimulated emission rates is not sufficient for operating a laser. To achieve that, or even a chemiluminescence experiment, we must characterize the competition between the radiative and collisional relaxation.

1.2 Molecular Rate Processes

The operation of the chemical laser is governed by the disequilibrium populations of the internal energy states. These in turn are determined by two factors, the primary distribution as determined by the pumping mechanism (e.g., the chemical reaction) and the diverse collisional and radiative transfer processes which populate and deplete the different levels. We can explore the role of such processes if we initiate the chemical reaction rapidly and then probe the temporal evolution of the populations. What we need to determine is the importance of collisions between the lasing species and other molecules in the gas, on the time scale of the photon emission.

As a concrete example, we consider the operation of the $F+H_2$ laser in a large excess of buffer gas at a moderate pressure (say, 50 Torr). A mixture of argon, H_2, and a source of fluorine atoms (say, CF_3I) is rapidly flash photolyzed (Fig.1.5). Vibrationally excited HF molecules are formed by the reaction sequence

$$\begin{aligned} &CF_3I \xrightarrow{h\nu} F + CF_2I \\ &F + H_2 \rightarrow H + HF^\dagger \quad . \end{aligned} \tag{1.2.1}$$

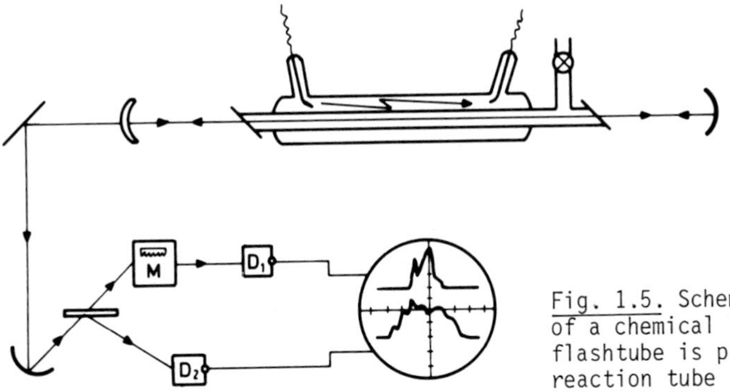

Fig. 1.5. Schematic illustration of a chemical laser apparatus. A flashtube is placed around the reaction tube (the vessel enclosed by the two angled windows). The laser "cavity" is bounded by the two curved mirrors. Upon repeated passages between the two mirrors, the initially spontaneous radiation increases in its intensity until a given transition reaches the lasing threshold condition. The stimulated emission is partially coupled out of the cavity (via the mirror on the left) and is detected either by D_1 after passing through a monochromator (recording the emission intensity of a given line) or without dispersion by D_2, yielding total emission intensity. The output of both detectors can be displayed (as a function of time). The temporal evolution of the individual lines, as monitored by D_1, is shown in Fig.1.6

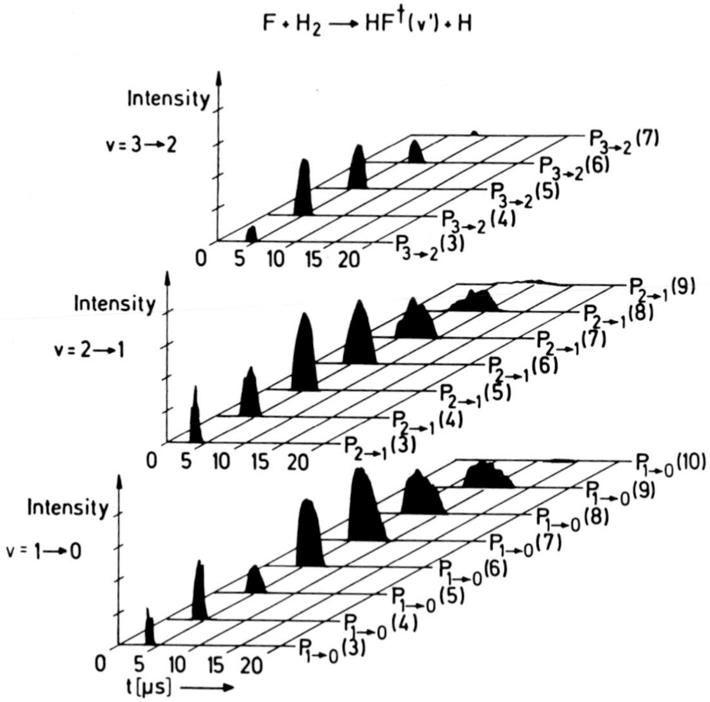

Fig. 1.6. Time-resolved pulse patterns in the $F+H_2$ laser. $P_{v\to v-1}(J)$ denotes the $v,J-1\to v-1,J$ transition (for experimental details, see [1.7]). Note the "J-shifting" phenomenon typical to lasers operating with rotationally equilibrated populations (Sects.4.2 and 4.4). The lasing within each vibrational band $v\to v-1$ is gradually shifted towards higher J values. (Adapted from [1.19])

When properly operated, the $F+H_2$ laser yields one to two coherent photons from each HF molecule formed in the pumping reaction. The detailed temporal evolution of the different transitions is shown in Fig.1.6. Laser emission is seen to start a few microseconds following flash initiation and the detectable emission is over within twenty microseconds. What collisional relaxation processes could effectively compete on this time scale? The buffer gas is in a large excess and hence most collisions are of the type

$$HF(v,J) + Ar \rightleftarrows HF(v',J') + Ar \quad . \tag{1.2.2}$$

We explicitly indicate the vibrotational state of the HF molecule to show that it may change as a result of the collision. How frequent are those collisions and how likely is a change of the internal state upon collision?

Estimates of the frequency of collisions in the gas phase are readily provided by the so-called "transport" phenomena which are governed by the mean free path λ traversed by the molecule between two successive collisions. If the speed of the molecule is $\underline{v}$, the number of collisions per unit time (the collision frequency) is $\omega = \underline{v}/\lambda$. In the dilute gas phase where the collisions are binary, involving two molecules at a time, we expect the collision frequency to be proportional to the (number) density ρ of the gas, $\omega \propto \rho$. The proportionality constant k,

$$\omega = \underline{v}/\lambda = k\rho \quad , \tag{1.2.3}$$

is the collision rate constant. We can express the same idea by considering the probability that the molecule will undergo a collision when it covers a short interval Δx during its otherwise free flight. By definition of the mean free path, this probability is $\Delta x/\lambda$, while for binary collisions $(\Delta x/\lambda) \propto \rho \Delta x$. The proportionality constant is known as the collision cross section, σ,

$$\lambda = (\rho\sigma)^{-1} \tag{1.2.4}$$

$$k = \underline{v}\sigma \quad . \tag{1.2.5}$$

The cross section (with dimensions of area) can be thought of as the effectived area presented by the molecule as it moves through the gas. It can be measured by considering the attenuation of a beam of molecules upon passage through a (dilute) gas. Even when only the average velocity of the molecules is known, it is customary to define an "effective" collision (or gas kinetic) cross section σ_{eff} by $\sigma_{eff} = k/\langle \underline{v} \rangle$, where the angular brackets denote an average[5], and $\langle \underline{v} \rangle = (8kT/\pi\mu)^{\frac{1}{2}}$.

In practical units $\omega = k(P/RT)$ or

$$P\tau = RT/k \quad , \tag{1.2.6}$$

where $\rho = P/RT$ is the density (in moles), k is the collision rate constant, and $\tau = \omega^{-1}$ is the time between successive collisions[6].

---

5 In practical units $k \simeq 8.76 \times 10^{11} (T/\mu)^{\frac{1}{2}} \sigma$ $cm^3 mole^{-1} s^{-1}$ where μ is the (reduced) mass of the colliding molecules in chemical mass units (^{12}C=12) and σ is in square angstroms. 1 Å = 10^{-8} cm.

6 $k(cm^3 mole^{-1} s^{-1}) = 2.445 \times 10^4 (T/298)/P\tau$ where $P\tau$ is in atm·s. 1 atm = 760 Torr. 1 Torr = 1 mm or mercury = 1.33×10^3 dyne·cm^{-2}. R = 82.06 cm^3atm·$mol^{-1}deg^{-1}$.

For HF in argon[7] at 50 Torr, taking an effective collision cross section of 30 $Å^2$, $\tau = RT/P \langle \underline{v} \rangle \sigma$ is about 10^{-9} s. An HF molecule suffers about 10^3 collisions per microsecond. The result (Fig.1.6) that laser emission is observed on a similar time scale implies that it takes on the average more than 10^3 collisions to bring a vibrationally hot HF molecule to thermal equilibrium.

Our first clue concerning the magnitude of energy transfer rate constants is again provided by the principle of detailed balance. Equating the equilibrium rates of the forward and backward processes in (1.2.2), we have

$$k(v',J'\to v,J)/k(v,J\to v',J') = (N_{v,J}/N_{v',J'})_{eq}$$

$$= [(2J+1)/(2J'+1)]\exp[-(E_{v,J}-E_{v',J'})/kT] \quad . \qquad (1.2.7)$$

Here $g_J = (2J+1)$ is the degeneracy (number of states of equal energy) of the level of rotational quantum number J. $E_{v,J}$ is the energy of the vibrotational level. For the purpose of a qualitative discussion it is often sufficient to use the rigid-rotor-harmonic-oscillator (RRHO) level scheme[8], $E_{v,J} = E_v + E_J$,

$$E_v = hcG(v) = hc\omega_e v \quad , \qquad (1.2.8)$$

$$E_J = hcF(J) = hcB_eJ(J+1) \quad . \qquad (1.2.9)$$

Here ω_e and B_e are the spectroscopic vibrational and rotational constants. Typically, ω_e is some two to three orders of magnitude larger than B_e, so that the spacing between adjacent vibrational levels far exceeds the rotational spacing.

If in the statement of detailed balance, (1.2.7), we take $v' > v$, say $v' = v+1$, then the exponent will contain the contribution $\exp(-hc\omega_e/kT)$ which, at ordinary temperatures, will be quite small. For changes of vibrational state the activation rate constant ($v'>v$) is usually much smaller compared to the deactivation rate. The same need not apply for rotations. Due to the small rotational spacing the exponential term will have a far less dominant role and, moreover, the degeneracy factors favor the higher J states. We can anticipate therefore some qualitative differences between the rates of the vibrational and rotational relaxation of HF (ω_e=4138.5 cm^{-1}, B_e=20.94 cm^{-1}).

---

7 The reduced mass μ is $\mu = 40.0 \times 20.0/(40.0+20.0) = 13.3$.

8 In (1.2.8) the vibrational energies E_v are measured from the ground level, i.e., we set $E_{v=0} = 0$.

In the collision between HF and Ar any energy removed out of (or deposited into) the vibrotational manifold of HF is provided by the relative[9] translational motion of Ar and HF. We can speak of the translational energy defect as the amount of energy provided by (or removed into) the translational motion. As a general rule, the larger is the translational energy defect the less efficient is the transfer process[10].

An illustration of the role of the energy defect is provided by the so-called V-V transfer processes, e.g.,

$$CO(v-1) + CO(v=1) \rightarrow CO(v) + CO(v=0) \quad . \tag{1.2.10}$$

In the harmonic approximation for the vibrational energy levels, the energy defect is zero. When anharmonicity (Sect.3.3) is taken into account, so that (1.2.8) is replaced by[11]

$$G(v) = \omega_e v - \omega_e x_e v(v+1) \quad ; \tag{1.2.11}$$

the energy defect $-\Delta E = hc[G(v-1)+G(1)-G(v)-G(0)] = 2hc\omega_e x_e(v-1)$ increases with v. The probability of the V-V transfer (per collision) is observed to be a decreasing function of $-\Delta E$ (Fig.1.7).

It is customary to express the efficiency of the different energy transfer processes in terms of their "collision number" Z. It is defined as the number of collisions required, on the average, to induce the particular transition. Thus, if k is the collision rate constant (1.2.3), then for (1.2.2)

$$Z(v,J\rightarrow v',J') = k(v,J\rightarrow v',J')/k \quad . \tag{1.2.12}$$

We can also think of the inverse of the collision number, Z^{-1}, as the probability that the particular change will occur in a single collision.

As a very rough rule of thumb, collision numbers for deactivation decrease exponentially with the translational energy defect. Thus, for vibrational deactivation of HF at 600 K, $\exp(-hc\omega_e/kT) = 5\cdot 10^{-5}$ and so we expect that more

---

9 It has to be the relative motion since the energy of the center of mass of the colliding pair cannot change during their encounter.

10 This is the collision theoretic analogue of the Franck-Condon principle. Large changes in the (relative) kinetic energy (more properly, in the momentum) of the colliding partners are disfavored.

11 In (1.2.11) as in (1.2.8) we use $G(0) = 0$.

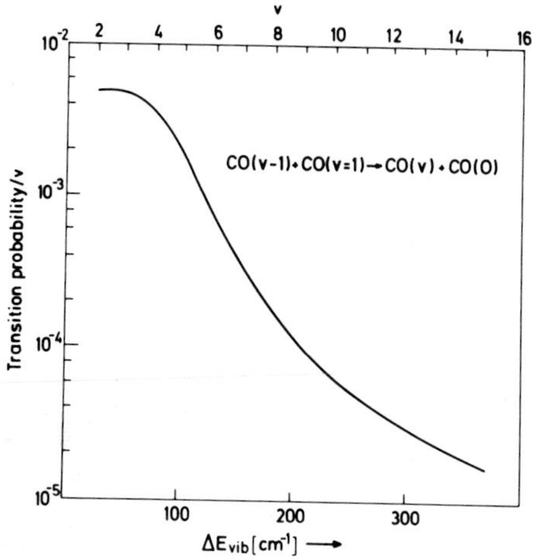

Fig. 1.7. Reduced probabilities (at T=300 K) for the exoergic V→V process $CO(v-1) + CO(1) \to CO(v) + CO(0)$ vs the energy defect (adapted from [1.20]). A qualitatively similar behavior is found for other V→V processes, e.g., $CO(v-1) + N_2(1) \to CO(v) + N_2(0)$

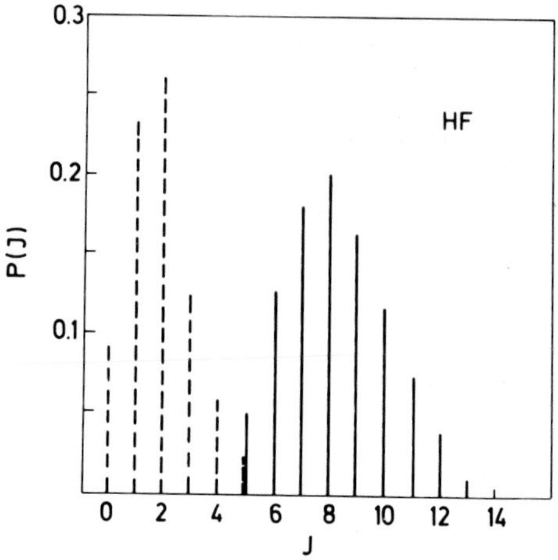

Fig. 1.8. Rotational state distribution of HF (in the v=1 manifold) immediately after the reaction $F + H_2 \to HF(v,J) + H$ (solid lines) and after complete rotational thermalization (broken lines) in a buffer gas at 300 K (nascent distribution adapted from [1.21])

than 10^5 collisions will be required, on the average, to cause vibrational deactivation when the energy is transferred to the translation. Direct experimental measurements verify this very rough estimate. The situation is far different for rotations. Figure 1.8 shows the nascent (and the thermally equilibrated) rotational state distribution for HF. For the $J = 8$ to $J' = 7$ transition (at the same v) $\Delta E = 16\ B_e$ [cf. (1.2.9)] and $\exp(-16hcB_e/kT) \simeq 0.2$. After some 10-10^2 collisions we expect that the HF will be in rotational equilibrium. If rotational relaxation is faster than the pumping reaction then at the beginning of the lasing the population of HF can be approximated by the functional form

$$N_{v,J} = N_v g_J \exp[-hcF(J)/kT]/Q_R \quad . \tag{1.2.13}$$

Here Q_R is the normalizing factor (the rotational "sum of states" or "partition function") which ensures that $\sum_J N_{v,J} = N_v$, i.e.,[12]

$$Q_R = \sum_J g_J \exp[-hcF(J)/kT] \quad . \tag{1.2.14}$$

12 In the rigid rotor approximation (1.2.9), where F(J) is independent of v, so is Q_R; $Q_R \simeq kT/hcB_e$.

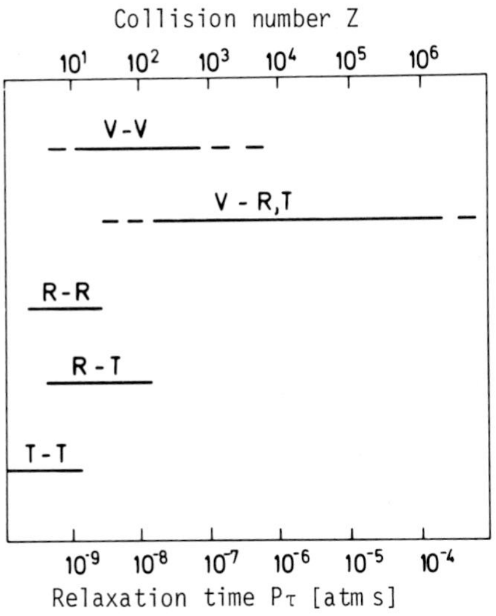

Fig. 1.9. Schematic survey of the typical ranges of energy-tranfer rates for thermal collisions. V,R and T refer to vibrational rotational and translational energies, respectively. The precise magnitude of the rate depends on the temperature and on the energy defect. (Collision numbers have been converted to $P\tau$ values assuming a total collision rate of about 2.4×10^{14} $cm^3mole^{-1}s^{-1}$)

N_v is the nascent population of the vibrational state v. Of course some vibrational relaxation has taken place, so that N_v is not precisely the nascent, initial population of the vibrations, but as a guide to the actual behavior, (1.2.13) is a realistic approximation.

A knowledge of the relative magnitude of the energy transfer rates has enabled us to offer a simple description of the population of HF at the beginning of the lasing stage. The number of molecules in any given vibrational state is essentially that produced by the chemical reaction. The distribution over the rotational levels (of the same vibrational state) is essentially completely relaxed — it is a thermal distribution. During any particular laser transition, the upper level is depleted and the lower level is gaining population. However, under the high buffer gas pressure condition corresponding to Fig.1.6 the duration of the pulse is sufficient to ensure that rotational energy-transfer collisions maintain the rotational thermal equilibrium by replenishing the emitting level and bleeding off the lower level.

These simple considerations cannot replace a more precise analysis (Sects. 4.2 and 4.4) but often provide reliable guidance to the observed behavior. Figure 1.9 provides a general survey of the order of magnitude of energy-transfer rates between different states. The numbers are quoted at '$P\tau$' values where the actual rate is connected to $P\tau$ as in (1.2.6). Note also that one can express the rates in terms of effective cross sections. Thus, the effective cross section for vibrational deactivation of HF in Ar at 300 K is several orders of magnitude smaller than the collision cross section. In other words, the collision cross section is a measure of the rate of all collisions[13] (irrespective of their outcome) while the cross section for vibrational deactivation is a measure of the rate of those collisions where a net decrease of the vibrational energy took place. Thus with every possible type of collision

$$M + AB(n) \rightarrow M + AB(m) \tag{1.2.15}$$

we can associate an 'effective' cross section $\sigma(n \rightarrow m)$ such that the collision rate is given by

$$d[AB(m)]/dt = k(n \rightarrow m)[M][AB(n)] \tag{1.2.16}$$

$$k(n \rightarrow m) = \langle \underline{v} \rangle \sigma(n \rightarrow m) \tag{1.2.17}$$

and the total collision rate $k(n \rightarrow)$ is

$$k(n \rightarrow) = \sum_m k(n \rightarrow m) \quad . \tag{1.2.18}$$

Of course, the rates quoted in Fig.1.9 are only meant as a rough guide. Moreover, in practice many other processes participate as well, say the V-R,T transfer process

$$HF(v,J) + H \rightarrow HF(v-1,J+L) + H \quad . \tag{1.2.19}$$

By adding HF (or other molecules) to the lasing mixture one can probe the removal (and also the final states) of vibrationally excited molecules by such mechanisms (cf. Fig.2.41). In particular, it is found that the vibrational energy mismatch is often made up by rotational energy (so-called V-R,T transfer) so that the net amount released to translation is minimal. This is particularly true for hydrides (cf. Sect.2.4.4).

Another possible relaxation process is via atom exchange, say

$$F + HF(v) \rightarrow FH(v') + F \quad . \tag{1.2.20}$$

Figure 1.10 shows the computed rates for such a process. The theoretical machinery of molecular collision dynamics is not limited to "rules of thumb"

13 By definition, the collision (or gas kinetic) cross section is the sum of the collision cross sections of the different possible processes.

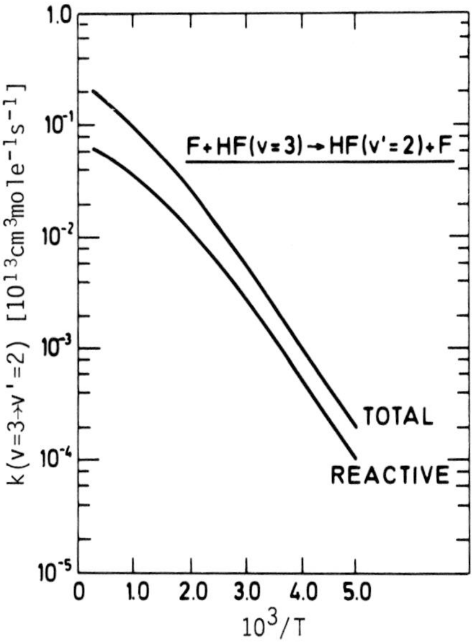

Fig. 1.10. Vibrational relaxation of HF. Rate constants as a function of temperature for the deexcitation process $v = 3 \rightarrow v' = 2$ by atom exchange and by both atom exchange and V-T transfer (adapted from [1.22]). Note that the V-T process in this (potentially reactive) system is quite efficient. This is often found to be the case. The computations also indicate that the $\Delta v = 2$ process is less probable than the $\Delta v = 1$ process by less than an order of magnitude. The plot of ln(k) vs 1/T serves to emphasize the 'Arrhenius' temperature dependence of the reaction rate; $E_a = -d\ \ln(k)/d(1/T)$ is only weakly temperature dependent

but, where necessary, can provide reliable estimates of both the primary energy disposal and the rates of subsequent energy-transfer processes.

Chapter 2 provides a detailed discussion of molecular processes and their macroscopic manifestations. Kinetic descriptions of chemical lasers are provided in Sect.4.4. Here we turn to the complementary aspect of our topic — laser-induced chemistry.

1.3 Photoselective Chemistry

Laser photochemistry is selective due to the preparation of the initial state. Currently available laser sources (Sect.3.4) provide photons over the entire range of the spectrum which is of chemical interest. Moreover, some lasers can provide high powers sufficient for the absorption of more than one photon per molecule, thereby preparing states that are not accessible by conventional light sources. In this section we examine the nature of the excited states of molecules that can be pumped by laser radiation and the types of dynamical processes that such excited states can engage in.

Lasers have contributed significantly to our understanding of both inter- and intramolecular dynamics not only by the preparation of the initial state but also by their use to probe its subsequent evolution. The availability of tunable laser sources, and of ultrashort laser pulses (all the way down to the picosecond region), provides powerful tools for this purpose. Examples of such applications are noted in this section. The primary division is however in order of increasing level of molecular excitation.

1.3.1 The Discrete Spectrum

The low-lying vibrotational states of molecules are discrete and comparatively sparse. The molecule absorbs light only at well-defined frequencies and the absorption spectrum is a series of sharp spikes. Our thinking about this region is conditioned by the level structure of diatomic or triatomic molecules, which is shown for HCl in Fig.1.11. Using a laser it is possible to excite a particular state. Except for spontaneous radiative decay [which in the infrared is quite slow; cf. (1.1.16)] such a prepared state can decay only via a bimolecular collisional mechanism.

Consider, say, the Cl + HI reaction

$$\mathrm{Cl} + \mathrm{HI} \rightarrow \mathrm{I} + \mathrm{HCl}(v) \quad , \tag{1.3.1}$$

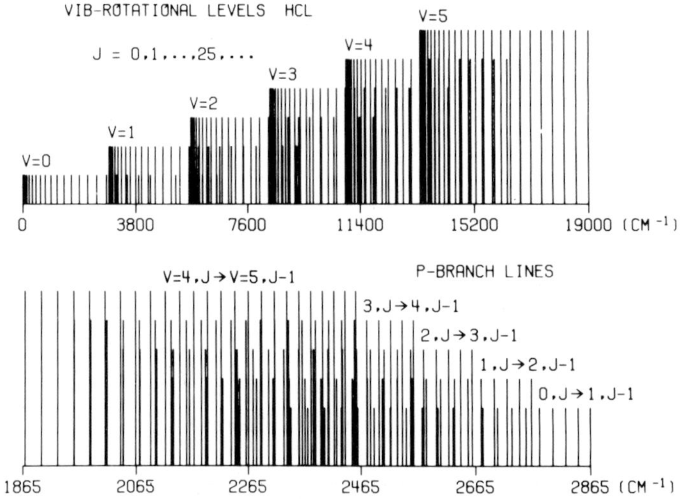

Fig. 1.11. Vibrotational energy levels and transition frequencies of HCl illustrating the discrete spectrum of small molecules. The upper trace shows the first twenty five rotational levels in each of the five vibrational levels. (For clarity, levels in different manifolds are designated by lines of different heights). The lower trace depicts the frequency distribution of the P-branch transitions, v,J→v+1,J-1. The widths of the transitions (cf. Sect.3.1.2) are much smaller than the average spacing between neighboring transitions

using thermal reactants which (cf. Fig.1.4) produces HCl molecules in the vibrational state v=3 in preference to v=2, 1 or 0. At equilibrium, the rate of formation of HCl(v) molecules by the reaction (1.3.1) equals the rate of their removal by the reversed reaction

$$I + HCl(v) \rightarrow Cl + HI \quad . \tag{1.3.2}$$

This, detailed balance, argument shows that the rate of the reaction (1.3.2) for equilibrium reactants is higher for v=3 than for v=2,1,0. At equilibrium, the concentration of HCl molecules in the v=3 state is pretty low, the vast majority being in v=0 (see Fig.1.1). Hence, we could considerably enhance the rate of the I + HCl reaction by pumping HCl molecules to higher vibrational states.

A quantitative discussion of the principle of detailed balance is given in Sect.2.1. It should however be clear that specificity of energy disposal in exoergic reactions, which provides a pumping mechanism for chemical lasers, implies that lasers can be used to pump reagents so as to selectively enhance

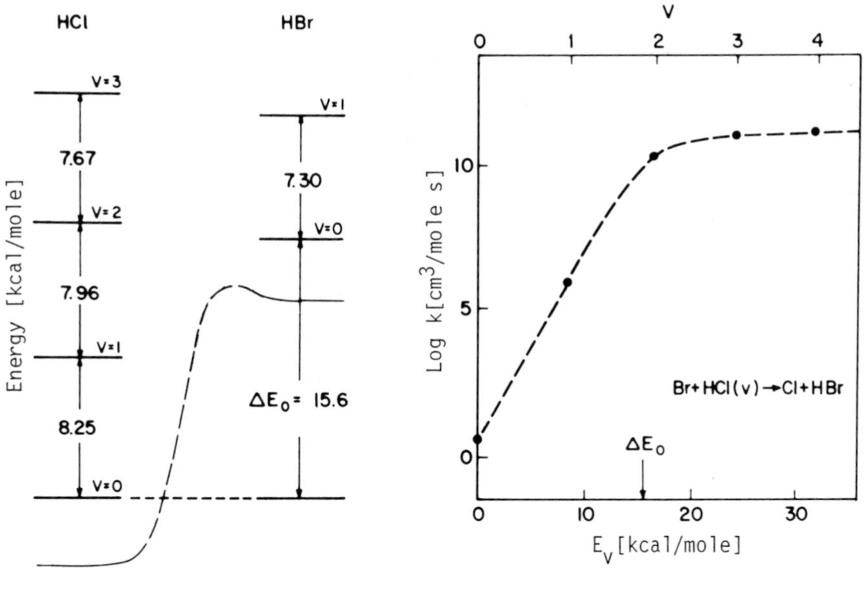

Fig. 1.12. Energy level diagram (*left*) and the rate constants of the Br + HCl(v) → Cl + HBr reaction vs the vibrational excitation of HCl (*right*). All energies in kcal/mole. Note the change from an exponential behavior in the endoergic regime to a moderate increase in the exoergic regime. (For experimental data see [1.18,23-25]; see also Sect.2.1)

the rate of endothermic reactions. Figure 1.12 shows the rate constant for the

$$Br + HCl(v) \rightarrow Cl + HBr \tag{1.3.3}$$

reaction. Relative rate constants for v=1-4 were determined using excited HCl molecules formed in the Cl + HI reaction (cf. Fig.1.4). The decrease in HCl(v) concentration was monitored by following the appropriate infrared luminescence bands [1.23]. Absolute rate constants for v=1 and 2 were determined using laser-excited HCl molecules by measuring the rate of appearance of Cl atoms [1.24], (see also [1.25]). Another method to obtain the endoergic rate constants is to convert the rates of the reverse exoergic reaction $Cl + HBr \rightarrow HCl(v) + Br$ via the detailed balance principle, as will be discussed in Sect.2.1. There we shall also examine the question of how to characterize the specific role played by the vibrational energy in enhancing the rate of the reaction.

One notable feature of Fig.1.12 deserves, however, an immediate comment. This is the much more moderate effect of HCl vibrational excitation past the v=2 level. We have already noted the rule of thumb (known as "the exponential gap rule") that direct molecular collisions distinctly disfavor changes in the translational energy. For the Br + HCl(v) reaction, as long as $v < 2$, the internal energy of the reactants is below the ground state energy of the products (cf. Fig.1.12). The energy required for the reaction is provided by the translation. Increasing the HCl vibrational excitation decreases the energy defect (makes the reaction less endothermic) and leads to an exponential enhancement of the reaction rate. For v=2 the process is about thermoneutral. The gap has been closed. For $v > 2$, the internal energy of the reactants exceeds the ground state of the products, and translational energy can be released. Since the rate does not decline upon increase of HCl vibrational excitation beyond v=2 it is reasonable to suggest that the excess HCl vibrational energy is preferentially released not as translation but rather as internal excitation of HBr.

The preferential conversion of reactant vibrational energy into vibrational, rather than translational, energy of the products is demonstrated in Fig.1.13 for the

$$F + HCl(v) \rightarrow Cl + HF(v') \tag{1.3.4}$$

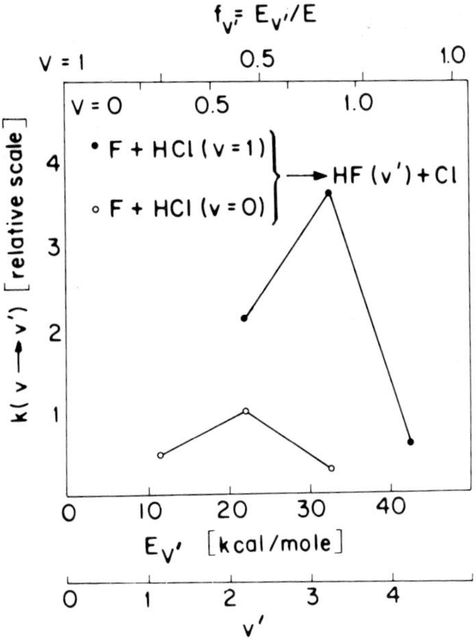

Fig. 1.13. Relative detailed rate constants of the reaction (1.3.4) for reactants in two different vibrational states. The rate constants (product state distributions) are shown vs the products vibrational energy $E_{V'}$, vibrational state v', and the fraction of the total reaction energy (E) appearing as products vibration $f_{V'}$. Note the increase in the vibrational energy content of the products upon vibrational excitation of the reactants. The overall reaction rate is also higher for v=1. (Adapted from [1.26])

reaction. The vibrational excitation of HCl leads to an increase in the mean vibrational energy of the HF molecules.

A vibrationally excited HCl molecule can be deactivated by V-R,T collisions. Such energy transfer side processes can compete with the reactive collisions. In other words, the rate of depletion of HCl(v) molecules due to collisions with a Br atom is due to two distinct processes,

$$HCl(v) + Br \begin{array}{l} \xrightarrow{k_r} Cl + HBr \\ \xrightarrow[k_{nr}]{} HCl(v') + Br \end{array} , \qquad (1.3.5)$$

only one of which leads to reaction. Given the measured vibrational energy disposal in the Cl + HBr reaction, detailed balance can be used to infer the dependence of k_r on the vibrational excitation of HCl. The enhancement of k_r with vibrational energy does not necessarily imply that k_{nr} is small. In fact such undesirable energy transfer processes can effectively compete with (e.g., Fig.1.9 and Sect.5.2) and reduce the efficiency of laser-driven processes.

To successfully compete with collisional deactivation it is thus necessary to achieve a high reaction rate which requires reagents excited past the threshold energy for reaction. Single-photon excitation from the ground (v=0) vibrational state to some high lying (v>1) state is only possible for anharmonic vibrations (Sect.3.3). Since near v=0 the anharmonicity is small, this so-called "overtone" absorption is weak but can be observed given a sufficiently high flux of photons. One way of realizing the required conditions is to perform the pumping inside a laser cavity.

Chemical reactivity is governed not only by energetic but also by steric requirements. Reaction may occur much more readily when the reagents approach along a preferred geometry [1.27]. Lasers can be used to prepare oriented molecules, as follows: The transition probability for absorption of light depends on the orientation of the molecular transition dipole with respect to the electrical field (i.e., as $\underline{\mu}\cdot\underline{E}_0$), being maximal when the two are parallel. Molecules which have absorbed polarized light are thus preferentially oriented.

The reaction

$$Sr + HF(v) \rightarrow SrF(v') + H \qquad (1.3.6)$$

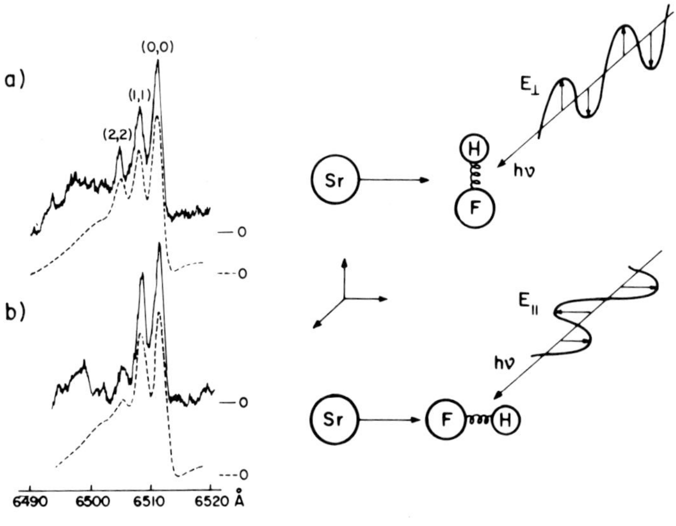

Fig. 1.14. The vibrational distribution of $SrF(X^2\Sigma^+)$ formed by the reaction Sr + HF(v=1,J=1), where HF molecules are preferentially aligned a) perpendicular, b) parallel to the approaching Sr atom. The alignment is achieved by selecting the polarization of the exciting HF laser (right panel). The left panel (solid line) shows the excitation spectrum of the $SrF(A^2\Pi_{3/2}-X^2\Sigma^+)$ transition. A dye laser excites a particular vibronic transition, and the total resulting fluorescence is monitored as a function of the excitation wavelength. Vibrational population ratios are obtained from known Franck-Condon factors. Computer simulation (dashed lines) shows that in case a) the initial vibrational distribution is $N_0:N_1:N_2 = 1.00:0.63:0.29$, and in case b) 1.00:0.72:0.13. A sideway attack thus favors higher vibrational excitation in the products. (Adapted from [1.28])

is endothermic by about 6.5 kcal/mole for v=0 and becomes exothermic by about 5 kcal/mole for v=1. The rate constant of (1.3.6) has been determined in a laser-molecular beam experiment (Sect.5.2), using an HF laser to excite the reagent molecule and the (dye) laser-induced-fluorescence method (LIF, Sect. 5.5) to monitor the SrF(v') products (Fig.1.14). It was found that while for HF(v=0) the reaction rate was negligible, a high yield was observed with HF(v=1).

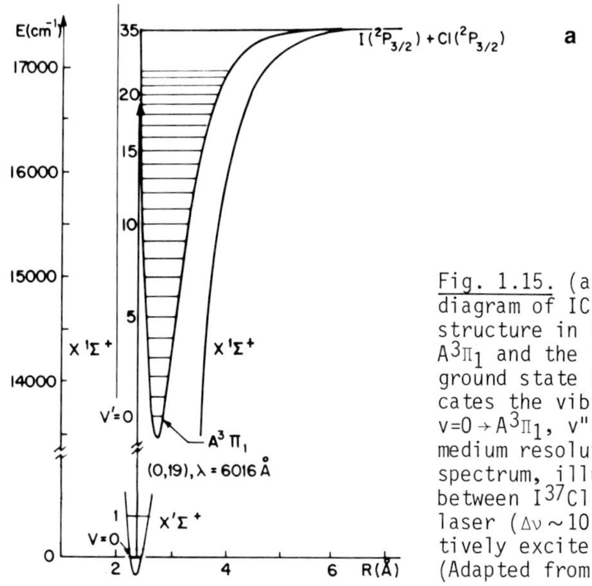

Fig. 1.15. (a) A partial potential energy diagram of ICl showing the vibrational structure in the excited electronic state $A^3\Pi_1$ and the two lowest levels in the ground state $X^1\Sigma^+$. The line (0,19) indicates the vibronic transition, $X^1\Sigma^+$, $v=0 \rightarrow A^3\Pi_1$, $v''=19$. (b) A portion of the medium resolution electronic absorption spectrum, illustrating the isotopic shift between $I^{37}Cl$ and $I^{35}Cl$. Even a broadband laser ($\Delta\nu \sim 10$ cm^{-1}) can be used to selectively excite one of the isotopic species. (Adapted from [1.29])

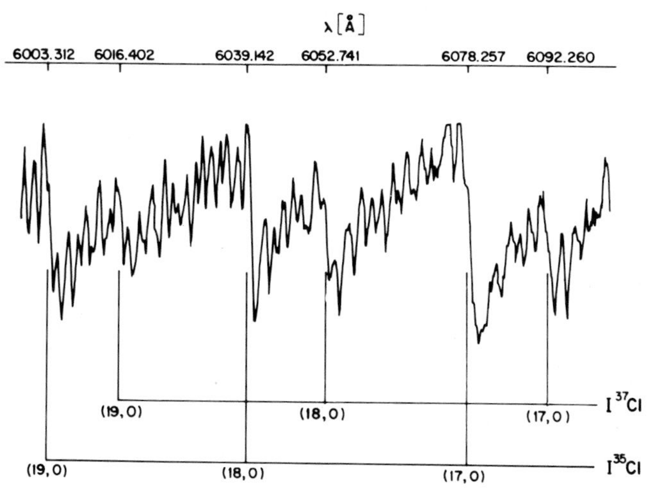

Using a polarized radiation of an (HF) laser, it is possible to prepare the HF(v=1) molecules in a selected orientation with respect to the direction of the Sr atom propagation. This is due to the fact that only molecules whose axis has a component parallel to the electric field vector ($\underline{E}$) can be excited. Thus, choosing the proper polarization, parallel ($\underline{E}_{\parallel}$) or perpendicular ($\underline{E}_{\perp}$) to the Sr beam, the Sr + HF(v=1) collision occurs preferentially along a collinear or a sideway approach (see Fig.1.14). The fraction of SrF(v'=2) products is larger in sideway collision.

In terms of the molecular reaction dynamics, electronically excited states of diatomic molecules often also belong to the classification of "discrete spectrum". The higher reactivity of electronically excited molecules has been used for, e.g., isotopic separation. Thus [1.29], ^{35}Cl and ^{37}Cl can be separated when I^{37}Cl is laser pumped to the $A^3\Pi_1$ state in the presence of scavengers which do not react with the ICl molecule in its ground electronic state. A portion of the visible absorption spectrum of ICl is shown in Fig. 1.15. It is evident that despite the heavy masses, the isotope shift is large enough, making even a broadband laser a suitable pump for preferential excitation of one isotopic species in the presence of the other.

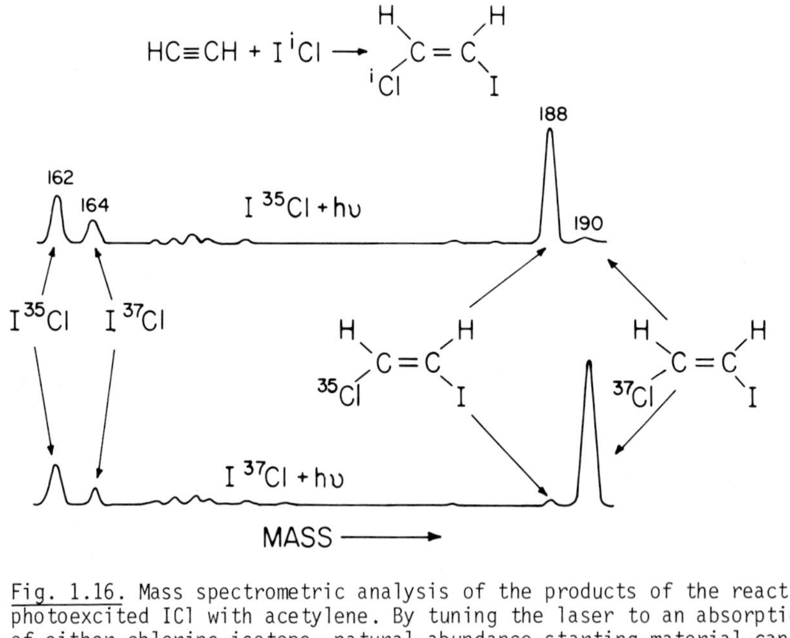

Fig. 1.16. Mass spectrometric analysis of the products of the reaction of photoexcited ICl with acetylene. By tuning the laser to an absorption line of either chlorine isotope, natural abundance starting material can be enriched to more than 97% purity in a single step. (Adapted from [1.31])

Energy transfer between different isotopic species is usually very rapid, tending to spoil the selectivity established in the initial excitation stage [1.30]. When the reaction rate with the scavenger is larger than the scrambling rate, high overall isotopic enrichment can be achieved, as shown in Fig.1.16.

1.3.2 The Quasicontinuum

For larger molecules and/or at higher excitation energies the number of molecular vibrotational states per energy interval increases very rapidly. Figure 1.17 illustrates the extreme congestion which is revealed upon improved resolution in a large but not giant molecule (SF_6) at a comparatively low (~1000 cm^{-1}, one quantum in the ν_3 mode) energy. Considerable simplification of the observed spectra is possible by cooling such molecules to the exceedingly low temperatures made possible by the use of supersonic beams. Under these conditions only the very lowest vibrotational states are populated, and hence most of the transitions possible at room temperature are of vanishingly small intensity. Even so, as the excitation energy is increased, so does the number of accessible final states[14]. Eventually the manifold of final states appears to be a continuum on the intrinsic energy resolution of the experiment (determined, for instance, by the exciting laser bandwidth). It is then referred to as the quasicontinuum. Note that the definition is an operational one – the quasicontinuum is reached when it is no longer possible to excite a single molecular eigenstate under the given experimental conditions.

Direct, single (typically visible) photon absorption into the quasicontinuum region of large molecules and emission from higher lying states into the quasicontinuum have been observed (Sect.2.3.7). Most eigenstates in the quasicontinuum have only very small optical transition probabilities from the ground state. The states prepared by optical excitation are a superposition of eigenstates, most of which have a character similar to the overtone spectra of diatomic molecules. In other words, the excitation is localized in a particular mode[15]. A superposition of eigenstates of somewhat different energies

---

14 In the classical limit, where internal energy can vary continuously, the number $\delta N(E)$ of excited vibrotational states in the energy range E to $E + \delta E$ increases with E as $\delta N(E) \propto E^{\alpha-1}\delta E$. Here $\alpha = s + r/2$, where s and r denote the number of vibrational and rotational degrees of freedom of the molecule, respectively. (See Appendix 2.A).

15 The superposition of delocalized vibrational eigenstates to describe a localized vibration is analogous to the superposition of delocalized molecular orbitals to described localized bond orbitals.

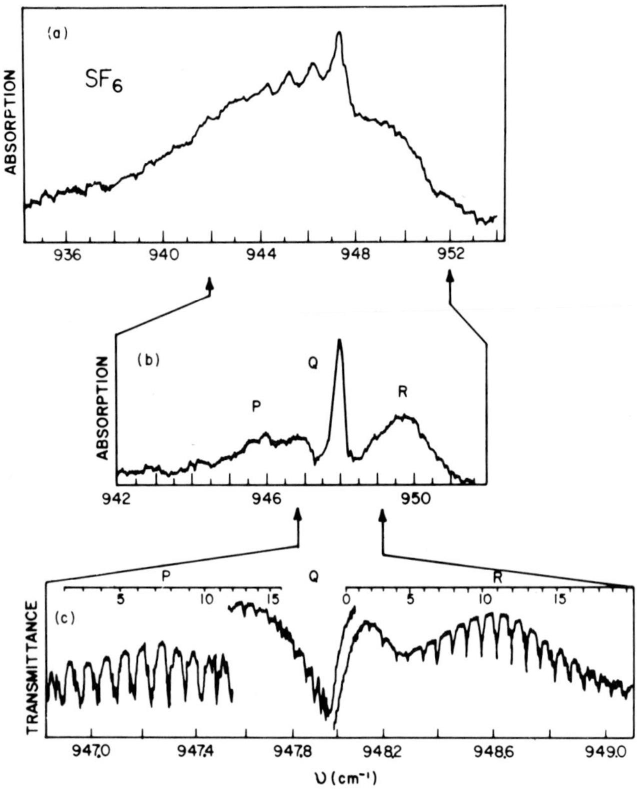

Fig. 1.17a-c. A portion of the vibrational absorption spectrum of SF_6 near 948 cm^{-1}, corresponding to single-photon absorption in the ν_3 mode (Fig.3.33). (a) Conventional low-resolution spectrum at room temperature. (b) A grating spectrum at 123 K. Some of the hot bands disappeared and P,Q,R branches are identified. (P, Q, and R branch transitions correspond to $v,J \to v+1,J'$ with $J' = J-1, J$ and $J+1$, respectively.) (c) A high-resolution spectrum of SF_6 at 123 K obtained with a tunable diode laser having a resolution of 0.001 cm^{-1} (adapted from [1.32], where the original experiments are cited; cf. Sect. 3.3). Each of the P, Q, and R "lines" can be shown to be further split into a cluster of lines, and is thus a manifold of lines. The splittings are mainly due to rotation-vibration interaction (Coriolis splitting). See also Fig.3.33

is no longer an eigenstate of the molecular Hamiltonian. Hence, the states prepared by optical excitation into the quasicontinuum are not stationary but are coupled (by the molecular potential) to other states. Energy can be pumped into the quasicontinuum but only indirectly, i.e., via these "doorway" states. This theme, that the optically prepared state in the quasicontinuum is typically nonstationary, will recur throughout this section.

Figure 1.18 shows the CH overtone absorption spectra of $CH_2=CH\text{-}CH_2NC$ (allyl isocyanide). The three overtone bands corresponding to the three types

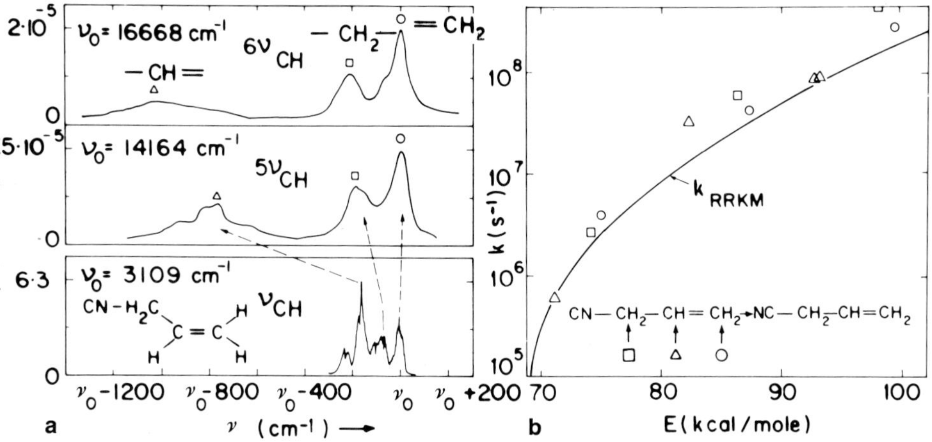

Fig. 1.18. (a) The fundamental ν_{CH}, the fifth overtone $5\nu_{CH}$, and the sixth overtone $6\nu_{CH}$ spectra of allyl isocyanide as measured in intracavity dye laser experiments [1.33] (Sect.5.5). The three types of CH stretching modes are identified. The overtone spectra are broadened and shifted with respect to the fundamental but the local mode structure is preserved even when the molecule is excited directly to the quasicontinuum. (b) The rate constant for the unimolecular photoisomerization of allyl isocyanide as a function of the excitation energy and the site of CH excitation (marked by □, △, and ○ for the three types of CH bonds). The solid curve is the prediction of the RRKM theory, which assumes that energy randomization in the molecule is complete (Sect.2.3.5). (Adapted from [1.33])

of CH sites (CH_2=, =CH-, and -CH_2-) are tentatively identified. Very many other states are present in this energy range, but because of very unfavorable transition probabilities (due to the need to simultaneously displace several modes from equilibrium), contribute only a very weak background absorption.

The overtone bandwidths are much larger than the rotational band contours in the fundamental CH stretch region (Fig.1.18). This is taken to reflect the nonstationary character of the optically prepared state. If we think of this state as n phonons (quanta of vibrational motion) localized in the CH stretch mode then it is not an eigenstate of the molecular Hamiltonian. It can undergo intramolecular transitions, say, to a state with n-1 quanta at the original site plus one CH stretch quantum on a different site. (The energy mismatch resulting from the anharmonicity of the vibration must be compensated by energy exchange with other modes.) Such states in turn are coupled to other states and ultimately the energy deposited by the optical excitation is spread throughout the molecule. The large bandwidth indicates (by the uncertainly principle) that the initial evolution of the prepared state will be rapid, of the order of few vibrational periods. It is therefore not necessarily the case that one can obtain bond selective chemistry using laser excitation.

Allyl isocyanide can isomerize to allyl cyanide.

$$CH_2{=}CHCH_2NC \rightarrow CH_2{=}CHCH_2CN \quad . \tag{1.3.7}$$

The available experimental results (Fig.1.18) do show however that while the rate constant for unimolecular isomerization (represented by the overall conversion efficiency) depends primarily on the total energy, it does show some dependence on the particular CH site where this energy was initially present. While the optically prepared state is not stationary, it does not, on the time scale of the unimolecular reaction (which is quite slow), reach a distribution of the excess energy which is independent of the site of the original excitation.

An observation that generated considerable interest is that polyatomic molecules can be pumped to the quasicontinuum region and ultimately to dissociation by high-power infrared lasers even under collision-free conditions. Moreover, the process can be species selective. It is possible to dissociate only those molecules corresponding to a particular isotopic composition by

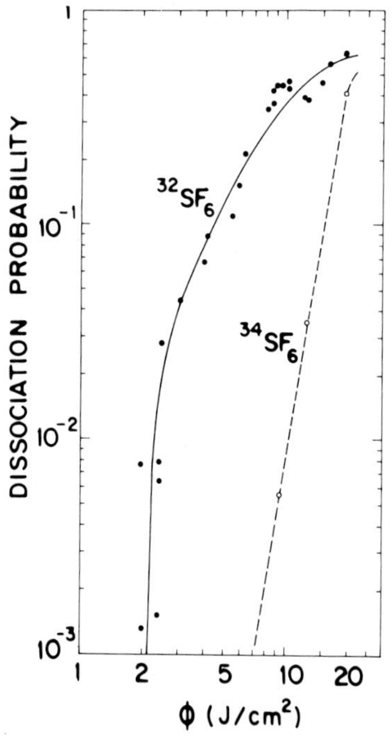

Fig. 1.19. The fraction of $^{32}SF_6$ and $^{34}SF_6$ molecules undergoing dissociation by multiphoton absorption of CO_2 laser radiation as a function of the flux ϕ of the laser radiation. [ϕ, often called the 'fluence', is the total pulse energy, per unit area of the laser beam (Sect.5.4)]. The data points were obtained by irradiating low-pressure (<1 Torr) samples of SF_6 static gas cells with the P(20) line (10.6 μm) of a pulsed CO_2 laser. The product compositions were analyzed in a mass spectrometer. Below 0.3 Torr the dissociation yields of the two isotopes were pressure independent. Above 1 Torr the isotopic selectivity is lost due to scrambling collisions. (Adapted from [1.34])

using the shifts in the absorption spectrum of different isotopic molecules. Figure 1.19 shows the fraction of SF_6 molecules which have undergone dissociation as a function of the laser energy. The laser energy determines the average number of laser quanta $<n>$ absorbed by the dissociating molecules. The process can be described as

$$SF_6 + nh\nu \to SF_5 + F, \quad SF_5 + mh\nu \to SF_4 + F, \quad \ldots \text{ etc.} \tag{1.3.8}$$

Species selectivity is achieved primarily due to absorption into the low-lying, discrete regime which serves as a bottleneck for entry into the quasicontinuum. This interplay between the two regimes has been studied in some detail and the following is only a brief introduction to the more extensive discussion in Sect.5.4.

Many infrared photons need be absorbed in order to dissociate a typical chemical bond. Since the laser photons have all the same frequency, the question is what mechanism allows their absorption by an isolated molecule. For a giant or large molecule at higher temperatures, the quasicontinuum provides the qualitative answer. The mean energy per (classical) vibration is kT. Hence if there are numerous modes and/or the temperature is high, the mean energy content of the molecule is high enough so that many states of the quasicontinuum are populated to begin with. For any initial state there will thus be a final state where their energy gap matches the photon frequency. Of course, this answer is somewhat too facile since it disregards the question of the magnitude of the transition probability (the "oscillator strength"), but it does provide for the more fundamental problem of a frequency matching.

For smaller molecules, the absorption of the first few photons requires a near coincidence between the laser frequency and some molecular transition. The coincidence should be close but need not be exact due to the phenomena of power broadening (Sect.3.1.3). Figure 1.20 shows the results of a pump and probe experiment on the ν_3 vibrational mode of SF_6. The P(18) line (of the $00^01 \to 10^00$ band) of a Q-switched high-power CO_2 laser (Sects.3.3 and 3.4) is used to pump SF_6 molecules from the ground to the first excited state of the ν_3 normal mode (cf. Sect.3.3). The pump line is in near resonance to some of the lines in the P(32) and P(33) manifolds of the $\nu_3 = 0 \to 1$ band of SF_6 (cf. Figs.1.17 and 3.33). A tunable, low-power infrared (diode) laser is then used to measure the absorption. Due to the pumping, there is a depletion of ground state molecules which can absorb light at the frequency of the pump laser. The absorption spectrum thus shows a transmission window, a phenomenon also termed "hole burning" or induced transparency. As the pump laser power is

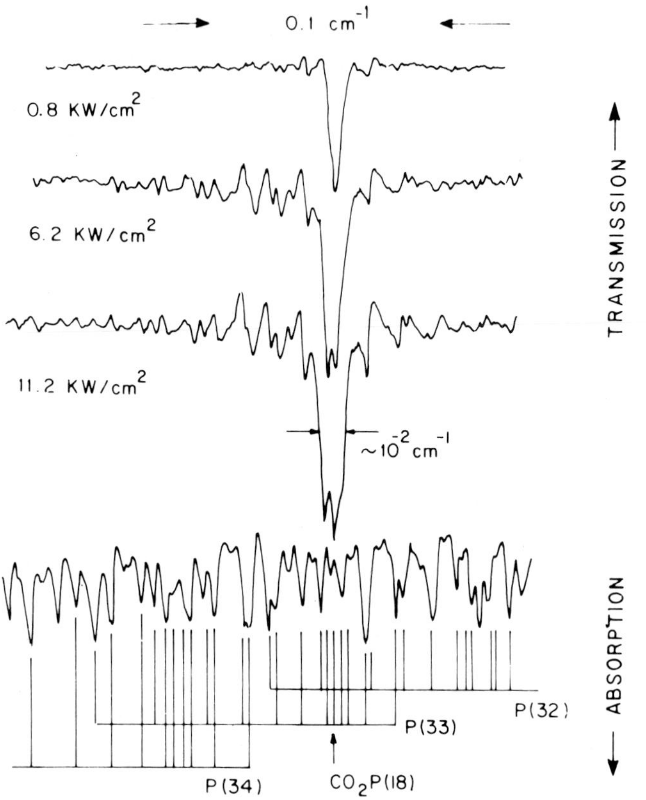

Fig. 1.20. Power broadening of the infrared double resonance spectrum of 0.08 Torr SF_6. The P(18) line of a high-power CO_2 laser is used to pump SF_6 molecules from the ground to the first excited vibrational level of the ν_3 mode. The pump line is in near resonance with some lines in the P(32) and P(33) manifolds of the $\nu_3 = 0 \to 1$ band of SF_6. The changes in the ground level population due to the pumping are reflected in the transmission spectrum of a continuous (cw), low-power infrared (diode) laser whose spectral width is less than 0.001 cm^{-1}. The bottom trace is a portion of the absorption spectrum of SF_6. The three upper traces are the transmission spectra corresponding to three different power levels of the pump laser. The range of frequencies excited increases with the power as is reflected by the widths of the dips ("holes") in the transmission spectrum. The width of these power-broadened dips is the order of the Rabi frequency, $\mu E_0/h$, and are thus proportional to $P^{1/2}$. μ is the transition dipole, E_0 is the electric field amplitude, and $P \propto E_0^2$ is the laser power (Sect.3.1). (Adapted from [1.35])

increased so does the width of the hole. Molecules which are not in exact resonance, can at high field strengths absorb energy from the field. The hole width at half dip is known as the Rabi frequency (Sect.3.1) and, depending on the magnitude of the transition probability can reach 1 cm^{-1} or even higher for currently available CO_2 lasers.

There are a number of different molecular states which correspond to a given rotational quantum number J (Sect.3.3). At high fields, some of these states are coupled by the high field strength of the intense laser light and can absorb. At higher pressures, collisions will also serve to replenish any state depleted due to absorption. The sublevels of a given J are very close in energy and are very effectively coupled by collisions. At somewhat higher pressures or over longer times, collisions will also couple states of different J's. The "double resonance" experiment of Fig.1.20 can be used to monitor this population transfer by introducing a delay between the pump and the probe lasers. The relaxation time is reflected by the rate of replenishing the hole burned by the pump laser. The rotational relaxation time due to SF_6-SF_6 collisions is about 43 ns-Torr corresponding to an "effective" collision cross section of 240 A^2.

The absorption of the second laser photon by SF_6 can follow a similar route: a $\nu_3 \rightarrow 2\nu_3$ transition where now, because of anharmonicity, a $J \rightarrow J$ (so-called "Q-branch") transition offers the closest frequency match. Fine tuning is again provided by the power broadening and the splitting of the upper rotational level. Alternatively, at high powers, the two photons can be simultaneously absorbed. Because the first photon is very nearly resonant with molecular transitions, the cross section for such a process can be significant. At high powers even the simultaneous absorption of three or more photons is possible (Sect.5.4).

Having absorbed three or more photons, the SF_6 molecule is essentially in the quasicontinuum region. One is then no longer worried about frequency matching but a new worry immediately takes over. In the quasicontinuum, states with, say, n quanta in the ν_3 mode are essentially at the same energy with states with n-1 quanta in ν_3 and the balance in other modes or with n-2 quanta in ν_3 and the balance in other modes, etc. Any one such state is coupled by the anharmonicity to the other states and is not stationary. The pumping moves the molecule up in the ladder of states. Due to anharmonicity the excitation at any given energy is being rapidly dispersed by the intramolecular coupling among the many accessible isoenergetic states. The two processes are in competition and state-selective pumping requires that the pumping-up rate be faster. One way of approaching this goal is to take larger steps — i.e., to use higher energy laser photons.

Past the lowest threshold for dissociation the up-pumping competes also against the depletion of molecules by dissociation. For a large molecule quite a few photons can be absorbed before the rate of dissociation is dominant. The fragments will then carry quite a bit of internal excitation. Such fragments may then be in the quasicontinuum and can themselves absorb the laser radiation, leading ultimately to secondary or even tertiary dissociation processes [cf. (1.3.8)]. Extensive fragmentation of polyatomic molecules has indeed been observed using high-power lasers.

The up-pumping of molecules beyond the lowest threshold is particularly evident when there are several reaction channels. An example is

$$C_2F_5Cl \rightarrow C_2F_5 + Cl \quad , \tag{1.3.9}$$

where the threshold is at about 83 kcal/mole and

$$C_2F_5Cl \rightarrow CF_3 + CF_2Cl \tag{1.3.10}$$

with a somewhat higher threshold (the C-C bond energy is, roughly 100 kcal/mole). At lower photon flux the first process dominates but at higher energy fluence the C-C bond rupture reaction becomes increasingly important (Fig. 1.21). Analysis of the products energy distribution verifies that the C_2F_5Cl molecule can indeed be pumped to energies considerably above its lowest threshold.

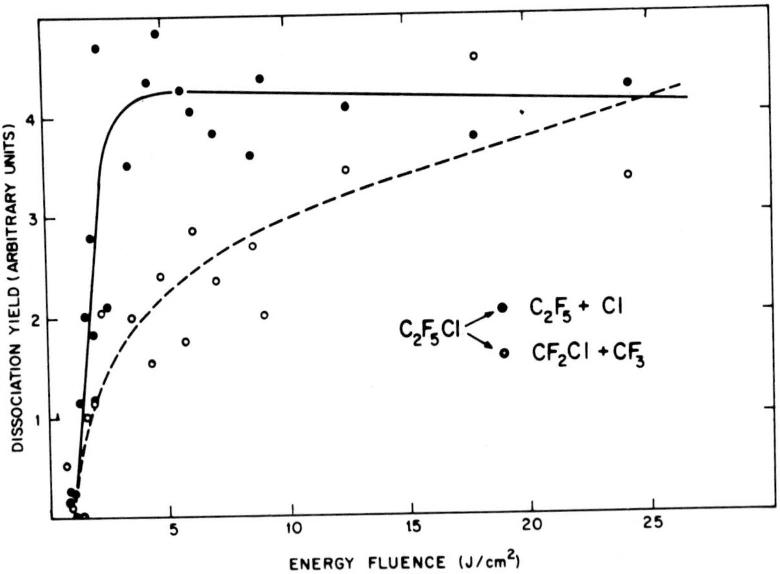

Fig. 1.21. Multiphoton dissociation of C_2F_5Cl by a CO_2 laser vs the laser energy fluence. At low fluence Cl elimination is the dominant pathway. At higher fluence the Cl elimination saturates while the rupture of the stronger C-C bond becomes increasingly important. (Adapted from [1.36])

The possibility of bottlenecks for intramolecular energy transfer, discussed in connection with Fig.1.18 (see also Fig.2.32) suggests that even multiphoton pumping can be used to achieve selectivity on a given potential energy surface. What is necessary is a rapid climbing up in the energy ladder. One example, discussed in Sect.5.4, is the isomerization vs fragmentation of cyclopropane. Multiphoton excitation in the region of the CH stretch (which requires higher, $\sim 3000\ cm^{-1}$, frequency photons) results essentially only in isomerization. Multiphoton excitation in the region of the CH_2 wag (photons of $\sim 1000\ cm^{-1}$) produces roughly equal yields of fragmentation and isomerization. If these results reflect the constraints on intramolecular relaxation of the CH stretch overtones then collisions should enhance the energy randomization. Pumping at the CH stretch frequency at higher pressures of an added inert gas does increase the yield of fragmentation. For a complete interpretation the effects of collisions during the light absorption need also be considered.

1.3.3 Radiationless Transitions

Above the ground potential energy surface lie the potential energy surfaces of electronically excited states. The novel feature associated with such surfaces for polyatomic molecules is that any vibrotational state of an excited surface is essentially isoenergetic with one or more states belonging to a lower surface[16]. If the minimal electronic excitation energy exceeds the dissociation energy on the ground surface then such a coincidence is assured: any state on the upper surface is exactly isoenergetic with a dissociative state of the lower surface. Since the (Born-Oppenheimer, Sect.3.3) separation into electronic and nuclear motion is only an approximation, such electronically excited states are not stationary. The single-photon transition probability from a low-lying vibrotational state to a dissociative state on the same surface is exceedingly small[17]. Hence optical, single-photon, excitation typically prepares a nonstationary state which is initially confined to the upper surface. The coupling to the lower surface will ultimately result in dissociation. This "predissociation" is further mentioned in Sect.1.3.4.

---

16 This energy matching is due to the quasicontinuum on the lower surface. For, say, a diatomic this coincidence is not necessarily the case.

17 This is another reflection of the extreme reluctance of the nuclei to undergo large changes in their momenta. We have previously seen this reluctance in the role of energy in collisions. The propensity for such changes that require the least change in momenta is known in spectroscopy as the Franck-Condon principle.

Often however the electronic excitation energy is sufficiently low so that the lower lying vibrotational states of the excited surface coincide in energy with the bound quasicontinuum region of the ground surface. This provides for an alternative route for entering the quasicontinuum and one which permits the potentially selective deposition of fairly large amounts of energy.

For a system at equilibrium all isoenergetic states are equally populated. In a large polyatomic molecule excited to the lower lying region of an excited electronic state there are overwhelmingly many more states in the quasicontinuum of the ground surface than there are in the same energy range in the excited surface. The intramolecular, interstate, relaxation will thus appear to be practically irreversible. The population in the excited surface will be severely depleted at a rate competitive with or even exceeding that of its fluorescence (hence the name "radiationless" transitions). Of course, unless the population in the quasicontinuum is further drained[18] by additional decay modes, the transition is not strictly irreversible. However, in a large molecule, the equilibrium is very much one sided.

Both the discrete and the onset of the quasicontinuum of the upper state and the coupling to the quasicontinuum of lower electronic states can be probed. Figure 1.22 shows the fluorescence excitation spectrum of pentacene, probing the vibrational structure of the first excited (S_1) singlet state. At relatively low frequencies, where the excitation is to the low-lying vibrational levels of S_1 the spectrum is discrete. Higher frequencies excite the molecules to the quasicontinuum region where the level density is very high and the identification of the transitions is intractable.

It is also possible to reach electronically excited states by multiphoton pumping to the quasicontinuum of the ground potential energy surface. Such transitions can be detected both by the appearance of electronically excited

18 Such will be the case if it is not a quasicontinuum but a real, dissociative continuum, or if collisional deactivation or dephasing takes place, etc.

Fig. 1.22. *Upper panel:* The onset of vibrational quasicontinuum in a large molecule. Shown is the excitation spectrum (total fluorescence vs excitation wavelength) of pentacene cooled by supersonic expansion. This cooling technique leads to population of low-lying vibrational and rotational states in S_0 avoiding spectral congestion (see lower panel). A dye laser with a bandwidth of 0.3 cm^{-1} was used for exciting the molecule from the ground vibrational level of the ground electronic state S_0 into the vibrational manifold of the first excited singlet state S_1. The discrete region (D, $\lambda > 5100$ Å) is characterized by well-separated lines. The vibrational quasicontinuum in S_1 (VQC, $\lambda < 5100$ Å) is experimentally observed by the onset of a congested structure, due only to the high density of vibrational levels in S_1. (Adapted from [1.37]). *Lower panel* (left): A low-resolution room temperature absorption spectrum of pentacene. (Right) A schematic energy level diagram

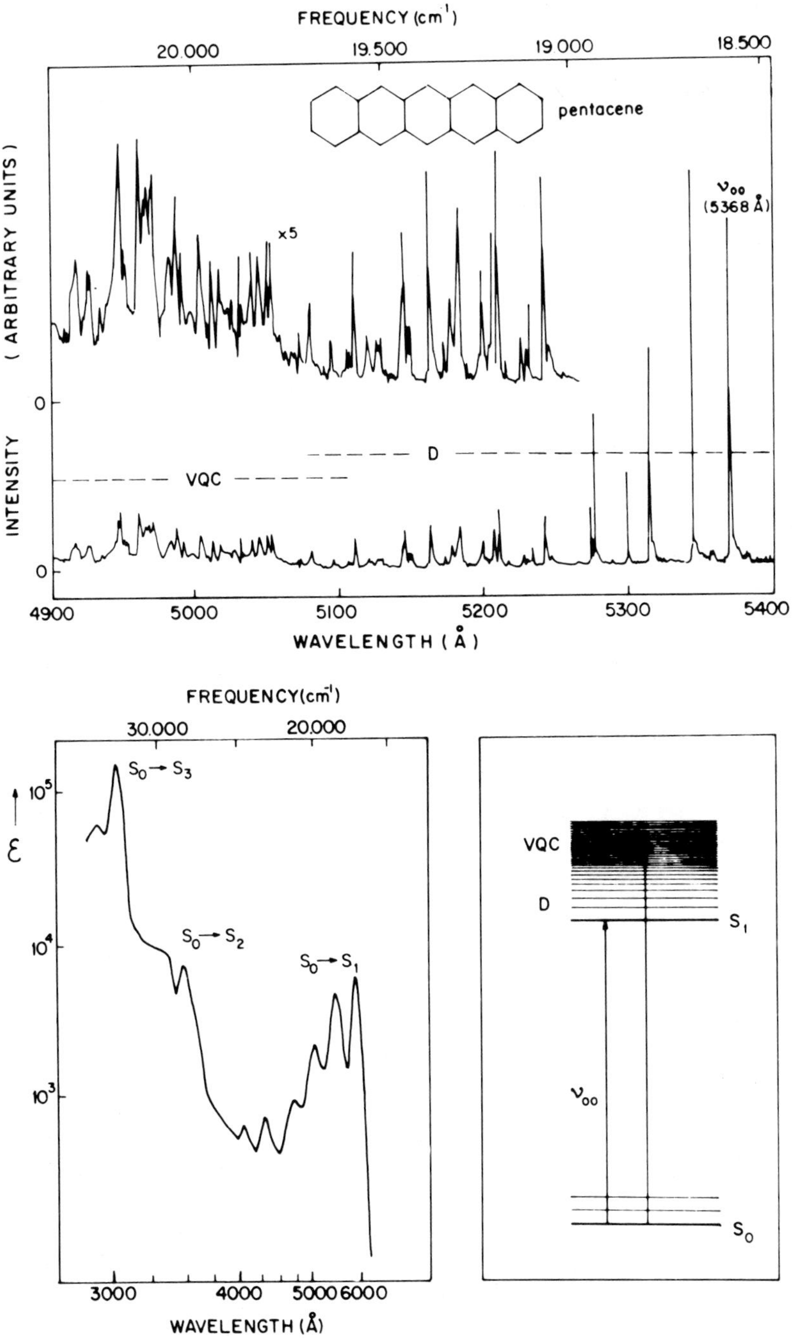

Fig.1.22. Figure caption see opposite page

fragments in multiphoton dissociation and by emission from low-lying electronically excited states of the parent molecule (Sect.5.4). As discussed before, it is preferable to use a smaller molecule with a low-lying electronic state if the process inverse to radiationless transitions is to be observed in practice.

1.3.4 Dissociative Continuum

High above the basin of a potential energy surface lies the continuum — a region where unbound motion of the nuclei is possible in at least one direction. A collision between any two reactants always takes place in the continuum of some potential energy surface. If the unbound motion is possible only in one direction, the collision cannot lead to any rearrangement but can of course be inelastic. If there is more than one exit channel then the corresponding collision can be reactive[19]. The two equivalent ways of thinking about the continuum — as a dissociative state of a parent molecule or as a state formed by collision of two fragments — are often summarized by referring to a dissociation as a "half collision".

If the potential energy surface has no basin, i.e., is purely repulsive, it has only states in the continuum[20]. Except if the potential is everywhere repulsive, states in the continuum of a polyatomic system can be isoenergetic with discrete states. If the surface has a deep basin these discrete states will form a quasicontinuum. As discussed in Sect.1.3.3, a molecule pumped to such a quasicontinuum will have a finite lifetime and need not dissociate within a vibrational period. Similarly, when two polyatomic fragments collide, they can be trapped in the multitude of bound states of the quasicontinuum and spend quite some time in the interaction region before they separate.

Photodissociation from an excited potential energy surface is possible when the optically prepared state is isoenergetic with some continuum. The dissociative continuum may belong to the same surface or to some other electronic state (predissociation). In the latter case, observable dissociation

---

19 Of course, a given potential energy surface may allow only inelastic collisions at lower energies but will allow reactive collisions at higher energies. This corresponds to a parent molecule having two dissociation modes. Reactive collisions are possible past the threshold of the higher energy mode.

20 Such surfaces are rare. It is typically the case that there is at least a shallow basin due to van der Waals forces. These interactions may be weak in comparison with chemical binding but are strong enough to ensure the existence of a bound spectrum even for such "inert" systems as a molecule colliding with a rare gas atom.

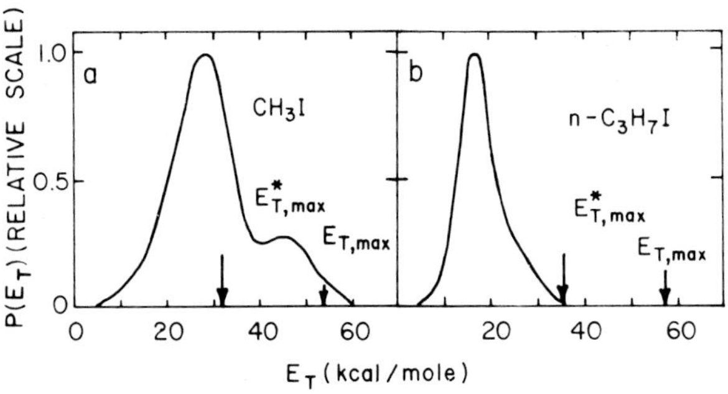

Fig. 1.23a,b. The distribution of relative translational energies of the products in the photodissociation of RI. (a) $R = CH_3$, (b) $R = n\text{-}C_3H_7$. In both cases the photon energy minus the bond dissociation energy $h\nu - D$ allows electronic excitation of the iodine atom, i.e., both $I(^2P_{3/2})$ and $I^*(^2P_{1/2})$ may be formed. The maximal translational energy possible is thus $E_{T,max} = h\nu - D$ or $E^*_{T,max} = h\nu - D - E^*$, depending on whether a ground state or excited iodine atom is formed. E^* is the electronic excitation. The experimental results show evidence of the formation of ground state I only for CH_3I. The sharp decline of $P(E_T)$ towards $E^*_{T,max}$ in the case of $n\text{-}C_3H_7I$ suggest that in this case most iodine atoms are electronically excited. [Indeed, $n\text{-}C_3H_7I$ is a good parent molecule for the iodine photochemical laser (Sect.4.3)]. (Adapted from [1.38])

requires that the coupling be not too weak. Besides purely electronic considerations the magnitude of the coupling depends, as always, on the changes in the momenta of the nuclei. Efficient crossing requires a minimal change. This condition can be realized when the two surfaces cross, i.e., at that configuration of the nuclei where the two (electronic) potential energy surfaces have the same value. The spectroscopy of dissociative states provides therefore valuable input towards the mapping of ground and excited potential energy surfaces.

Due to surface crossing the dissociation products can appear in more than one electronic state. In the photodissociation of CS_2 (Fig.1.2) the initial optical excitation is to a predissociating state S_3 crossed by two dissociative states, one leading to $CS(X^1\Sigma^+)$ and $S(^1D)$ and the other to $CS(X^1\Sigma^+)$ and $S(^3P)$.

The photodissociation of alkyl iodides to yield electronically excited I atoms was mentioned in Sect.1.1 as a pumping mechanism for the iodine atomic laser. Figure 1.23 records the observed translational energy distribution in the photodissociation of CH_3I and of $n\text{-}C_3H_7I$. The possible pathways are

$$RI \xrightarrow{h\nu} \begin{cases} R + I^*(^2P_{1/2}) \\ R + I(^2P_{3/2}) \end{cases} . \qquad (1.3.11)$$

Due to the electronic excitation carried by the I^* atom, the available energy (for distribution among translation and vibrotational excitation of R) is lower in the first path. The experimental results show that for CH_3I, about 20% of the products are formed with more translational energy than is consistent with the first path. Not so for n-C_3H_7I. There are practically no products with translational energy in excess of that allowed by the first path. This does not rule out the second path (since the n-C_3H_7 radical can soak up the energy not in translation). It does however suggest that I* formation is more dominant for n-C_3H_7I than for CH_3I. The ratio of I^*/I production is a function of the pumping wavelength.

1.3.5 Ionization

Highest on the photochemical energy scale is ionization – the photodetachment of electrons from neutral molecules. Since the resulting molecular ion is often formed with high internal excitation, ionization is frequently followed by fragmentation.

The final state in the ionization process is, as in dissociation, a continuum state, here, a state corresponding to the unbounded motion of an electron and a molecular ion. For polyatomic molecules such states are typically isoenergetic with different quasicontinua of bound states. These include states where the electron is bound by an internally excited ion. (The energy required for the ejection of the electron is provided by the deexcitation of the ion.) There may also be bound states of other, stable[21], electronic surfaces. Optical excitation is often to one or more of these nonstationary quasicontinua, known as "autoionizing" states. In short, direct photoionization is not necessarily the rule in polyatomics.

Using high-power lasers in the visible or near UV range, one can provide the energy required to ionization by multiphoton absorption. Each photon provides a substantial energy and, particularly at high powers, the up-pumping can compete with the lateral decay processes which deplete molecules which could otherwise absorb. The fragmentation pattern of benzene (Fig.1.24) and in particular the appearance of ions which require very high energies (such as C^+, whose minimal appearance requires 26 eV or 8 photons), indicates the considerable extent of up-pumping that can be achieved. Note also that (Fig. 1.24) the fragmentation pattern is qualitatively different from that observed following ionization by (70 eV) electron impact.

21 These are, of course, only nominally stable. The coupling between different electronic states enables them to ultimately ionize.

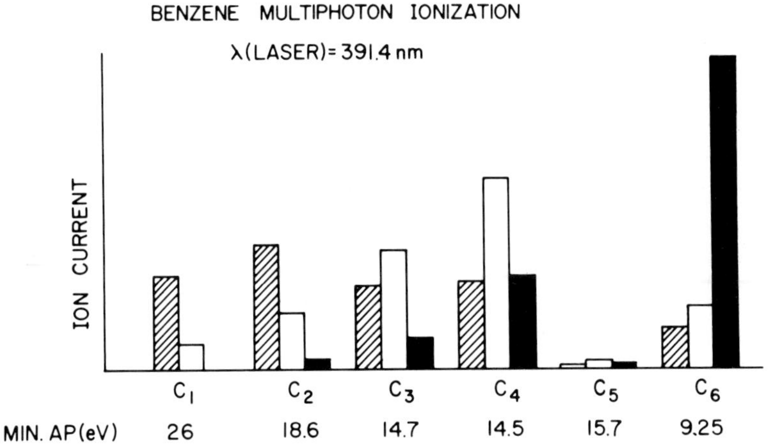

Fig. 1.24. Formation of fragment ions by dye laser resonance enhanced (Sect.5.4) multiphoton ionization of benzene. The bars show the relative abundance of fragments containing the indicated number of carbon atoms. Black bars are for electron impact ionization at 70 eV. Blank bars for weak focusing of the laser beam, and dashed bars for tight focusing. Fragmentation is much more extensive in the case of multiphoton ionization than in the case of electron impact. Note that C^+, absent from the electron impact spectrum, can be made the most abundant fragment by using high enough laser power. Minimum appearance potentials are indicated for each fragment. (Adapted from [1.39])

1.4 The Road Ahead

The purpose of this introductory chapter was only an overview of the observed molecular processes in a system of molecules and radiation in disequilibrium. In particular, we have examined the necessary (but not sufficient) conditions for lasing and the establishing of these by chemical reactions, the acquisition of energy by molecules as determined by the level structure and the distribution of oscillator strengths, and energy disposal by intramolecular and intermolecular processes, both reactive and of the energy transfer type. The latter, while acting primarily as the dissipation mechanism, can also be harnessed and put to good use.

The following chapters will examine many of the experiments, the phenomena, and the concepts in some depth. As is already evident, lasers and chemical change is an interdisciplinary subject and requires the synthesis of ideas and techniques (both experimental and theoretical) from a number of fields. We begin, in Chap.2, with the discussion of disequilibrium on both the molecular and the macroscopic levels. We address such questions as what do we mean by selectivity[22] and how do we characterize it in a simple quantitative fashion? Other topics include an introduction to the dynamics of molecular collisions, the exponential gap and how it can be bridged, and the implications of the selectivity and specificity of the microscopic processes to the macroscopic evolution. Chapter 3 is again technique oriented. The basic concepts of light-matter interaction are introduced. Laser physics is discussed, with an emphasis on the rate equation approach, used in the analysis of chemical lasers in Chap.4. A survey of atomic and molecular spectroscopy serves mostly to help the reader through Chaps.4 and 5 without extensive reference to auxiliary texts. Finally, several laser systems of interest to chemists are discussed.

Chapters 4 and 5 are problem oriented. What have we learned about chemical lasers and photoselective chemistry and what would we like to do and can do, in the near future. The accent in these chapters is on the practice. The operation and applications of chemical lasers is discussed in Chap.4 with special reference to representative examples. Case studies in laser chemistry including the design of the experimental arrangements and the interpretation of the results are analyzed in Chap.5, which concludes with the prospects for large-scale laser chemistry.

---

22 Ans: When all isoenergetic states are not equally populated.

2. Disequilibrium

Our concern in this volume is with molecular systems that have been displaced from thermal equilibrium, often to a considerable extent. Lasers and chemical lasers in particular operate because such displacements are feasible and are used, in turn, as means for driving other systems away from equilibrium. The first task is thus to examine those concepts and theoretical constructs of chemical physics that provide the framework for a discussion of disequilibrium on the molecular level. This chapter will center attention on the means for describing the system, means for the characterization of the phenomena in a compact fashion. The need for such a chapter stems from the traditional concern of physical chemistry with the simpler case of systems in thermal equilibrium. When disequilibrium was introduced it was often discussed as a macroscopic phenomenon. An extension of the traditional conceptual machinery is now required. In particular, the two mainstays of physical chemistry, chemical kinetics and chemical thermodynamics, need to be discussed from a molecular or state-to-state point of view.

2.1 Specificity and Selectivity of Chemical Reactions

2.1.1 Overview: Microscopic Disequilibrium

Traditionally, chemical reactions were studied under conditions which ensured thermal equilibrium; the reaction was maintained at a rate slow compared to that of collisional energy transfer. Any momentary depletion of excited reagents due to their high reaction rate was immediately restored by collisional activation. The reactants could be characterized as being in an effective thermal equilibrium at all times. The only control over the energy of the reactants was via their temperature. To characterize the variation of the reaction rate it was sufficient to study the temperature dependence of the rate constant. Similarly, the energy released by the reaction could be measured by the exothermicity, the mean excess of energy of the (thermally equilibrated) reactants over that of the products.

One can however consider probing chemical reactions on a more "molecular" level [2.1]. Here one overcomes the limitations imposed by thermal equilibrium and considers such questions as "How would the reaction rate be affected if energy is specifically pumped into the reagent vibration?" or "Immediately after the reaction, and before any collisional relaxation, how much of the available energy is in vibrational excitation of the newly formed products?" Ultimately one could even link the two questions: "How does reagent selective vibrational excitation influence the vibrational energy disposal in the products?" [2.2-5].

Consider, for example, the influence of reagent excitation on the rate of reactions [2.1-14]. Figure 2.1 shows the rate constants $k(v\rightarrow;T)$ for reactions of vibrationally excited HCl molecules with the different halogen atoms

$$I + HCl(v) \rightarrow Cl + HI \tag{2.1.1}$$

$$Br + HCl(v) \rightarrow Cl + HBr \tag{2.1.2}$$

$$F + HCl(v) \rightarrow Cl + HF \quad . \tag{2.1.3}$$

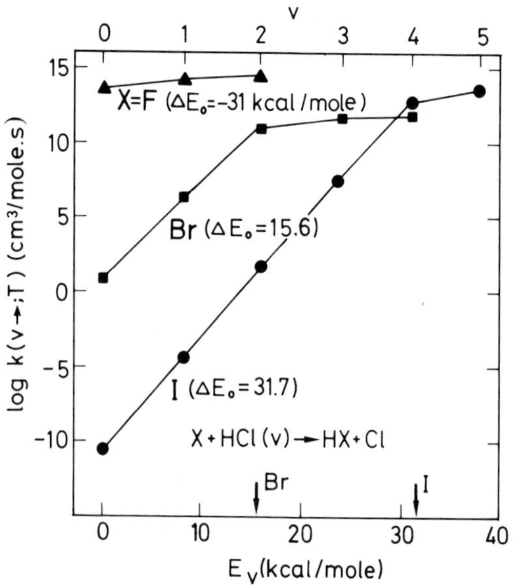

Fig. 2.1. The reaction rate coefficients $k(v\rightarrow;T)$ vs the vibrational energy of HCl for endoergic (X=I,Br) and exoergic (X=F) reactions. ΔE_0 is the difference between the ground state energies of the products and the reagents. In these reactions only the vibrational energy of HCl is specified. The other degrees of freedom (rotation of HCl, relative translation of X and HCl) have a thermal distribution. The arrows indicate the point where the (mean) energy of the reagents equals the endoergodicity (ΔE_0). Experimental results for X = F from [2.14] (relative rate constants) and [2.15] (absolute total rate constant); X = I from [2.16] (relative rates measured in the exoergic direction Cl + HI → HCl(v) + I, and converted by detailed balance, see below); X = Br from [2.17] (relative rates). Absolute rates for X = I,Br from [2.18]

The notation $k(v\rightarrow;T)$ serves as a reminder that the rate constant is for a process where the reagent diatomic is in a definite vibrational state while the distribution of the energy in the other degrees of freedom (rotation, translation) is that of thermal equilibrium. A far greater effect is evident for the endoergic I + HCl reaction than for the exoergic F + HCl reaction.

The activation energy E_a is defined as a measure of the increase of the reaction rate with the temperature of the reagents. For the reactions of vibrationally selected reagents one can similarly define $E_a(v)$ by

$$E_a(v) = -R d \ln[k(v\rightarrow;T)]/d(1/T) \quad . \tag{2.1.4}$$

$E_a(v)$ has the (rigorous) interpretation as the mean energy of those molecules that react minus the mean energy of all molecules [2.1]. We would thus expect that $E_a(v)$ would decrease with increasing vibrational excitation of the reagents (Fig.2.2).

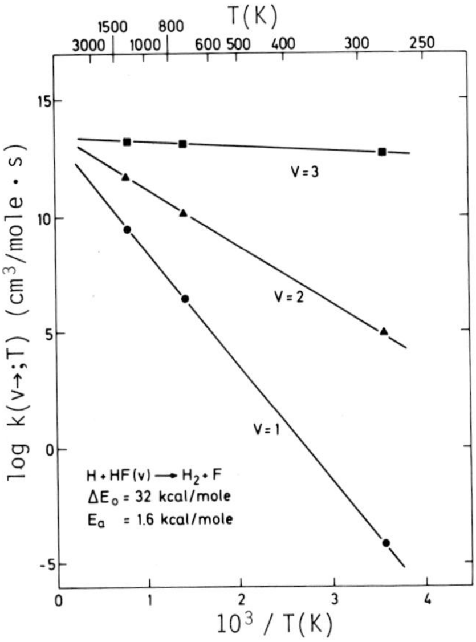

Fig. 2.2. Temperature dependence of the reaction rate constant from specified reagent vibrational states for the endothermic H + HF(v) reaction. The "Arrhenius" plot serves to show that the "preexponential" factor (the $T\rightarrow\infty$ limit) is nearly independent of v and the increase of the endothermic rate constant with reagent vibration is primarily due to the decrease in the activation energy. Experimental results adapted from [2.19]

For an endoergic reaction, the reagents must, on thermodynamic grounds, have a minimal excess energy (equal to the reaction endoergicity) before reaction can take place. Thus, endoergic reactions will always have an energy of activation and their reaction rates would be greatly enhanced by any excitation of the reagents. Even so, the dependence of the activation energy on E_v as evident in Fig.2.2 and in other reactions cannot be completely accounted for in this fashion. The decrease of $E_a(v)$ with increasing E_v is with a slope that exceeds unity [2.12], indicating that reagent vibrational excitation is particularly effective in promoting the endoergic reaction when all other degrees of freedom have a thermal distribution. The comparison of the efficacies of vibrational and translational energies is given in Sect.2.2.8.

The potential practical implications of the enhancement of reaction rates by reagent excitation requires that one considers two important loss mechanisms. The first is the intramolecular migration of excitation energy in a polyatomic molecule. How specific is the effect? Can the internal excitation be locked in a specific mode or, if it cannot, can the collision or the unimolecular rearrangement under consideration compete with the intramolecular relaxation? Evidence bearing on this point will be discussed in Sect.2.3.7. Even when the excitation energy is rapidly redistributed the effective "heating" of the internal degrees of freedom that is achieved by a quantum of vibrational motion is significant. The second loss mechanism is relaxation by collisions. This will be particularly serious in liquids where the density is typically three orders of magnitude higher than in the gas phase. Collisional loss of vibrational energy in the gas phase can however be fairly inefficient, particularly so for nonpolar diatomic molecules. The presently available experimental evidence suggests that the rule of thumb for the liquid phase is that one can scale the gas phase results to the liquid by simply allowing for the change in the density, $\tau_{liquid} = (\bar{v}_{liquid}/\bar{v}_{gas})\tau_{gas}$, where $\bar{v}$ is the molar volume. The range of measured vibrational relaxation times (in pure liquids) spans the same considerable range that is found in gases (Fig.1.9). From upwards of 1 s for liquid N_2 [2.20] down to the subnanosecond range for both intramolecular and intermolecular vibrational energy transfer of polyatomic molecules [2.21, 22], for example, the V-T relaxation time from the ν_3 mode of SF_6 is 27 (±7) 10^{-12} s in liquid O_2 or 160 (±40) 10^{-12} s in liquid Ar [2.22]. The scaling law does correlate these rates with the results obtained in the gas phase. At the lower end of the scale, the lifetime is just long enough to compete with intramolecular rearrangements or intermolecular encounters with a scavenger.

Reaction rates can thus be enhanced by selective reagent excitation. Variations are also evident in the energy disposal of chemical reactions. Figure 2.3 shows the populations of the vibrational states of HCl formed in a series of

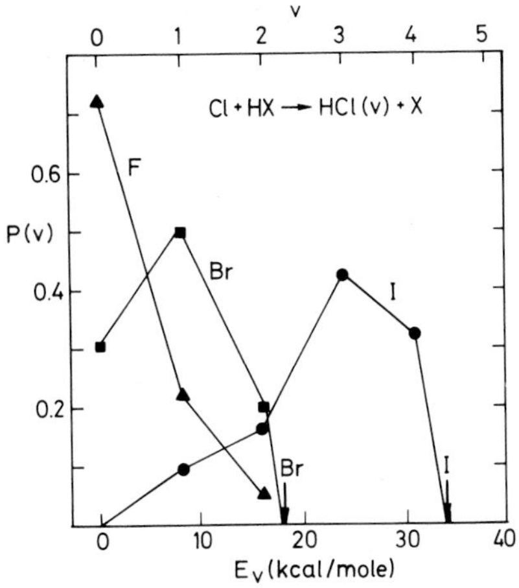

Fig. 2.3. P(v), the fraction of HCl in the vibrational state v, produced in the Cl + HX reaction vs v. Experimental results from [2.16,17]. The arrows indicate the magnitude of the exothermicity (the reaction is endothermic for X = F). The higher is the exothermicity the more extensive is the population inversion

$$Cl + HX \rightarrow X + HCl(v) \tag{2.1.5}$$

reactions: from extensive inversion in the highly exoergic Cl + HI reaction to a total lack of inversion in the endoergic Cl + HF reaction.

Comparison of Figs.2.3 and 2.1 suggests that selectivity in energy consumption and specificity in energy disposal must be closely related. The Cl + HI → I + HCl(v) reaction preferentially populates the higher vibrational states of HCl and the rate of the I + HCl(v) → Cl + HI reaction is very strongly enhanced by HCl vibrational excitation. For all reactions selectivity and specificity go hand in hand. Our task is to make this correlation explicit and to devise a common measure to characterize it. We shall then seek to determine the origin of these effects from the point of view of the dynamics of these collisions.

2.1.2 The Detailed Rate Constant

For an elementary bimolecular reaction the reaction rate constant is defined by writing down the reaction rate as a typical second-order kinetics law. Thus for the reaction (Fig.2.2)

$$H + HF(v) \rightarrow F + H_2 \quad , \tag{2.1.6}$$

the rate of disappearance of H atoms due to their reaction with HF molecules in the vibrational state v is expressed as

$$-d[HF(v)]/dt = k(v\rightarrow;T)[H][HF(v)] \quad , \tag{2.1.7}$$

where the square brackets denote concentrations. We have assumed here, as before, that the rotational states of HF and the kinetic energy have a thermal distribution at the temperature T. Hence the rate constant $k(v\rightarrow;T)$ can depend only on the vibrational state of HF and on the temperature. As in ordinary chemical kinetics, the fact that the reaction is an elementary bimolecular one ensures that $k(v\rightarrow;T)$ defined by (2.1.7) is independent of the concentrations.

Chemical kinetics with state-selected reactants is no different than ordinary chemical kinetics. We need simply regard HF molecules in a specified vibrational state as a distinct chemical species (always provided that the other, nonselected degrees of freedom of the reagents remain in thermal equilibrium).

For completely thermal reactants, H atoms can be consumed by their reaction with HF molecules in different vibrational states. The total reaction rate is then just the sum of the rates of the separate processes that lead to disappearance of H atoms

$$-d[H]/dt = -d[HF]/dt = -\sum_v d[HF(v)]/dt = \sum_v k(v\rightarrow;T)[H][HF(v)]$$

$$= \left\{\sum_v k(v\rightarrow;T)[HF(v)]/[HF]\right\}[H][HF] \quad . \tag{2.1.8}$$

Hence the thermal reaction rate $\vec{k}(T)$ is related to $k(v\rightarrow;T)$ by

$$\vec{k}(T) = \sum_v p(v|T)k(v\rightarrow;T) \quad . \tag{2.1.9}$$

Here $p(v|T)$,

$$p(v|T) = [HF(v)]/[HF] \quad , \tag{2.1.10}$$

is the fraction of HF molecules that are in the vibrational state v at thermal equilibrium.

The thermal rate constant is obtained as an average of the detailed rate constants over the distribution of initial vibrational states. It is clear however that nothing in our discussion here limits it to vibrational states. v can equally well stand for any specification of the initial state. Hence the general rule is: When there is a distribution of the states of the reagents, the total rate constant is obtained as a weighted sum (i.e., an average) over the rate constants for the different initial states.

The asymmetric character of the definition of the rate constant is made explicit when we consider a resolution of the states of the products. Consider, say, the reaction

$$H + HF(v) \rightarrow F + H_2(v') \quad . \tag{2.1.11}$$

The rate of consumption of HF molecules in the state v to form H_2 molecules in the state v' is expressed as $k(v\rightarrow v';T)[H][HF(v)]$. But clearly the rate of consumption of HF(v) is the sum of the rates to form $H_2(v')$ in the different vibrational states v' so that

$$k(v\rightarrow;T) = \sum_{v'} k(v\rightarrow v';T) \quad , \tag{2.1.12}$$

and, using (2.1.9)

$$\vec{k}(T) = \sum_{v} p(v|T) \sum_{v'} k(v\rightarrow v';T) \quad . \tag{2.1.13}$$

The dichotomy between reagents and products states evident in (2.1.13) is summarized by the "canon": Sum over final states, average over initial states.

The form (2.1.13) remains valid also for more elaborate methods for preparing an ensemble of reactants, e.g., by laser excitation of the sample. There is however a very significant practical difference. In the general case, where for "preparation X"

$$\vec{k}(X) = \sum_{i} p(i|X) \sum_{f} k(i\rightarrow f;X) \tag{2.1.14}$$

we do not necessarily know the weights p(i|X) of the different initial states i in the mixture prepared by "method X". This is of course not the case when "X" is "thermal equilibrium". In Sect.2.1.7 we discuss an approach whereby such weights like p(i|X) can be inferred.

2.1.3 Detailed Balance

The reaction

$$F + H_2 \rightarrow H + HF(v) \tag{2.1.15}$$

is simply the reaction (2.1.6) in the reverse direction. The rate constant $k(\rightarrow v;T)$ for formation of HF in the vibrational state v

$$d[HF(v)]/dt = k(\rightarrow v;T)[F][H_2] \quad , \qquad (2.1.16)$$

may well therefore be related to the rate $k(v\rightarrow;T)$ for consumption of HF(v) by the reversed reaction (2.1.6).

The principle of detailed balance is a statement about rates (not rate constants) of collisions which enables us to provide the connection. It states that *at equilibrium* the rate of a detailed process is equal to the rate of the reversed process. Hence, using equilibrium concentrations in (2.1.7,16)

$$k(v\rightarrow;T)[H][HF(v)] = k(\rightarrow v;T)[F][H_2] \quad , \qquad (2.1.17)$$

where we emphasize that in (2.1.17) the concentrations are at equilibrium. Rearranging (2.1.17), we have

$$k(\rightarrow v;T)/k(v\rightarrow;T) = K(\rightarrow v;T) \quad . \qquad (2.1.18)$$

Here $K(\rightarrow v;T)$,

$$K(\rightarrow v;T) = [H][HF(v)]/[F][H_2] \qquad (2.1.19)$$

is the equilibrium constant for the $F + H_2 \rightarrow H + HF(v)$ reaction, at *thermal equilibrium*. The concentrations in (2.1.19) as in (2.1.17) are those that obtain at equilibrium. Again we see that if we regard HF(v) as a distinct chemical species then (2.1.18) is just the usual result that for an elementary process the equilibrium constant is the ratio of the forward and reverse rate constants.

The considerable utility of expressing the principle of detailed balance in the form (2.1.18) stems from our ability to compute the equilibrium constant without any kinetic input. The simplest route to computing $K(\rightarrow v;T)$ is to relate it to $K(T)$,

$$K(T) = \overleftarrow{k}(T)/\overrightarrow{k}(T) = [H][HF]/[F][H_2] \qquad (2.1.20)$$

the equilibrium constant of the thermal

$$F + H_2 \rightarrow H + HF \qquad (2.1.21)$$

reaction. Using the definitions (2.1.10,19,20)

$$K(\rightarrow v;T) = p(v|T)K(T) \quad . \qquad (2.1.22)$$

Standard thermochemical tables enable one to compute $\Delta G^0(T)$, the standard free energy change of the reaction, in terms of the standard free energies of the reactants and products; $\Delta G^0(T) = -RT \ln[K(T)]$.

Alternatively, one can compute K(T) in terms of the partition functions of the reactants and products,

$$K(T) = (\mu'/\mu)^{3/2}(Q_{HF}/Q_{H_2}) \exp(-\Delta E_0/kT) \quad . \tag{2.1.23}$$

Here μ' and μ are the reduced masses of the products $\mu' = m_H m_{HF}/(m_H+m_{HF})$ and reactants, respectively, and the Q's are the partition functions for the *internal* degrees of freedom of the diatoms.

The previous discussion and its extension in Sect.2.1.5 are based on the principle of detailed balance as derived from the need to balance the rates of any two opposing processes, at equilibrium [2.23]. One can however reach the same conclusions, starting from a (theoretical) examination of the collision under the operation of time reversal. The procedure is then referred to as "microscopic reversibility". The final, working equations are however the same, irrespective of the starting point [2.24,25].

2.1.4 Energy Disposal and Energy Consumption

The principle of detailed balance cast in the form (2.1.17) clearly implies the essence. Those vibrational states that are preferentially populated in the forward reaction would react with a high rate in the reversed reaction. To compare the reactivity of two different vibrational states, say, n and m we can write, using (2.1.10,17)

$$p(n|T)k(n\rightarrow;T)/p(m|T)k(m\rightarrow;T) = k(\rightarrow n;T)/k(\rightarrow m;T) \quad . \tag{2.1.24}$$

By definition, $p(n|T)$ is the fraction of HF molecules in the vibrational state n at thermal equilibrium. Thus $p(n|T)k(n\rightarrow;T)$ is the actual rate of the reaction from the nth vibrational state of HF (at unit concentration) at thermal equilibrium. The relative rates of reactions of two different states equals the relative rates of populating the two states by the reversed reaction. If we now allow a more general interpretation of the index n, as any state (or group of states) of the reactant molecule, then (2.1.24) remains valid.

Rather than considering the relative rates we can examine P(v), the fraction of reactive collisions which populate the HF vibrational state v in the reaction (2.1.15),

$$P(v) = k(\to v;T)/\overleftarrow{k}(T) \qquad (2.1.25)$$

$$\overleftarrow{k}(T) = \sum_v k(\to v;T) \qquad (2.1.26)$$

$$\sum_v P(v) = 1 \quad . \qquad (2.1.27)$$

The principle of detailed balance can now be reexpressed by noting that P(v) is also the fraction of reverse collisions which proceed from HF in the vibrational state v,

$$P(v) = p(v|T)k(v\to;T)/\overrightarrow{k}(T) \qquad (2.1.28)$$

$$\overrightarrow{k}(T) = \sum_v p(v|T)k(v\to;T) \quad . \qquad (2.1.29)$$

The two expressions (2.1.25) and (2.1.28) do not appear to be equivalent. This is merely a reflection of the asymmetric canon: Average over initial states sum over final states. By substituting (2.1.18,22) into (2.1.28) one readily verifies that (2.1.25) = (2.1.28).

The same probability distribution, P(v) (Fig.2.3), characterizes the vibrational energy disposal in the forward reaction and the vibrational energy requirements in the reverse reaction. The results displayed in Figs.2.1 and 2.3 are thus identical in their content. In fact, only one set of measurements was employed by us to draw both figures, using the equivalence of (2.1.25)

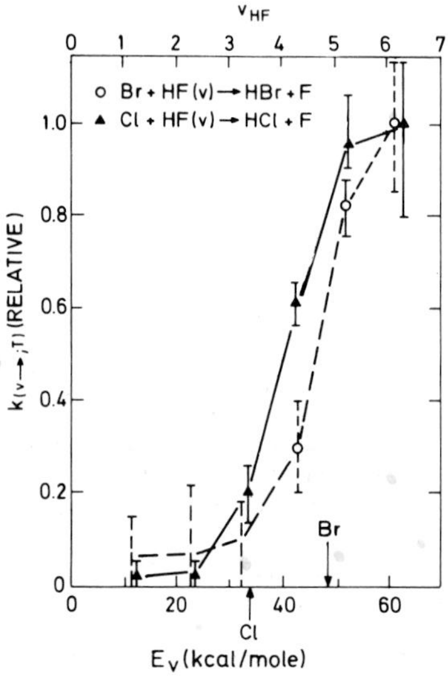

Fig. 2.4. Experimentally determined rate constants for endoergic reactions. The arrows mark the onset of exothermic transitions ($E_V-\Delta E_0 \geq 0$). The error bars demonstrate the difficulty in measuring (or computing) reaction rates for endothermic transitions (adapted from [2.17])

and (2.1.28). Even so, one cannot conclude that the two routes to P(v) are equivalent on practical grounds [2.12]. Consider the exothermic $F+H_2$ reaction (2.1.15). The measured [2.19] product vibrational state distribution has but a moderate v dependence. Hence from (2.1.28), the exponential v dependence of $p(v|T)$ implies that $k(v\rightarrow;T)$ for the endothermic $H+HF(v)$ reaction (2.1.6) will increase essentially exponentially with v (Fig.2.2). The exponential dependence is so dominant that it tends to wash out the more moderate v dependence of P(v). Thus it is far easier to determine the v dependence of P(v) via the energy disposal [i.e., using (2.1.25)] than through the energy consumption [i.e., using (2.1.28)] (similar conclusions apply to the role of reagent translational energy). This is reflected in the large error bars in direct measurements of endothermic rate constants (Fig.2.4). Once E_v is sufficiently large so that $E_v - \Delta E_0 \gtrsim 0$, it becomes easier to measure the rate coefficients for energy consumption, as is evident from the figure.

2.1.5 The Reaction Probability Matrix

We have thus far considered the simpler situation where either the state of the reagents was selected or the state of the products was probed. Experimental and computational studies show that reagent excitation not only affects the magnitude of the overall reaction rate (Fig.2.2), but also modifies the state distribution of the products (Fig.2.5). When reagent state selection is carried out it is not possible to consider "the" product state distribution. The initial state must also be specified.

We can relate the rate constants for the forward and reverse reactions, say, $F+H_2(v) \rightleftarrows H+HF(v')$, by appealing again to the principle of detailed balance. Thus

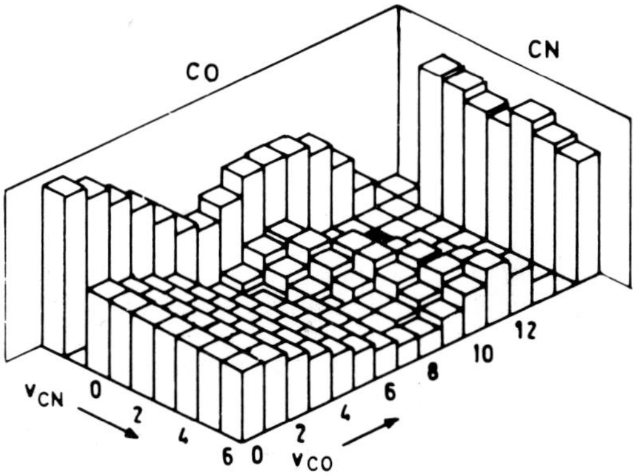

Fig. 2.5. The distribution of products vibrational states as a function of the reagents vibrational states, for the $O+CN(v)\rightarrow N+CO(v)$ reaction (adapted from [2.26]). The histogram along the CN axis is the nascent CN vibrational distribution following photolysis of $(CN)_2$. The CO vibrational distribution which results is shown along the CO axis

$$k(v\to v';T)/k(v'\to v;T) = K(v\to v';T) = [H][HF(v')]/[F][H_2(v)] \quad . \tag{2.1.30}$$

Here $K(v\to v';T)$ is the equilibrium constant for the process. As in the proof of (2.1.22) one concludes that

$$K(v\to v';T) = p(v'|T)K(T)/p(v|T) \quad . \tag{2.1.31}$$

For a nonreactive collision, e.g., $Cl + HCl(v) \rightleftarrows Cl + HCl(v')$, we have $K(T) = 1$ and

$$K(v\to v';T) = \exp[-(E_v - E_{v'})/kT] \quad . \tag{2.1.32}$$

In general, for atom-diatom collisions

$$K(v\to v';T) = (\mu'/\mu)^{3/2}(Q_R'/Q_R)\exp[-(E_v - E_{v'} - \Delta E_0)/kT] \quad , \tag{2.1.33}$$

where Q_R is the rotational partition function. Clearly, (2.1.33) and (2.1.30) are the state-to-state analogues of the thermal expressions (2.1.23) and (2.1.20), corresponding to the reactions (2.1.11) and (2.1.21), respectively.

Combining our three key results (2.1.20), (2.1.30), and (2.1.31) we can write the principle of detailed balance as

$$\begin{aligned} P(v,v') &\equiv p(v|T)k(v\to v';T)/\vec{k}(T) \\ &= p(v'|T)k(v'\to v;T)/\overleftarrow{k}(T) = P(v',v) \quad . \end{aligned} \tag{2.1.34}$$

For inelastic (energy transfer) collisions where $\vec{k}(T) = \overleftarrow{k}(T)$ the results (2.1.34) and (2.1.25) = (2.1.28) read

$$p(v|T)k(v\to v';T) = p(v'|T)k(v'\to v;T) \tag{2.1.35}$$

and

$$p(v|T)k(v\to;T) = k(\to v;T) \quad . \tag{2.1.36}$$

These results will be used extensively in Sect.2.5. Since $\vec{k}(T)$ is the thermal reaction rate constant of the forward reaction it follows from (2.1.13) that the numbers $P(v,v')$ are normalized

$$\sum_v \sum_{v'} P(v,v') = 1 \quad . \tag{2.1.37}$$

The role of vibrational energy in a process can thus be represented by an array $P(v,v')$. It is a direct implication of detailed balance that the array corresponding to the reversed process is simply the transpose of the one for the forward process. The same array thus suffices to characterize both the forward and the reverse processes.

Less detailed information is readily extracted from the array or "matrix" $P(v,v')$. We first examine the significance of the sums of the elements along a given row or given column. Taking the sum of elements along the row v' we obtain $P(v')$, using (2.1.34),

$$
\begin{aligned}
P(v') &= \sum_{v} P(v,v') = \sum_{v} p(v|T)k(v \to v';T)/\vec{k}(T) = k(\to v';T)/\vec{k}(T) \\
&= \sum_{v} p(v'|T)k(v' \to v;T)/\overleftarrow{k}(T) = p(v'|T)k(v'\to;T)/\overleftarrow{k}(T) \quad . \qquad (2.1.38)
\end{aligned}
$$

By examining (2.1.38) we recognize $P(v')$ as the distribution of HF vibrational states in the $H_2 + F$ reaction using thermal reactants or the relative reaction rates in the reversed $H + HF(v')$ reaction (irrespective of the state of the products). Similarly, for the column sums we obtain

$$
\begin{aligned}
P(v) &= \sum_{v'} P(v,v') = p(v|T)k(v\to;T)/\vec{k}(T) \\
&= \sum_{v'} p(v'|T)k(v' \to v;T)/\overleftarrow{k}(T) = k(\to v;T)/\overleftarrow{k}(T) \quad . \qquad (2.1.39)
\end{aligned}
$$

The normalized distribution $P(v)$ can thus be interpreted either as the relative reactivity in the $F + H_2(v)$ reaction or as the distribution of the product vibrational states in the $H + HF$ reaction. Probability matrices for other reactions will, of course, have a corresponding interpretation.

Detailed balance, in itself, is sufficient to ensure that, say, the distribution of HF vibrational states in the $F + H_2$ reaction equals the different relative reaction rates in the $H + HF(v)$ reaction. It relates the energy disposal of the forward reaction to the energy consumption of the reverse reaction. To relate the energy consumption in the forward reaction [i.e., $P(v)$] to the energy disposal in the same, forward, reaction we must proceed via the $P(v,v')$ matrix. Detailed balance, in itself, does not suffice.

Finally, we note that there was nothing in our discussion in this section that was unique to vibrational states. The symmetry (2.1.34) remains valid if v stands for any specification of the reagent energy states and similarly for v' and the products. Note however that any degree of freedom of the *reagents* that is not specified must be in thermal equilibrium.

2.1.6 Measures of Specificity and Selectivity

For a system at thermal equilibrium it is shown in every textbook in statistical mechanics that the fraction of molecules $p(n|T)$ in the quantum state n is given by the one parameter distribution

$$p(n|T) = \exp(-E_n/kT)/Q \quad . \tag{2.1.40}$$

Q, the partition function, ensures that $p(n|T)$ is normalized

$$Q(T) = \sum_n \exp(-E_n/kT) \tag{2.1.41}$$

and is hence a function of T. The magnitude of the parameter T, known as "the temperature", is determined in principle in terms of the mean energy $\langle E \rangle$,

$$\langle E \rangle = \sum_n E_n p(n|T) \quad , \tag{2.1.42}$$

of the system. In other words, the value of T is that found by equating the right-hand side of (2.1.42), regarded as a function of T, to the value of $\langle E \rangle$. The purpose of this and the following section is to discuss a point of view where the equilibrium distribution appears as a special case of a more general approach. This point of view is not limited to the discussion of macroscopic systems. Hence we shall be able to provide a unified framework for the description and analysis of systems in disequilibrium valid for both the macroscopic and microscopic domains. The same machinery that will be used in Sect.2.2 to discuss specificity of energy disposal and selectivity of energy consumption (disequilibrium in a single collision) will also be used in Sect.2.5 to discuss macroscopic disequilibrium, where the bulk concentrations differ from their equilibrium value.

As was emphasized in Chap.1, this is a volume about systems in disequilibrium and hence the unified approach is particularly useful. In a limited space we cannot however do full justice to the different aspects of this formalism. The general approach [2.27-29] with special reference to macroscopic systems in equilibrium is discussed in several textbooks [2.30-32]. There are also a number of reviews [2.13,29,33-35] dealing in detail with microscopic disequilibrium. The formalism is thermodynamic-like but adopts its central theme from information theory [2.36-38]: That the entropy of a distribution is the measure of the missing information.

A distribution P(n) over quantum states n is assigned the entropy

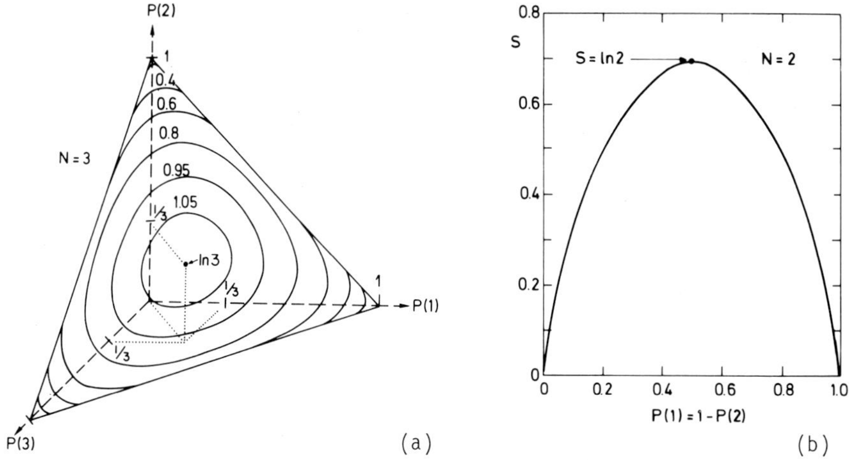

Fig. 2.6. (a) Contour plots of the entropy for a distribution over three states. The three Cartesian axis specify P(1), P(2), and P(3). All possible normalized [P(1)+P(2)+P(3)=1] distributions correspond to points located in the plane bounded by the triangle shown. The entropy is maximal [ln(3)] at the midpoint [P(1)=P(2)=P(3)=1/3] of the triangle and decreases as one moves towards the vertices. The contours along an edge (where one probability is zero) represent the entropy for a two-state system (figure prepared by J. Manz). (b) The entropy for a two-state system P(1) + P(2) = 1

$$S = -\sum_n P(n)\ \ln[P(n)] \quad . \tag{2.1.43}$$

As defined in (2.1.43), the entropy is dimensionless. To obtain the entropy in thermodynamic units it is necessary to multiply the sum by Boltzmann's constant k. As a check, one readily verifies that for the thermal equilibrium distribution (2.1.40) the entropy is

$$S = k\ \ln(Q) + \langle E \rangle / T \tag{2.1.44}$$

which is the familiar expression.

We shall regard (2.1.43) as defining the entropy for any distribution over quantum states. Figure 2.6 shows the entropy for a system of three and of two states. The entropy is seen to attain a (unique) maximal value for the uniform (all states are equiprobable) distribution. We shall show below that the same holds for any number of states: The (unique) maximal value of the entropy of a distribution over quantum states is for the uniform distribution. To compute the entropy one needs the distribution over quantum states. Experiments will often provide only a lower resolution. For example, in the study of energy disposal, only the product vibrational state distribution may be known. For an isolated A + BC collision, the total energy is the sum of the vibrational and rotational energies plus the translational energy of the relative motion. Even when the vibrational energy is given, the balance of the energy can be partitioned in various ways between rotation and translation. Specifying the vibrational state of BC does not therefore specify a unique quantum state for the isolated collision. Rather, it specifies a group of different quantum states each of which corresponds to the same vibrational state of BC. Let there be g(v) quantum states in the group v (the counting of states is discussed in Appendix 2.A). When all the quantum states within the group are equiprobable,

$$P(n) = P(v)/g(v) \tag{2.1.45}$$

and (2.1.43) becomes

$$S = -\sum_n [P(v)/g(v)]\ln[P(v)/g(v)] \quad . \tag{2.1.46}$$

The terms in the square brackets in (2.1.46) are constant for all the states which belong to the same group and one can sum over n within each group separately, leading to

$$S = -\sum_v P(v)\ln[P(v)/g(v)] \quad . \tag{2.1.47}$$

It is shown in Appendix 2.A that the final result (2.1.47) remains valid even if all the states within the group are not equally probable [in which case ln[g(v)] is the entropy of the distribution of quantum states within the group v]. It should also be clear that (2.1.47) is valid for any grouping of quantum states where v is the group index.

The prior distribution $P^0(v)$ is defined as that (unique) distribution over the groups v for which the entropy is maximal. We first state the result,

$$P^0(v) = g(v)/\sum_v g(v) \quad , \tag{2.1.48}$$

and then proceed to prove it. With the definition (2.1.48), the entropy (2.1.47) can be written as

$$S = S_{max} - DS[P|P^0] \tag{2.1.49}$$

$$DS[P|P^0] = \sum_v P(v)\ln[P(v)/P^0(v)] \tag{2.1.50}$$

and $S_{max} = \ln\left[\sum_v g(v)\right]$. For the special case of systems at a given total energy when g(v) is the number of states in the group v, S_{max},

$$S_{max} = \ln\left[\sum_v g(v)\right] = \ln\left[\sum_n 1\right] \quad , \qquad (2.1.51)$$

is the logarithm of the number of accessible quantum states.

The measure DS will play a central role in our discussion. We first show that $DS[P|P^0]$ is nonnegative and equals zero if and only if $P(v) = P^0(v)$ for all v. Hence S_{max} is the unique maximal value of the entropy which can be realized only if $P(v) = P^0(v)$. In particular, if there is no grouping, g(v) = 1, then our proof shows that the entropy is maximal only for a uniform distribution over quantum states.

The proof is based on the inequality $\ln(x) \leqq x-1$ or, equivalently, by putting $y = 1/x$ that $\ln(y) \geqq 1-1/y$ with equalities holding if and only if $x = y = 1$ (Fig.2.7). It follows that, since both P(v) and $P^0(v)$ are normalized,

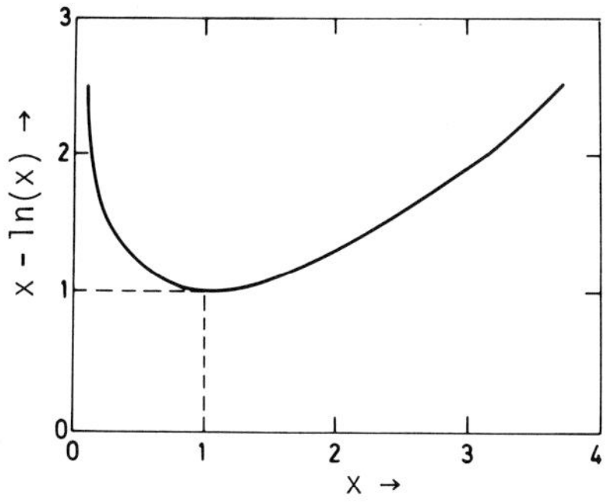

Fig. 2.7. The inequality $\ln(x) \leqq (x-1)$ with equality only at x = 1

$$DS[P|P^0] = \sum_v P(v)\ln[P(v)/P^0(v)] \geqq \sum_v P(v)[1-P^0(v)/P(v)] = 0 \quad . \qquad (2.1.52)$$

$DS[P|P^0]$ is thus positive except if and only if $P(v) \equiv P^0(v)$, when it equals zero.

$DS[P|P^0]$ is the difference between the maximal and actual value of the entropy (2.1.49). For this reason it is known as "the entropy deficiency". It is an overall measure of the deviance of the observed distribution from the prior one.

The qualitative concept of specificity of energy disposal, or selectivity of energy consumption, is thus given a quantitative definition by comparing the observed distribution with the prior one (where all isoenergetic quantum states are equally probable). Appendix 2.C justifies the definition of the prior distribution on dynamical grounds. The overall deviance from the prior is measured by the entropy deficiency while the deviance of a particular result (the "local" deviance) is characterized by the "surprisal",

$$I(v) = -\ln[P(v)/P^{0}(v)] \quad . \qquad (2.1.53)$$

In practice, it is often the case (Sect.2.2) that the surprisal has a simple structure. The rest of this section motivates the choice of the entropy as the measure of deviance from the prior distribution and provides a functional form for the surprisal.

2.1.7 The Maximum Entropy Formalism

Consider a quantal system at a given value of the energy (and of other conserved quantities). For a system of more than one degree of freedom, there can be a number of degenerate quantum states which are consistent with the stated conditions. The only description of such a system which is a solution of the quantal equations of motion, which is invariant under time evolution and which has the same form in any basis set, is the uniform – so-called "microcanonical" – distribution. This symmetry consideration dictates the choice of the reference distribution – all degenerate quantum states are equiprobable. Any other choice implies at least a partial removal of the degeneracy and hence implies that some additional information is available. Given the choice of the reference distribution, one can still question the use of the functional form (2.1.43) as the measure of disequilibrium. After all, there are very many alternative functions of the probabilities P(n) (so-called "convex functions"), all of which share the property that their unique maximal value is for the uniform distribution. One such example is $\sum_n P^2(n)$ but there are many others. The property that singles out the entropy was already pointed out by Boltzmann and Planck. It is the connection between the entropy and the probability of an entire distribution [2.29,30,34]. The more probable is a particular distribution of outcomes – the higher is its entropy. The derivation of this result has been often discussed and will not be repeated here.

Among all possible distributions over quantum states, the prior distribution (whose entropy is maximal) is the most probable one. It may well be the case however that attention should be limited to a subset of all possible distributions. The most familiar example is thermal equilibrium where the mean value of the energy of the system is given and one only seeks the most probable distribution among the set of distributions whose mean energy is given. The additional conditions that limit the range of distributions under consideration will be referred to as "constraints". In the presence of constraints the most probable distribution is not necessarily the prior one. Rather, among the set of distributions consistent with the constraints it is the one of maximal entropy.

Applications of this point of view for specifying a probability distribution are discussed in Sects.2.2 and 2.5 and elsewhere [2.27-35]. Here we are concerned with the functional form of such a distribution of maximal entropy. Consider the constraints as the mean values of m quantities $\langle A_r \rangle$,

$$\langle A_r \rangle = \sum_v P(v) A_r(v) \quad , \quad r = 1,\dots,m \quad , \tag{2.1.54}$$

where $A_r(v)$ is the value of A_r for the group v. The mean value of the energy, (2.1.42), is a particular example of a constraint. The distribution which is normalized, consistent with the m values $\langle A_r \rangle$, $r = 1,\dots,m$, and of maximal entropy is given by

$$P^{ME}(v) = P^0(v) \exp\left[-\lambda_0 - \sum_{r=1}^{m} \lambda_r A_r(v)\right] \quad . \tag{2.1.55}$$

The functional form (2.1.55) contains $m+1$ (Lagrange) parameters λ_0 and λ_1, $\dots,\lambda_m$. Their values are determined by the $m+1$ conditions that $P^{ME}(v)$ is normalized and reproduces the m values $\langle A_r \rangle$. In general, they need to be determined numerically [2.39]. It is sometimes convenient however to impose normalization explicitly by making λ_0 a function of $\lambda_1,\dots,\lambda_m$ via the normalization condition

$$\exp(\lambda_0) = \sum_v P^0(v) \exp\left[-\sum_{r=1}^{m} \lambda_r A_r(m)\right] \quad . \tag{2.1.56}$$

The structure of (2.1.56) suggests that $\exp(\lambda_0)$ be referred to as the "sum over states" or, using the more common terminology as the "partition function", $Q = \exp(\lambda_0)$.

If there are no constraints (beyond normalization) $m = 0$, and, from (2.1.56) since $P^0(v)$ is normalized, $\lambda_0 = 0$. Hence, in the absence of constraints the distribution of maximal entropy $P^{ME}(v)$ is just the prior distribution. Otherwise, to show that $P^{ME}(v)$ is indeed the (unique) distribution of maximal entropy from the set of P(v)'s satisfying the constraints consider the inequality

$$DS[P|P^{ME}] = \sum_v P(v)\ln[P(v)/P^{ME}(v)] \geqq 0 \tag{2.1.57}$$

with equality if and only if $P(v) = P^{ME}(v)$ (cf. Fig.2.7). From (2.1.57) and (2.1.54) and putting $A_0(v) = 1$

$$\begin{aligned} DS[P|P^{ME}] &= \sum_v P(v)\ln[P(v)/P^{o}(v)] - \sum_v P(v)\ln[P^{ME}/P^{o}(v)] \\ &= DS[P|P^{o}] + \sum_{r=0}^{m} \lambda_r \sum_v P(v)A_r(v) \\ &= DS[P|P^{o}] + \sum_{r=0}^{m} \lambda_r \langle A_r \rangle \\ &= DS[P|P^{o}] - \sum_v P^{ME}(v)\ln[P^{ME}(v)/P^{o}(v)] \\ &= DS[P|P^{o}] - DS[P^{ME}|P^{o}] \geqq 0 \quad . \end{aligned} \tag{2.1.58}$$

The entropy deficiency of any distribution P(v) which is consistent with the constraints either exceeds the entropy deficiency of the distribution $P^{ME}(v)$ or equals it, in which case $P(v) \equiv P^{ME}(v)$.

The thermal equilibrium of quantum states (2.1.40) is a particular example of a distribution of maximal entropy. In general, we define equilibrium as the special case where the constraints are conserved quantities and, furthermore, are additive for independent systems (e.g., the energy of two separate systems is the sum of the energies of the two systems). The constraints for chemical equilibrium, for example, are the numbers C_r of the atoms of the different kinds that are in the system [2.40], while the unknown distribution is the numbers C_j of molecules of different kinds which can be formed from these atoms

$$C_r = \sum_j C_j A_r(j) \quad , \tag{2.1.59}$$

where $A_r(j)$ is the number of atoms of the type r per molecule of the type j. The number m of constraints is smaller (or, at most equal) to the number of unknowns. The new feature when we come to systems in disequilibrium is the nature of the constraints. One can show that also out of equilibrium the constraints retain the character of a conserved quantity, but of a novel type [2.41]. The functional form of the surprisal remains however the same: For a system with m constraints it is [cf.(2.1.55), $A_0(v)=1$],

$$I(v) = -\ln[P^{ME}(v)/P^{o}(v)] = \sum_{r=0}^{m} \lambda_r A_r(v) \quad . \qquad (2.1.60)$$

Equation (2.1.60) is the surprisal of the distribution of maximal entropy subject to m specified constraints. We are, however, often faced with the reversed problem: An observed distribution P(v) is given and one needs to identify the constraints. In other words one wishes to fit the surprisal $I(v) = -\ln[P(v)/P^{o}(v)]$, by

$$-\ln[P(v)/P^{o}(v)] \simeq \lambda_0 + \sum_{r=1}^{m} \lambda_r A_r(v) \quad . \qquad (2.1.61)$$

Some practical details [2.39,42] of this "surprisal analysis" procedure are discussed in Appendix 2.B. We would like to recommend that this appendix (or another source, e.g., [2.35]) be consulted before attempting an analysis of some new data. The text continues with examples of the results of such an analysis.

2.2 Surprisal Analysis

2.2.1 The Prior Distribution

In a collision of an atom with a diatomic molecule the total available energy E is the sum of the product vibrational, rotational, and translational energies

$$E = E_V + E_R + E_T \quad . \qquad (2.2.1)$$

(In Sect.2.2.5 we shall recognize the possibility of electronic excitation). The energy not deposited in the vibration, $E - E_V$, is necessarily either in the translation or the rotation. If E_V is high, less energy is available for the other degrees of freedom and vice versa. When we specify E and E_V there may be many collisions with the specified values but with different magnitudes of E_R and E_T (provided only that $E_R + E_T = E - E_V$). There are many possible outcomes of the collision, all of which correspond to the same given vibrational state but which differ in the way the rest of the energy, $E - E_v$, is partitioned.

In interpreting the distribution P(v) we need to allow for the fact that many different collisions can result in a specified vibrational state. The first thing we thus need to find out is the fraction of all possible final states where the vibrational state is v. We designate this fraction by $P^0(v)$. If all states of the products are equally probable, i.e., if all states are formed with the same rate, then $P(v) = P^0(v)$. If there is some specific disposal of energy then the rate of populating the state v will not be simply proportional to $P^0(v)$ and the two distributions will not be identical. $P^0(v)$ serves as a reference against which P(v) is compared. The definition of $P^0(v)$ is discussed from a collision-theoretic point of view in Appendix 2.C.

The details of state counting for atom-diatom collision are given in Appendix 2.A. When the total energy is in the range E, E + dE, the number of product states (per unit volume) when the vibrational and rotational states are specified is $\rho(v,J;E)dE$,

$$\rho(v,J;E) = (2J+1)A_T(E-E_{v,J})^{\frac{1}{2}} \quad , \tag{2.2.2}$$

where A_T is a (unit-bearing) constant depending on the masses of the colliding molecules (Appendix 2.A). Less detailed results are obtained by appropriate summations. If only the vibrational state is specified the number of final states is $\rho(v;E)dE$,

$$\rho(v;E) = \sum_J \rho(v,J;E) \tag{2.2.3}$$

and the total number of states $\rho(E)dE$ is given by

$$\rho(E) = \sum_v \rho(v;E) = \sum_v \sum_J \rho(v,J;E) \quad . \tag{2.2.4}$$

In (2.2.3,4) summation is restricted to values of J and v such that $E_J \leqq E - E_v$ and $E_{v,J} \leqq E$, respectively. With these definitions the prior distribution is [cf. (2.1.48)]

$$P^0(v) = \rho(v;E)/\rho(E) \tag{2.2.5}$$

and, by an obvious extension

$$P^0(v,J) = \rho(v,J;E)/\rho(E) \quad . \tag{2.2.6}$$

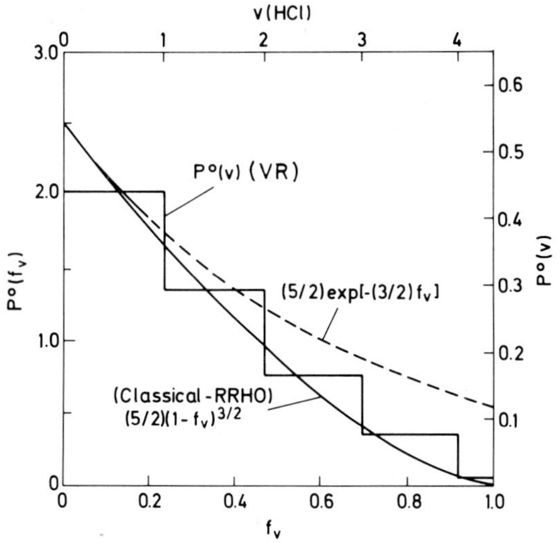

Fig. 2.8. Prior vibrational product state distributions. Shown are the classical RRHO approximation (2.2.9), where f_v is treated as a continuous variable, and the more accurate VR approximation (scale on the right) where v is discrete and $E_v = f_v E$ is given by the vibrating rotor level scheme (Appendix 2.A). Also shown (dashed curve) is a Boltzmann-like approximation appropriate only for $f_v \ll 1$; see (2.2.21). The prior distributions shown are for HCl formed in the $Cl + HI \rightarrow HCl + I$ reaction. The experimental P(v) distribution for this reaction is shown in Fig.2.3

Simple (but not completely reliable) closed expressions are obtained by considering the diatomic molecule as a rigid rotor and harmonic oscillator (RRHO) and replacing the summations over discrete quantum numbers by integration over continuous ("classical") energy levels. One then finds for $P^O(E_v) = P^O(v)(dv/dE_v)$

$$P^O(E_v) = (5/2)(E-E_v)^{3/2}/E^{5/2} \qquad\qquad (2.2.7)$$

and other results listed in Appendix 2.A.

In the following we shall see that it is often instructive to represent product state distributions, at a given total energy E, in terms of the reduced energy variables $f_x = E_x/E$ where $x = v,R,T$ and according to (2.2.1)

$$f_v + f_R + f_T = 1 \quad . \qquad\qquad (2.2.8)$$

For example, in this representation the classical RRHO expression (2.2.7) transforms to $P^O(f_v) = P^O(E_v)dE_v/df_v = P^O(E_v)E,$

$$P^O(f_v) = (5/2)(1-f_v)^{3/2} \quad . \qquad\qquad (2.2.9)$$

This, and other approximations for the prior distribution, are shown in Fig.2.8.

The prior distribution of rotational states within a given vibrational manifold $P^0(J|v)$ is given by

$$P^\circ(J|v) = P^0(v,J)/P^0(v) = \rho(v,J;E)/\rho(v;E)$$
$$\propto (2J+1)(E-E_v-E_J)^{1/2}/(E-E_v)^{3/2} \qquad (2.2.10)$$

and shown in Fig.2.14.

Strictly speaking, (2.2.5-10) apply only at a well-defined energy E. For thermal reactants one requires an averaging over the energy available to the products. However, for highly exoergic reactions, $-\Delta E_0 \gg kT$, the major contribution to E is $-\Delta E_0$ and the fractional spread in E is small. It is then (but only then) legitimate to use (2.2.1-10) with E replaced by its mean value $\langle E \rangle$. For other situations, an actual averaging is required and will be discussed in Sect.2.4.

It should be stressed that the prior distribution, e.g., $P^0(v,J)$, is defined to serve as a reference for comparison with the corresponding experimental distributions, e.g., P(v,J). In chemical kinetics such distributions are always defined as relative rates. Thus [cf. (2.1.25)] P(v,J) is the fractional rate into the v,J products state. In technical terms, P(v,J) is a "flux density". If we place a detector some distance away from the reaction center, P(v,J) is the fraction of the flux reaching the detector where the final state is v,J. (See, in particular, Ref.2.1 [Sects.3.1.6 and 6.2.2].) One can also define a "number density" N(v,J), as the fraction of molecules in the state v,J that reach the detector. Note that while N(v,J) is not a relative rate constant, some detection methods (mass spectrometry, laser-induced fluorescence) actually measure the number rather than the flux density. For products with a well-defined relative velocity the number density times the velocity is the flux density. Hence, the prior distribution for the number density is [cf. (2.2.2,6)]

$$N^0(v,J) = (2J+1)/\sum_v \sum_J (2J+1) \quad . \qquad (2.2.11)$$

2.2.2 The Surprisal

The prior distribution is the most probable result in the absence of any constraints (cf. Sects.2.1.7 and 2.2.7). The observed energy disposal or energy requirements of chemical reactions will often markedly differ from the "democratic" behavior discussed in the previous section. To characterize this deviance we employ the surprisal. Thus, say, for vibrational energy disposal

$$I(v) = -\ln[P(v)/P^0(v)] \quad . \qquad (2.1.53)$$

The surprisal clearly vanishes when our prior expectations are borne out by the experiment, i.e., when $P(v) = P^0(v)$.

An important feature of the surprisal is that it has precisely the same magnitude for energy disposal in the forward reaction as for energy requirement in the reverse reaction. This is an immediate implication of detailed balance when expressed in the form (2.1.25) = (2.1.28) (cf. Sect.2.1.4). This symmetry is also valid for the prior distribution (Sect.2.4.1) and hence also for the surprisal.

To forge a link between the concept of the surprisal and the dynamics of the process recall that (e.g., Sect.3.1.1) the rate of any transition can be expressed as

$$R = (\text{dynamical factor}) \cdot (\text{density of states factor}) \qquad (2.2.12)$$

or, in a more compact quantal notation [2.24,25]

$$R(i \to f) = (1/h)|T_{if}|^2 \rho_f \quad , \qquad (2.2.13)$$

where T is the transition operator and ρ_f is the density of final states. If one is interested in transitions to a group Γ of final states

$$R(i \to \Gamma) = \sum_f{}' R(i \to f) = (1/h)\left(\sum_f{}' |T_{if}|^2 \rho_f \Big/ \sum_f{}' \rho_f\right)\rho_\Gamma \quad , \qquad (2.2.14)$$

where $\rho_\Gamma = \sum' \rho_f$ is the density of states of the group Γ and the prime on the summation symbol indicates that it is restricted to states f in the group Γ.

The result (2.2.14) is also of the form (2.2.12) and shows that the dynamical factor has invariably the interpretation of an elementary dynamical factor $\left(|T_{if}|^2\right)$ averaged over the final states. The considerable advantage of the factorization (2.2.12) will become evident throughout the rest of this volume. The first thing that one wants to know about a proposed transition is "What is the density of the final states?" Only then does he turn to consider the dynamical factor.

Surprisal analysis seeks to formalize this common attitude. Up to normalization of P^0 [recall that $P^0(\Gamma)=\rho_\Gamma/\sum_\Gamma \rho_\Gamma$] the surprisal is simply the logarithm of the dynamical factor.

For a class of processes with a common mechanism there are indeed special names for the dynamical factor. The most obvious example are transitions between different electronic states, e.g., radiationless transitions (cf. Sect. 2.4.9). The magnitude of T_{if} is then largely governed by the overlap between the initial and final vibrational states. This is the so-called "Condon" approximation and the dynamical term in (2.2.12) is then referred to as the "Franck-Condon factor" (see also Sect.3.3.2).

In the weak coupling approximation it is possible to replace the elements of the T matrix by the elements of the potential responsible for the transition. Then (2.2.13) reads

$$R(i \to f) = (1/h)|V_{if}|^2 \rho_f \quad . \tag{2.2.15}$$

This approximation is often referred to as "the golden rule". As long as one is not actually computing matrix elements there is no reason not to use the exact expression (2.2.12).

In Appendix 2.C we derive the resolution (2.2.12) and the concept of the surprisal from purely collision-theoretic considerations. For simplicity, classical mechanics is employed so as to make the appendix reasonably accessible.

2.2.3 Vibrational Surprisal

The definition of the surprisal merely allows us to replace one set of numbers, $P(v)$, $v=0,1,\ldots$, by another set, the surprisals, $I(v)$. When we examine the dependence of the surprisal on v (Fig.2.9) we find that it varies simply and monotonically with v. Additional examples of surprisal plots for product vibrational state distribution are shown in Figs.2.10 and 2.11. Surprisal plots for vibrational state distributions for several other reactions leading to (or consuming) electronically excited species have been discussed [2.29, 33,50-53].

For the examples shown in Figs.2.9-2.11 (and others) one can approximate the surprisal by a single v-independent parameter λ_v,

$$I(v) = \lambda_0 + \lambda_v(E_v/E) = \lambda_0 + \lambda_v f_v \quad , \tag{2.2.16}$$

where E is the (mean) energy available to the products. Inserting (2.2.16) in the definition (2.1.53), we obtain a simple representation

$$P(v) = P^0(v) \exp(-\lambda_0-\lambda_v f_v) \tag{2.2.17}$$

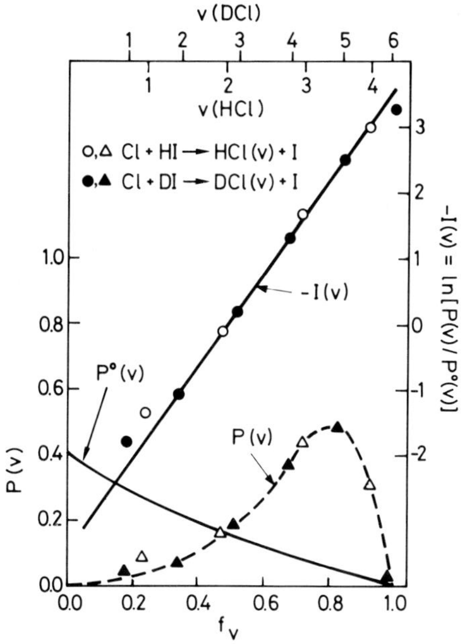

Fig.2.9

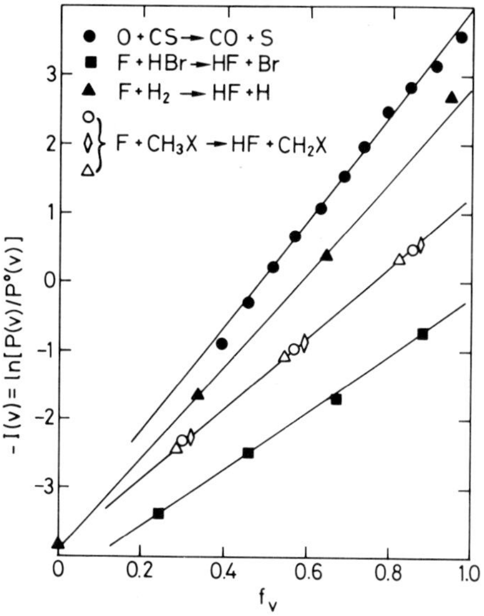

Fig.2.10

Fig. 2.9. Surprisal analysis of the product vibrational state distribution in the exothermic $Cl + HI \to I + HCl(v)$ (and Cl+DI) reactions. *Bottom part:* Prior $P^0(f_v)$ and experimental $P(f_v)$ (adapted from [2.16]), normalized distributions vs $f_v = E_v/\langle E \rangle$. *Top part:* The surprisal vs f_v (adapted from [2.43]). Note that the slope λ_v of the surprisal plot is essentially invariant with respect to isotopic substitution. This is not the case if v, the vibrational quantum number, is employed as the independent variable

Fig. 2.10. Surprisal analysis for product vibrational state distributions in exoergic reactions. Experimental results and analysis of $F + CH_3X \to HF + CH_2X$ (o for X=Br, ◊ for Cl, and Δ for I from [2.44]. Results and analysis of the $F + H_2 \to HF + H$ reaction from [2.45]. Experimental results for $F + HBr \to HF + Br$ from [2.46] and for $O + CS \to CO + S$ from [2.47] with analysis from [2.48]). The exothermicities of the reactions O + CS, F + HBr, $F + H_2$, and $F + CH_3Cl$, Br, I in units of kcal/mole, are about 85.0, 50.0, 34.6 and 39.0, 38.0, 37.0, respectively

for P(v). Since $P^0(v)$ is a monotonically decreasing function of f_v [cf. (2.2.9) and Fig.2.8], λ_v is negative when population inversion occurs. Table 2.1 lists some typical values of λ_v for a series of reactions. Reactions of the same family and, in particular, reactions which differ by isotopic substitution often have very similar λ_v parameters. Not listed in the table is λ_0, the reason being that (2.2.17) is a strictly one-parameter representation. If λ_v is known, λ_0 can be immediately computed from the condition that P(v) is a normalized distribution $\sum_v P(v) = 1$ or [cf. (2.1.56)]

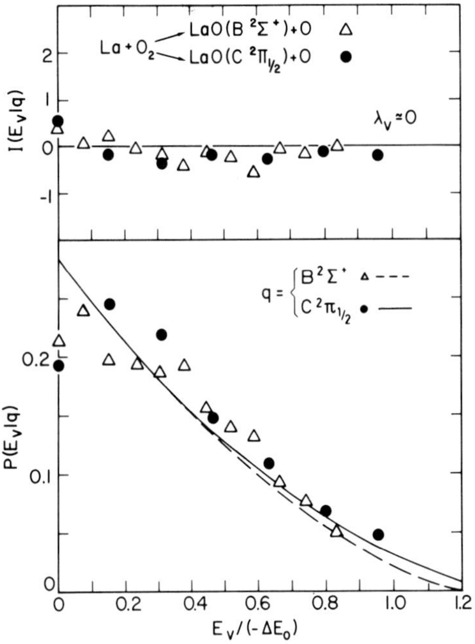

Fig. 2.11. Experimental and prior vibrational distributions (bottom panel) and surprisal plots (top panel) for the reactions $La + O_2 \rightarrow La(q,v) + O$. q denotes the electronic state of the diatomic products; both $q = B^2\Sigma^+$ and $C^2\Pi_{1/2}$ are excited electronic states. The respective exoergicities are $\Delta E_0 = -21.3$ and -7.6 kcal/mole. The circles and triangles in the bottom panel are the experimental results. The prior distributions (properly averaged over the thermal spread in reagent energy) are represented by the smooth lines. The vibrational populations are not inverted and the surprisal is correspondingly small (adapted from [2.49]; see also [2.29])

Table 2.1. Measures of specificity of energy disposal

Reaction	$\langle f_v \rangle$	$-\lambda_v$	DS[P|P^0] [e.u.]	E [kcal/mole]
$Cl+HI \rightarrow HCl+I$	0.70	8.0	3.54	34.0
$CI+DI \rightarrow DCl+I$	0.69	8.0	3.25	34.0
$F+H_2 \rightarrow HF+H$	0.67	6.9	3.35	34.6
$F+D_2 \rightarrow DF+D$	0.60	5.7	2.27	34.7
$F+HD \rightarrow HF+D$	0.59	6.7	2.52	33.8
$F+DH \rightarrow DF+H$	0.55	5.5	2.10	35.5
$O+CS \rightarrow CO+S$	0.68	7.7	2.9	87.0
$F+HCl \rightarrow HF+Cl$	0.52	4.8	1.8	34.8

$$\exp(\lambda_0) = \sum_v P^0(v)\exp(-\lambda_v f_v) \quad . \tag{2.2.18}$$

The characterization of the distribution P(v) in terms of the magnitude of λ_v is essentially equivalent to using the magnitude of the mean value of $f_v = E_v/E$ as a measure. To see this we first note that, at a given E, $P^0(v) \propto (1-f_v)^{3/2}$ and hence, using (2.2.9,18)

$$\langle f_v \rangle = \sum_v f_v P(f_v) = \sum_v f_v(1-f_v)^{3/2}\exp(-\lambda_v f_v)\Big/\sum_v(1-f_v)^{3/2}\exp(-\lambda_v f_v) \quad . \tag{2.2.19}$$

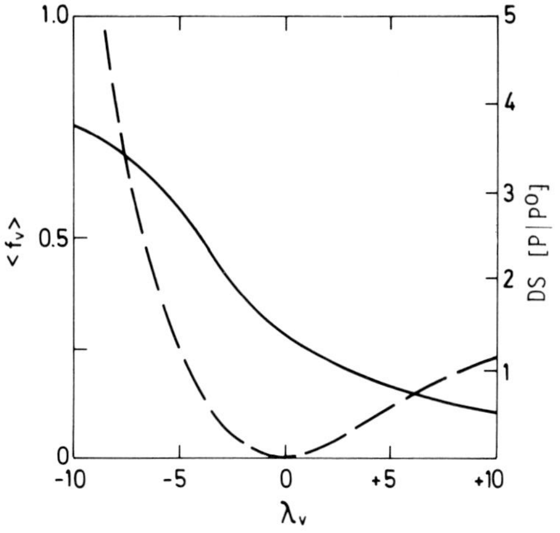

Fig. 2.12. The role of λ_V as a measure of the specificity of vibrational energy disposal. *Left scale:* The mean fraction of the available energy released as product vibrational excitation $<f_V>$ vs λ_V (solid curve: monotonic dependence). *Right scale:* The entropy deficiency (2.1.52) of the product vibrational state distributions vs λ_V (dashed curve). In the prior limit ($\lambda_V=0$) the entropy deficiency is zero and $<f_V> = 2/7$

Given $<f_V>$ we can thus compute λ_V from (2.2.19). In the approximation in which the discrete summation over v is replaced by integration over f_V, the solution of (2.2.19) in graphical form is shown in Fig.2.12.

As has been previously discussed, the surprisal analysis of P(v) characterizes not only the vibrational energy disposal in the forward reaction but also the role of reagent vibrational excitation in the reverse reaction.

A prior vibrational distribution can, of course, be determined also for a collision with polyatomic products, e.g., [2.42]

$$F + RH \rightarrow R + HF(v) \quad , \tag{2.2.20}$$

where R is an organic radical. The rapid increase of the density of states of a polyatomic molecule with internal excitation (cf. Appendix 2.A) implies that $P^O(v)$ will be a particularly rapidly declining function of v. More explicitly $P^O(v) \propto (E-E_V)^\gamma$ where E is the total collision energy and $\gamma = s + \alpha/2-1$ with s denoting the number of vibrational degrees of freedom of R and α the number of translational-rotational degrees of freedom of both products in the center of mass system. [For the atom-diatom case corresponding to R being an atom, e.g., R = H, we have s = 0, $\alpha = 5$; hence $\gamma = 3/2$ in accordance with (2.2.7).] If the number s of vibrational degrees of freedom of R is large enough $P^O(v)$ is large only for $E_V << E$ and can thus be approximated (Appendix 2.A) as

$$P^O(v) \propto \exp(-E_V/kT_V) \tag{2.2.21}$$

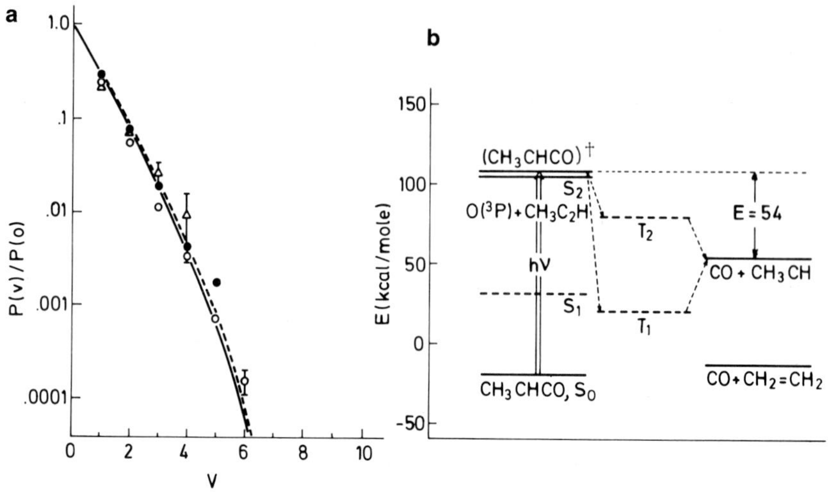

Fig. 2.13. Vibrational energy disposal (a) and schematic energy diagram (b) for C_3H_4O (adapted from [2.55]). At the wavelength of λ=226 nm ($h\nu$=126 kcal/mole) the photoexcited C_3H_4O carries about the same total energy as that energy-rich molecule formed in the $O(^3P)$ + methylacetylene collision. (a) CO vibrational state distribution. Triangles and open circles: photoexcitation experiments; (•) $O + C_3H_4$ reaction; (---) prior distribution

with $kT_v \equiv E/\gamma$. This is an example of the familiar result that for a small system (HF) that can exchange energy with a large system (R), the equilibrium distribution is the canonical one [2.54]. Clearly the approximation (2.2.21) becomes poorer, especially for $E_v \sim E$, as $s \to 0$, as demonstrated in Fig.2.8.

The use of polyatomic reagents also offers examples of product vibrational state distributions that do conform to the prior one (see Fig.5.41). Figure 2.13 shows the CO vibrational distributions obtained [2.55] by preparing the energy-rich CH_3CHCO molecule either via photoexcitation from the ground state or by the

$$O(^3P) + CH_3C \equiv CH \to (CH_3CHCO)^† \to CH_3CH + CO(v) \qquad (2.2.22)$$

reaction. The distribution is essentially independent of the mode of formation of $(CH_3CHCO)^†$ and, as therefore expected, agrees with the prior.

2.2.4 The Rotational State Distribution

The distribution of both vibrational and rotational states of the product molecules P(v,J) is defined, like P(v), as the relative reaction rate into the specified state. Thus if $k(\to v,J;T)$ is the rate constant of, say, the

$$F + H_2 \rightarrow H + HF(v,J) \tag{2.2.23}$$

reaction (using thermal reactants) and $\vec{k}(T)$ is the thermal (total) reaction rate $\vec{k}(T) = \sum_v \sum_J k(\rightarrow v,J;T)$, then

$$P(v,J) = k(\rightarrow v,J;T)/\vec{k}(T) \quad . \tag{2.2.24}$$

Rotational relaxation is typically much faster than vibrational relaxation (cf. Fig.1.9). Hence considerable care and ingenuity are required to measure the nascent distribution of rotational states and thereby to obtain the $P(v,J)$ distribution. The initial state of the rotational relaxation is often limited to transitions between the different rotational states of the same vibrational manifold (Sect.2.5.2. See however Sect.2.4.4, Fig.2.41 in particular). Hence $P(v)$,

$$P(v) = \sum_J P(v,J) \quad , \tag{2.2.25}$$

could be measured even if considerable rotational relaxation occurs. It is thus convenient to express $P(v,J)$ as

$$P(v,J) = P(J|v)P(v) \quad , \tag{2.2.26}$$

where $P(J|v)$ is the distribution of rotational states within a given vibrational manifold. Using (2.2.24-26), [see also (2.1.25)],

$$P(J|v) = k(\rightarrow v,J;T)/\sum_J k(\rightarrow v,J;T) = k(\rightarrow v,J;T)/k(\rightarrow v;T) \quad . \tag{2.2.27}$$

Of the two factors in (2.2.26), $P(J|v)$ is difficult to determine with accuracy.

Figure 2.14 shows the detailed rate constants $k(\rightarrow v,J;T)$ for the $F + H_2$ reaction. Shown for comparison is the prior distribution, (2.2.10), and the thermal (i.e., relaxed) distribution of the rotational states in the different vibrational manifolds. On inspection it is evident that both $P(J|v)$ and $P^0(J|v)$ differ from the thermal distribution in one important respect. The maxima of both $P(J|v)$ and $P^0(J|v)$ shift to lower J values as v is increased. This observation is of particular interest because, as will be argued later (Sect.2.5.2), the position of the maximum is only little affected by the rotational relaxation.

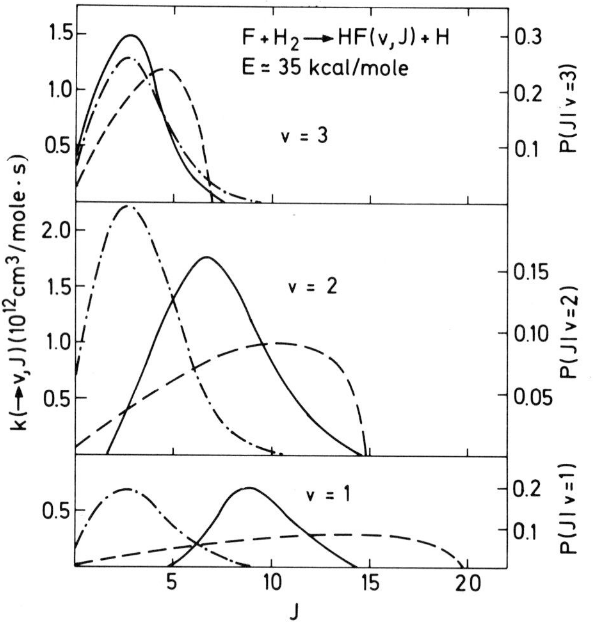

Fig. 2.14. Detailed vibrotational rate constants (left scale) and distributions (right scale) in the $F + H_2 \rightarrow H + HF(v,J)$ reaction (adapted from [2.56]). The solid curves represent the results obtained in infrared chemiluminescence-arrested relaxation experiments [2.57]. The dashed curves are the prior distributions $P^0(J|v)$, (2.2.28). The dashed-dotted curves are thermal rotational distributions describing an intermediate stage in the relaxation of the bulk system. The rotational state distribution is already in equilibrium with the buffer gas while the vibrational distribution is still nonrelaxed. The areas under the curves are proportional to P(v).

The first clue to the shape of P(J|v) is provided by the observation that not only the position of the maximum but also the range of states populated by the collision diminished as v is increased. This latter aspect is easy to understand from considerations of conservation of energy.

$$P^0(J|v) \propto (2J+1)(E-E_v-E_J)^{1/2}/(E-E_v)^{3/2}$$

$$\cong E_R^{1/2}[1-E_R/(E-E_v)]^{1/2}/(E-E_v) = [g_R(1-g_R)]^{1/2}/(E-E_v)^{1/2} \quad .(2.2.28)$$

Here we have introduced g_R [2.58]

$$g_R = E_R/(E-E_v) = f_R/(1-f_v) \quad (2.2.29)$$

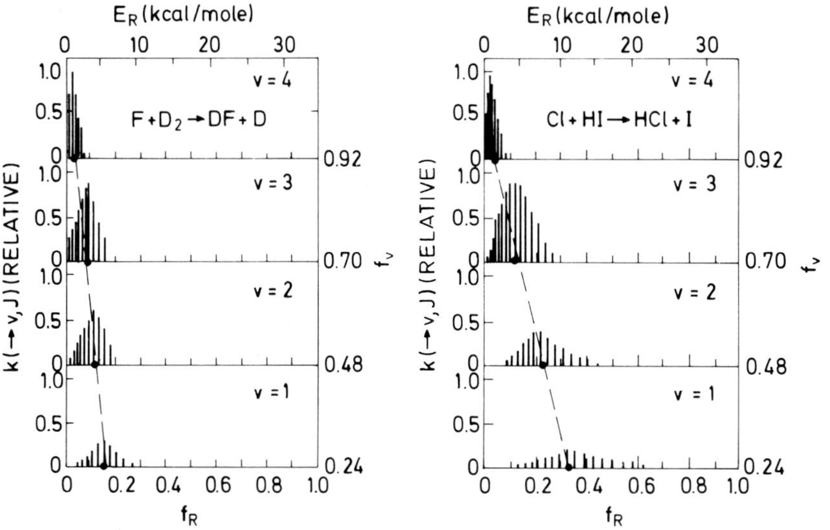

Fig. 2.15. Experimental vibrotational distributions for two reactions shown as "stick" diagrams vs f_R (adapted from [2.57] and [2.16]). The height of each bar is proportional to the magnitude of the rate constant $k(\rightarrow v,J)$. The points correspond to $\hat{f}_R(v) = \hat{g}_R(1-f_v)$, i.e., to the peak of the rotational distributions in different vibrational manifolds. The linearity of the (----) lines connecting these points indicate that $\hat{g}_R = d\hat{f}_R(v)/d(1-f_v)$ is independent of v; $\hat{g}_R = 0.45$ and 0.21 for the Cl + HI and F + D_2 reactions, respectively

as the fraction of available energy (in the vibrational manifold v) which is present as rotational excitation of the products, $0 \leqq g_R \leqq 1$ [we have used $E_R=E_J=BJ(J+1) \simeq B(J+\frac{1}{2})^2$ so that $2J+1=2(J+\frac{1}{2}) \propto E_R^{\frac{1}{2}}$].

Could it be that not only the prior but also the actual distribution [and hence also the surprisal $I(J|v)$] can be expressed as a function only of g_R?

As a first examination of this conjecture we note from (2.2.28) that when we plot $P^0(J|v)$ vs g_R the maximum occurs at $g_R = 0.5$, independent of v. Hence, we expect that when we plot $P(J|v)$ vs g_R (or E_R) the maximum should occur at the same value of g_R, say, $\hat{g}_R$, for all the vibrational manifolds. Figure 2.15 shows the results for the $F + D_2 \rightarrow DF(v,J) + D$ and $Cl + HI \rightarrow HCl(v,J) + I$ reactions. Shown is the shift of $\hat{f}_R$ for different values of v. Clearly for these reactions g_R accounts well for the shift in the $P(J|v)$ distribution to lower J values as v increases. It also accounts for the isotope effect in this and other reactions, in analogy to the transformation from E_v to f_v (cf. Fig.2.9). For most reactions $\hat{g}_R \leqq 0.5$. For some reactions (e.g., Cl + HI, $\hat{g}_R$=0.45) it is closer to 0.5 than for others (e.g., F + H_2, $\hat{g}_R$=0.21), as is shown in Fig.2.15. The main kinematic reason for the spread in $\hat{g}_R$ values is discussed in [2.1].

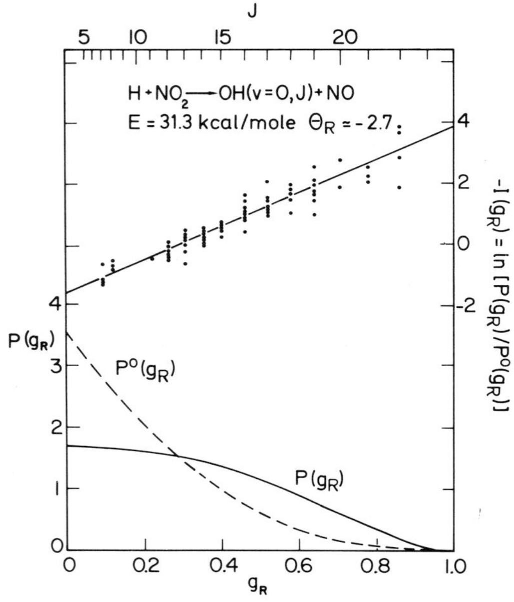

Fig. 2.16. Rotational surprisal plot for the OH(v=0,J) rotational state distribution from the $H + NO_2 \rightarrow NO + OH(v',J')$ reaction [2.59]. The linear surprisal parameter g_R is defined via
$I(g_R) = -\ln[P(g_R)/P^o(g_R)] = \theta_0 + \theta_R g_R$

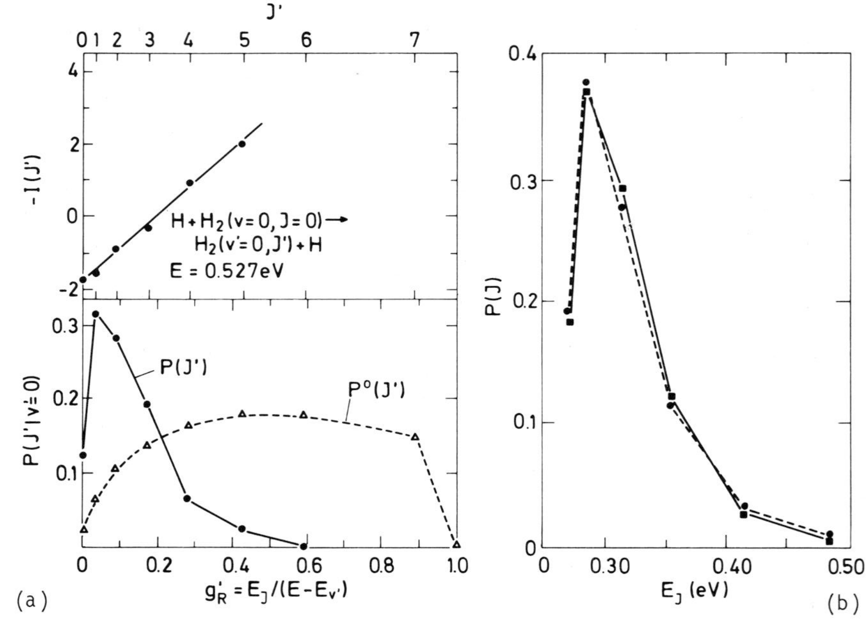

Fig. 2.17a,b. Rotational distributions and surprisal analysis for the reactive $H + H_2$ collision. (a) Quantal computation P(J') and prior distribution $P^o(J')$ for the $H + H_2(v=0,J=0) \rightarrow H_2(v'=0,J') + H$ reaction plotted vs the reduced energy variable g_R (bottom). The rotational surprisal plot is linear (top) (adapted from [2.60]). (b) A comparison of an exact quantal calculation (——) and a distribution of maximal entropy (----) subject to a single constraint $\langle E_J \rangle$, implying linear surprisal (adapted from [2.61])

Linear rotational surprisal plots have been noted for both experimental (Fig.2.16) and quantum mechanical computations for both reactive (Fig.2.17) and inelastic collisions. Deviations from linearity (vs g_R) are however noted when a large fraction of the available energy is converted into rotational excitation.

2.2.5 Electronic Excitation

Exoergic reactions can promote the products to excited electronic states. On prior grounds, the occupation of the qth electronic state would be proportional to $g_q\rho_q(E)$. Here g_q is the electronic degeneracy of the state while $\rho_q(E)$ is [cf. (2.2.4)] the total density of states for products in the qth electronic state. The currently available experimental results [2.29,33,49-53,62-66] suggest that often the deviations from the prior distribution are not large. Table 2.2 shows the experimental and prior branching ratios for the reaction

$$M + O_2 \rightarrow MO(q) + O \quad , \quad M = Sc,Y \quad . \qquad (2.2.30)$$

The multitude of surface crossings (as discussed below) that is often present for such reactions is one reason why there is a distinct lack of specificity in such processes. Indeed, the vibrational state distribution (Fig. 2.11) is also often not very specific. When this is not the case, one can offer dynamic explanations for the deviance from the prior distribution [2.66].

Table 2.2. Experimental (Γ_q) and prior (Γ_q^o) branching ratios (relative rate constants) for the reactions $M + O_2 \rightarrow MO(q) + O$; M = Sc,Y. (Adapted from [2.66])

Product molecule	q	Γ_q	Γ_q^o
ScO	$A^2\Pi_{1/2}$	1.23	1.19
	$A^2\Pi_{3/2}$	1.00	1.00
YO	$A'^2\Delta_{3/2}$	-	3.86
	$A'^2\Delta_{5/2}$	-	3.19
	$A^2\Pi_{1/2}$	1.74	1.37
	$A^2\Pi_{3/2}$	1.00	1.00

2.2.6 Polyatomic Molecules

For polyatomic reagents there is the possibility that the collision proceeds via an energy-rich intermediate which will survive for many vibrational periods. The reason (cf. Sect.2.3.5) is that the excitation energy can be partitioned among very many vibrational modes. Hence the mean energy per mode is low and the probability that there is sufficient energy in the exit mode is small. One could therefore expect that the energy in the dissociation products will be equipartitioned, as in the prior distribution (Fig.2.13). This is not always the case.

Figure 2.18 shows the surprisal (curve labelled III) of the HF vibrational distribution [2.44] produced in the

$$F + CH_2O \rightarrow CHO + HF(v) \tag{2.2.31}$$

reaction. The surprisal is very large since on prior grounds (curve labelled III [cf. (2.A.19) and the approximation (2.2.21)]) high vibrational excitation

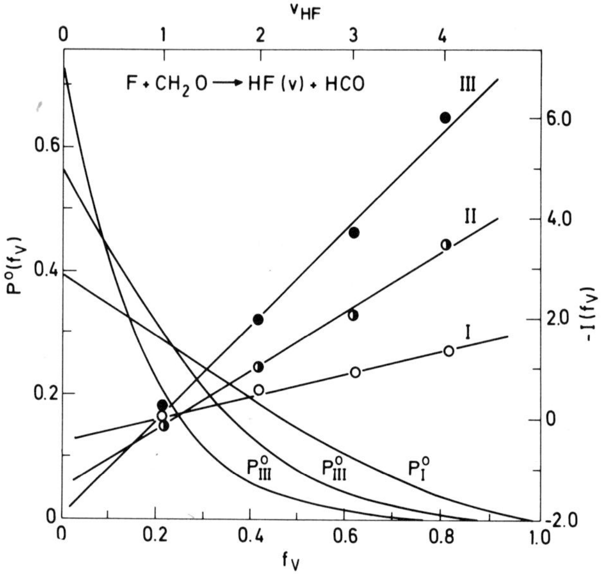

Fig. 2.18. Surprisal analysis of the HF vibrational distribution in the $F + CH_2O$ reaction. Three choices of the prior distribution (bottom part) are shown, leading to three surprisal plots (upper part) (adapted from [2.44]). The curves labelled III correspond to the usual choice of P^o: all final quantum states are equally probable. Curves II and I exclude from the counting the vibrational and both the vibrational and rotational states of CHO, respectively

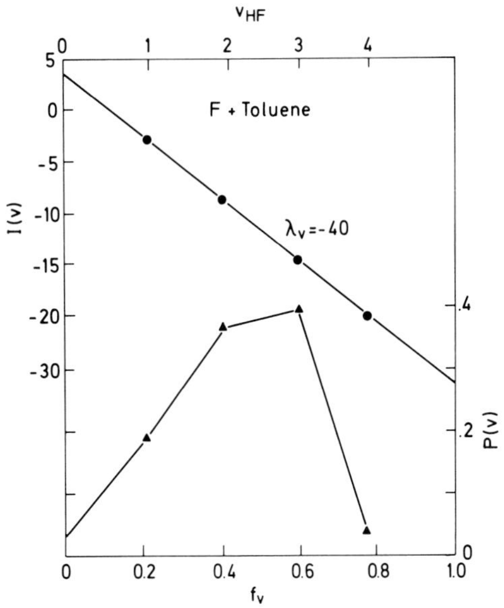

Fig.2.19

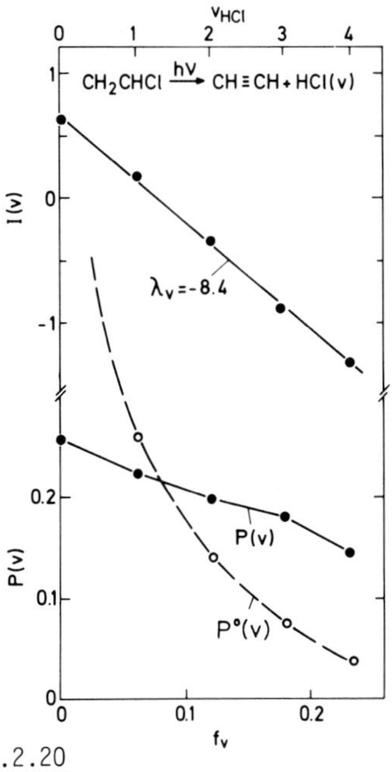

Fig.2.20

Fig. 2.19. Surprisal analysis of the HF vibrational state distribution for the reaction (2.2.32) (adapted from [2.44]). *Bottom part:* Observed distribution; *top part:* the surprisal using the usual prior distribution

Fig.2.20. Surprisal of HCl vibrational distribution from photoelimination of chloroethylene (experimental results from [2.70]) (adapted from [2.71])

of HF is very improbable. It is then tempting to argue that the origins of the surprisal are the vibrational modes of CHO which, on prior grounds, can be populated but which may be only very weakly coupled to the reaction center. Say we freeze them out, treating CHO like a rigid rotor. The HF distribution (curves labelled II in Fig.2.18) is still quite surprising. Even treating CHO as a rigid mass point (curves labelled I) does not reduce the surprisal to zero. A similar behavior is found for other H abstraction processes, e.g., (2.2.20). Figure 2.19 shows the results for the

$$F + C_7H_8 \rightarrow C_7H_6 + HF(v) \qquad (2.2.32)$$

reaction. While the number of vibrational modes is large, the distribution is very deviant from the prior one. Photoelimination and photodissociation processes also often show inverted product vibrational distribution [2.67,68] and have indeed been used (Chap.4) as pumping reactions for laser action. Figure 2.20 shows the surprisal of HCl vibrational distributions for chloroethylene photoelimination processes. Even photoelimination induced by multiple absorption of IR photons (cf. Sects.1.3.2 and 5.4.2), e.g., [2.69]

$$CH_3CF_3 \xrightarrow{CO_2 \text{ laser}} (CH_3CF_3)^{\dagger} \rightarrow CH_2CF_2 + HF(v) \quad , \qquad (2.2.33)$$

leads to nonstatistical vibrational state distributions of both HF and of CH_2CF_2.

Energy-rich CH_3CF_3 can also be produced via chemical activation [2.67],

$$CH_3 + CF_3 \rightarrow (CH_3CF_3)^{\dagger} \rightarrow CH_2CF_2 + HF(v) \quad , \qquad (2.2.34)$$

where the source of the excitation is the energy released by the formation of the C-C bond. Here too the HF vibrational population is very deviant from prior expectations and is similar to that obtained in the photoelimination reaction (2.2.33) and in the chemical activation reaction

$$H + CH_2CF_3 \rightarrow (CH_3CF_3)^{\dagger} \rightarrow CH_2CF_2 + HF(v) \quad . \qquad (2.2.35)$$

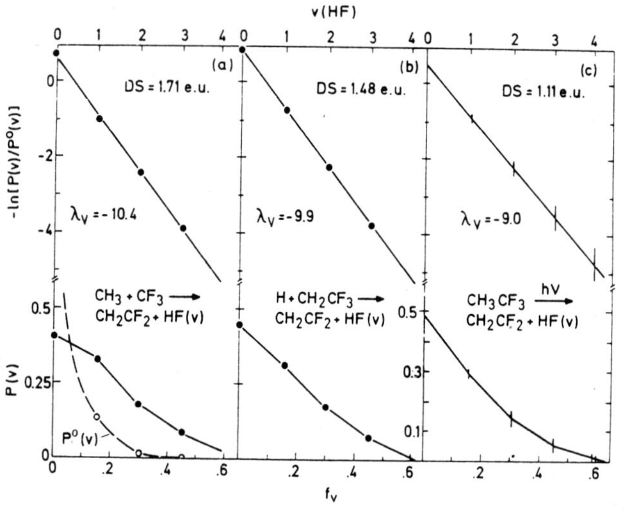

Fig. 2.21a-c. Surprisal analysis of HF vibrational distribution produced in the dissociation of $(CH_3CF_3)^{\dagger}$ formed via (a) chemical activation $CH_3 + CF_3$, (b) chemical activation $H + CH_2CF_3$, and (c) multiphoton excitation of CH_3CF_3. The surprisal (and distribution) for case (c) were computed using (2.2.16,19). The error margin reflects the experimental uncertainty in $\langle f_v \rangle$ (adapted from [2.71]). An energy level diagram is provided in Fig.2.30. Note however that the energy-rich CH_3CF_3 molecules produced by chemical activation may have been somewhat collisionally relaxed

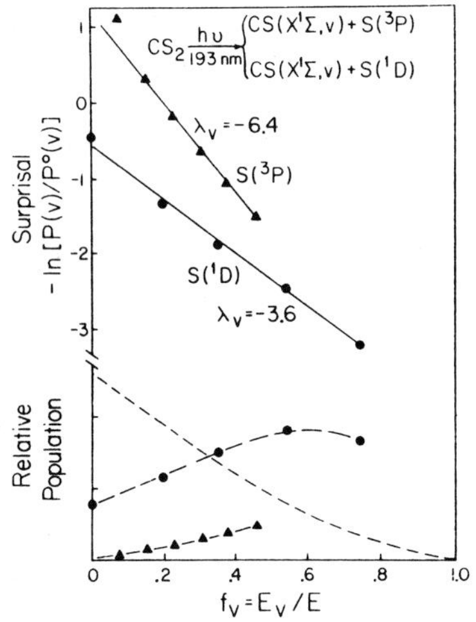

Fig. 2.22. Surprisal analysis of the CS vibrational distribution in the two channels observed in the photodissociation of CS_2. The dashed line is the prior distribution (cf. Fig.1.2 for further details)

Figure 2.21 is a surprisal plot of the HF vibrational distribution from the dissociation of $(CH_3CF_3)^\dagger$ produced by the three excitation methods (2.2.33-35). For the process (2.2.33) only $\langle E_V \rangle$ was reported and the surprisal was calculated from it as in Sect.2.2.3, (2.2.16,19) in particular.

Photoelimination reactions studied in a molecular beam-laser configuration (photofragment spectroscopy, Sects.5.3 and 5.5) confirm the nonstatistical nature of the product state distribution in four center photoelimination reactions [2.72]. In sufficiently small molecules, even simple bond fission, e.g., reactions (1.1.2) and (1.1.3), can yield vibrationally inverted distributions. Figure 2.22 shows the results of surprisal analysis for the photodissociation of CS_2 (cf. Fig.1.2).

2.2.7 Surprisal Analysis and Collision Dynamics

The results of surprisal analysis imply that molecular collisions are neither entirely statistical (so-called [2.1] "complex" or "compound" dynamics) nor entirely "direct". Rather, they have the characteristics of both extreme limits. The collision is maximally statistical subject however to constraints. The nature and magnitude of the constraints depend not only on the system but also on the details of the initial state. The use of state-selected reagents in bimolecular collisions or selective activation of reagents in unimolecular processes enhances the constraints. Similarly, increasing the total energy of the system beyond the threshold for reaction will result in a more direct [2.1], and hence more constrained collision. Figure 2.23 illustrates this behavior for the

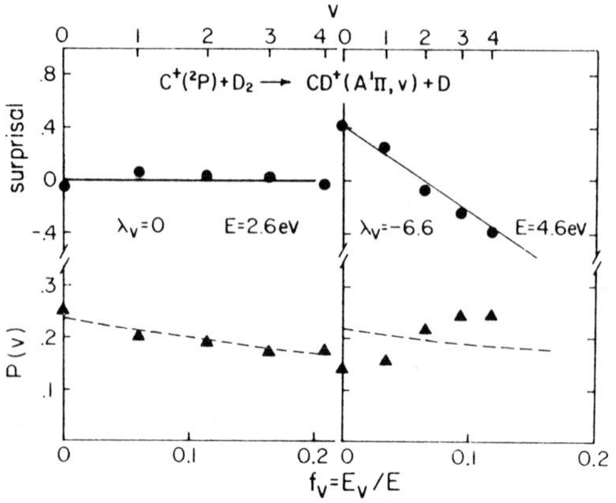

Fig. 2.23. *Bottom panels:* CD^+ vibrational state distribution (▲) at lower and higher energies and the prior distribution (---) vs $f_v = E_v/E$. *Top panels:* Surprisal plots. λ_v is essentially zero up to collision energies at about 3 eV. At higher energies there is extensive population inversion [2.73]

$$C^+(^2P) + D_2 \rightarrow CD^+(A^1\Pi,v) + D \tag{2.2.36}$$

collision which at low energies proceeds via a long-lived CD_2^+ intermediate.

How do we arrive at this picture of "as statistical as possible in the presence of constraints" collision dynamics? Consider first the final state distribution after the collision, e.g., in the reaction (2.2.34), and say that the vibrational distribution of HF was subjected to surprisal analysis (Fig.2.21) leading to a linear surprisal. For a total energy in the range E to $E + \delta E$, there will be many quantum state of the products, which correspond to HF being in the vibrational state v. These correspond to all possible partitioning of the available energy $(E-E_v)$ among all the degrees of freedom of the products except for the vibration of HF. Let n be a set of quantum numbers that specify a single quantum state where the energy $E-E_v$ is so partitioned. A quantum state of the products is then specified by the pair of indices v,n and the distribution of final states is $P(v,n)$. The observed vibrational distribution of HF is given by

$$P(v) = \sum_n P(v,n) \quad , \tag{2.2.37}$$

where summation in (2.2.37) is over such states n whose energy is in the range $E-E_v$, $E + \delta E - E_v$.

The entropy of the distribution of final states is

$$S[v,n] = -\sum_v \sum_n P(v,n) \ln[P(v,n)] \quad . \tag{2.2.38}$$

When the entropy is maximized subject to no constraints [except of course that $P(v,n)$ must be normalized] we conclude that [cf. (2.1.55)]

$$P^{ME}(v,n) = \exp(-\lambda_0^0) \tag{2.2.39}$$

and normalization implies

$$\exp(-\lambda_0^0) = \sum_v \sum_n 1 \quad . \tag{2.2.40}$$

Hence (2.2.39) is just the prior distribution. All final states are equally probable. Say now that the mean fraction $\langle f_v \rangle$ [cf. (2.2.19)] of energy present in HF vibration is imposed as a constraint

$$\langle f_v \rangle = \sum_v (E_v/E)P(v) = \sum_v \sum_n (E_v/E)P(v,n) \quad . \tag{2.2.41}$$

The distribution of maximal entropy subject to $\langle f_v \rangle$ as a constraint is now of the form [cf. (2.1.55)]

$$P^{ME}(v,n) = \exp(-\lambda_0 - \lambda_v f_v) \quad , \tag{2.2.42}$$

where now [cf. (2.1.56)]

$$\exp(\lambda_0) = \sum_v \sum_n \exp(-\lambda_v f_v) \quad . \tag{2.2.43}$$

The predicted vibrational distribution of HF is

$$P^{ME}(v) = \sum_n P^{ME}(v,n) = P^0(v)\exp[-\lambda_v f_v - (\lambda_0 - \lambda_0^0)] \quad . \tag{2.2.44}$$

Here $P^0(v)$ is the prior distribution

$$P^0(v) = \sum_n \exp(-\lambda_0^0) = \sum_n 1/\sum_v \sum_n 1 \quad . \tag{2.2.45}$$

The prediction (2.2.44) accords with the data (Fig.2.21). It hence shows that the observed distribution is constrained and is not the prior one where all final states are equally probable. However, even in the presence of the constraint, all final quantum states which correspond to HF being in a particular vibrational state v remain equally probable. In other words, $P(v,n)$ is independent of n for a given v. Indeed, if we define $P^0(n|v)$,

$$P^{0}(n|v) = 1/\sum_{n} 1 \quad , \tag{2.2.46}$$

as the prior distribution of n for a given v then the constrained distribution (2.2.42) can be written as

$$P^{ME}(v,n) = P^{0}(n|v)P^{ME}(v) \quad . \tag{2.2.47}$$

Except for the deviance enforced by the constraint, the distribution of final states is as statistical as possible.

Surprisal analysis examines the terminal results of the dynamics — that of the well-separated and no longer interacting products. How can one extrapolate back to probe the dynamics during the evolution when all degrees of freedom are interacting? To do so it is necessary to introduce the procedure of maximal entropy from collision theoretic considerations. A start in this direction is made in Appendix 2.C. A much more complete account can be found elsewhere [2.41,74,75].

The major conclusion is that a distribution of maximal entropy remains one throughout the time evolution. Hence it is possible to extrapolate back and conclude that the dynamics of the collision (or dissociation) can be described by: the state of the system is maximally statistical (i.e., of maximal entropy) subject however to constraints.

Say we extrapolate all the way back to the initial state. Then it too is a state of maximal entropy subject to constraints. But surely it is the experimentalist who selects the initial state. It thus follows that in preparing the initial state the experimentalist preordains what constraints will be present throughout and hence also after the collision. By exercising initial state selection one builds more constraints into the initial state and hence ensures a more constrained (i.e., less statistical) time evolution and product distribution.

The control that can be exercised by initial state selection is familiar from bimolecular collision experiments and will be extensively discussed in Sect.2.4. The general rule is that the more imbalance in the energy content of different modes of the initial state, the more specific is the final state distribution. The available evidence (Sect.2.2.6, Figs.2.20 and 2.21 in particular) is that the same holds for dissociation processes, even when the molecule is activated by multi IR photon absorption. What is still missing is demonstrations of product species specificity following laser excitation. Figure 2.24 shows such a specificity for the bimolecular reaction

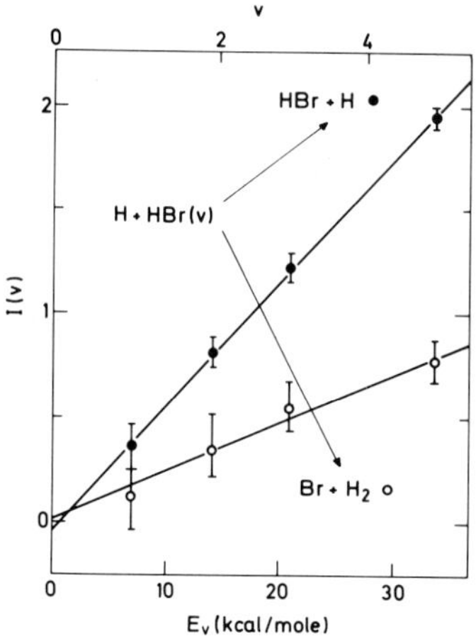

Fig. 2.24. Surprisal analysis of the reagent vibrational energy dependence of the two reaction paths in (2.2.48) (adapted from [2.11])

$$HBr(v) + H \begin{cases} \rightarrow H_2 + Br \\ \rightarrow H + BrH \end{cases} . \qquad (2.2.48)$$

The branching ratio is quite dependent on the initial vibrational excitation of HBr and the surprisal of the energy consumption is quite different for the two different reaction paths.

The discussion of Sect.2.2.7 was largely limited to selectivity and specificity on a given potential energy surface. Examples of selectivity following laser excitation of vibrational degrees of freedom in polyatomics were already mentioned in Chap.1 and will be taken up again in Sect.2.3.6 and in Chap.5. That constraints can be altered (or imposed) by reagent electronic excitation has long been familiar to photochemists. Chemical lasers are possible and infrared laser photoselective chemistry of polyatomics may be possible because constraints are also operative for intra (electronic) state dynamics, and not only for inter (electronic) states. Not all isoenergetic states are equiprobable and the surprisal is the measure of the deviances.

2.2.8 On the Role of Reagent Translation

We have paid particular attention to those degrees of freedom of the reagents that can be selected using lasers. In principle, however, one can also control the translational energy of the reagents, e.g., by using a molecular beam configuration with velocity selection [2.4]. It is then possible to envisage different partitions of a given total energy. It can be placed mostly in translation with little internal (vibrational/rotational/electronic) excitation or vice versa, etc. These different partitions will differ in the magnitude of the reaction rate constant. What then is the efficacy of the different partitions?

As a concrete example, consider the

$$\mathrm{Br} + \mathrm{HCl}(v) \rightarrow \mathrm{HBr} + \mathrm{Cl} \tag{2.2.49}$$

reaction (Figs.1.12, 2.1, and 2.35). At a given total energy E how can we best enhance the reaction rate, by increasing v or by increasing the energy in the other degrees of freedom[1] (which in practice means the translation)?

What we are thus after is the v dependence of $k(v\rightarrow;E)$ [cf. (2.1.14)]. Note that this rate constant is a different quantity than $k(v\rightarrow;T)$. The Br + HCl reaction is endoergic; hence for E below the endoergicity ($E<\Delta E_0, \Delta E_0>0$), $k(v\rightarrow;E)$ is zero. Not so for $k(v\rightarrow;T)$, where the tail of the Boltzmann distribution (cf. Fig.2.35) can always ensure that enough energy is available for the reaction.

Above the energetic threshold for reaction we can determine $k(v\rightarrow;E)$ using detailed balance [Appendix 2.C, (2.C.14) in particular]

$$\rho(v;E)k(v\rightarrow;E) = \rho'(E-\Delta E_0)k(\rightarrow v;E) \quad . \tag{2.2.50}$$

Here $\rho(v;E)$ is the density of states of the reagents in a given vibrational level v and total energy E [cf. (2.2.3)] and $\rho'(E-\Delta E_0)$ is the total, (2.2.4), density of states of the products (where the total energy is $E-\Delta E_0$).

Hence, using (2.1.25,2.2.50)

$$k(v\rightarrow;E) = \frac{\rho'(E-\Delta E_0)}{\rho(v;E)} \cdot k(\rightarrow v;E) = \frac{\rho'(E-\Delta E_0)}{\rho(E)} \cdot \frac{\rho(E)}{\rho(v;E)} \cdot P(v)\overleftarrow{k}(E) \tag{2.2.51}$$

or, limiting attention to the v-dependent terms

$$k(v\rightarrow;E) \propto P(v)/P^0(v) = \exp[-I(v)] = \exp[-\lambda_0-\lambda_v f_v] \quad . \tag{2.2.52}$$

where the final form is only valid for reactions where the surprisal is linear.

Since I(v) is typically above -5, the range of variation of $k(v\rightarrow;E)$ with v will seldom span more than two orders of magnitude. In other words, at a

---

1 Note that since the total energy is constant, any increase in vibration is at the expense of the other degrees of freedom and vice versa.

given total energy the efficacy of vibrational vs translational reagent excitation will not very much exceed 10^2, except for cases where extreme population inversion is observed in the reverse reaction. It is also important to emphasize that $k(v\rightarrow;T)$ will be much more v dependent since, as discussed in Sect.2.1, the analogous relation to (2.2.52) is [cf. (2.1.28)]

$$k(v\rightarrow;T) \propto P(v)/p(v|T) \tag{2.2.53}$$

and at about room temperatures $p(v|T)$ is a very rapidly varying function of v.

2.3 Molecular Reaction Dynamics

We have thus far discussed the characterization of the role of energy in molecular collisions. Section 2.1 explored the (rigorous) implications of detailed balance to relate the group to group rate coefficients in the forward and reverse reactions. It enabled us to conclude that if the forward reaction is, say, highly specific in its vibrational energy disposal then the reverse reaction will be highly selective in its vibrational energy consumption. Alternatively, if the specificity is low, so is the selectivity. To characterize the discrimination in the dynamics we have introduced the surprisal in Sect.2.2. A measure of the specificity (or the selectivity) was thereby obtained. It is still desirable however to relate this measure to more fundamental descriptions of the collision. In particular, it is of interest to correlate the dynamics with the structure. The present section notes the essential ingredients in such a program. It is however not meant to replace the more systematic accounts of molecular collision dynamics that are available [2.1,68,75-80]. In particular, those aspects that are thoroughly discussed in [2.1] will not be replicated.

2.3.1 Computational Studies

Considerable attention has been given to accounting for the product state distribution (or the reactant state utilization) using the method of classical trajectories [2.1,81-86]. One cannot overestimate the contribution of computational studies to our present level of understanding. Yet, one must recognize their limitations. First, such computations necessarily require as input the intermolecular interaction potential which will be further discussed in this section. While the art of tailor-making such surfaces semi-empirically is quite advanced, we still lack sufficient examples of rigorous and accurate quantum mechanical computations of such potentials. The problem is particularly acute for processes involving a change in the electronic state, i.e., where electronic excitation energy can be consumed (or released) during the collision (Sects.2.2.6 and 2.4.7). The second limitation of computational studies is that, by their very nature, they provide a quantitative rather than a qualitative picture of the collision dynamics. The highly refined computational procedure tends to mask the essential physical features of the problem. To obtain these we turn to models, approximate descriptions of the dynamics which retain the bare minimal elements of reality to simulate the special characteristics of the particular reaction under study. Of course, the computational studies themselves often suggest starting points for simple models. Particularly useful in this respect is the ability, provided by the method of classical trajectories, to follow the interatomic distances during the collision.

Figure 2.25 shows typical trajectories and the resulting product vibrational distribution (determined by running many trajectories) for the $O+CN(v=4)$ collision. The initial vibrational excitation is quite evident in the oscillatory motion in the reactant region. The two (empirical) potential energy surfaces shown correspond to the formation of $N(^2D)$ which proceeds via the NCO intermediate, and to the formation of $N(^4S)$, respectively, where there is only a small barrier between the entrance and exit valleys. In the former case, even though the NCO molecule formed in the collision has sufficient energy to dissociate, the classical trajectory shows that it does have a fleeting existence.

Noncomplex forming or "direct" reactions that occur on a time scale of a vibrational period have common characteristics. The trajectories computed for such reactions show a continuous (monotonic) tightening of the newly formed bond and a simultaneous loosing up of the old bond. Within very few vibrational periods the order[2] of the new bond has increased to unity. Computational studies helped to establish a correlation between the direct nature of the collision and the considerable specificity or selectivity it exhibits. The more dominant is the contribution from "snarled" trajectories the less discriminatory is the role of the energy in the collision. This is quite reasonable. By the time the energy has been exchanged several times between the several vibrational modes all memory of the initial disposition of the exci-

2 The bond order is a useful measure of the bond strength as further discussed below in connection with the BEBO method.

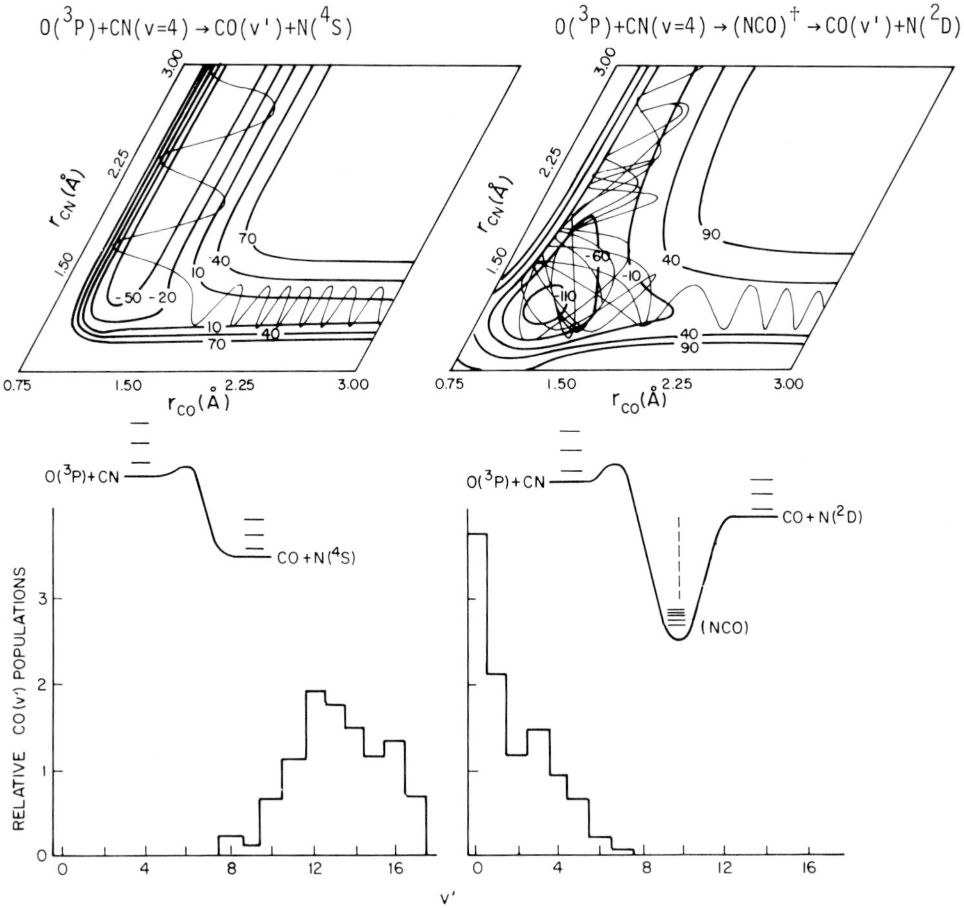

Fig. 2.25. Reactive trajectories for the O + CN(v=4) collision on potential energy surfaces corresponding to the formation of N(^{2}D) and N(^{4}S), respectively. The bottom panels show the resulting CO vibrational energy distribution (and, in the inset, the profile of the potential energy along the reaction coordinate) (adapted from [2.87])

tation has been erased and all final partitionings of the energy are equally probable. The differences in energy disposal between direct and compound collisions are quite evident in the computational results shown in Fig.2.25 and also in the corresponding experimental results (cf. Fig.2.5).

Computational studies of energy-rich polyatomic molecules, or of collisions of greater complexity than diatom-diatom, are currently limited because of the "Monte Carlo" nature of the method of classical trajectories. The more complex is the system, the more trajectories need to be run to obtain good statistics. The same considerations imply that the accuracy of the method will diminish whenever very detailed product state distributions are required. In general, the smaller the rate constant the less reliable is the method of classical trajectories (unless special steps are taken).

Finally, the method is subject to two peculiar classical limitations. The standard algorithms as currently in use do not ensure that quantal detailed balance is satisfied. Rate coefficients generated in this fashion and used in a kinetic (or master) equation program may fail to bring the system to thermal equilibrium. A related aspect is the manner of "quantization" of the classical product energy distribution. This needs to be done with some care if the detailed product state distribution is desired.

Some of the very same weaknesses of the classical trajectory method are also its strengths. The Monte Carlo selection of initial conditions enables one to carry the computation for any desired manner of reagent preparation. The product mean energy (and higher moments thereof) is typically well accounted for. In particular, the ease of executing the procedure has made it the standard way to survey (in a semiquantitative fashion) the role of such features as the overall topography of the potential energy surfaces or different manners of reagent excitation on the collision dynamics.

2.3.2 Potential Energy Surface(s)

The potential energy surface is the electronic energy of the system as a function of the interatomic distance (see also Sect.3.3.2). There will in general be a multitude of electronic states corresponding to a given configuration of the nuclei. Only in favorable circumstances can one assume that the motion of the system can be confined to a single surface. As discussed in Sect.2.4.7, surface "hopping" becomes important whenever the time required to the change in electronic state (essentially $h/\Delta E_{el}$) is comparable to the time spent by the system in the region where the gap is small.

Potential energy surfaces can be computed using the methods of molecular quantum chemistry [2.83,88,89]. Semiempirical procedures have however played an important role, particularly in interpreting trends in a family of related reactions and in allowing for systematic variations in the key features of the interaction. The different but related potentials are then used in computational studies thereby allowing us to correlate the topography of the potential with the nature of the collision dynamics. We have already mentioned one such aspect. If the interaction is such that the polyatomic species formed during the collision is potentially stable, the collision will typically not be direct, at least not at low energies. As the energy is increased, the details of the underlying potential matter less, and the collision becomes more direct.

Semiempirical procedures are particularly important for polyatomic collisions, where the surface is in actuality a hypersurface, a function of very many variables. It is important therefore that one develops means for representing the important regions of the surface in terms of the properties of the reagents and products. The BEBO (bond energy bond order) method [2.90-94], to be discussed below, is a semiempirical one aimed at using as much as possible of the accumulated empirical wisdom in the construction of the potential. Another extensively used method, with slightly firmer theoretical foundations, is the LEPS (London Eyring Polanyi Sato) method [2.81,83]. In practice this method is also used as a means of interpolation from the reagent to the product potentials, and is typically employed outside the range of its theoretical validity as a convenient empirical form.

The central concept that emerges from the consideration of potential energy surfaces is that of "concerted" reaction. The energy required for, say, an atom transfer, is typically far below a bond-breaking energy because the new bond is being formed simultaneously as the old bond loosens. The potential energy barrier separating reagents from products is much reduced when part of the energy required to break the old bond is provided by the simultaneous formation of the new bond.

The relations between the energy and the length of the bond can be conveniently summarized by two empirical relations [2.90]. The first is Pauling's definition of bond order n for a bond of length R

$$R = R_{eq}(n=1) - a \ln(n) \quad . \tag{2.3.1}$$

Here $R_{eq}(n=1)$ is the equilibrium bond length for a single bond and a is the Pauling parameter ($\simeq 0.5$ a.u.). The energy in a bond of order n is given by the Johnston-Parr relation [2.90]

$$E(n) = -Dn^p \quad , \tag{2.3.2}$$

where $D \equiv E(n=1)$ is the dissociation energy of a single bond and p is an empirical constant (the "bond index", which may be determined from spectroscopic data [2.91] and does not differ much from unity.)

On the basis of the empirical relation between bond energy and bond order, intuition suggests that along the path of minimal potential energy in an $A + B\text{-}C \rightarrow A\text{-}B + C$ reaction the sum of the two bond orders n_{BC} and n_{AB} will remain constant

$$n_{BC} + n_{AB} = 1 \quad . \tag{2.3.3}$$

To locate the crest in the potential energy profile vs bond order (or vs interatomic distances) one now needs to express the energy of A-B-C (including A-C repulsion) as a function of the bond orders [2.90]. Rather than do so explicitly, we appeal to a qualitative rule (sometimes known as Hammond's principle [2.93,94]) which states that the crest occurs early along the minimal energy path for exoergic processes and conversely for endoergic ones. In other words, for exoergic processes the order of the old bond is not much reduced below unity while for endoergic processes the order of the old bond is far lower and the order of the new bond is nearly unity at the "transition state".

2.3.3 Bond-Tightening Models

Armed with the concept of a direct reaction we shall seek to obtain simplified models which highlight the essential features of the problem. Consider, for example, the concept of the tightening of the newly formed bond. At the start of the act of changing partners, the bond to be formed has a rather low order which rapidly increases to unity. A simple assumption is that this increase in bond strength is sudden; that it occurs on a time scale shorter than a vibrational period during which it is also unperturbed by the departing atom. If so, the distribution of final vibrational states is very readily computed. It is simply given by the Franck-Condon overlap factor between the vibrational wave functions of the loose oscillator and the fully tightened one, the latter corresponding to the product final vibrational state. Figure 2.26 shows the wave functions and the corresponding vibrational state distribution for the $F + H_2$ reaction. In this ultrasimple picture the dynamics are primarily governed by the bond order of the loose oscillator or, equivalently, by its (extended) separation R as given by (2.3.1).

A class of processes for which the picture of a sudden bond tightening may be particularly appropriate is photodissociation (cf. Sect.5.3.1). In the excited state of, say, CO_2 the CO bond will be compressed (because of its higher bond order) as compared to the equilibrium separation in CO electronically excited. If we imagine a sudden increase in the bond order during the dissociation we can determine the distribution of final vibrational states in terms of the overlap between the vibrational wave functions in the two vibrational potentials (as in Fig.2.26).

A particularly important feature of the potential is the location of the transition state or "point of no return". The model of sudden tightening of

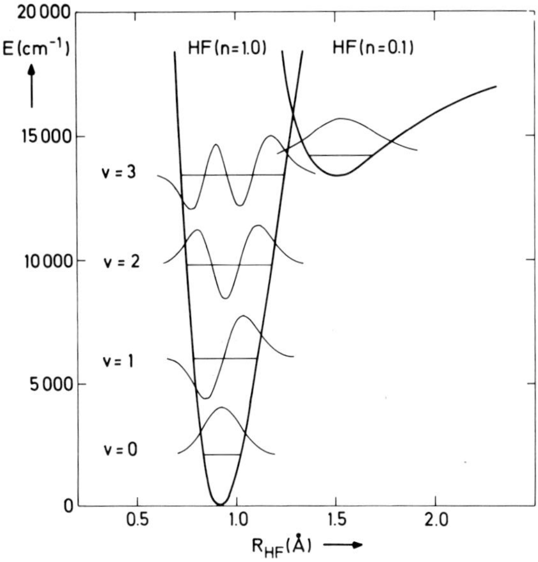

Fig. 2.26. Wave functions for the loose (low bond order, n=0.1) and tightened (n=1) HF oscillators in the $F + H_2 \rightarrow HF + H$ reaction (adapted from [2.95]). The loose oscillator corresponds to the HF bond in the "point of no return". The tightened oscillator is the stable product HF molecule. The product vibrational distribution of HF(v) molecules is determined by the overlap between the wave functions describing the two oscillators. Maximal overlap obtains here for HF(v=3)

the new bond shows why. If the bond order is low at that point, there is considerable change in the equilibrium bond distance. The Franck-Condon factors will then favor overlap of the loose oscillator with high vibrational states of the tightened one. The products will be preferentially formed with high vibrational excitation. On the other hand, if at the point of no return the bond order is already high, a large fraction of the exoergicity will be released as translation while the vibrational excitation of the products will be lower. These considerations are illustrated in Fig.2.27.

We can thus summarize. For an exoergic reaction we expect an "early" transition from reactants to products, with consequent substantial vibrational excitation. Detailed balance now implies that for the reversed endoergic reaction one expects a "late" transition state. Turning to the role of reagent excitation we note that for an exoergic reaction the reagent bond order is nearly unity at the point of no return. Reagent vibrational excitation is thus not available to overcome the barrier separating the reactants from products. For endoergic reactions the transition state is late, the reagent bond order is well below unity, and the initial vibrational energy is available

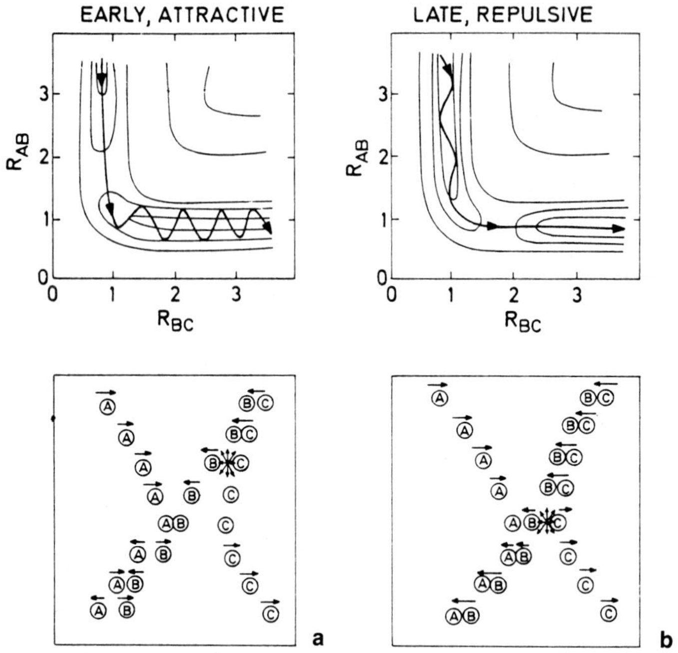

Fig. 2.27. Schematic illustration of (a) attractive energy release on a potential surface with early barrier and (b) repulsive energy release on a potential surface with late barrier [2.81]. In (a) the reaction exoergicity is released when the A atom and the BC molecule are still approaching each other, leading to vibrational excitation. In (b) the energy is released only when the atom A has approached BC and is almost at rest or already on its way out. As a result, the AB product molecule rebounds and most of the energy is released as product translation

Table 2.3. Structure discrimination relations in direct reactive collisions

Location of transition state	Typical class of reactions	Reagent energy consumption	Product energy disposal
Early	Exoergic	Translation-enhancement Vibration-detraction	Vibration
Late	Endoergic	Translation-detraction Vibration-enhancement	Translation

for surmounting the potential hump. Experiments and trajectory computations on a series of model potential energy surfaces [2.1,81-86] lend considerable support for this simple picture which is summarized in Table 2.3. It is in the nature of chemistry that there is sufficient diversity to provide an exception to every rule. Yet the simple considerations of Table 2.3 provide a reasonable overall view of the correlation between structure and energy discrimination in direct elementary reactions.

2.3.4 Kinematic Models of Collision Dynamics

Thus far we have adopted a structural point of view, but collision-oriented considerations lead to the same general conclusions. Figure 2.27 draws the distinction between an early and a late release of the collision exoergicity. For an early release the reaction exoergicity is made available as kinetic energy while the incident atom is still on its approach motion. The transferred atom B is set in motion on a collision course towards the approaching atom A. The considerable kinetic energy of the relative motion of A and B during the collision is the considerable vibrational energy of AB past the collision. For a late release, C first rebounds elastically from B. Only as C is on the way out does the BC bond snap, sending B in the same direction of A. The relative velocity of A and B is now smaller and much of the energy is in the motion of the center of AB, i.e., in product translational energy.

To make the picture more quantitative we examine the changes in the momenta of the atoms as a result of a reactive collision, e.g. Cl + HI → HCl + I. Let $\underline{v}_i = \underline{v}_{HI} - \underline{v}_{Cl}$ denote the relative velocity. Conservation of linear momentum implies that in the center of mass system $\underline{p}_{HI} = -\underline{p}_{Cl}$. We can thus speak of the initial relative momenta $\underline{p}_i = \underline{p}_{HI} = \mu_i \underline{v}_i$ where μ_i is the reactants' reduced mass. The initial momentum of the I atom is its fraction, $\gamma_i = m_I/m_{HI}$, of the initial momentum $\underline{p}_i$ of HI. Say now that reaction took place upon collision and the I atom departs stripped of its partner. Let the final momentum of the I atom be $\underline{p}_f$. If the heavy and cumbersome iodine atom suffered no change in its momentum during the collision then

$$\underline{p}_f = \gamma_i \underline{p}_i \quad . \tag{2.3.4}$$

In terms of the change in translational energy we note that the initial relative translational energy of the reactants is given by $E_T = \mu_i \underline{v}_i^2/2 = \underline{p}_i^2/2\mu_i$, where $\mu_i = m_{Cl} m_{HI}/(m_{Cl} + m_{HI})$. Similarly the final translational energy E_T' is given by $E_T' = \underline{p}_f^2/2\mu_f$, where $\mu_f = m_{HCl} m_I/(m_{HCl} + m_I)$. Thus (2.3.4) is equivalent to

$$E_T' = E_T \cos^2\beta \quad . \tag{2.3.5}$$

Here, $\cos^2\beta = \gamma_i^2(\mu_i/\mu_f) < 1$ is a frequently occurring mass ratio. In this simple picture any energy released by the reaction is necessarily deposited in the internal (vibrational and/or rotational) degrees of freedom of the newly formed molecule. The translational motion cannot accommodate it.

To allow for the change of the momentum of the I atom under the action of the interparticle potential we can modify (2.3.4) to

$$\underline{p}_f = \gamma_i \underline{p}_i + \underline{q} \quad , \tag{2.3.6}$$

where $\underline{q}$ is the impulse imparted to the iodine atom during the collision. The change in the energy of HI corresponding to such an impulse is $R = \underline{q}^2/2\mu_{HI}$ and (2.3.5) for $E_T' = \underline{p}_f^2/2\mu_f$ is now modified to[3]

$$E_T' = E_T \cos^2\beta + R \sin^2\beta \quad . \tag{2.3.7}$$

For H atom transfer between heavier atoms $\sin^2\beta$ is very small and the correction is not significant. It does represent however the effect of, among others, the change in HI energy during the collision (e.g., due to the decrease of the bond order from unity). The later is the release of energy, the larger is R and the higher is the fraction of exoergicity that is released as product translation.

Two further refinements can be easily included in the present model. The first concerns the relative force constants of the old and new bonds. If the force constant of the new bond is very high the newly formed diatomic product molecule recoils as a rigid body. The energy R is then released as the relative energy of the departing atom and the product diatomic molecule. In this case the second term in (2.3.7) is simply R, $R = \underline{q}^2/2\mu_f$. The second modification brings in the mechanism of rebound reactions (Fig.2.27). Here reaction follows the rebound of the incident atom from the reactant diatomic. This has implications for both terms in (2.3.7). To begin with, R is released either as the relative energy of the departing reactant atom and the reactant diatomic $R = \underline{q}^2/2\mu_i$ or as the kinetic energy in the old bond, $R = \underline{q}^2/2\mu_{HCl}$, depending on the magnitude of the relative force constants. In addition, even if R is zero, one must take cognizance of the fact that an elastic encounter took place, i.e., that the initial relative velocity is also the final relative velocity or $\underline{p}_f = \mu_f \underline{v}_i = (\mu_f/\mu_i)\underline{p}_i$ when $R = 0$. Thus, if $R = \underline{q}^2/2\mu_i$ we have for rebound reactions

3 We drop the cross term obtained by squaring (2.3.6) since the orientation of $\underline{q}$ relative to $\underline{p}_i$ is random so that on the average $\langle \underline{q}\cdot\underline{p}_i \rangle = 0$.

$$E'_T = E_T(\mu_f/\mu_i) + R(\mu_i/\mu_f) \quad . \tag{2.3.8}$$

The present, dynamical, point of view is closely related to the structural considerations discussed in the previous section. To see this one should note that the Franck-Condon overlap factors are large for those transitions that involve little change in the momenta of the nuclei [2.96,97]. This concept, that efficient transitions are those that involve the least change in momenta, except for the impulses imparted by the potential, would serve as our guiding principle when we interpret state-to-state transitions in Sect.2.4.

2.3.5 Unimolecular Processes — The RRK Approach

The multiple absoprtion of infrared photons was considered over sixty years ago as a possible mechanism for the formation of energy-rich polyatomic molecules. At the time it was dismissed in favor of the Lindemann mechanism [2.98-100] of collisional activation. This mechanism assumes that the energy-rich molecule does not dissociate immediately but can survive for some time, and be deactivated by collisions instead of dissociating. Statistical considerations [known in this context as the Rice-Ramsperger-Kassel (RRK) theory [2.98-100]; see Appendix 2.C] readily account for the longer lifetime of polyatomic molecules. If, at a given energy, all states of the system are equally probable, then due to the high number of vibrational modes, the mean energy per mode is not high, even though the total energy is above the dissociation threshold. The fraction of molecules where enough energy is localized in the mode along which the fragments recede during the dissociation is thus low. Counting the states as if the molecule consists of s harmonic oscillators of equal frequency ν, one obtains [2.24,98-100] that for a molecule containing n vibrational quanta the degeneracy, i.e., the number of states with energy $E = nh\nu$, is

$$G(n) = \frac{(n+s-1)!}{n!(s-1)!} \simeq \frac{1}{(s-1)!}\left(\frac{E}{h\nu}\right)^{s-1} = G(E) \quad , \tag{2.3.9}$$

where the second equality is valid only in the classical limit $n \gg s$. The total number of states with energy less than or equal to E is

$$N(n) = \sum_{n'=0}^{n} G(n') = \frac{(n+s)!}{n!s!} \simeq \frac{1}{s!}\left(\frac{E}{h\nu}\right)^{s} = N(E) \quad . \tag{2.3.10}$$

The classical analogue of the relation $G(n) = N(n) - N(n-1)$ is $G(E) = N(E) - N(E-\delta E) = \rho(E)\delta E$, where $\delta E = nh\nu - (n-1)h\nu = h\nu$ and $\rho(E)$ is the density of states

$$\rho(E) = \frac{dN(E)}{dE} = \frac{G(E)}{h\nu} = \frac{1}{(s-1)!h\nu}\left(\frac{E}{h\nu}\right)^{s-1} \quad . \tag{2.3.11}$$

Say now that $E_0 = mh\nu$ is the minimal energy that needs to be placed in an oscillator so that it will break. The RRK theory assumes that the dissociation is the rate determining step, i.e., that on the time scale of the dissociation there is ample time for energy equipartitioning. In simple terms, if $A^\dagger$ is an energy-rich molecule and $A^\ddagger$ is a molecule with energy in excess of E_0 in the critical mode, then the (microcanonical) equilibrium is rapidly established and is unperturbed by the slow depletion of $A^\ddagger$ molecules due to dissociation. The fraction of critically excited molecules is thus given by the detailed balance ration $[A^\ddagger]/[A^\dagger] = G^\ddagger(E)/G(E)$, where $G^\ddagger(E)$ is the total number of states of $A^\ddagger$ molecules. The rate of dissociation of $A^\dagger$ molecules $R = R(E) = -d[A^\dagger]/dt$ is proportional to $[A^\ddagger]$, hence to $[G^\ddagger(E)/G(E)][A^\dagger]$. The dissociation rate constant $k_d(E) = R(E)/[A^\dagger(E)]$ is thus

$$k_d(E) = a[G^\ddagger(E)/G(E)] = aG^\ddagger(E)/h\nu\rho(E) \quad , \tag{2.3.12}$$

where a is a proportionality factor which is assumed to be energy independent. Noting that the number of states of a molecule with energy in excess of $E_0 = mh\nu$ in one oscillator is equal to the number of states with energy up to $E-E_0 = (n-m)h\nu$ in the remaining s-1 oscillators, we find, using (2.3.10)

$$G^\ddagger(E) = N^\ddagger_{s-1}(E-E_0) = [(E-E_0)/h\nu]^{s-1}/(s-1)! \quad . \tag{2.3.13}$$

Combining (2.3.11-13), we get the RRK result

$$k_d(E) = a[(E-E_0)/E]^{s-1} \quad , \quad E \geqq E_0 \quad . \tag{2.3.14}$$

For $E \leqq E_0$, $k_d(E) \equiv 0$.

Molecular interpretations to the parameters a and s ("the effective number of oscillators") and elaborations of the simple form (2.3.14) have been given by the RRKM (M=Marcus) theory which casts the basic RRK assumptions into transition state theory (TST) [2.24,98-100] (cf. Appendix 2.C). The general expression for the dissociation rate constant is then

$$k_d(E) = N^\ddagger(E-E_0)/h\rho(E) \quad , \tag{2.3.15}$$

where $N^\ddagger(E-E_0)$ is the number of states of molecules in the transition state region, excluding the contribution of the dissociating degree of freedom (the motion along the reaction coordinate). In the RRK approximation, where the energy-rich molecule is regarded as a collection of strongly coupled degenerate oscillators $N^{\ddagger}$ and ρ are given by (2.3.13) and (2.3.11); hence

$$k_d(E) = \nu[(E-E_0)/E]^{s-1} \quad , \qquad (2.3.16)$$

which is the original RRK expression, (2.3.14), with a identified as ν.

Powerful infrared lasers have been used to induce multiphoton dissociation of polyatomic molecules (Sect.5.4). Given a mixture of several components one can identify one (or more) wavelengths where a particular species will have a far higher absorption. When a high-power laser line at that frequency is available, one can selectively photolyze a particular component of the mixture. The method is thus very species selective. A question of both practical and theoretical interest is however that of energy selectivity. That is, what is the population distribution of energy-rich molecules which is created by the multiphoton light absorption and, moreover, how is the energy distributed within the molecule. In other words, there are really two questions. The first question reflects the fact that when a sample absorbs the laser light not all the molecules have absorbed the same number of photons. Some may have absorbed more than their average share and others may have absorbed less. In principle, this depends on the absorption cross section and its dependence on the energy content of the absorbing molecule.

The question here is whether after the laser pulse the distribution of total energy among the different molecules is at all similar to that which obtains for a system at thermal equilibrium (but at a high temperature). The second, but related, question concerns a subgroup of molecules, those that have absorbed a given number of photons. Here the question is whether the excess energy is distributed among all the modes or is it more or less localized. If it is, then one might be able to achieve mode-specific excitation.

To answer such questions we consider first the situation where the answer to both questions is in the negative. Any deviations will then indicate that selectivity is operative. Consider therefore a sample of molecules which has absorbed a given mean number of photons. Such an ensemble is characterized by its mean energy – just as in a system at thermal equilibrium (cf. Sect. 2.1.7). The fraction of molecules which are in a given quantum internal state n of the energy E_n is determined by the procedure of maximum entropy to be the canonical distribution (2.1.40). The probability of a group v of g_v degenerate states is

$$p(v|T) = \sum_n{}' p(n|T) = \sum_n{}' \exp(-E_n/kT)/Q = g_v \exp(-E_v/kT)/Q \quad , \tag{2.3.17}$$

where the summation is restricted to those states with $E_n = E_v$ and Q is the partition function

$$Q = \sum_n \exp(-E_n/kT) = \sum_v g_v \exp(-E_v/kT) \quad . \tag{2.3.18}$$

The temperature T is determined by the given value of the mean excitation energy $\langle E \rangle$,

$$\langle E \rangle = \sum_n p(n|T)E_n = \sum_v p(v|T)E_v \tag{2.3.19}$$

or, using (2.3.17,18)

$$\langle E \rangle = -k\partial \ln[Q(T)]/\partial(1/T) \quad . \tag{2.3.20}$$

For the polyatomic molecules of interest the density of internal quantum states is sufficiently high to be regarded as a continuum. Let $\rho(E)$ be the internal density of states so that the number of vibrational states at an energy range E, E + dE is $\rho(E)dE$. According to (2.3.17) all these vibrational states are equiprobable. Hence the fraction $\rho(E)dE$ of molecules with energy in the range E, E + dE is given by

$$P(E) = \rho(E) \exp(-E/kT)/Q \quad , \tag{2.3.21}$$

where the definitions of T and Q remain unchanged. One can however break the sums over n in (2.3.18,19) to sum first over all those states with energy in the range E, E + dE followed by a sum (i.e., integral) over E. Carrying out the first sum enables us to write, for example, (2.3.18) as

$$Q = \int dE \rho(E) \exp(-E/kT) \quad . \tag{2.3.22}$$

The two questions we started with are manifested in the two terms of (2.3.21). Since no constraints on the intramolecular energy distribution were imposed, the maximal entropy solution was that all quantum states of (about) the same energy are equally probable. In (2.3.21) $\rho(E)dE$ is the number of quantum states in the energy range E, E + dE. If not all degenerate vibrational states are in fact populated by the excitation process, the first factor would be lower. Since no constraints on the uptake of photons were imposed, the de-

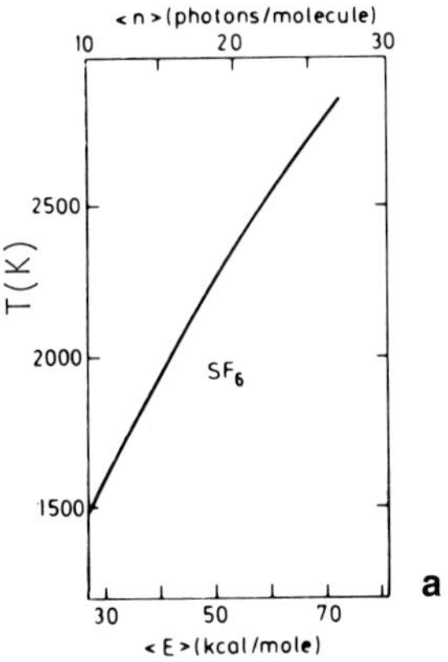

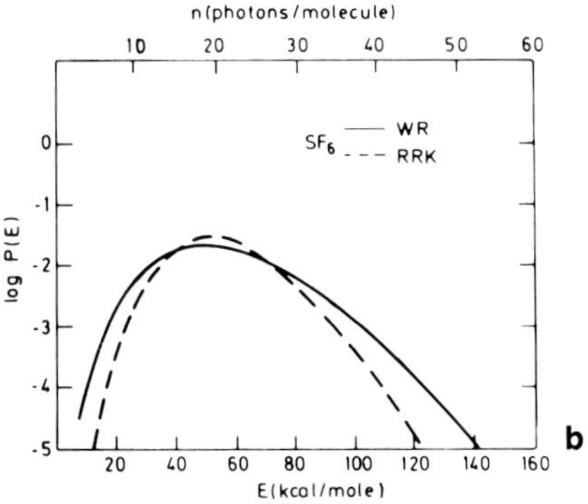

Fig. 2.28. (a) The temperature T vs $<E>$ and $<n> = <E>/h\nu$ for SF_6. $<n>$ is the average vibrational energy of the molecule measured in CO_2 laser quanta (~2.7 kcal/mole). Calculated using the Whitten-Rabinovitch (WR) representation of $\rho(E)$. (b) The canonical distribution, (2.3.21), vs n (or E) for $<n> = 20$ using the WR and the improved RRK expression (2.A.21) for $\rho(E)$ (adapted from [2.101])

pendence on the total energy is governed simply by the mean excitation energy $<E>$, or equivalently, by the temperature T. The presence of constraints on the absorption process will thus be manifested by a non-Boltzmann second factor in (2.3.21) while lack of rapid intramolecular energy sharing among all the isoenergetic states will be reflected in deviations from the preexponential factor in (2.3.21).

The temperature T which characterizes the canonical distribution P(E), (2.3.21), provides a measure for the average energy per mode. In the simplest approximation where the molecule is regarded as a collection of s degenerate classical oscillators so that $\rho(E) \propto E^{s-1}$ [cf. (2.3.11)] it is easy to show using (2.3.20) and (2.3.22) that $kT = <E>/s$. Hence, for a large polyatomic molecule the temperature is a slowly increasing function of the absorbed energy. This is illustrated in Fig.2.28 for SF_6 using the more realistic Whitten-Rabinovitch expression for $\rho(E)$ (cf. Appendix 2.A). The figure shows also P(E) for an energy-rich molecule. The fast increase of the density of states factor and the fast decrease of the Boltzmann factor as a function of E are responsible for the sharp peak of $P(E) \propto \rho(E)\exp(-E/kT)$ around the most probable value E_{mp}. In the simplest approximation, $\rho(E) \propto E^{s-1}$, we get $E_{mp} = (s-1)kT$ implying that for large s, $E_{mp} \simeq <E>$.

The dissociation rate constant corresponding to an energy distribution P(E) of the dissociating molecules is $k_d = \int dE P(E) k_d(E)$, where $k_d(E)$ is the

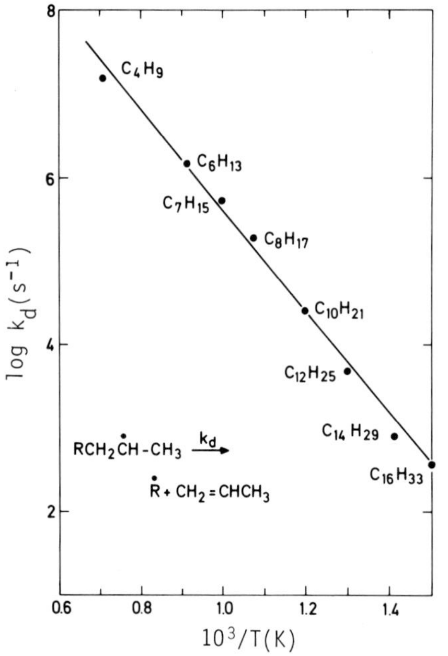

Fig. 2.29. $\log(k_d)$ vs $1/T$ for the series of reactions (2.3.24) (adapted from [2.104]). The activated alkyl radical is identified in the figure. The temperature T is defined as discussed in the text, in terms of the energy content of the activated molecule, i.e. $kT = \langle E \rangle/s$ is the average energy per oscillator. Note that E_0 (the C-C bond energy) is essentially the same for all the reactions. Similarly, $\langle E \rangle$ – the C-H plus C-C bond energies minus the C=C bond energy – is essentially the same for all reactions. Hence $1/T$ is a direct measure of s

dissociation rate constant for molecules with energy E [cf. (2.3.14-16)]. For the canonical distribution (2.3.21), with $\rho(E)$ and $k_d(E)$ given by the simplified expressions (2.3.11) and (2.3.16), one gets

$$k_d(T) = \int_{E_0}^{\infty} dEP(E)k_d(E) = \nu \exp(-E_0/kT) \quad . \tag{2.3.23}$$

The exponential factor $\exp(-E_0/kT)$ is simply the fraction of molecules where energy in excess of E_0 is localized in one oscillator. [This follows immediately from the fact that the energy distribution of a single oscillator is $P(E) \propto \exp(-E/kT)$.]

An alternative derivation of (2.3.23) showing that this form is insensitive to the details of the energy distribution can be given as follows: In the energy range where $k_d(E)$ is significant (equivalently when $\langle E \rangle$ is significantly above E_0) we can write $k_d(\langle E \rangle) = \nu[1-sE_0/\langle E \rangle] = \nu(1-E_0/kT) \simeq \nu \exp(-E_0/kT)$ where $kT = \langle E \rangle/s$ is the average energy per oscillator. Thus $k_d(E)$ increases exponentially as s diminishes. This behavior is illustrated in Fig.2.29 for the series of chemical activation reactions [2.102-104]

$$RCH_2CH{=}CH_2 + H \rightarrow [RCH_2\dot{C}H\text{-}CH_3]^{\dagger} \xrightarrow{k_d} \dot{R} + CH_2{=}CH\text{-}CH_3 \quad . \tag{2.3.24}$$

For all the molecules shown in Fig.2.29 $\langle E \rangle$ is roughly the same (the C-H plus C-C bond energies minus the C=C bond energy, ~38 kcal/mole), but s changes from 33 to 135. Since $kT \simeq \langle E \rangle / s$ the temperature ranges over a wide range, leading to a dramatic variation in the rate, as implied by (2.3.23).

The considerable range of k_d values correlated in Fig.2.29 provides additional support for the notion that on the slow time scale considered (well above a nanosecond) there is ample time for the molecule to act as its own heat bath.

What one means by this often repeated phrase is that the fraction of molecules where energy in excess of E_0 is localized in the reaction coordinate can be computed as if the reaction coordinate has come to equilibrium with a heat bath at the temperature T. Since the reaction coordinate can only exchange energy with the other s-1 modes of the (isolated) molecule, these s-1 modes act as the heat bath for the mode of interest.

The detailed studies supporting the concept of efficient intramolecular relaxation were carried out using chemical activation [2.100,102-104]. Here, by collision of suitable reagents a new and strong chemical bond is formed. The bond energy is thereby made available and a weaker bond elsewhere in the molecule can dissociate, as for example in (2.3.24) (Fig.2.29), or (2.2.34,35) (Fig.2.21). Another example is the ring closure process followed by isomerization [2.102,103],

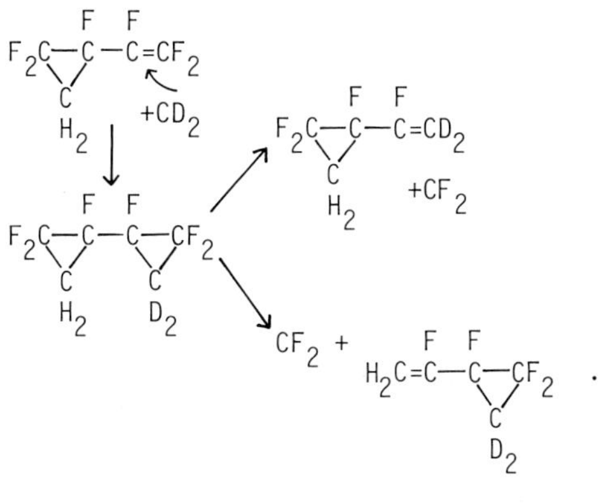

(2.3.25)

The use of CD_2 is simply to label the new and presumably initially "hot" ring. Internal energy migration spreads the excess energy to the old and initially cold ring. At very high pressures the energy-rich molecules are rapidly deactivated by collisions and hence only those molecules which have been newly formed have a chance to dissociate. Indeed decomposition of the initially hot new ring is somewhat favored by high pressures. At low pressures, where the interval between collisions is long, the decomposition is independent of the site of the initial excitation. The use of collisional deactivation as a clock suggests that for such highly excited (~110 kcal/mole) species the intramolecular relaxation rate is of the order of 10^{11}-10^{12} s^{-1}, i.e., 10-10^2 vibrational periods, which is quite rapid. It is therefore to be expected that only at high pressures can collisions compete with this process (Figure 1.9 shows the frequency ranges of different energy transfer collisions in units of Atm•s).

It should be noted that while chemical activation experiments provide the most extensive studies thus far available on intramolecular relaxation, the evidence is indirect. In particular, the conclusion hinges on the assumption of strong collisions, i.e., that the primary role of collisions is to deactivate the energy-rich molecule. One must note however that an at least equally important role is that of collision-induced intramolecular energy transfer, a process that is well documented for small polyatomic molecules [2.6,105]. Indeed we shall argue below (Sect.2.4.5) that in an atom large polyatomic collision, the distribution of internal energy computed as if the energy has been completely randomized by the collision (the "prior" distribution), is in accord with the available experimental data.

There is both experimental and theoretical evidence that the rate of intramolecular energy transfer is very significantly lowered as the energy is decreased. Near the bottom of the well it is of course possible to excite molecules to specific vibrational states where, in the absence of collisions, fluorescence with unit quantum yield will be observed. Collisions deplete the fluorescence but populate other specific states which can be monitored by their fluorescence [2.6,105,106]. Indeed, the selectivity with respect to the initial state and the specificity of the population of the final states demonstrate the lack of complete intramolecular scrambling (Sect.2.5.5). As the energy increases, so does the density of states, (2.3.11). The more degrees of freedom in the molecule, the faster the increase. Past the onset of the vibrational quasicontinuum resonant intramolecular energy transfer is invariably feasible and is limited only by dynamic constraints.

2.3.6 Unimolecular Processes – Selectivity and Specificity

The statistical point of view adopted in the previous paragraph is more restricted than the more general statement of Sect.2.2.6. There we concluded that the intramolecular coupling processes result in the most random distribution of states subject however to constraints. Of course, the purely statistical limit (i.e., no constraints) is included in this description as a

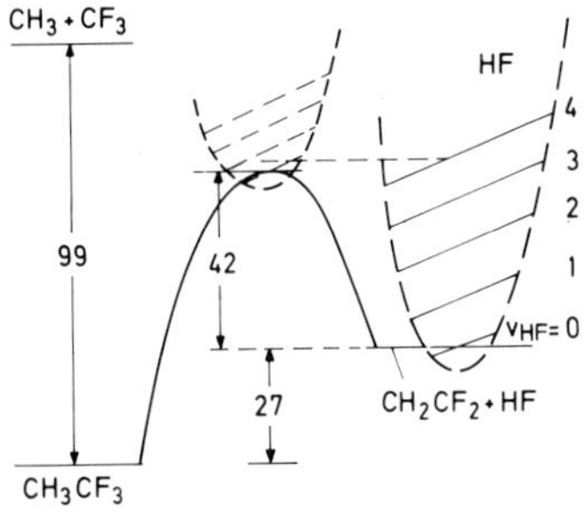

Fig. 2.30. Schematic representation of the potential energy profile for the elimination reaction (2.2.34). The energies are in kcal/mole (adapted from [2.69]). The dashed line on top of the barrier is the HF potential energy at the point of elimination. The lower potential curve is that of free HF. For a surprisal analysis of the HF vibrational state distribution see Fig.2.21

special case. The point is, however, that analysis of experimental results, e.g., elimination following chemical activation (Sect.2.2.6), leads to the conclusion that constraints are not invariably absent and that constraints can be locked in by suitable preparation of the initial state. Many examples of direct, bimolecular collision processes where the constraints on the product distribution reflect the details of the initial state selection will be discussed in Sect.2.4. In unimolecular reaction theory the extreme limit where the vibrational energy is strictly confined to the mode in which it was initially placed is known as the Slater model [2.24,98-100]. Surprisal analysis indicates that the actual state of affairs is intermediate between the Slater and the RRK models.

It is possible to relate the constraints observed in unimolecular dissociation processes to the collision dynamics. The considerable extent of vibrational excitation of HX molecules following four-center elimination (Fig. 2.21) can be discussed in reference to the potential energy profile for the process (Fig.2.30). There are two points to note. One is that the HF bond length at the point of elimination (on the top of the barrier) is quite a bit longer than the equilibrium bond length of HF. Hence the bond tightening model (Sect.2.3.3) will predict considerable vibrational excitation in the products. In addition, as the products separate they are strongly repelled by the downhill barrier and hence can exchange energy, performing a "half inelastic collision". This will, on prior grounds, tend to reduce the HF vibrational excitation and to enhance the vibrational excitation of $CH_2{=}CF_2$ (which is indeed observed to be vibrationally excited).

Refined statistical models (e.g., [2.107]) recognize the need to introduce constraints into the simple RRK picture and do so by restricting the range of states that are assumed to be strongly coupled and hence equiprobable. We have seen, however (Figs.2.18 and 2.20), that it is not always possible to explain away the surprisal by freezing out some states.

The point of view of the information theoretic approach of Sect.2.2 is that intramolecular dynamics, even on a single potential energy surface, is as statistical as possible, subject to constraints [2.75]. Such constraints are revealed by the observation that not all isoenergetic states are equally probable. Much of the current evidence for the dependence of the constraints on the initial preparation comes however either from experimental studies on tri- and tetra-atomic systems (discussed in detail in Sect.2.4) or from theoretical [2.41] or computational studies. In particular, trajectory computations of energy-rich molecules past the dissociation threshold show that one can prepare initial states where energy is rapidly shared only among a subset of the possible states (and in Fig.2.32 we examine some of the corresponding experimental evidence). Trajectory computations also show that at excess excitation, i.e., well past the threshold, the role of the constraints is much more dominant. Both computations, experiments, and intuitive reasoning suggest that excitation energy placed initially in a mode which requires a large change in the nuclear momenta for its deexcitation will not be readily transferred. In Sect.2.4 we shall extensively document this conclusion for intermolecular events. It is however equally valid for intramolecular evolution. A simple example is shown in Fig.2.31 and another in Fig.2.50.

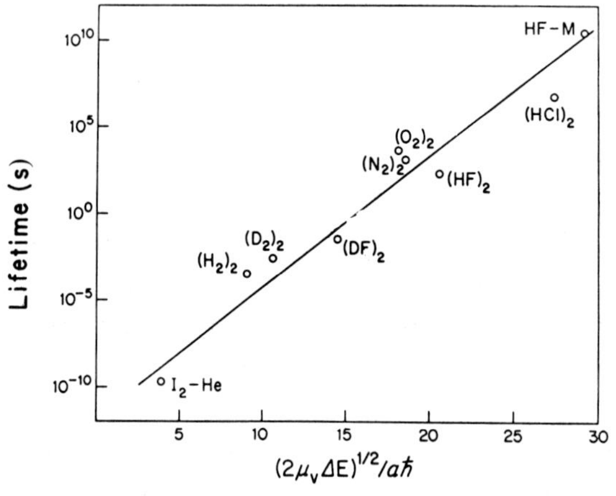

Fig. 2.31. The lifetimes for unimolecular dissociation of van der Waals bound molecules, following vibrational excitation of the chemical bond. The dissociation requires energy transfer from the higher frequency molecular bond, e.g., the H-F, to the much weaker (and hence lower frequency) van der Waals bond (e.g., the HF-HF bond). The lifetimes (inverse rate) are plotted vs $\Delta p = (2\mu_V \Delta E)^{\frac{1}{2}}/ah$, the change in the momentum of the dissociating bond. ΔE is the translational energy of the fragments and μ_V their reduced mass (a is the exponential parameter of the Morse potential used to describe the van der Waals bond). M is a heavy ether molecule (adapted from [2.108])

2.3.7 Preparing the Initial State

Intramolecular energy transfer reflects not only the properties of the molecules but also the nature of the excitation process. It occurs because the initially prepared state is not a stationary state of the system. The simplest example of a stationary state is an eigenstate of the Schrödinger equation. For a polyatomic system, the high density of excited states makes the excitation of a definite single eigenstate unlikely. A state which is a linear combination of eigenstates (where the coefficients of the different eigenstates bear definite phase relations to one another) is not, in general, a stationary state. To see this let ϕ_n be an eigenstate of energy E_n and let the prepared initial state at the time $t=0$ be $\psi(0)$,

$$\psi(0) = \sum_n C_n \phi_n \quad . \tag{2.3.26}$$

At the time t after the preparation the initial state $\psi(0)$ evolved into $\psi(t)$,

$$\psi(t) = \sum_n C_n \exp(-iE_n t/\hbar)\phi_n \quad . \tag{2.3.27}$$

If D is the mean spacing between the eigenstates so that D^{-1} is the level density, then, roughly $E_n = \langle E \rangle + nD$, $(n=0,\pm1,\pm2)$, so that the state recurs, $\psi(t)=\psi(0)$, for times roughly of the order of h/D. The recurrence time can be exceedingly long ($h \simeq 3.15 \cdot 10^{-12}$ $cm^{-1}s$ so that already at a density of 10^3 levels per cm^{-1}, the recurrence time is in the nanosecond range and collisional effects can compete).

We are specifically concerned here with a situation where there are many terms in the sum in (2.3.26) so that the range δE of energies of the states E_n, which contribute to the sum, while narrow is still large compared to the mean spacing D between two adjacent eigenstates. In other words, the time $h/\delta E$ required to prepare the initial state is quite short compared to h/D. The emission of light by the excited state at the time t to return to the ground state is governed by the magnitude of the squared matrix element $|\langle g|\underline{\mu}|\psi(t)\rangle|^2$ where g is the ground state and $\underline{\mu}$ is the dipole moment operator (cf. Sects.1.1 and 3.1.1). Introducing the explicit form of $\psi(t)$ from (2.3.27), we find

$$|\langle g|\underline{\mu}|\psi(t)\rangle|^2 = \sum_{n,m} C_n^*(t) C_m(t) \langle \phi_n|\underline{\mu}|g\rangle\langle g|\underline{\mu}|\phi_m\rangle \quad . \tag{2.3.28}$$

Here $C_n(t) = C_n \exp(-iE_n t/\hbar)$. Due to the different energies, terms of the type $C_n^*(t)C_m(t)$ tend to rapidly oscillate or, in technical terms, to dephase. It is only for time of the order of h/D that all contributions to the sum in (2.3.28) will have their original phase. For shorter times, the emission spectrum governed by $|<g|\underline{\mu}|\psi(t)>|^2$ may be quite distinct from the absorption spectrum governed by $|<g|\underline{\mu}|\psi(0)>|^2$. The larger the range of energies of the eigenstates that contribute to the initial state $\psi(0)$, the faster is the dephasing. This disappearance of the initial state is the phenomenon of radiationless transitions, further discussed in Sects.2.4.9 and 5.3. If D is sufficiently large (say $>10^{-3}$ cm) then even at very low pressures collisions with distant molecules[4] will enhance the dephasing thereby ensuring that the initial state never recurs.

A more general example of a stationary state is a mixture of eigenstates, e.g., an equilibrium distribution. Such a state cannot be described by a wave function but requires the concept of a density matrix [2.24,54]. The matrix whose elements are the products $C_n^*(t)C_m(t)$ as previously discussed is an example of a density matrix. The diagonal elements of such a matrix are the probabilities of the different eigenstates, and a stationary density matrix is diagonal. The probabilities can be computed for example by the procedure of maximum entropy subject to constraints which are conserved quantities (e.g., energy). The probabilities are then time independent. If one determines the state by the procedure of maximum entropy using constraints which are not conserved quantities then one obtains a nonstationary mixture. Here, the values of the constraints change with time and hence so do the values of the Lagrange parameters. Since, typically, one uses reagents which themselves are not in a definite eigenstate, such a nonstationary mixture is the typical initial state in a unimolecular process.

Visualizing a nonstationary state is more immediate in classical mechanics [2.109,110]. One can readily think of a triatomic molecule where all the vibrational energy is initially placed in one normal mode. Because the exact potential is not strictly harmonic, this state is not stationary. (It would be, were the harmonic approximation exact.) On solving the equations of motion for such an initial state energy transfer to the other modes will take place. To be sure, as in quantum mechanics, the initial state will recur. This behavior characterizes the low energy (or "quasiperiodic" regime where most

---

4 Any collision, even if too distant to cause any energy transfer, will lead to a phase shift of the molecular eigenstates. Such collisions are termed "elastic" but they suffice for dephasing and thus broadening of the energy levels (cf. Sect.3.1.2).

classical trajectories are periodic [2.109-112]). However, at higher energies where the anharmonicity is sufficient to ensure what is known technically as "mixing" the probability of finding the system in any given finite region of its phase space (including the initial one) will ultimately become independent of the initial mode of excitation. Mixing therefore destroys selectivity. When computing classical trajectories, this loss of initial conditions is exemplified by those trajectories which originate in neighboring regions of phase space yet proceed to execute markedly different motions [2.111,112]. This irregular behavior has a quantum analogue. For a molecule which is strictly harmonic each eigenstate corresponds to a particular number of quanta in each one of the normal modes. The eigenstate can thus be labelled by the occupation numbers of the different modes, i.e., by a series of numbers $n_1, n_2, \dots$ (see, e.g., Fig.1.17). For a real molecule, the normal modes are no longer exact but the labelling of the eigenstates can still be used. In the irregular region [2.113] eigenstates which are adjacent in terms of their energy do differ, in an irregular fashion, in their labelling. Despite the similarity of the total energy the same mode may have quite different occupation numbers in two neighboring eigenstates.

The special role of the eigenstates of the molecular Hamiltonian is that except for radiative transitions they are stationary. A state prepared by optical excitation is thus either an eigenstate or a linear combination of eigenstates. Due to the high density of states, the latter is invariably the case for excitation into the quasicontinuum. Because of the irregular nature of the spectra the description in terms of eigenstates [cf. (2.3.26-28)] while correct, is not necessarily the most intuitive one. Rather, one may obtain a more physical picture by considering an initial state [denoted by $\psi(0)$ in (2.3.26)] which is not stationary and does evolve in time even in the absence of the radiation field. The nature of this initial state is defined in principle by its resolution into eigenstates, i.e., by the spread of the transition probabilities from the ground state. In practice one often identifies the optically prepared nonstationary initial state by physical reasoning (i.e., based on the Franck-Condon, Born-Oppenheimer, and similar considerations, what is the likely character of the state). An example discussed in Sects.1.3.3 and 2.4.9 is the state prepared by visible or UV excitation past the lowest electronic transition. While optical transitions directly into the quasicontinuum of the ground electronic state are, in principle, possible, considerations of the Franck-Condon factors suggest that the initially prepared state is very largely confined to the upper electronic surface. In the irregular region of the spectrum, the local modes [2.114-116]

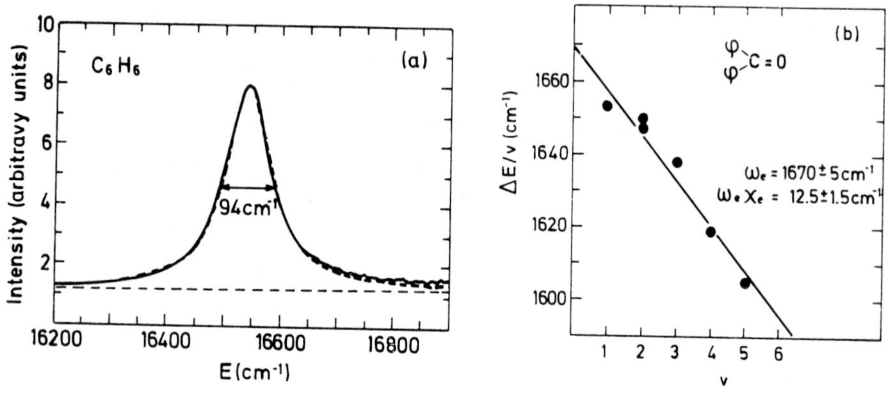

Fig. 2.32. (a) The spectrum of the 5th overtone of the CH stretch of vapor benzene. The dashed line is a Lorentzian fit superimposed on a linear background (adapted from [2.117]). (b) A plot of $\Delta E/v = (E_v - E_{v-1})/v$ for the CO stretch of benzophenone (in solid matrix at 2 K). The near linearity of the plot indicates that the CO bond is a local mode with the characteristics of an anharmonic oscillator. The spectroscopic constants correspond to the best fit linear curve (adapted from [2.119])

have been successfully used to interpret characteristic features in the absorption at high vibrational excitation. The local modes are bond vibrations (or linear combinations of equivalent bond vibrations)..

One of the best-known examples of a local mode is the CH stretch of benzene and various other molecules. The level structure of this mode of vapor benzene was detected by measuring (using dye laser intracavity absorption; cf. Sect. 5.5) its high overtone spectrum [2.117]. The spectrum appears similar to that of a single oscillator but with Lorentzian broadening of the order of 100 cm^{-1}, which can be attributed to dephasing and relaxation into the dense background of other vibrational modes [2.117,118] (Fig.2.32). Further evidence to the locality of the CH excitation stems from the fact that essentially the same overtone spectrum was found also in the deuterated compounds C_6D_5H, p-$C_6D_4H_2$ and C_6DH_5. For instance, the 5th overtone of benzene in these deuterated analogues is centered at 16550 ± 10 cm^{-1}. Bond localization has also been noted in condensed phases. For instance, the overtone spectrum of the CH stretch in liquid benzene (and deuterated analogues) displays very similar features to those observed in the gas phase, except for a larger broadening ($\sim$300 cm^{-1}) [2.116].

Another example, measurements [2.119] of the overtone emission spectrum of benzophenone $(C_6H_5)_2C{=}O$ isolated in a solid matrix at 2 K, reveal that the level structure of the C=O stretch is very nearly like that of an anharmonic (Morse) oscillator (cf. Sect.3.3.2). That is, the vibrational energies are given as $E_v = hc\omega_e(v+\frac{1}{2}) - hc\omega_e x_e(v+\frac{1}{2})^2$ so that $\Delta E = E_v - E_{v-1} = 2hc\omega_e x_e v$ (Fig.2.32b).

Examples where by using laser excitation of nonstationary states it is possible to achieve products specificity were noted in Chap.1 and are further discussed in Chap.5. The first is multiphoton ionization. The second (Fig. 1.18) is that of mode specificity. The rate constant for unimolecular isomerization is seen to depend not only on the total energy but also on the particular mode which was laser pumped to create the energy-rich molecule. Moreover and in contrast to expectations based on the RRK approach (cf. Sect. 2.3.5), the observed unimolecular rate constant is not a monotonic function of the total energy.

2.4 State-to-State Processes

This section examines the prior state-to-state rate coefficients, the deviation of the actual rates from the prior forms, and the dynamic reasons for these deviances. Particular attention will be given to the simplest functional form for representing a dynamic bias, the so-called "exponential gap" representation and to its physical interpretation. In addition to the detailed technical results this section is meant to reinforce the message of Sects. 2.2.7 and 2.3.6: By suitable initial state selection it is possible to govern the specificity of the product state distribution.

2.4.1 The Prior Detailed Rate Constant

The discussion in Sect.2.2 was largely restricted to highly exoergic (or highly endoergic) processes, where the thermal spread in the collision energy could be neglected. Here we want to consider the general case. We need to define the prior detailed rate constant, $k^0(v \to v';T)$. We can then consider not only the effects of energy consumption and energy disposal but also the temperature dependence. The prior rate constant $k^0(v \to v';T)$ is defined such that any deviation of the actual rate constant from the prior represents special features of the dynamics. It follows that detailed balance (and other rigorous symmetries) must hold for the prior rate constants as well. The definition of $k^0(v \to v';T)$ must be consistent with the explicit statement of detailed balance

$$k^{0}(v\rightarrow v';T)/k^{0}(v'\rightarrow v;T) = K(v\rightarrow v';T) \quad , \tag{2.4.1}$$

where $K(v\rightarrow v';T)$ is the equilibrium constant [cf. (2.1.31)]. The actual rate constants are also related as in (2.4.1) [cf. (2.1.30)]. The deviation between the actual and prior rate constants is thus identical for the forward and reverse processes,

$$k(v\rightarrow v';T)/k^{0}(v\rightarrow v';T) = k(v'\rightarrow v;T)/k^{0}(v'\rightarrow v;T) \quad . \tag{2.4.2}$$

The considerations of the previous paragraph are by no means unique to vibrational states. The result (2.4.2) remains true if we interpret v (and v') as any specification of the internal states. The considerations of Sect.2.1.2 will now be recognized as a particular case of (2.4.2) when v stands for "no state selection", and v' continues to represent the vibrational state of the products of the forward reaction. Similarly, if both v and v' stand for "no state selection" (2.4.2) reads

$$\vec{k}(T)/\vec{k}^{0}(T) = \overleftarrow{k}(T)/\overleftarrow{k}^{0}(T) \quad . \tag{2.4.3}$$

It is advantageous to derive the prior rates in a manner consistent with our previous assumption (Sect.2.1 and Appendix 2.A) that for collisions at a given total energy, all final states are, a priori, formed with the same rate. When the distribution of kinetic energy is thermal, different collisions will have different energies even though the internal energy of the reactants may be sharply specified. The collision theory canon now provides the procedure; average the rate constant at a given energy over the distribution of energy in the reactants.

To derive the prior rates we start with the results at a given energy. There, the prior rate into any group of product states does not depend on the precise disposition of the energy in the reactants; it only depends on the magnitude of the total energy. The rate constant is then simply proportional to the number of product states at the total energy range (Appendix 2.A and Sect.2.2). For collisions at a given total energy E

$$k^{0}(v\rightarrow v';E) \propto \rho(v';E) \tag{2.4.4}$$

and there is no explicit dependence on v. Only the total energy matters. Other prior rates are similarly defined by allowing a more general interpretation of v and v'.

The rate constant at a given temperature is obtained by averaging over the distribution of energy in the reagents

$$k^0(v \to v';T) = \langle k^0(v \to v';E) \rangle \quad . \tag{2.4.5}$$

The prior $k^0(v \to v';T)$ can be evaluated in closed form. Details are given in Appendix 2.A and elsewhere [2.11,35,120,121]. Here we turn to examine the qualitative predictions provided by the prior rates. Only the deviation from such behavior can be taken as evidence for discriminatory dynamics.

All the detailed properties which we shall derive are just elaborations on the following statement: The prior rate constant $k^0(v \to v';T)$ depends on the initial and final energies only through its dependence on the reduced variable

$$\Delta_{v,v'} = (E_v - E_{v'} - \Delta E_0)/2kT \quad , \tag{2.4.6}$$

where ΔE_0 is the exoergicity of the reaction. The definition (2.4.6) applies also, as a special case, for nonreactive collisions in which case $\Delta E_0 = 0$. $k^0(v \to v';T)$ is a monotonically increasing function of $\Delta_{v,v'}$ (Fig.2.33). The increase is exponential, $\exp(-2\Delta_{v,v'})$, for negative $\Delta_{v,v'}$, and changes to a much more moderate power law, $\Delta^m_{v,v'}$, for positive $\Delta_{v,v'}$. (The power m depends on the number of degrees of freedom of the collision products and on the specification of the initial and final states v and v', cf. Sect.2.4.5 and Appendix 2.A.)

The entire story of the prior rate is thus told by Fig.2.33. Changes in the exoergicity of the reaction, of the temperature, of the vibrational energy of the reagents (or of the products) all correspond to altering the magnitude of $\Delta_{v,v'}$, and hence are all represented on the same graph. The qualitative behavior shown in Fig.2.33 remains valid even if we reinterpret the symbols v and v' as any specification of the internal states. Consider, say, the effect of reagent vibrational energy on the reaction rate. Then v retains its meaning as the reagent vibration and v' is now any product state, from the ground state up. In this case we set $E_{v'} = 0$. As long as $E_v - \Delta E_0 < 0$ then, on prior grounds, the reaction rate increases exponentially with increasing E_v (decreasing Δ). Once however the reagent excitation is high, i.e., once $E_v - \Delta E_0 > 0$, the effect of further increasing E_v is considerably less. This is just the behavior found experimentally or in computational studies (cf. Figs.2.1 and 2.34).

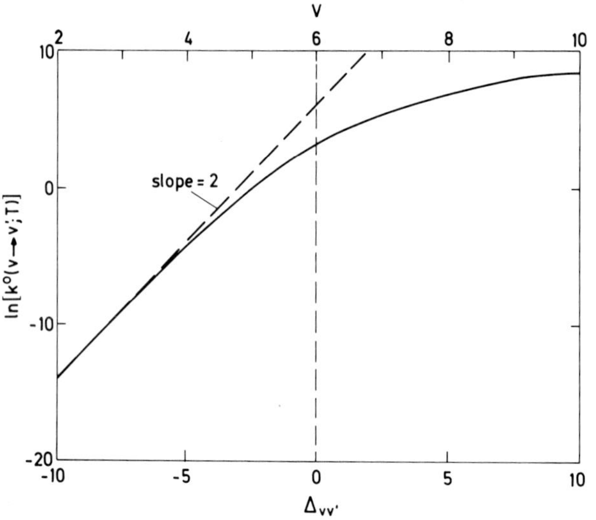

Fig. 2.33. The thermal prior rate constant $k^0(v\to v';T)$ for an atom-diatom exchange reaction as function of the reduced variable $\Delta_{v,v'} = (E_v - E_{v'} - \Delta E_0)/2kT$ (adapted from [2.11]). The prior rate constant increases exponentially for $\Delta_{v,v'} < 0$ and moderately, as a power law for $\Delta_{v,v'} > 0$. By plotting the rate vs the reduced variable one accounts for any changes in the exoergicity, temperature, and initial or final states. For example, the top scale for v corresponds to a typical endoergic reaction with $\Delta E_0 = 30$ kcal/mole, T = 500 K, $E_v - E_{v-1} = 5$ kcal/mole, and $E_{v'} = 0$. Note the Arrhenius-like behavior; $\ln[(k^0)(v\to v';T)] \propto -2\Delta_{v,v'}$ in the highly endothermic region

The temperature effect is also given in Fig.2.33. $\Delta_{v,v'}$ decreases as T increases. Changes in T will thus affect endothermic transitions ($\Delta_{v,v'} < 0$) much more so than exothermic ones. The strong temperature effect in the endothermic regime has already been noted in Fig.2.2. Figure 2.34 provides another example contrasting the T dependence of the rates of the

$$Cl + HCl(v) \to H + Cl_2 \tag{2.4.7}$$

reaction for v = 5 and 8 corresponding to E_v below and above ΔE_0, respectively.

How can the prior rates contain so much chemistry? The answer is that the results we discussed are really very obvious. Essentially all that we are seeing is conservation of energy at work and, of course, we have designed[5] the

5 This was achieved by computing first the reaction rate at a given (total) energy and then averaging over the distribution of total energy [cf. (2.4.5)]. All the algebra of Appendix 2.A is necessary to achieve the correct averaging. We are now reaping the benefits. An alternative, seemingly simpler route is to "forget" about conservation of energy. But then we would have been surprised by the effects we just considered.

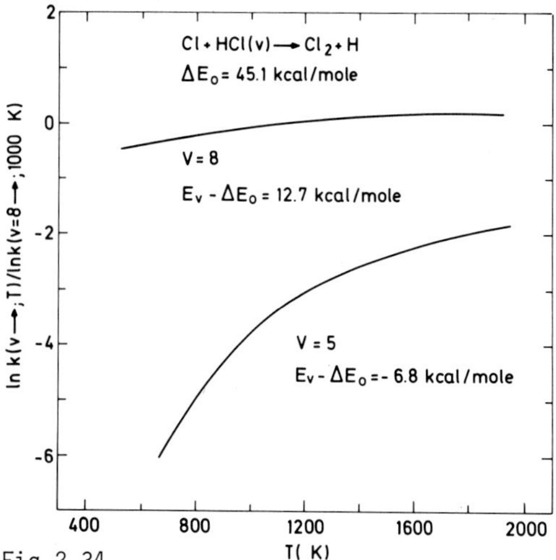

Fig.2.34

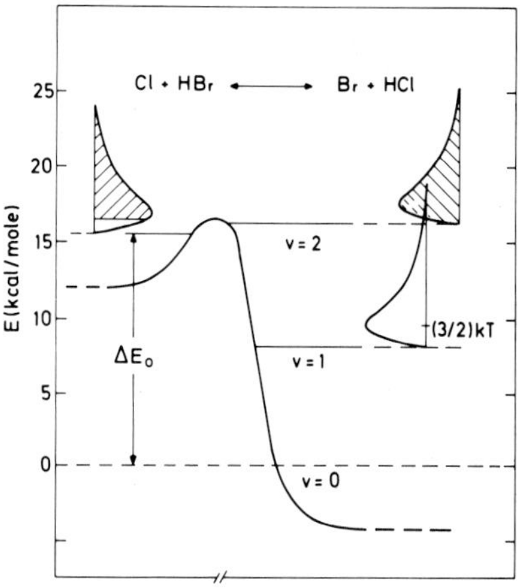

Fig.2.35

Fig. 2.34. The temperature dependence of the reaction rate constant $k(v\to v';T)$ for an endoergic reaction in the endothermic (v=5) and exothermic (v=8) regimes (trajectory computations by E. Pollak). As is the case for the prior rates (Fig.2.33) the temperature has a far greater role in the endothermic regime

Fig. 2.35. The effect of reactant vibrational excitation on the rate of endoergic reactions ($Br+HCl \to HBr+Cl$, $\Delta E_0 = 15.8$ kcal/mole is a typical example). While practically all the HCl(v=2) molecules have sufficient energy to react with Br atoms to form Cl+HBr products, only a very small fraction of HCl(v=1) molecules can do so. The ratio between the number of HCl(v=1) and HCl(v=2) which are capable of reaction is reflected by the ratio of the shaded areas of the corresponding thermal ($T \simeq 500$ K) rotational-translational distributions; $(3/2)kT$ is the most probable translational-rotational energy. The figure also serves to illustrate that in the exoergic direction, $Cl+HBr \to Br+HCl$, most collisions are potentially reactive

prior to know that energy is conserved. Consider an endoergic process where $E_v - \Delta E_0 < 0$. Product formation is only possible in those collisions which occur with high translational energies. The fraction of such collisions is exponentially small (Fig.2.35). As E_v (or T) is increased more (and exponentially so) collisions are potentially reactive, i.e., have the necessary energy for reaction to take place. Now consider the opposite extreme $E_v - \Delta E_0 > 0$. All collisions are potentially reactive. Increasing E_v or T helps little in this respect. Then why are exoergic processes at all sensitive to initial conditions? Part of the answer is that by increasing the energy the number of possible final states is also increased. To see this in detail, note from (2.4.4, 5) that $k^0(v\to v';T)$ is simply the average of $\rho(v';E)$ over the distribution of

E. Now $E = \varepsilon + E_v - \Delta E_0$, where[6] $\varepsilon = E_T + E_R$. If $E_v - \Delta E_0 \gg kT$ one can (very approximately) replace the average of $\rho(v';E)$ by $\rho(v';\langle E\rangle)$. Hence, if $\langle E\rangle > E_{v'}$

$$k^0(v \to v';T) \propto \rho(v';\langle E\rangle) = A_{v'}(\langle E\rangle - E_{v'})^{3/2} \quad . \tag{2.4.8}$$

The second equality is only valid for product vibrational states and we used equation (2.2.3) for $\rho(v';E)$. In (2.4.8)

$$\langle E\rangle = \langle \varepsilon\rangle + E_v - \Delta E_0 \quad , \tag{2.4.9}$$

where $\langle\varepsilon\rangle$ is the mean translational-rotational energy of those reagents that react (the "reactive reactants", Fig.2.35). If no reagent vibrational state selection is carried out so that the reactants are thermal, we need to average $k^0(v \to v';T)$ over the distribution of initial vibrational states. The correct procedure is given in Appendix 2.A. Here we carry the averaging by replacing E_v by $\langle E_v\rangle$ in (2.4.9). Then

$$k^0(\to v';T) \propto A_{v'}(\langle \varepsilon + E_v\rangle - \Delta E_0 - E_{v'})^{3/2} \quad . \tag{2.4.10}$$

Equations (2.4.8,10) are quite useful (e.g., Sect.2.2.3) in analyzing the product vibrational state distribution for exoergic reactions.

A similar approximation will not work in the opposite, endoergic limit where $\langle E\rangle < E_{v'}$. Now however we can invoke detailed balance to compute $k^0(v' \to v;T)$ from (2.4.1). Introducing the thermodynamic equilibrium constant $K(T) = \vec{k}(T)/\overleftarrow{k}(T) = \vec{k}^0(T)/\overleftarrow{k}^0(T)$, we find, for highly endoergic processes

$$k^0(v' \to v;T) \propto [p(v|T)/p(v'|T)K(T)]A_{v'}(\langle E\rangle - E_{v'})^{3/2} \quad , \tag{2.4.11}$$

where $E_{v'} + \Delta E_0 < E_v + \langle\varepsilon\rangle$ (note that now ΔE_0 refers to the reversed reaction). The major temperature dependence of (2.4.11) is given by $p(v|T)/p(v'|T) \times K(T)$ or, very approximately

$$k^0(v' \to v;T) \propto \exp[-(E_v - E_{v'} - \Delta E_0)/kT] \quad . \tag{2.4.12}$$

We have already seen (Sect.2.1.1) that even this simple approximation correctly tells a major part of the story: reagent vibrational excitation serves to reduce the activation energy (Fig.2.2).

---

6 Henceforth, for notational brevity, we use the symbol ε to denote the sum of translational and rotational energies.

2.4.2 The Exponential Gap Representation

The simplest collisions where changes of internal state can be monitored are energy transfer processes, $\Delta E_0 = 0$. One typically finds that the representation

$$I(v,v';T) = -\ln[k(v\rightarrow v';T)/k^0(v\rightarrow v';T)] = I_0(T) + \bar{\lambda}_v|E_v - E_{v'}|/kT \quad (2.4.13)$$

with $\bar{\lambda}_v$ and I_0 functions only of T offers a convenient approximation to the surprisal. (In Sect.2.2 we have used λ_v as the Lagrange parameter conjugated to the dimensionless variable $f_v = E_v/E$. The symbol $\bar{\lambda}_v$ is the Lagrange parameter conjugated to the dimensionless energy variable $|E_v - E_{v'}|/kT$. Note again that for both λ_v and $\bar{\lambda}_v$ the subscript v stands for "vibration" and not for the vibrational quantum number.)

Figure 2.36 shows the results for the inelastic process

$$Li_2^*(v) + M \rightarrow Li_2^*(v') + M \quad . \quad (2.4.14)$$

Here M is a rare gas atom and the Li_2^* molecule is in an electronically excited state so that the populations of the different vibrational states can be monitored via their fluorescence [2.122]. The surprisal parameters cor-

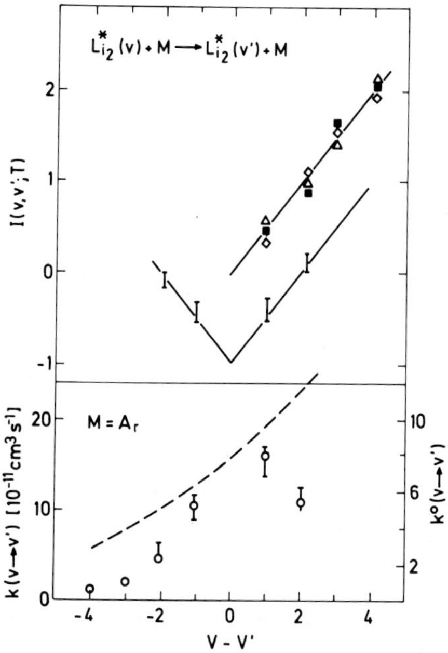

Fig. 2.36. Surprisal synthesis of the distribution of final states and the role of the collision partner in vibrational energy transfer. *Upper panel:* Surprisal for the excitation and deexcitation of $Li_2^*(v=2)$ by Ar (I) and of the deactivation of $Li_2^*(v=4)$ by Xe (Δ), Ar (■), and He (◊) (experimental results from [2.122]). The solid lines are not fits but predictions (adapted from [2.121]). *Lower panel:* Experimental (I) and synthetic (o) rates for the $Ar + Li_2^*(v=2)$ vibrational energy transfer. The synthetic rates correspond to the solid surprisal lines in the upper panel. (The dashed line represents the prior rates)

responding to the solid lines in the upper panel were synthesized according to a procedure which will be described in Sect.2.5.6. The behavior displayed in Fig.2.36 and implied by the exponential gap representation (2.4.13) is quite general: The higher is the amount of internal energy exchanged in an inelastic collision the less likely it is to occur. There is thus a qualitative difference between the prior and actual distribution of final states. The prior favors disposing of as much as possible of the initial internal energy into product translation. The dynamic bias tends to strongly discourage this. The final internal energy is restricted to a narrow range (of width kT/λ_V) about its initial value (Fig.2.36). On one aspect both the prior and the dynamic bias agree. Activation collisions, where the final internal energy exceeds the initial one, are strongly disfavored. Activation rates are thus small and not easily measured (or computed). Where results are available, they conform to the representation (2.4.13). An alternative route to rates of activation is thus through the application of detailed balance (2.1.30-32). The rate of a $v \to v'$, $E_v < E_{v'}$, collision is thereby expressed in terms of the rate of a $v' \to v$ collision $E_v < E_{v'}$.

The functional representation (2.4.13) is a particular example of the general requirement of detailed balance that, for energy transfer collisions, $I(v,v';T)$ is a symmetric function of v and v'. A more flexible form is obtained when the constant λ_V in (2.4.11) is replaced by a symmetric function $\lambda_V(v,v') = \lambda_V(v',v)$. The simplest improvement is $\lambda_V(v,v') = \lambda_V + C(v+v')$. This form offers a more accurate account of the detailed rates.

Sensitive tests of the exponential gap behavior are provided by considering processes with small gaps, such as V-V transfer (Fig.2.37). This includes processes where the gap is due to collisions between different molecules (Fig.2.37a) or when it is entirely due to anharmonicity, e.g.,

$$CO(v) + CO(0) \to CO(v-1) + CO(1) \quad . \qquad (2.4.15)$$

Figure 2.37b shows the surprisal for this and two other V-V processes involving CO(v) as one of the collision partners. In all these cases the energy exchanged in a collision decreases linearly with v. In one case, the CO(v) + NO(0) collision, the energy defect changes sign so that $|\Delta E|$ first decreases and then increases again.

Another, larger, class of processes involving small energy gaps is rotational energy transfer; typically the reduced energy gap $\Delta_{J,J'} = (E_J - E_{J'})/2kT$ is small compared to unity. Figure 2.38 shows the behavior for the two collisions with HD in the v=0 level. Different transitions (induced by different

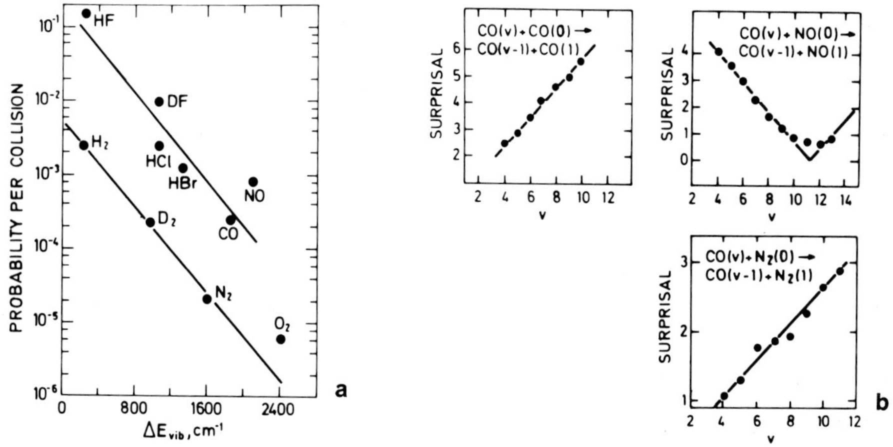

Fig. 2.37a,b. The dependence of V-V energy transfer on the energy gap. (a) Energy transfer probabilities for HF(v=1) + AB(v=0) → HF(v=0) + AB(v=1) collisions, except for the data points marked H_2 and HF which represent the probabilities for $H_2(1)$ + HF(0) → $H_2(0)$ + HF(1) and HF(1) + HF(1) → HF(0) + HF(2), respectively (adapted from [2.123]). Note the higher transfer probabilities to polar partners. (b) The dependence of the surprisal for three V-V processes (T=300 K) on the energy gap which is due to anharmonicity and is thus proportional to v (adapted from [2.121])

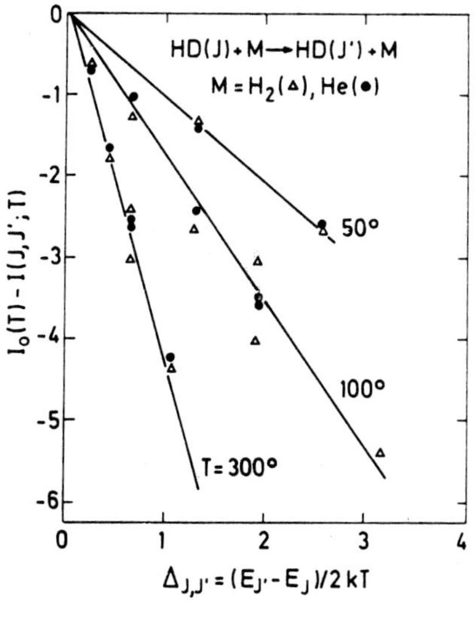

Fig. 2.38. Surprisal analysis for rotational energy transfer (adapted from [2.124]). Computed (quantal) rate constants for $M=H_2$ (Δ), and for M=He (●). Note that θ_R (9.6, 7.25, and 5.9 for T=300, 100, and 50 K, respectively) is insensitive to the collision partner

collision partners) can be characterized (on a surprisal plot) by the magnitude of $(E_J-E_{J'})/kT$. The rotational surprisal parameter θ_R is the analogue of λ_v in (2.4.13) with E_v replaced by E_J.

Rotational transitions in molecules with lower moments of inertia (including hydrides) are already in the $\Delta_{J,J'} < 1$ range. Their rates can thus be well approximated by using the "low Δ" form of the prior rate, i.e., for $J > J'$, $k^0(J \to J';T) \propto (2J'+1)\exp(\Delta_{J,J'})$, leading, through (2.4.13) to [2.124]

$$k(J \to J';T) \propto (2J'+1)\exp\left\{(E_J - E_{J'})/2kT - \theta_R|E_J - E_{J'}|/kT)\right\} \quad . \tag{2.4.16}$$

Experimental data for rotational relaxation of, for example, HCl in rare gas collisions can be well represented in this fashion (cf. Sect.2.5.2).

2.4.3 Reactive Collisions

The exponential gap representation is not limited to vibrational or rotational energy transfer. For a reactive collision the form (2.4.13) remains equally useful provided that the initial and final vibrational energies refer to a common origin. If the energy of the product ground vibrational state is taken as the reference point we have that

$$k(v \to v';T) = k^0(v \to v';T)\exp[-I_0(T) - \lambda_v|E_v - E_{v'} - \Delta E_0|/kT] \quad . \tag{2.4.17}$$

Figure 2.39 shows the results for the $Cl + HF(v) \to HCl + F$ reaction.

Experimental or computational results for other reactions show a similar behavior. In particular this dependence is found for energy transfer via atom exchange collisions, e.g., $F + HF(v) \to FH(v') + F$ [2.121], where even though $\Delta E_0 = 0$ the collision is a reactive one.

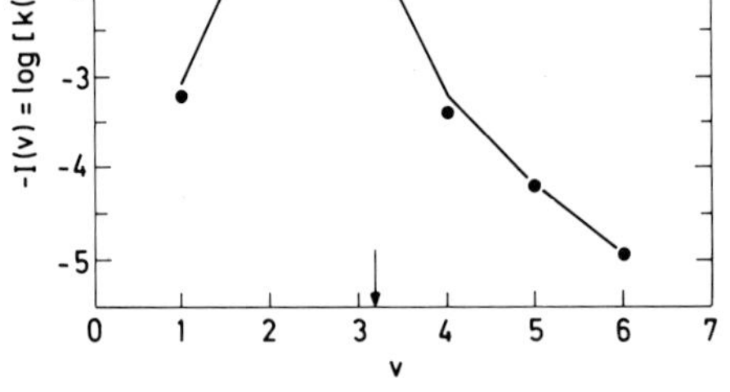

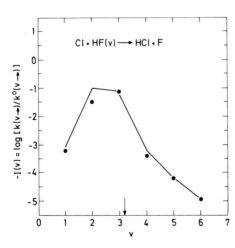

Fig. 2.39. Deviation of experimental rate constants $k(v\to;T)$ from prior expectations $k^0(v\to;T)$ as function of reagent vibrational excitation. The arrow indicates the endoergicity limit. The figure illustrates the opposing trends of the surprisal $I(v)$ in the endoergic [$I(v)$ decreases] and exoergic [$I(v)$ increases] regimes. Experimental data from [2.2,17]. The solid curve is a theoretical fit, based on an exponential gap representation, (2.4.17), for $k(v \to v';T)$. Summing over v', one obtains the v dependence shown as the solid line. The break in $I(v;T)$ is a direct reflection of the exponential gap dependence of the detailed rates (adapted from [2.10])

A particularly interesting example is that of the

$$O + CN(v) \rightarrow N + CO(v') \qquad (2.4.18)$$

reaction [2.26] (Figs.2.5 and 2.25). The reaction can proceed either via the energy-rich, longer lived NCO intermediate [leading to the excited $N(^2D)$ nitrogen atom] or through a direct mechanism [leading to the ground state $N(^4S)$ nitrogen atom]. For the direct process λ_v is finite and the shift of the most probable vibrational state of CO with increasing vibrational excitation of CN is quite evident in Figs.2.5 and 2.25. For the compound process $\lambda_v \simeq 0$, and the CO vibrational distribution is practically independent of the CN excitation (as it should for an exoergic process on prior grounds), (cf. Fig.2.33).

The dynamic point of view of Sect.2.3 (and of Sect.2.4.4) readily accounts for the break in the surprisal plot at $E_v = E_{v'} + \Delta E_0$ in (2.4.17). This is the transition that requires the least change in the momenta of the heavy nuclei during the transfer of the light H atom. In general, when a repulsive release of the energy is possible (Sect.2.3.4) we expect the break to occur when the final (product) translational energy is at (or near to) its most probable final value, as computed by the kinematic considerations. In general, we can express this most probable value as $E_T' = \alpha E_T + \chi R$, where R is the energy of the repulsive release. Now $E = E_T + E_I - \Delta E_0 = E_T' + E_I'$, where E_I is the internal energy. Hence, the most probable value of E_I' is $E_I' = E - \alpha E_T - \chi R = E_I - (\Delta E_0 + \chi R) + (1-\alpha)E_T$. For H atom transfer along the approach motion, $\alpha = \cos^2\beta \simeq 1$, $\chi = \sin^2\beta \simeq 0$, [cf. (2.3.7)], and $E_I' = E_I - \Delta E_0$.

The representation (2.4.17) also accounts for the role of reagent vibrational excitation. Consider an endoergic reaction where the vibrational energy of the reactants does not suffice to overcome the endoergicity, $\Delta E_0 - E_v > 0$. The necessary energy for reaction is then partly provided by the thermal energy of the reagents. In this case $|E_v - E_{v'} - \Delta E_0| = E_{v'} - (E_v - \Delta E_0)$ and the reaction rate has an exponential dependence on E_v. Using (2.4.12) and (2.4.17) we see that both $k(v \rightarrow v';T)$ and $k(v \rightarrow;T)$ will depend on E_v as $\exp[-(1-\lambda_v) \times (\Delta E_0 - E_v)/kT]$ as indeed is found to be the case (Fig.2.2). When the reagent vibrational energy exceeds ΔE_0, initial vibrational energy can exceed the vibrational energy of the products. The energy mismatch $E_v - \Delta E_0 - E_{v'}$ can be of either sign depending on $E_{v'}$. $k(v \rightarrow v';T)$ is no longer exponentially increasing with increasing E_v. For such transitions that $E_v > \Delta E_0 + E_{v'}$, reagent vibrational excitation detracts the reaction rate even though the reaction from ground state reagents is endoergic.

The correlation between reagent state selection and product state distribution as discussed in this and previous sections illustrates our claim that constraints can be built into the initial state so as to keep the excitation "locked in" a given mode. A simple dynamical interpretation of the constraint governing this behavior is discussed next.

2.4.4 The Adiabaticity Parameter

The exponential gap representation relates the detailed rate constant to the energy defect ΔE of the collision. The rate constant is found to deviate from prior expectations in an exponential dependence on $|\Delta E|$

$$-\ln(k/k^0) = I_0(T) + \lambda|\Delta E|/kT \quad . \tag{2.4.19}$$

The quantitative role of $|\Delta E|$ in restricting the range of product states which are effectively populated clearly depends on the magnitude of λ/kT. For the rate constant for a transfer of energy ΔE to be significant it is necessary that $\lambda|\Delta E|/kT < 1$. The range of probable final states is restricted to those whose vibrational energy lies within a band of width kT/λ about the vibrational energy of the initial state. (Similar comments apply to rotational energy. It is the interconversion of internal and translational energy that carries exponential levies.) In the limit where λ/kT is large the product energy is narrowly confined to the initial energy while if λ is small the distribution closely follows the prior limit which favors the conversion of vibrational to translational-rotational energy [cf. (2.2.7) and Appendix 2.A]. Figure 2.40 shows, for a vibrational energy transfer collision, the mean change in the vibrational quantum number $\langle\Delta v\rangle = \langle v'-v\rangle \simeq \langle\Delta E_v\rangle/hc\omega_e$ vs the reduced variable $\Lambda_v = (hc\omega_e/kT)\lambda_v \equiv u\lambda_v$, computed for HF(v) + M collisions using (2.4.13). We note that the higher is Λ_v the less is the initial state perturbed by the collision. The inset to the figure shows the range of typical Λ_v values in the high- and low-temperature limits.

The dynamical point of view is that the propensity for energy transfer in a collision is governed by the "adiabaticity" of the collision. To examine this interpretation we consider a simple model of a collinear collision of a particle A with a diatomic molecule B-C. Let A approach towards the atom B. If the duration of the A-B collision is fast compared to the period of the oscillation of B-C, atom B will be set in motion relative to C. During the fast or "sudden" collision, energy is readily imparted to the B-C molecule. Rather than considering the duration of the collision, we can examine the B-C force constant. If the B-C spring is very soft, it is very easy to dis-

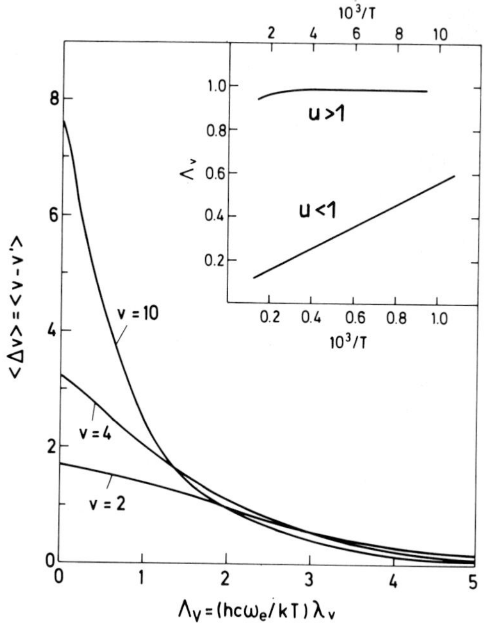

Fig. 2.40. Average transfer of vibrational quanta as function of $\Lambda_V = (hc\omega_e/kT)\lambda_V$ [cf.(2.4.13)]. As Λ_V increases the collision becomes more adiabatic and the probability of energy transfer decreases. In the prior limit $\lambda_V = 0$, most of the initial vibrational excitation of the diatomic molecule is converted to translational and rotational energy, i.e., $<\Delta v> \simeq v$. The inset shows the typical values of Λ_V for different temperatures as obtained by analyses of experimental and computational data or using sum rules [cf. Sect.2.5.6] (adapted from [2.12]). At the low-temperature limit, $u = hc\omega_e/kT > 1$, λ_V becomes proportional to T and $\Lambda_V \to 1$ (computed for HF, ω_e=4138.5 cm^{-1}; T^{-1} scale on the upper abscissa). At the high-temperature limit, $u < 1$, λ_V is nearly independent of T and Λ_V varies linearly with T^{-1} (computed for Li_2^*, ω_e=269.6 cm^{-1}, T^{-1} scale on lower abscissa)

place B relative to C. On the other hand, if the B-C spring is stiff or, equivalently, if the A-B collision is slow, it is difficult to displace B relative to C. Any force exercised on B is rapidly communicated to C and the BC molecule responds "as a whole". The molecule adjusts adiabatically to the perturbation.

The adiabaticity, i.e., the propensity to resist changes in the vibrational energy, is thus conveniently measured by ξ,

$$\xi = \tau_c/\tau_v \quad , \tag{2.4.20}$$

where τ_c is the duration of the collision and τ_v is the period of the internal motion (τ_v is inversely proportional to the stiffness of the "spring"). A high ξ means an adiabatic collision, one that is rather unlikely to change the state of the molecule.

For a collision where the energy mismatch is ΔE we can estimate τ_v as $h/\Delta E$ so that $\xi = \tau_c \Delta E/h$. This form of ξ lends itself to a simple quantal interpretation. Consider a collision of finite duration τ_c. Such a collision cannot be described in terms of states of well-defined kinetic energy. Rather one must use a wave packet where the kinetic energy has a spread of about h/τ_c. Changes in the kinetic energy of up to h/τ_c are thus quite probable.

Larger changes correspond to the wings of the wave packet and are thus less likely. Rather than invoke the time-energy uncertainty relation, we can employ the position-momentum one. Consider a potential of range L. To localize a wave packet to within this range, one needs a spread $\Delta p \sim h/L$ in the momentum. The change in relative momentum from p to p' during the collision is likely if $|p-p'| < h/L$. The long-range part of the potential cannot induce large changes in the momentum. It is the short-range part of the potential that has to carry the arduous taks of providing the hefty impulse required for a large change in the momentum.

The two formulations of the uncertainty principle are readily seen to be equivalent by noting that one can write

$$\xi = (p-p')L/h = (p^2-p'^2)L/h(p+p') = 2\mu L\Delta E/h(p+p') = \Delta E\tau_c/h \quad , \qquad (2.4.21)$$

where $\tau_c = L/[(p+p')/2\mu] = L/\bar{v}$, with $\bar{v} = (v+v')/2$ denoting the arithmetic mean of the velocities before and after the collision. The consideration of the propensity for momentum transfer also helps to bring into focus the relative importance of the short- and long-range parts of the intermolecular potential. The long-range forces cannot bring about large transfers of energy into or out of the translation. However, when ΔE is small, so that they can contribute, their role is particularly important since their long range implies a large cross section. (The cross section being an effective area tends to increase as the square of the range of the interaction.) V-T transfer is typically a short-range process. V-R,T; V-V, R-R, and R-T transfers can also be induced by the long-range potential and hence, typically, are characterized by larger cross sections than V-T.

We have thus far stressed the transfer of energy to (or from) the translation. The rotations of the molecules provide an alternative source (or sink) for vibrational energy. An estimate for the propensity of V-R transfer in, say, A+BC collision, can be obtained using (2.4.21) but with p representing the tangential momentum of the rotor AB instead of the relative momentum of A+BC. For a (classical) rigid rotor of equilibrium internuclear distance R_e, we have that $p = \mu_{BC}v_R = \mu_{BC}\omega R_e = (h/2\pi)J/R_e$, where $\mu_{BC} = m_B m_C/(m_B+m_C)$ is the reduced mass of the rotor, v_R is the tangential velocity, ω is the angular rotational velocity, and $(h/2\pi)J$ is the angular momentum of the rotor so that its energy is $E_R = (h/2\pi)J\omega$. From (2.4.21) we have $\xi_{V-R} = \Delta EL/h\bar{v}_R = \Delta EL/h\bar{\omega}R_e$ compared to $\xi_{V-T} = \Delta EL/h\bar{v}$ for V-T transfer. The decision to transfer an amount of vibrational energy to rotation or translation will thus depend on the ratio $\xi_{V-T}/\xi_{V-R} = \bar{\omega}R_e/\bar{v}$. If the collision velocity v is less than the rotational velocity ωR_e, V-R transfer will be favored. At the same collision energy, heavier projectiles (i.e., lower $\underline{v}$'s) will favor V-R transfers. Also since $J(h/2\pi) = I\omega$, V-R transfer is favored for higher initial J's. Similarly, for a given ΔE, diatomics with high rotational constants (i.e., hydrides) will require a smaller change in J (i.e., $\xi_{V\text{-}R}$ will be smaller) to accommodate ΔE and hence will accept it more readily.

Often, both the translation and the rotation have a thermal distribution at the temperature T. In that case the relative efficiency of V-R and V-T transfers is governed largely by the masses [2.125]. Consider an A+BC collision at the temperature T. $\langle E_R \rangle = kT = \langle I\omega^2/2 \rangle = \langle \mu_{BC} R_e^2 \omega^2/2 \rangle$, where $I = \mu_{BC} R_e^2$ is the moment of inertia of BC. Hence $\langle R_e \omega \rangle = (2kT/\mu_{BC})^{\frac{1}{2}}$. Similarly $\langle \underline{v} \rangle = (\langle 2E_T \rangle/\mu_{A,BC})^{\frac{1}{2}} \simeq (3kT/\mu_{A,BC})^{\frac{1}{2}}$. Hence $\xi_{V\text{-}T}/\xi_{V\text{-}R} = \langle R_e \omega \rangle / \langle \underline{v} \rangle \simeq (\mu_{A,BC}/\mu_{BC})^{\frac{1}{2}}$. For hydrides $\mu_{BC} \approx 1$ while $\mu_{A,BC} \gtrsim 1$; hence $\xi_{V\text{-}R} < \xi_{V\text{-}T}$. For a given diatomic molecule, $\xi_{V\text{-}T}/\xi_{V\text{-}R}$ increases as m_A increases. Indeed dynamical computations for X+HY collisions (X and Y halogen atoms) show that V-R is the important mechanism while in H+HY collisions V-R and V-T are about equally important.

A particularly interesting illustration of the propensity for V-R transfer in collisions involving hydrides is lasing on pure rotational transitions in HF [2.126] (and other hydrides like OH and NH; cf. Sect.4.1). Figure 2.41 shows those transitions on which lasing was observed. The highest gain transitions are identified by dots. The experimental output pattern suggests that in addition to the direct chemical pumping the initial states of such transitions may also be populated by the V-R processes $M + HF(v,J) \rightarrow M + HF(v\text{-}n,J')$,

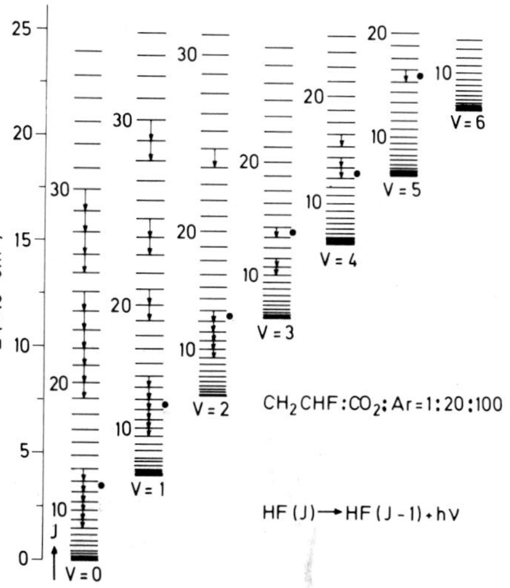

Fig. 2.41. Pure rotational transitions on which laser action in HF (produced by photoelimination from CH_2=CHF) was observed. The highest gain transitions are identified by a dot (adapted from [2.126])

where $n = 1$, 2, or 3 and the final rotational state J' is such that $\Delta E_R \simeq -\Delta E_V$ [i.e., $\psi \simeq 1$ in (2.4.23)].

The adiabaticity factor for V-R,T processes can be expressed as

$$h\xi = |\Delta E_T|\tau_c + |\Delta E_R|\tau_R \tag{2.4.22}$$

where the first factor reflects the inertia of the linear momentum $\tau_c \simeq L/<\underline{v}>$, while the second factor accounts for the inertia of the angular momentum $\tau_R \simeq L/<R_e\omega>$. For a given $v \to v'$ transition, i.e., for a given ΔE_V, (2.4.22) can be expressed as $h\xi = |\Delta E_V - \Delta E_R|\tau_c + |\Delta E_R|\tau_R$ which shows that (for a given ΔE_V) the most probable rotational energy transfer is

$$\Delta E_R = -\psi\Delta E_V \quad , \tag{2.4.23}$$

where $\psi = \tau_c/(\tau_R+\tau_c) \simeq 1/[(<\underline{v}>/<R_e\omega>)+1]$. Equation (2.4.23) accounts for the often observed propensity rules for V-R,T transfer. I) The most probable ΔE_R is proportional to ΔE_V. II) The proportionality constant ψ is nearly energy independent and is governed primarily by the masses. Indeed we find $\psi \simeq 1/[(\mu_{A,BC}/\mu_{BC})^{\frac{1}{2}}+1]$. III) The shift increases with increasing initial rotational state J of the diatomic (because $\tau_R \propto J^{-1}$). Not immediately evident from (2.4.23) is IV) that the range of significantly populated final rotational states (i.e., the width of the J' distribution) increases with increasing ΔE_V. To see this we note that the width σ^2 is defined by putting $h\xi_{V-R} = (J-J')^2/2\sigma^2$ so that $\sigma^2 \propto (J-J') \propto E_R/(J+J')$.

Even for simple R-T transfer, (2.4.22) indicates two distinct sources of dynamic bias. One, $|\Delta E_R|\tau_c$, reflecting the inertia to changes in the linear momentum would dominate when $\psi \to 1$, and the other $|\Delta E_R|\tau_R$ would be characteristic of collisions with high rotational inertia $\psi \to 0$.

The immediate implications of our discussion of the exponential gap for state-to-state processes are clear. The most efficient transitions are those that minimize the momentum transfer gap[6a] (cf. Sect.2.3.4) $p' - \gamma p' - q$ (where q is the impulse imparted by the interaction) and the angular momentum transfer gap J-J'-T (where T is torque imparted during the duration of the collision). By selecting the reagent initial state to conform to these requirements one can considerably enhance the reaction rate. Similarly, by selecting reactions

6a For potential energy surfaces with an early release of the exoergicity one expects that $q \sim 0$, and for H atom transfer $\gamma \simeq 1$ so that the momentum transfer gap and the vibrational energy gap are equivalent.

where the gap for the formation of product ground state is large, one can achieve preferential population of excited states.

The role of reagent vibrational excitation is thus to narrow the energy gap [cf. (2.4.13)]. Any excess vibrational energy (over the minimal amount that suffices to bridge the gap) would be converted to internal excitation of the products and would not dramatically enhance the magnitude of the overall reaction rate [2.10,11]. Similarly population inversion is to be expected not only for exoergic reactions but also for highly excited reagents in nominally endoergic processes.

2.4.5 Polyatomic Molecules

Typically, larger amounts of energy are transferred in a collision with a polyatomic molecule, particularly when it is highly vibrationally excited [2.103,127]. In contrast to the case of small (e.g., atom-diatom) colliding systems where the dynamical bias is large the final energy partitioning in collisions involving polyatomics can be explained, to a large degree, on prior grounds [2.101]. Due to the many vibrational degrees of freedom of polyatomic molecules the prior expectations are that most of the collision energy will be found in the dense manifold (the quasicontinuum) of vibrational levels, as opposed to the case of, say, atom-diatom collisions where the vibrational levels are discrete and, on prior grounds, the translation and rotation share most of the energy (Sect.2.2.1). To quantify these arguments we shall next explicitly evaluate the prior final energy distribution in a collision between an atom and a molecule with s vibrational degrees of freedom. The differences between large and small molecules will be reflected by the value of s.

Consider for example a collision between a (nonlinear) polyatomic and an atom, e.g., $SF_6 + Ar$. Let E_V and $\varepsilon = E_R + E_T$ denote the initial vibrational energy of the polyatomic molecule and the sum of the initial rotational and (relative) translational energies, respectively. For a collision at a given total energy $E = E_V + \varepsilon$ we have $\Delta\varepsilon = -\Delta E_V = \varepsilon' - \varepsilon = E_V - E_V'$, where E_V' and ε' are the vibrational and translational-rotational (T-R) energies after the collision. The prior distribution $P^0(\Delta\varepsilon;E_V,E)$ for the change $\Delta\varepsilon$ at given E_V and E is proportional to the product of the vibrational and T-R densities of states (cf. Appendix 2.A),

$$P^0(\Delta\varepsilon;E_V,E) \propto \rho_{TR}(\varepsilon')\rho_V(E_V') = \rho_{TR}(\varepsilon+\Delta\varepsilon)\rho_V(E_V-\Delta\varepsilon) \quad . \tag{2.4.24}$$

For an atom-polyatomic collision $\rho_{TR}(\varepsilon) \propto \varepsilon^2$, (in the more general case $\rho_{TR}(\varepsilon) \propto \varepsilon^{\alpha/2-1}$ where α is the number of relevant translational-rotational degrees of freedom, Appendix 2.A). The simplest approximation for $\rho_v(E_v)$ is the RRK expression (2.3.11). Thus, approximately

$$P^o(\Delta\varepsilon;E_v,E) \propto (\varepsilon+\Delta\varepsilon)^2(E_v-\Delta\varepsilon)^{s-1} = \varepsilon'^2(E-\varepsilon')^{s-1} \tag{2.4.25}$$

where s is the number of vibrational degrees of freedom. Using (2.4.25) one can show that (for a given E) the most probable and average values of $\varepsilon' = E - E_v'$ are given by $\varepsilon'_{mp} = [2/(s+1)]E$ and $<\varepsilon'> = [3/(s+3)]E$, respectively. Thus, most of the energy is expected to be found in the vibrations of the polyatomic molecule, as opposed to the case of atom-diatom collisions where, on prior grounds, most of the energy appears in the translational and rotational motions of the departing species [cf. (2.2.7)].

For a thermal distribution of T-R energies (2.4.25) should be averaged over the initial distribution $p(\varepsilon|T) \propto \varepsilon^2 \exp(-\varepsilon/kT)$. Noting that $P^o(\Delta\varepsilon;E_v,E) = P^o(\Delta\varepsilon;E_v,\varepsilon)$, $(E=\varepsilon+E_v)$, the thermal average of (2.4.25) is

$$P^o(\Delta\varepsilon;E_v,T) = \int_0^\infty d\varepsilon p(\varepsilon|T)P^o(\Delta\varepsilon;E_v,\varepsilon)$$

$$\propto \left[1+\frac{|\Delta\varepsilon|}{2kT}+\frac{1}{3}\left(\frac{|\Delta\varepsilon|}{2kT}\right)^2\right]\left(1-\frac{\Delta\varepsilon}{E_v}\right)^{s-1} \times \begin{cases}\exp(\Delta\varepsilon/2kT) & \Delta\varepsilon < 0 \\ 1 & \Delta\varepsilon > 0\end{cases} \tag{2.4.26}$$

Figure 2.4.2 shows the prior distribution (2.4.26) for $Ar + SF_6$ collisions (s=15) for three different initial vibrational energies. Also indicated are the average energy transfers $<\Delta\varepsilon> = \int d\varepsilon\Delta\varepsilon P^o(\Delta\varepsilon;E_v,T)$, which are in good agreement with available experimental data [2.101].

Highly endoergic transitions, $\Delta\varepsilon/kT < -1$ are strongly disfavored due to the dominance of the exponential factor $\exp(\Delta\varepsilon/2kT)$ in (2.4.26). However, as $\Delta\varepsilon \to 0$ the probability increases rapidly due to the vibrational density of states factor $(1-\Delta\varepsilon/E_v)^{s-1}$. (This factor is 1 for atom-diatom collisions, s=1, and the probability of $\Delta\varepsilon \sim 0$ remains small). In the exoergic regime $P^o(\Delta\varepsilon;E_v,T)$ first increases due to the first factor in (2.4.26) but then decreases rapidly as soon as $\Delta\varepsilon \to E_v$ due to the dominance of the density of states factor. A quick estimate for the location of the maximum can be obtained using (2.4.25) with $E = E_v + \varepsilon$ replaced by $E_{mp} = E_v + \varepsilon_{mp} = E_v + 2kT$. This

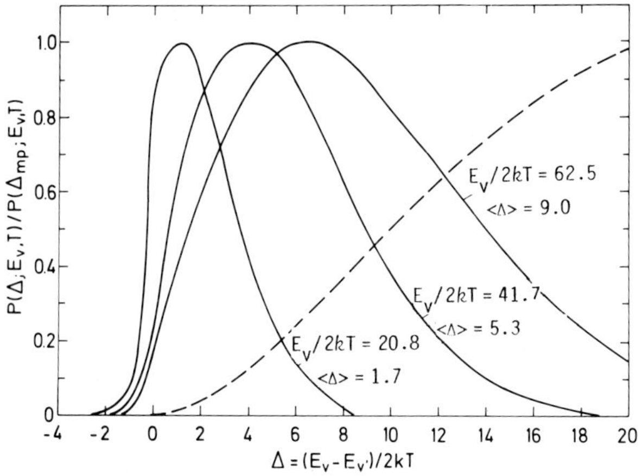

Fig. 2.42. The prior distribution (2.4.26) for vibrational energy transfer in SF_6 + Ar collisions at room temperature vs the reduced energy transfer $\Delta = -\Delta E_V/2kT = \Delta\varepsilon/2kT$, for three levels, E_V, of initial SF_6 excitation. $\langle\Delta\rangle$ is the average (reduced) energy transfer per collision. Shown for comparison (dashed line) is the prior distribution for atom-diatom collision, (data based on Fig.2.33 with $E_V = 30$ kcal/mole, i.e., $E_V/2kT = 25$). (After [2.101])

yields $\Delta_{mp} = (\varepsilon'_{mp} - \varepsilon_{mp})/2kT = (E_v - E_{v'mp})/2kT = E_v/(s+1)kT - (s-1)/(s+1)$. For the E_v values in Fig.2.42, i.e., $E_v/2kT = 62.5$, 41.7 and 20.8, this approximation leads to $\Delta_{mp} = 6.9$, 4.3 and 1.7, respectively; in good agreement with the exact values.

2.4.6 Temperature Dependence

We have thus far examined the exponential gap behavior from the point of view of the distribution of collision products. Another, if indirect, evidence comes from the overall temperature dependence of the rates [2.128-130].

The vibrational relaxation time τ is observed in shock tube studies to decrease with temperature according to the Landau-Teller dependence

$$\ln(P\tau) = A + BT^{-1/3} \quad , \tag{2.4.27}$$

where P is the pressure (Fig.2.43). Measurements at lower temperatures often show deviation from a simple $T^{-1/3}$ dependence, most notably a rise, or levelling off, at low temperatures (see Fig.2.44). To determine the temperature

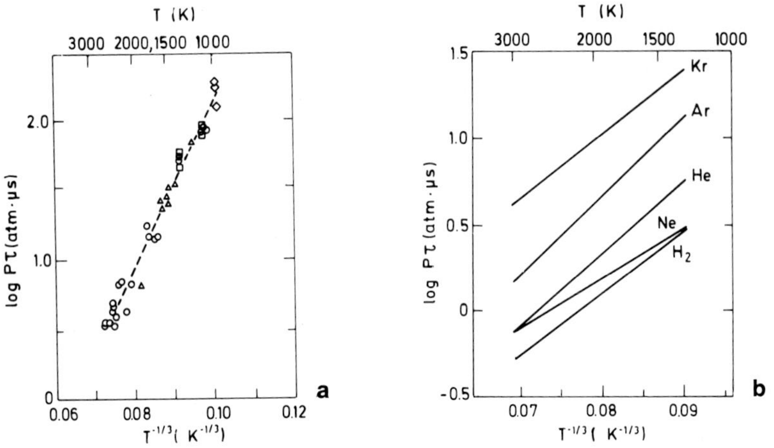

Fig. 2.43a,b. Landau-Teller plots (vibrational relaxation times, as $P\tau$, vs $T^{-1/3}$). (a) Highly diluted DF in Ar-He mixtures (adapted from [2.131]). (b) Highly diluted H_2 in a series of buffer gases with increasing mass. Except for Ne, the mass dependence is that suggested by the simple theory (2.4.31) (adapted from [2.132])

dependence of the relaxation time we consider first the co-called "adiabatic limit". Here λ_v (or ξ) is sufficiently large that only $v \to v' = v \pm 1$ transitions need to be considered as effectively contributing. In this case one can solve explicitly for the vibrational relaxation time τ ([2.128-130]; see also Sects.2.5.1 and 2.5.6). Taking ρ as the density of the buffer gas

$$1/\rho\tau = k(1\to 0;T)[1-\exp(-h\nu/kT)] \quad , \tag{2.4.28}$$

where $h\nu$ is the energy of the $v = 1 \to 0$ transition and

$$k(1\to 0;T) = \langle k(1\to 0;E)\rangle \approx \langle \exp(-\xi)\rangle = \langle \exp(-\underline{v}_0/\underline{v})\rangle \quad . \tag{2.4.29}$$

Here the brackets denote a thermal average and $\xi = \nu L/\underline{v} \equiv \underline{v}_0/\underline{v}$. The fraction of collisions with velocity in the range $\underline{v}$, $\underline{v}+d\underline{v}$, decreases exponentially [as $\exp(-\mu\underline{v}^2/2kT)$] as $\underline{v}$ increases, while $\exp(-\xi)$ increases exponentially as $\underline{v}$ increases. The main contribution to the integral comes therefore from the velocity range where the integrand is maximal. This range is centered at the velocity $\underline{v}_\beta$ which is the solution of the implicit equation

$$d[-\mu\underline{v}^2/2kT-\underline{v}_0/\underline{v}]/d\underline{v} = 0 \tag{2.4.30}$$

Table 2.4. Comparison of experimental and theoretical parameters for V-T transfer (adapted from [2.130])

System	$B(K^{1/3})$ experimental	ν [cm^{-1}]	μ^a (m_p)	L^b [A]	$B(K^{1/3})$ calculated^c
O_2-O_2	126	2143	16.0	0.202	156
O_2-H_2	36		1.88	0.186	73
O_2-D_2	63		3.55	0.185	90
O_2-He	60		3.55	0.175	86
O_2-Ar	161		17.8	0.199	161
CO-CO	160	1556	14.0	0.224	130
CO-He	87		3.5	0.185	72
CO-Ne	142		11.7	0.191	110
CO-Ar	182		16.5	0.209	131
CO-Kr	188		23.0	0.215	149

[a] m_p = proton mass

[b] $L = \sigma_{ab}/17.5$; $\sigma_{ab} = (\sigma_a+\sigma_b)/2$. σ is the Lennard-Jones range parameter.

[c] $B \propto \mu^{1/3}\nu^{2/3}L^{2/3}$. Normalized such that B experimental = B calculated at O_2-Ar.

or $\underline{v}_\beta = (\underline{v}_0 kT/\mu)^{1/3}$. The rate is then given by

$$k(1\rightarrow 0;T) \propto \exp(-\mu\underline{v}_\beta^2/2kT-\underline{v}_0/\underline{v}_\beta)$$

$$= \exp\left[-3\underline{v}_0^{2/3}\mu^{1/3}/2(kT)^{1/3}\right] \equiv \exp[-BT^{-1/3}] \qquad (2.4.31)$$

with $B \propto \mu^{1/3}\underline{v}_0^{2/3} \propto \mu^{1/3}\nu^{2/3}$ [2.130]. Table 2.4 compares experimental results and theoretical predictions for B as obtained for CO and O_2 relaxation in the rare gases. Even in the temperature range where the $T^{-1/3}$ dependence obtains, the $B \propto \mu^{1/3}\nu^{2/3}$ correlation may fail (Fig.2.44). The vibrational spacing of HCl is about 1.4 that of DCl, yet both are relaxed at about the same rate by collisions with He or Ar. The considerations of Sect.2.4.4 suggest that this behavior is due to the importance of transfer to rotation rather than to translation. Thus we need to replace ξ in (2.4.29) by $\xi_{V-R} = \Delta E \cdot L/hv_R$ where $v_R = \omega R_e$ is the rotational velocity during the collision. The thermal average now is carried out on the thermal distribution of rotational energy. One finds [2.125] using similar considerations $B \propto \mu_{BC}^{1/3}\nu^{2/3}$ where BC is the diatom, of vibrational frequency ν. Since however $\nu \propto \mu_{BC}^{-1/2}$, B is roughly independent of the vibrational spacing, provided that V-R transfer is the dominant mechanism.

The $\exp(-BT^{-1/3})$ increase of the relaxation rate with increasing temperature is typical of the adiabatic regime, where transition probabilities are small. When long-range forces contribute significantly to the transition

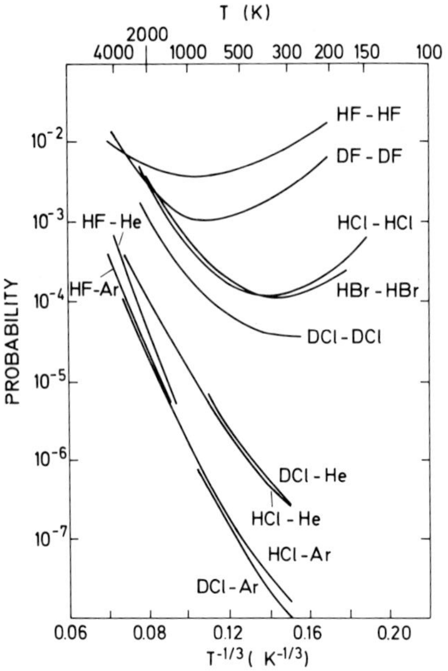

Fig. 2.44. Probabilities of vibrational relaxation vs $T^{-1/3}$ for hydrogen (deuterium) halides by He, Ar, and other hydrogen halides. The near absence of isotope effect in HCl(DCl) + M(=He,Ar) collisions, despite the considerable difference between ν_{HCl} and ν_{DCl} suggests that V-R transfer is the dominant relaxation mechanism. The importance of long-range forces in collisions between two polar molecules in the low-temperature regime is quite evident. (Data adapted from [2.133-137])

and/or when Δv (or ΔJ) changes of more than unity are not improbable, one encounters deviations (Fig.2.44). At low temperatures, the possibility of compound, long-duration encounters (cf. Sect.2.3.5) increases the probability of energy transfer as energy tends to be equipartitioned during such collisions. In general, the contribution from long-range forces tends to diminish as the temperature increases. Often therefore there will be a minimum in the magnitude of the probability of energy transfer as a function of T. This reflects an intermediate velocity range, one where the effect of the long-range forces is already diminishing yet the mechanisms which are possible due to the short-range forces are not yet fully operative. A somewhat analogous situation is evident for the far simpler situation of elastic collisions [2.1]. Intermediate between the glancing collisions which pull the molecules "in" under the influence of the attractive, long-range, force and the "hard" nearly head-on collisions which cause the molecules to rebound is a region leading to practically no *net* deflection. The long- and short-range forces act in opposite directions, essentially cancelling one another [2.138].

2.4.7 Electronic Energy Transfer

Electronic excitation energies are typically even larger than vibrational spacings. The exponential gap law would thus suggest that, at ordinary velo-

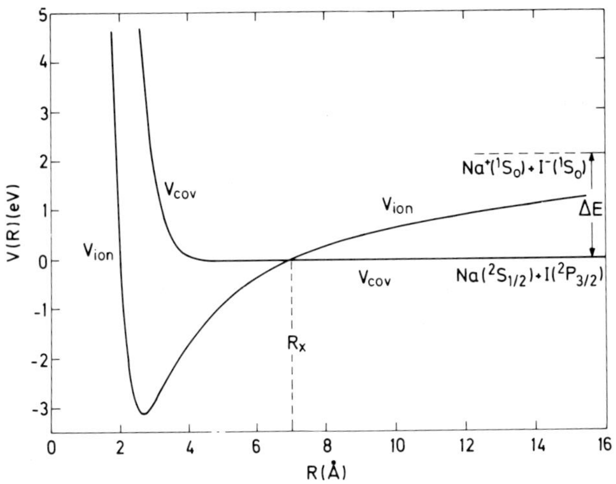

Fig. 2.45. Potential energy curves for the interaction of Na and I. (Adapted from [2.140])

cities, electronic energy transfer or quenching would be restricted to such processes that are nearly resonant, where the energy mismatch is small and little energy needs to be put into or taken out of the translational motion. We have often remarked how chemical reactions tend to preferentially populate internally excited states rather than dump the exoergicity into translation, and the same applies to electronic excitation. Processes which are nearly resonant do indeed occur with significant rates. Even so, examples of efficient conversion of translational into electronic energy (or vice versa) are well known, even though the magnitude of $\Delta E/\underline{v}$ is far too large. A simple example is the so-called chemi-ionization collisions, e.g.,

$$Na + I \rightarrow Na^+ + I^- \quad , \qquad (2.4.32)$$

where the difference ΔE between the ionization potential of the metal and the electron affinity of the halogen is provided by the initial translational energy. Other examples abound in the reactions of electronically excited rare gas atoms, which are quite similar to the alkali metals because of their low ionization potential [2.139]. Figure 2.45 shows a typical set of potential energy curves. At large separations the energy of the ions exceeds that of the neutral atoms by ΔE. As the ions approach they readily gain energy because of their Coulomb attraction. There is thus a critical separation R_x, where the interaction between the ions about equals the interaction between

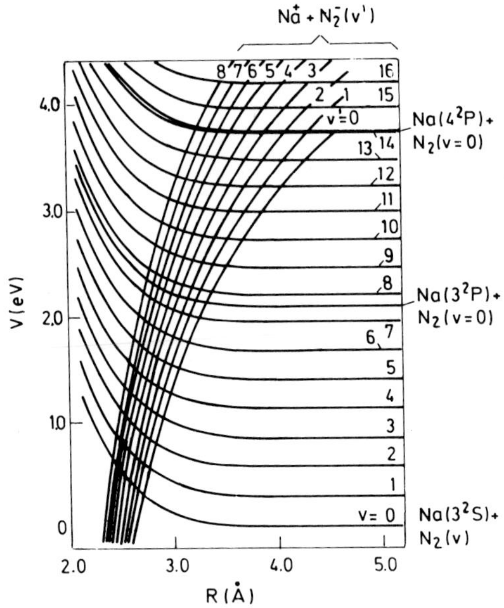

Fig. 2.46. The maze of crossing electronic potential energy curves in the $Na+N_2(v) \rightarrow Na^+ + N_2^-(v')$ system (adapted from [2.141]). The curves show the interaction energy as a function of the Na-N_2 distance for the different vibrational states of N_2 (the vibrational quantum number of N_2 is indicated on the right) and of N_2^- and for Na in different lower lying electronic states and for Na^+

the neutral atoms. For $R < R_X$ the ions are more stable. As two neutral atoms approach they reach the point R_X where the electronic state can change without any significant change in the kinetic energy. At R_X the reactants can cross from the covalent to the ionic state.

An estimate of the crossing distance is readily provided by the approximation which neglects the covalent potential as compared to the ionic one. Then, at R_X, the Coulombic interaction e^2/R_X is just sufficient to balance ΔE or, in practical units,

$$R_X[\text{in Å}] = 14.35/\Delta E[\text{in eV}] \quad . \tag{2.4.33}$$

The crossing can thus occur at large interactomic separations. The valence electron is used as a "harpoon" to bring in the other atom. Once the crossing takes place, the Coulombic interaction brings the two atoms together.

When the atom collides with a diatomic molecule ΔE depends on the energy content of the molecule. The more excited is the molecule, the lower is the additional energy required and the higher is R_X. One then has a maze of curve crossings rather than an isolated region, as shown in Fig.2.46.

The possibility of curve crossing provides a mechanism for a change in electronic state which proceeds with a significant change in the kinetic energy. To estimate the crossing probability as $\exp(-\xi)$, where ξ is the adiabaticity parameter, we need to redefine ξ since now the actual change in the kinetic energy at the crossing point is much reduced compared to the nominal value ΔE.

The crossing between the ionic and covalent curves is brought about by a coupling potential U(R). Examination of a large number of systems [2.142] suggests that U(R) decreases exponentially with R,

$$U(R) \propto \exp(-R/L) \quad , \tag{2.4.34}$$

reflecting the decline, with R, of the overlap between the wave functions on the two centers. The change in kinetic energy due to crossing at R_x is $U(R_x)$. For any other separation the change in energy is about $U(R_x) + \Delta V(R)$, where $\Delta V(R)$ is the difference between the ionic and covalent potentials, $\Delta V(R_x) = 0$. The range of the interaction is thus the range $\underline{a}$ over which V(R) remains small. We estimate $\underline{a}$ by expanding $\Delta V(R)$ in a Taylor series about R_x; $\Delta V(R) \simeq (R-R_x)\partial\Delta V(R)/\partial R$, where the derivative is evaluated at R_x. We now take $\underline{a}$ such that $\Delta V(\underline{a}) = U(R_x)$ or $\underline{a} = U(R_x)/|\partial\Delta V(R)/\partial R|$ and $\xi = \underline{a}U(R_x)/h\underline{v}$, where $\underline{v}$ is the velocity at R_x. We thus expect that the crossing probability P has the, by now familiar, $\exp(-\underline{v}_0/\underline{v})$ velocity dependence, except that here

$$\underline{v}_0 = |U(R_x)|^2/h|\partial\Delta V(R)/\partial R| \simeq R_x^2|U(R_x)|^2/e^2h \quad , \tag{2.4.35}$$

where $\Delta V(R)$ is approximated by the ionic term e^2/R. The simple estimates (2.4.33) and (2.4.35) offer a qualitative interpreation of the role of the asymptotic energy gap ΔE. If ΔE is very large (say more than 4 eV), R_x and hence $\underline{v}_0$ are small and the transition is not very efficient. On the other hand, for small ΔE (but still larger than kT), R_x is large, but then $U(R_x)$ and hence $\underline{v}_0$ are small. The transition is again not very efficient. Curve crossing is an important mechanism for electronic energy transfer when the states differ, but not too much, in their energy.

In an actual collision, once the atoms transverse the crossing region their relative separation continues to decrease until thery reach the inner, repulsive core of the potential. From then on, R increases until the atoms separate. In every collision the atoms go through the crossing region twice, once on the way in and once on the way out. If the collision is to result in a change in electronic state, then either a crossing does take place on the way in and not take place on the way out or vice versa. The actual transition probability in atom-atom collision is thus of the form $2P(1-P)$ where $P = \exp(-\underline{v}_0/\underline{v})$ is the single passage transition probability.

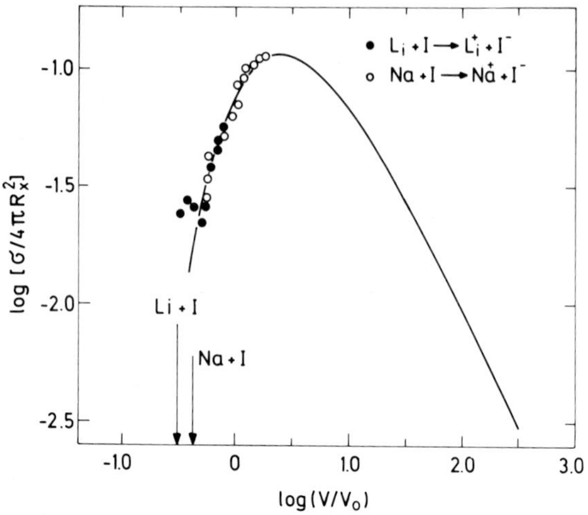

Fig. 2.47. Reduced plot of the curve crossing cross sections vs the reduced velocity $\underline{v}/\underline{v}_0$ (adapted from [2.143]). The solid curve is the Landau-Zener cross section. The dots are experimental results from [2.144]. Arrows indicate threshold energies for ionization

The cross section computed in terms of the P(1-P) transition probability, known as the Landau-Zener model, is found to be a universal function of the reduced velocity $\underline{v}/\underline{v}_0$, Fig.2.47. The maximal value is at $\underline{v} = 2.36\ \underline{v}_0$. In contrast to the simple exponential gap law the transition probability does not necessarily increase as ΔE decreases. The reasons why are easy to see. As ΔE decreases, R_x increases [cf. (2.4.33)]; consequently $U(R_x)$ decreases [cf. (2.4.34)], and so does $\underline{v}_0 \propto R_x^2 \exp(-2R_x/L)$. For the collision to take place it is necessary that $\underline{v}$ exceeds the threshold velocity v_{th}; $\mu v_{th}^2/2 = \Delta E$. Hence, unless $v_{th} < 2.36\ \underline{v}_0$, the cross section will be past its peak (Fig. 2.47). In other words, if $U(R_x)$ is too small, the coupling between the covalent and ionic states is too weak to allow for efficient transitions. An example is provided by the quenching of electronically excited alkali atoms by diatomic molecules.

Even in polyatomic systems the crossing regions can be well separated, in which case one can use the method of classical trajectories augmented by the introduction of crossing probabilities P, which allow a given trajectory to bifurcate at a crossing point [2.145]. If, however, there is a maze of crossings and plenty of energy, the actual distribution of final electronic states will closely resemble a prior distribution.

The discussion thus far centered on the interconversion of electronic and translational energies. When at least one of the reagents is a molecule, there arises the possibility of the interconversion of electronic and internal (vibrational) energy. Indeed, the process may occur not only during a collision but also in an intramolecular fashion ("internal conversion") as discussed in Sect.2.4.9. Collisional processes of this type include, for example,

$$NF(a\ ^1\Delta, v=0) + HF(v=3,J) \rightarrow NF(b\ ^1\Sigma^+, v=0) + HF(v=1,J\pm1) \quad , \qquad (2.4.36)$$

where HF vibrational energy was converted to NF electronic excitation or

$$Br(^2P_{1/2}) + HCN(0,0,0) \rightarrow Br(^2P_{3/2}) + HCN(0,0,1) \quad , \qquad (2.4.37)$$

where the Br electronic excitation is converted to the asymmetric stretch vibrational excitation (cf. Sect.3.3) of HCN. Indeed, as in an exoergic chemical reaction, $E\rightarrow V$ processes lead to specific product vibrational excitation and laser emission was observed [2.146] on the $001\rightarrow100$ and the $001\rightarrow010$ bands of HCN [compare to the CO_2 laser (Sect.3.4.4) where CO_2 is pumped to the 001 state by V-V transfer. Indeed, high gain is observed on the $001\rightarrow100$ transition of CO_2 following the quenching of $Br(^2P_{1/2})$ by CO_2].

2.4.8 A Laser Bridge for the Exponential Gap

By selective excitation of the reagents it is possible to bridge the exponential gap. In a bulk system the prepared reagents may also undergo energy transfer collisions which will tend to degrade the excitation and hence reduce the efficiency. Can one therefore prepare the excited reagents during the actual collision and are there any additional advantages to doing so?

An example of a laser-induced bridging of the exponential gap is that of electronic energy transfer in the

$$Sr(5s5p\ ^1P^0) + Ca(4s^2\ ^1S) \xrightarrow{4977\ \text{Å}} Sr(5s^2\ ^1S) + Ca(4p^2\ ^1S) \qquad (2.4.38)$$

atomic collision [2.146]. Figure 2.48 shows a scheme of the energy levels and the efficiency of the energy transfer as a function of the laser frequency. The rapid decline about the gap matching frequency provides an explicit experimental demonstration of the steepness of the "resonance function", $\exp(-\xi)$.

A simple way to visualize the role of the laser is to view the function of the photon as that of changing the potential energy of the reactants so as to more closely match it to that of the products (Fig.2.49) [2.147-149].

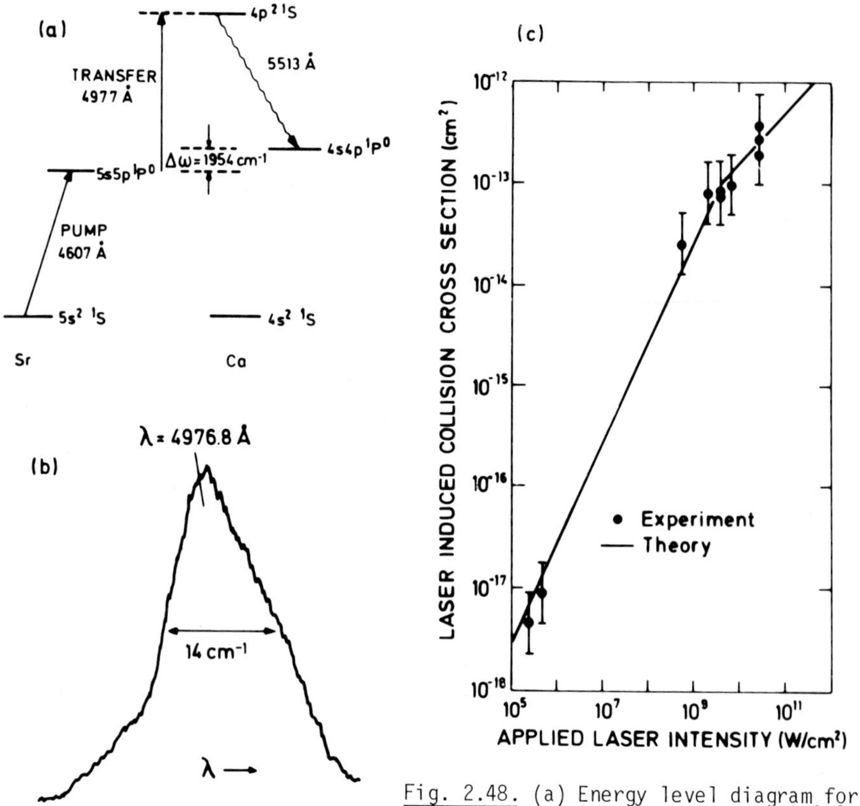

Fig. 2.48. (a) Energy level diagram for the reaction (2.4.38). The Sr(5s5p $^1P^0$) state is prepared by excitation (at 4607 Å) from the ground state, using a dye laser. The Ca($4p^2$ ^1S) population is monitored by its fluorescence (at 5513 Å). (b) The cross section as a function of the transfer wavelength for the reaction (2.4.38). (c) The dependence of the cross section on the transfer laser power (adapted from [2.147])

The laser photon thus serves to reduce (or even close) the potential energy gap, enabling the transition to take place with least change in the momenta of the collision partners. In technical terms, in the presence of the strong laser field the molecule is "dressed" by the photons. The dressed ground state can then cross an excited state.

The analysis in Fig.2.49 suggests that two opposing considerations govern the efficiency of the transfer. The larger is the separation R_x at which the closing of the gap occurs, the larger will be the cross section. Opposing this effect however is the expected rapid decline of the coupling between the two states [cf. (2.4.34)] as a function of R. In general it will therefore be advantageous to tune the laser frequency.

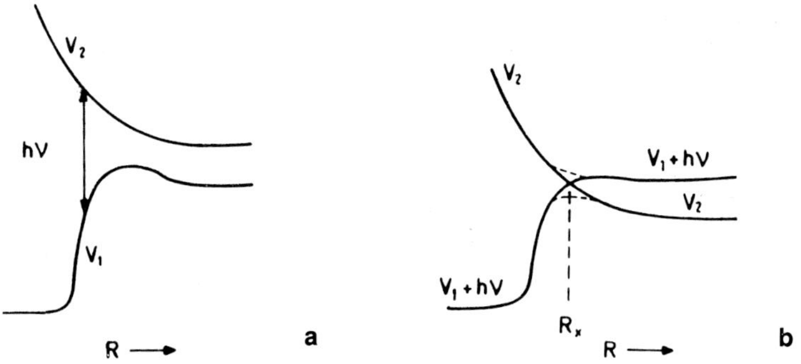

Fig. 2.49. A sketch of two potential energy profiles (a), and the same curves shifted by the energy of the laser photon (b) (in technical terms these are known as "dressed states"). At the crossing between $V_1(R)+h\nu$ and $V_2(R)$, the gap is closed and an efficient transition between the two states can take place, accompanied by an absorption of the photon. One can also view the process as a (vertical) absorption of the photon at the indicated separation, R_x (adapted from [2.148])

Having noted the considerable potential of this method a strong caveat is in order. At the high fields required to dress the ground state, the higher electronic states will also be dressed. Hence multiphoton absorption leading to dissociative or autoionizing states is possible. Such processes, using visible photons, have indeed been observed as has been discussed in Sect. 1.3.5. These processes result in a considerable energy uptake (cf. Fig.1.21) and represent an example where laser pumping does effectively compete with intramolecular decay processes (cf. Sect.2.3.5). However, they imply that when dealing with polyatomic reagents it is not going to be easy to ensure that just one-photon transitions take place.

2.4.9 Intramolecular Electronic to Vibrational Energy Transfer: Radiationless Transitions

When a large polyatomic molecule is electronically excited by a single-photon absorption, the quantum yield of the fluorescence is typically below unity and can be quite small even in the absence of deactivation by collisions. The same is of course true for a diatomic molecule provided that it can dissociate (or predissociate). In the case of the polyatomic molecule however the quantum yield is below unity even in the absence of any dissociation or isomerization. The energy is not radiated but dissipated via an intramolecular transfer or, in a descriptive language, the molecule acts as its own heat bath [2.150-153]. Several detailed studies which document the impact of lasers on this problem are discussed in Sect.5.3.3.

At the energy required for electronic excitation the density of states of a polyatomic molecule is exceedingly high. Besides the manifold of the vibrotational states of the electronic state which was excited by the dipole-allowed absorption from the ground electronic state there is the very dense set of highly excited vibrotational states of the ground state and one (or two) vibrotational manifolds belonging to electronic triplet states (see Fig.3.29). Due to the much higher densities of states in these alternative manifolds the a priori probability of finding the molecule in one of these other electronic states (which do not however have dipole-allowed transitions from the ground states) is much higher. These highly dense manifolds act as the equivalent of the continuum of states provided by the translation in a dissociation or in a collision-induced quenching process. Indeed we have already commented on the fact (cf. Fig.2.42) that, in comparison to the translation, the high density of vibrational states of a polyatomic molecule can be quite competitive.

The considerations above centered on the "density of final states" factor in (2.2.12). We turn next to the "dynamical" factor for intramolecular energy transfer from an optically excited electronic state. Since the electrons move faster than the nuclei it is possible to recognize two separate dynamical factors (the so-called Condon approximation). The first is that transitions to alternative electronic states depend on the electronic coupling between them (spin-orbit for singlet-triplet or "intersystem crossing" transitions, electronic-nuclear for singlet-singlet or "internal conversion" transitions). The second is that such transitions imply the conversion of large amounts of electronic energy into the nuclear degrees of freedom. They should thus be governed by an exponential gap rule as previously discussed. The adiabaticity parameter (Sect.2.4.4) is now the amount ΔE of electronic energy which can be converted in a period $\tau = 1/h\nu$ of the vibrational mode which accepts this energy $\xi = \Delta E/h\nu$. The higher the vibrational frequency, the more active is the vibrational mode. The dependence of the rate of intersystem crossing on the energy gap between the first excited singlet state and the lowest triplet state (Fig.2.50) and the marked slowdown of this rate when the high-frequency C-H modes are replaced by lower frequency C-D modes offer support for this further example of the exponential gap law.

The special role of the C-H vibrations in accepting the excess energy identifies them as "local modes" of the type discussed in Sect.2.3.5. The high vibrational frequency of these modes which makes them the preferred acceptors is also responsible for their ineffective mixing with other bond vibrations of lower frequencies.

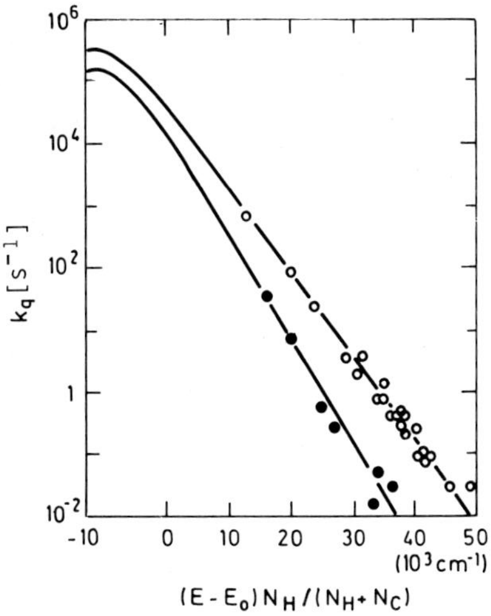

Fig. 2.50. Rate constants k_q for the decay rate of the triplet state of aromatic hydrocarbons as a function of the triplet energy times the C-H or C-D ratio. N_H is the number of H atoms per molecule, etc. Open circles refer to C_nH_m and closed circles to C_nD_m compounds (adapted from [2.152])

The rate of very fast radiationless transitions can be monitored by competition with collisional deactivation as discussed in Sect.2.3.7. Indeed, the discussion of dephasing there is just as applicable here. Due to the higher excitation energy the mean level spacing D is quite small, while the range of eigenstates that are excited is governed by the strength of the coupling between the two electronic states. By pumping very cold molecules which are effectively in their ground state it is possible to excite specific vibrotational levels of the excited electronic state. The decay rates of these states do depend on the specific initial vibrotational state, indicating that the rate of intramolecular electronic to vibrational energy transfer is faster than the rate of intramolecular vibrational energy transfer within the excited (singlet) electronic state. As the excitation energy is increased, the decay rates are less and less dependent on the initial state.

The considerable dependence of the rate of depletion of the excited singlet state on the particular initial quantum state (cf. Sect.5.3.3) implies that the "rate constant" is going to depend on the details of the preparation process, except for a fully state-selected experiment. To see this recall the general expression, (2.1.14), which now reads

$$k_{RT}(X) = \sum_i p(i|X) k_{RT}(i) \quad . \tag{2.4.39}$$

Here $k_{RT}(i)$ is the rate of radiationless transitions for the initial quantum state i and $p(i|X)$ is the fraction of molecules excited to the state i in the particular experiment X. The energy spread in the excitation pulse is the primary variable that determines the distribution $p(i|X)$. In the limit of broadband excitation very many terms contribute to the sum in (2.4.39), and only a smooth, averaged behavior will be observed. However, using narrower lines (and particularly so at energies just above the minimal excitation energy of the electronic excited states where the density of states i is still not high), the operational measured rate constant $k_{RT}(X)$ will reflect not only the properties of the molecule but also the properties of the light source, except when $k_{RT}(i)$ is only weakly dependent on i.

2.5 Macroscopic Disequilibrium

Various microscopic mechanisms can carry a molecular system to equilibrium. These include for example bimolecular collisions, spontaneous and stimulated radiation, unimolecular isomerization and dissociation processes, and intramolecular radiationless transitions. In the previous sections of this chapter we have studied the dynamical aspects of such events. Now we turn to discuss the mutual, statistical, effects of many microscopic events on the approach to equilibrium of macroscopic molecular systems, with particular emphasis on collisional relaxation. We shall begin the discussion with the kinetic description of spontaneously relaxing systems, continue in describing their thermodynamic characteristics, and end up analyzing work processes in nonequilibrium molecular systems, referring to chemical lasers as examples of special interest. The kinetic analysis of chemical lasers will be discussed in Sect.4.4 since it requires Sect.3.2 as a prerequisite.

2.5.1 The Master Equation: Relaxation of Harmonic Oscillators

As a way of introducing the basic concepts of molecular relaxation consider a very simple example: vibrationally excited diatomic molecules, e.g., HF, highly diluted in a monoatomic host gas like Ar. Neglecting the influence of the infrequent HF-HF collisions, radiative decay, and wall effects, the relaxation is caused by collisions with the buffer gas

$$HF(v) + Ar \rightleftarrows HF(v') + Ar \quad . \tag{2.5.1}$$

As in Sect.2.1.2 the time rate of change of the vibrational populations is governed by the kinetic equation, known commonly as the master equation [2.154

$$\frac{dP(v,t)}{dt} = \rho \sum_{v'} [k(v' \to v;T)P(v',t) - k(v \to v';T)P(v,t)] \quad . \tag{2.5.2}$$

Here $P(v,t) = [HF(v)]/[HF]$ is the fraction of HF molecules in level v at time t, $\rho = [Ar]$ is the buffer gas density, and T is the buffer gas, or "heat bath", temperature. This temperature also characterizes the translational and rotational degrees of freedom of the diatomic molecules whose relaxation is assumed to be instantaneous on the time scale of vibrational relaxation (see Sects.1.2 and 2.5.3).

The positive terms in (2.5.2) represent the rate of population transfer into level v, while the negative terms, the sum of which is $\rho k(v\to;T)P(v,t)$, describe the depletion rate of this level. At equilibrium ($t\to\infty$) these opposing contributions balance each other and $dP(v,\infty)/dt = 0$ or $P(v,\infty) = p(v|T)$. This property is ensured by the condition of detailed balancing [see (2.1.35)].

The master equation (2.5.2), derived here on the basis of ordinary kinetic considerations, provides the mathematical framework for describing the irreversible[7] approach of $P(v,t)$ to $P(v,\infty) = p(v|T)$. In general, the solution of a master equation requires numerical integration. One of the very few cases for which closed-form solutions are available is (2.5.2) with the $k(v\to v';T)$ predicted by the Landau-Teller (LT) model (Sect.2.4.6). Besides the $T^{-1/3}$ dependence of $\ln[k(1\to 0;T)]$ [cf. (2.4.27)] the LT model implies that only near neighbor $v\to v\pm 1$ transitions are allowed and

$$k(v\to v-1;T) = v\, k(1\to 0;T) \quad . \tag{2.5.3}$$

The reversed, excitation, rate constants are fixed by the requirement of detailed balance (2.1.35). In the harmonic oscillator approximation $E_v = v\hbar\omega$, we have $k(v-1\to v;T) = \alpha k(v\to v-1;T) = \alpha v\, k(1\to 0;T)$, where

$$\alpha = p(v|T)/p(v-1|T) = \exp(-\hbar\omega/kT) \quad . \tag{2.5.4}$$

In this notation the equilibrium distribution of the oscillators is

$$p(v|T) = \alpha^v \Big/ \sum_{v=0}^{\infty} \alpha^v = \alpha^v(1-\alpha) \quad , \tag{2.5.5}$$

7 For derivations of the master equation from the reversible equations of motions describing the exact time evolution of all the particles constituting a macroscopic system, see for example the collection of articles in [2.154].

where $\sum \alpha^v = (1-\alpha)^{-1} = Q$ is the vibrational partition function [see, e.g., (2.3.18)].

Using (2.5.3) and the appropriate reversed rate constants $k(v-1 \rightarrow v;T)$ and noting that $v' = v \pm 1$, the master equation (2.5.2) reads

$$\frac{dP(v,t)}{dt} = \frac{1}{\tau(1-\alpha)} \left\{ \alpha v P(v-1,t) - [v+\alpha(v+1)]P(v,t) + (v+1)P(v+1,t) \right\} , \qquad (2.5.6)$$

where, for reasons that will soon be clarified

$$\tau = [(1-\alpha)\rho k(1 \rightarrow 0;T)]^{-1} \qquad (2.5.7)$$

will be called the vibrational relaxation time [see also (2.4.28)].

An example demonstrating the time evolution of $P(v,t)$ according to (2.5.6) is shown in Fig.2.51. The highly excited initial distribution $P(v,0)$ is of the type which often characterizes the nascent vibrational populations of exoergic reaction products (see for example Figs.2.3,14). The figure illustrates the short memory of the initial distribution as reflected by the gradual shift of the most populated level towards $v=0$. Note also the narrowing of $P(v,t)$ as time evolves. This is due to the increase of $k(v \rightarrow v-1;T)$ with v which implies that the rear edge (high v's) of $P(v,t)$ relaxes faster than the front edge, leading to effective narrowing of $P(v,t)$.

This behavior is even more pronounced for anharmonic oscillators where a significant enhancement of $k(v \rightarrow v-1;T)$ as a function of v arises from the fact that the energy gaps $E_v - E_{v-1}$ decrease with v (see Sect.2.5.4).

While $P(v,t)$ is usually a rather complicated function of $P(v,0)$ the mean vibrational energy per oscillator $\langle E_v(t) \rangle$ displays a very simple time evolu-

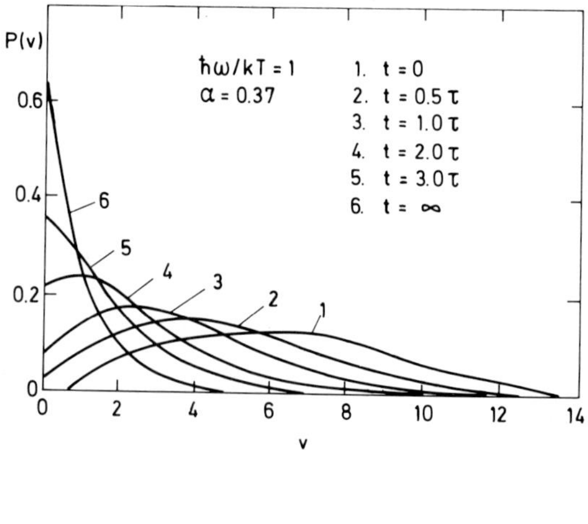

Fig. 2.51. Relaxation of harmonic oscillators, (2.5.6). (For detailed discussions and analytic solutions see [2.155])

tion, independent of the initial distribution. For the special case of harmonic oscillators

$$<E_v(t)> = \sum_v E_v P(v,t) = \hbar\omega \sum_v vP(v,t) = \hbar\omega <v(t)> \quad , \tag{2.5.8}$$

where $<v(t)>$ is the average number of quanta per oscillator at time t. At equilibrium, i.e., as $t\to\infty$, $P(v,t)\to P(v,\infty) = p(v|T)$ and, using (2.5.5), we find

$$<E_v(\infty)> = \sum_v E_v p(v|T) = \hbar\omega(1-\alpha) \sum_v v\alpha^v = \hbar\omega\alpha(1-\alpha)^{-1} = \hbar\omega <v(\infty)> \quad , \tag{2.5.9}$$

where we have used $\sum v\alpha^v = \alpha(\partial/\partial\alpha)\sum\alpha^v = \alpha(\partial/\partial\alpha)(1-\alpha)^{-1}$.

The rate of change of $<E_v(t)>$ is given by

$$\frac{d<E_v(t)>}{dt} = \frac{d}{dt}\sum_v E_v P(v,t) = \sum_v E_v \frac{dP(v,t)}{dt} \quad . \tag{2.5.10}$$

Multiplying both sides of (2.5.6) by $E_v = \hbar\omega v$, summing over v, and using (2.5.7-9) we find after some algebraic manipulations

$$\frac{d<E_v(t)>}{dt} = \frac{1}{\tau}[<E_v(\infty)> - <E_v(t)>] \quad . \tag{2.5.11}$$

This equation can be immediately integrated to yield

$$<E_v(t)> = <E_v(\infty)> + [<E_v(0)> - <E_v(\infty)>]\exp(-t/\tau) \quad . \tag{2.5.12}$$

Hence the mean vibrational energy decays exponentially to its equilibrium value $<E_v(\infty)>$ with a single relaxation time τ, (2.5.7), independent of the initial distribution P(v,0). As could be expected, the relaxation time is inversely proportional to $\rho k(1\to 0;T)$ which measures the rate of inelastic collisions.

The master equation (2.5.2) can be rewritten in the form

$$\frac{dP(v,t)}{dt} = \sum_{v'} A(v,v')P(v',t) \tag{2.5.13}$$

with A(v,v'), the "relaxation matrix", defined by

$$A(v,v') = \rho[k(v'\to v;T) - \delta_{vv'}k(v\to;T)] \quad , \tag{2.5.14}$$

where $\delta_{vv'}$ is the Kronecker symbol. The first (gain) term in A(v,v') represents the transfer of molecules from v' to v while the second (loss) term accounts for the reversed process.

Equation (2.5.13) has the formal solution [2.154]

$$P(v,t) = P(v,\infty) + \sum_{m=1}^{\infty} C_m(v)\exp(\lambda_m t) \quad , \qquad (2.5.15)$$

where the negative constants λ_m are the eigenstates of the relaxation matrix and the $C_m(v)$'s are functions of the corresponding eigenvectors $P_m(v)$, i.e., $\sum_{v'} A(v,v')P_m(v') = \lambda_m P_m(v)$. The equilibrium solution $P(v,\infty) = p(v|T)$ is an eigenvector of A(v,v') with the eigenvalue $\lambda_0 = 0$ since $dP(v,\infty)/dt = \sum_{v'} A(v,v') \times P(v',\infty) \equiv 0$. This is a direct consequence of the requirement for detailed balancing.

For most relaxation problems the eigenvalue solution (2.5.15) requires numerical evaluation of the λ_m's and C_m's and therefore does not present a special advantage over direct numerical integration of the master equation. For the case of harmonic oscillators governed by the master equation (2.5.7) the eigenvalues of A(v,v') are quite simple [2.155]

$$\lambda_m = -m/\tau = -(1-\alpha)\rho k(m \to m-1;T) \quad . \qquad (2.5.16)$$

An important consequence of this result is that as time evolves the number of terms in (2.5.15) that contribute significantly to P(v,t) is constantly decreasing. In particular, near equilibrium or more precisely when $t/\tau = t(\lambda_1-\lambda_2) \gg 1$, we find

$$P(v,t) = P(v,\infty) + C_1(v)\exp(-t/\tau) \qquad (t\to\infty) \quad . \qquad (2.5.17)$$

Thus as $t\to\infty$ all the average (macroscopic) observables $\sum A_v P(v,t) = \langle A_v(t)\rangle$ display an exponential decay to their equilibrium value $\langle A_v(\infty)\rangle$, with a single common relaxation time τ. This property characterizes many relaxing systems besides harmonic oscillators and is valid whenever the lowest nonzero eigenvalue λ_1 is well separated from the other λ_m's [2.154].

The master equation (2.5.6) exhibits an interesting property known as "canonical invariance" implying the following: Given a canonical initial distribution P(v,0) then the canonical form is preserved throughout the approach to equilibrium, i.e.,

$$P(v,t) = \exp[-E_v/kT(t)]/Q[T(t)] \quad , \qquad (2.5.18)$$

with $T(0)=T(t=0)$ characterizing the initial distribution. In other words (2.5.18) is a solution of (2.5.6) at all times $0 \leq t \leq \infty$, provided of course that $P(v,0)$ is also canonical.

The property of canonical invariance is not limited to harmonic oscillators. (Also, it can be shown that for systems with this property the energy relaxation, $<E_v(t)>$, is exponential as in (2.5.12), [2.154].) For harmonic oscillators, canonical invariance implies a particularly simple form for $P(v,t)$. Noting that (2.5.18) is equivalent to (2.5.4) with $\alpha=\alpha(t)=\exp[-\hbar\omega/kT(t)]$ we can extend (2.5.9) to $<E_v(t)>/\hbar\omega = <v(t)> = \alpha(t)[1-\alpha(t)]^{-1}$. Hence, $\alpha(t) = <v(t)>/(1+<v(t)>)$. Now using (2.5.5) with $\alpha(t)$ we obtain

$$P(v,t) = <v(t)>^v/(1+<v(t)>)^{v+1} \quad , \tag{2.5.19}$$

showing that a single macroscopic observable $<v(t)> = <E_v(t)>/\hbar\omega$ [whose explicit time dependence is given by (2.5.12)] uniquely determines $P(v,t)$ throughout the relaxation. This important result is consistent with the conclusions of Sects.2.1.7 and 2.1.8 where we showed that a canonical distribution of the type (2.5.19) is the distribution which maximizes the entropy $S=-\sum P(v)\ln[P(v)]$ subject to the single constraint $<E_v> = \sum E_v P(v)$. In the present case the constraint and consequently the Lagrange parameter $1/kT$ are time dependent.

Canonical invariance is a rather special property obeyed only by certain types of master equations. Moreover, it applies only for canonical initial distributions $P(v,0)$. However near equilibrium, i.e., as $t\to\infty$, the canonical representation does often provide a very good approximation for $P(v,t)$ regardless of $P(v,0)$ and of the specific form of the rate constants $k(v\to v';T)$. This behavior will be discussed in more detail in Sect.2.5.6.

Two reasons have motivated the choice of the V-T relaxation of harmonic oscillators as our introductory example. First, it displays some important qualitative characteristics of V-T relaxation of diatomic molecules. Second, the master equation for this problem could be solved analytically. Using again the harmonic oscillator model one can obtain analytic solutions also for the relaxation of a pure diatomic gas, where both V-V and V-T collisions are important. However, some of the most interesting relaxation features of diatomic (and polyatomic) gases are direct consequences of vibrational anharmonic effects, in particular the anharmonicity affects the V-V relaxation pattern. These effects will be discussed in Sect.2.5.4.

2.5.2 Rotational Relaxation

Due to the smaller energy changes associated with rotationally inelastic collisions, such as

$$HF(J) + Ar \rightarrow HF(J') + Ar \quad , \tag{2.5.20}$$

rotational relaxation is usually much faster than vibrational relaxation (see Fig.1.9). A qualitative difference between vibrational and rotational relaxation is associated with the fact that vibrational energy spacings are nearly constant, or more precisely, slowly decreasing with v due to anharmonicity, whereas rotational energy spacings increase with J, e.g., $\Delta E_J = BJ(J+1) - BJ(J-1) = 2BJ$. Thus, according to the exponential gap law (Sects.2.4.2-4), the rate constants $k(J \rightarrow J')$ which determine the population transfer rates decrease (exponentially) as J increases. Consequently, the rate at which the population ratios $P(J,t)/P(J+1,t)$ approach their equilibrium proportions is fast in the low J regime where typically $\Delta E_{JJ'} < kT$, and relatively slow in the high J regime [2.156-160]. This behavior is quite different from that displayed in V-T relaxation where the high v levels relax faster.

Figure 2.52 shows the solutions of the rotational master equation

$$\frac{dP(J,t)}{dt} = \rho \sum_{J'} [k(J' \rightarrow J;T)P(J',t) - k(J \rightarrow J';T)P(J,t)] \tag{2.5.21}$$

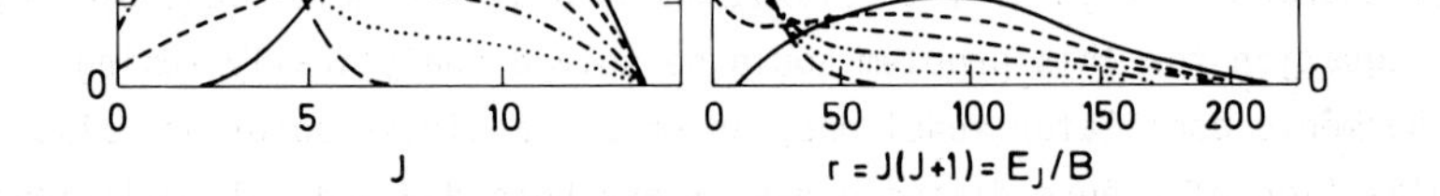

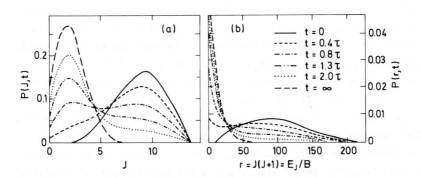

Fig. 2.52. (a) Rotational relaxation patterns P(J,t) describing the solutions of (2.5.21) with exponential gap type rate constants (2.4.16); $\theta_R = \frac{1}{2}$. (b) The same distributions after transformation to the (reduced) energy variable $r = E_J/B = J(J+1)$; $P(r,t) = P(J,t)dJ/dr = P(J,t)/(2J+1)$. The time constant τ is the energy relaxation time defined by $[\langle E_J(\tau)\rangle - \langle E_J(\infty)\rangle]/[\langle E_J(0)\rangle - \langle E_J(\infty)\rangle] = 1/e$ (adapted from [2.160]). Comparison with Fig.2.51 reveals that P(r,t), as opposed to P(J,t), is not so markedly different from $P(v,t) \propto P(E_v,t)$. Of course, the vibrational relaxation time is typically much larger due to the much larger spacings between vibrational levels

describing the relaxation of HF by Ar. The rate constants used in the solution of (2.5.21) are of the form (2.4.16) [2.157], with $\theta_R = \frac{1}{2}$ (this value agrees with available results from infrared chemiluminescence [2.156], laser fluorescence [2.158], and double resonance [2.159] experiments). The initial distribution P(J,0) is similar to the nascent product distribution in the $F + H_2 \rightarrow HF(v=1,J) + H$ reaction (cf. Fig.2.14).

Comparison of Fig.2.51 with Fig.2.52a reveals the differences mentioned above between P(v,t) and P(J,t). While the peak of P(v,t) is gradually shifted towards v=0 the rotational distribution in the high J region nearly preserves its shape (though not its magnitude), reflecting the weak coupling between high J levels. The decline of the original peak is accompanied by a simultaneous rise of the Boltzmann peak at the low J region, giving rise to the double peak structure of the partly relaxed P(J,t). These intermediate distributions can often be well approximated (e.g., in the case of Fig.2.52a) as a superposition of the initial P(J,0) and final $P(J,\infty)$ distributions [2.157,29]. That is

$$P(J,t) = P(J,\infty) + [P(J,0)-P(J,\infty)]\exp(-t/\tau) \tag{2.5.22a}$$

or, equivalently,

$$-\frac{d}{dt}[P(J,t)-P(J,\infty)] = [P(J,t)-P(J,\infty)]/\tau \quad . \tag{2.5.22b}$$

These forms[8] imply that all the absolute level populations approach their equilibrium values at the same rate. It should be noticed that this conclusion does not contradict our previous notion that the equilibration among neighboring rotational levels, i.e., the approach of P(J,t)/P(J+1,t) to $P(J,\infty)/P(J+1,\infty)$ is faster in the low J regime. Consider for example the relaxation pattern of Fig.2.52a for which the representation (2.5.22) provides a good approximation. Here, $P(J,0) \gg P(J,\infty)$ in the region of high J's ($J \sim 10$), whereas $P(J,\infty) \gg P(J,0)$ in the thermal region ($J \sim 2$). Thus, except very near to equilibrium ($t \gg \tau$) where all $P(J,t) \rightarrow P(J,\infty)$, we find, using (2.5.22a), that $P(J,t)/P(J+1,t) \simeq P(J,0)/P(J+1,0)$ for high J's and $P(J,t)/P(J+1,t) \simeq P(J,\infty)/P(J+1,\infty)$ for low J's. The "long memory" of the initial distribution in the high J regime is a reflection of the relatively slow population trans-

8 The adequacy of the approximation (2.5.22) depends on the initial distribution P(J,0) and the eigenvalues of the relaxation matrix A(J,J') (cf. Sect.2.5.1). It is exact, regardless of P(J,0), when all the eigenvalues are degenerate.

fer among the widely spaced rotational levels. The immediate buildup of the low level populations according to their equilibrium proportions is due to the fast population exchange among closely spaced levels. In Sect.4.4 we shall see that rotational nonequilibrium effects are more clearly observed in the high J regime.

The exponential gap law provides a common dynamical basis for interpreting the markedly different vibrational and rotational relaxation patterns of Figs.2.51 and 2.52. One may therefore suspect that the difference between the relaxation patterns of $P(v,t)$ and $P(J,t)$ are only apparent and may largely disappear by transforming to the corresponding energy distributions $P(E_v,t)$ and $P(E_J,t)$. Applying the standard transformation rule $P(y) = P(x)dx/dy$ from a general distribution function $P(x)$ over the variable x to the distribution $P(y)$ over $y = f(x)$, we find: I) For harmonic oscillators $E_v = \hbar\omega v$, $P(E_v) = (\hbar\omega)^{-1}P(v)$; hence $P(E_v)$ and $P(v)$ are simply proportional to one another so that Fig.2.51 describes both $P(v,t)$ and $P(E_v,t)$. II) For rotors, $E_J = BJ(J+1)$, $P(E_J) = P(J)/(2J+1)$, and $P(E_J)$ [equivalently $P(r)$, $r=J(J+1)$] is quite different from $P(J)$ as shown in Fig.2.52. Indeed $P(E_J,t)$ appears more similar to $P(E_v,t)$ than to $P(J,t)$ from which it was derived.

In Sect.2.2 it was pointed out that presenting product state distributions as functions of the reduced energy variables f_V, f_R, or f_T reveals the unifying aspects of apparently different phenomena. For instance $P(f_V)$, as opposed to $P(v)$, is the same for isotopic reactions like $Cl + HI(DI) \rightarrow HCl(DCl) + I$ (Fig.2.9) and $P(g_R)$ as opposed to $P(J|v)$ is the same for nascent rotational distributions in different vibrational manifolds (Fig.2.15). The transformation from $P(J)$ to $P(E_J)$ is yet another example demonstrating the importance and power of this procedure. Another application is that the parameter θ_R in (2.4.16) can be computed by the same procedure used to determine λ_V in (2.4.17) which is discussed in Sect.2.5.6.

2.5.3 Separation of Time Scales

Vibrational deactivation is usually accompanied by rotational excitation (Sects.2.4.5 and 2.4.6). Namely, in collisions like

$$HF(v,J) + Ar \rightarrow HF(v',J') + Ar \qquad (2.5.23)$$

a change in v, $v' \neq v$, often implies a simultaneous change in J, $J' \neq J$. (The converse is not true.) Nevertheless, in our discussion of vibrational relaxation in Sect.2.5.1 we have apparently disregarded the possible changes in J [cf. (2.5.2)]. Also, in describing rotational relaxation in the previous section we have ignored the vibration [cf. (2.5.21)]. Treating vibrational and rotational relaxations as two separate problems is justified when these phenomena occur on different time scales. More generally, the existence of distinct time scales for different rate processes implies a considerable conceptual and mathematical simplification of the relaxation problem. In this section we present a few examples which illustrate this argument and are of interest in laser chemistry and chemical laser research. As our first example consider the vibrotational relaxation process (2.5.23).

The master equation describing (2.5.23) is equivalent to (2.5.2) [or (2.5.21)] with $P(v,J;t)$ replacing $P(v,t)$, $[P(J,t)]$. Omitting the time symbol for notational brevity and using the familiar representation $P(v,J) = P(v) \times P(J|v)$ [cf. (2.2.26)], we get

$$\begin{aligned}
\frac{dP(v,J)}{dt} &= P(v)\frac{dP(J|v)}{dt} + P(J|v)\frac{dP(v)}{dt} \\
&= \rho P(v)\sum_{J'}[k(J'\to J|v;T)P(J'|v) - k(J\to J'|v;T)P(J|v)] \\
&\quad + \rho\sum_{v\neq v'}[k(v',J'\to v,J;T)P(v')P(J'|v') \\
&\quad - k(v,J\to v',J';T)P(v)P(J|v)] \quad ,
\end{aligned} \tag{2.5.24}$$

where on both sides of this equation we have separated between the R-T terms, $v,J \to v,J' \neq J$, and the V-R,T terms $v,J \to v' \neq v,J'$ and used the notation $k(J\to J'|v;T) = k(v,J\to v,J';T)$. In general (see Fig.1.9)

$$k(v,J\to v',J';T)/k(J\to J'|v;T) \sim \tau_{R-T}/\tau_{V-R,T} \ll 1 \quad ,$$

where τ_{R-T} and $\tau_{V-R,T}$ are the characteristic times for rotational and vibrational relaxation, respectively. (Deviations from this behavior are observed at the high J regime; see for example Fig.2.41.) At times $t < \tau_{R-T} \ll \tau_{V-R,T}$ only $dP(J|v)/dt \neq 0$ while $P(v)$ is practically constant; $dP(v)/dt = 0$. Hence, for this time interval we can drop the second V-R,T terms on both sides of (2.5.24) in order to regain the rotational master equation (2.5.21). At somewhat later times $\tau_{R-T} < t \ll \tau_{V-R,T}$ rotational relaxation has been completed so that $P(J|v) = p(J|v,T)$ or $P(v,J) = P(v)p(J|v,T)$, where $P(v) = P(v,0)$. Such intermediate distributions which characterize the populations after the fast process has been completed whereas the slow process has not yet begun are often called "quasiequilibrium" distributions. As time evolves, $\tau_{R-T} < t \sim \tau_{V-R,T}$, $P(v)$ starts changing. That is, $dP(v)/dt \neq 0$ while $P(J|v) = p(J|v;T)$

remains equilibrated, i.e., $dP(J|v)/dt = 0$. At this time interval (2.5.24) reduces to the vibrational master equation (2.5.2) with the rotationally avaraged rate constants

$$k(v \to v';T) = \sum_J p(J|v;T) \sum_{J'} k(v,J \to v',J';T) \qquad (2.5.25)$$

in accordance with the "canon" of collision theory [cf. (2.1.13,14)]. The averaged rate constants $k(v \to v';T)$ are not only fewer but also much easier to measure than the detailed rate constants $k(v,J \to v',J';T)$.

Sections 2.5.4 and 2.5.5 describe two additional relaxation problems in which the separation of time scales plays a major role.

2.5.4 Vibrational Anharmonicity and V-V Up-Pumping

Electrical CO lasers involve two consecutive pumping stages (Sect.4.3). First, inelastic electron-CO collisions displace the vibrational CO populations from thermal equilibrium and then fast V-V processes elevate molecules from intermediate ($v \sim 2$-7) to highly excited vibrational levels ($v \sim 7$-15), thereby creating the population inversion necessary for the lasing process. Let us examine the V-V pumping stage in more detail.

Immediately after the initial excitation inelastic collisions

$$CO(v) + CO(u) \rightleftarrows CO(v') + CO(u') \qquad (2.5.26)$$

start carrying the vibrational populations back to equilibrium. At the low v regime the relaxation is dominated by the near-resonant V-V collisions, $(v'+u'=v+u)$. Due to the anharmonic level spacings the rate of the V-V processes decreases while that of the V-T processes $(v+u \neq v'+u')$ increases as a function of v. Consider for example the single quantum exchange transitions $v,u \to v+1,u-1$ $(v \geqq u)$ and $v,u \to v-1,u$ as representatives of the (exoergic) V-V and V-T processes, respectively. Using the anharmonic oscillator level scheme

$$E_v = E_{v=0} + hc\omega_e v - hc\omega_e x_e v(v+1) \qquad (2.5.27)$$

we find that $\Delta E = -2hc\omega_e x_e(v-u+1)$ for the $v,u \to v+1,u-1$ transition and $\Delta E = -hc\omega_e + 2hc\omega_e x_e v$ for the $v,u \to v-1,u$ transition. Thus for a given u, $|\Delta E|$ increases with v in the V-V process and decreases with v in the V-T process.

These features are demonstrated for CO and HF in Fig.2.53 (see also Fig. 2.37). The rate constants for CO-CO collisions are based on a modification

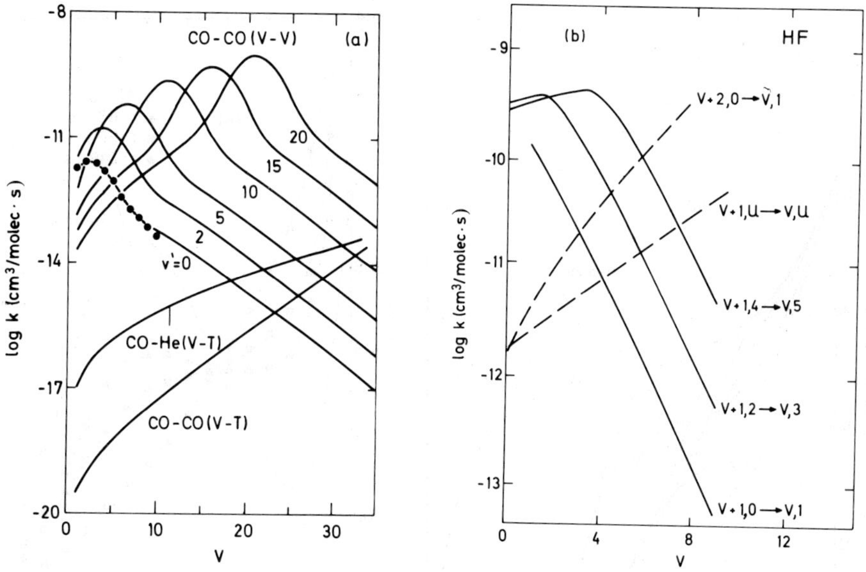

Fig. 2.53a,b. V-V and V-T rate constants for CO and HF, at 300 K. Due to the anharmonicity the energy gaps associated with V-V transfer increase with v thereby leading to a decrease of the corresponding rate constants. An opposite behavior is displayed by the V-T rate constants. (a) Rate constants for the single quantum transfer V-V processes $CO(v)+CO(v')\to CO(v-1)+CO(v'+1)$ for $v'=1,2,5,10,15$, and 20, calculated according to a modified SSH model [2.161] (see also Fig.1.7). The dots are experimental data for $v'=1$ [2.162]. Also shown are computed V-T rate constants of $CO(v)+M\to CO(v-1)+M$ for M=CO [2.161] and M=He [2.163]. All rate constants are of the exponential gap type. (b) Exponential gap type rate constants [cf. (2.4.17)] for $HF(v)+HF(u)\to HF(v')+HF(u')$ collisions. The solid and dashed lines correspond to V-V, $(u+v=u'+v')$, and V-T $(u+v\neq u'+v')$ transfer, respectively. The rate constants were "synthesized" on the basis of partial kinetic information as described in Sect.2.5.6, (see [2.164])

[2.161] of the SSH model [2.128,129] which takes into account the influence of long-range interactions. [The SSH (Schwartz, Slavsky, Hertzfeld) model is an extension of the Landau-Teller model and, likewise, allows only $\Delta v=\pm 1$ transitions.] The HF rate constants are of the type (2.4.17) with $\bar{\lambda}_v=\frac{1}{2}$. This value was determined on the basis of partial kinetic information as described in Sect.2.5.6. The trends predicted by the two approaches are quite similar since both incorporate the exponential gap dependence as a major ingredient.

Due to the anharmonicity the V-V "up-pumping" or "splitting" transitions $v,u\to v+1,u-1$ $(v\geqq u)$ are exoergic and thus faster than the reversed transitions $v+1,u-1\to v,u$. Hence the V-V processes tend to spread the vibrational populations over the vibrational ladder so as to elevate a considerable portion of the molecules to high lying vibrational states. This tendency is the origin

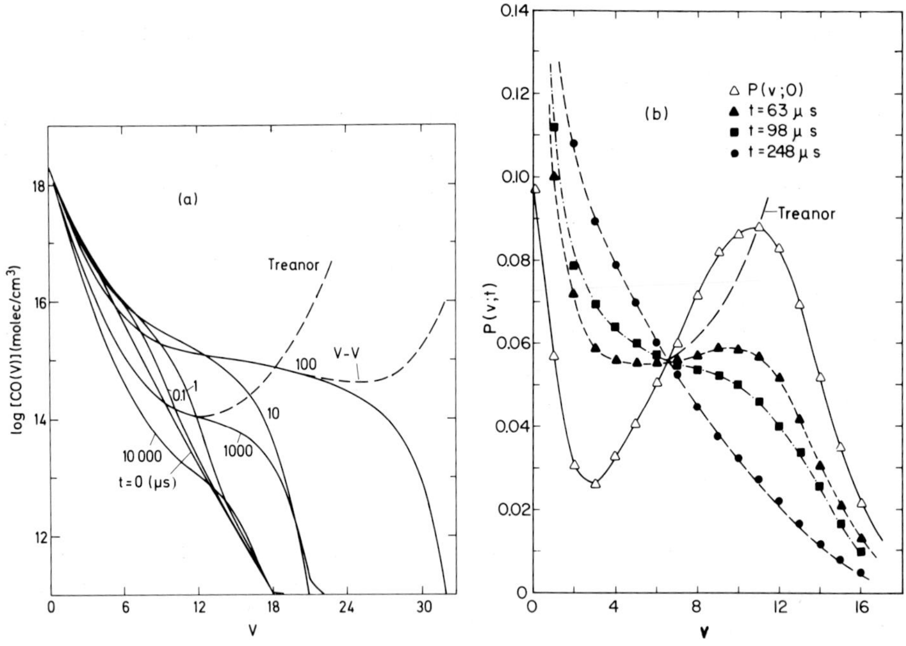

Fig. 2.54a,b. Stages in the time evolution of CO vibrational populations. (a) Numerical solutions of the vibrational master equation for a system containing 10% CO and 90% He at a total pressure of 1 atm ($2.7\cdot10^9$ mole/cm^3) and T = 300 K (computed for this book by A. Ofir using the rate constants of Fig.2.53a). The initial distribution P(v,t=0) is a hot (3000 K) Boltzmann distribution. The V-V up-pumping mechanism at the early stages of the relaxation is evident. The broken curves, which are the solutions of the master equation obtained by keeping the V-V terms and deleting the V-T terms, demonstrate more clearly this trend and the approach towards the Treanor distribution. (b) Experimental results for the O_2-CS_2-He system [2.166]. The CO molecules are mainly produced in the O + CS → CO(v) + S reaction (cf. Fig.4.22). Some of the intermediate distributions here appear similar to those in (a) (adapted from [2.29])

of the so-called "V-V up-pumping" mechanism (often referred to as the Treanor, Rich and Rehm [2.165] mechanism). Clearly, this mechanism is operative only in the V-V dominated region and only for times $\tau_{V-V} \sim t < \tau_{V-T}$; it is not effective in the high v regime dominated by the V-T processes (Fig.2.53). In the time and energy intervals governed by the V-V processes they tend to establish a quasiequilibrium, vibrationally inverted distribution $P_q(v)$ known as the Treanor distribution [2.165] (Fig.2.54). More precisely, $P_q(v)$ is the stationary solution of the V-V part (v'+u'=v+u) of the master equation describing (2.5.26), that is,

$$\frac{dP(v)}{dt} = \rho \sum_{\substack{u,v',u' \\ (v'+u'=v+u)}} [k(v',u' \rightarrow v,u;T)P(v')P(u')-k(v,u \rightarrow v',u')P(v)P(u)] \quad , \tag{2.5.28}$$

where $\rho = [CO]$ and T is the instantaneous translational-rotational temperature. In other words, $P_q(v)$ is the equilibrium distribution, $dP_q(v)/dt = 0$, in a system governed by V-V processes only. Equivalently, $P_q(v)$ is the distribution which maximizes the entropy $S = -\sum P(v) \ln P(v)$ subject to the (time-independent) constraints governing such a system (cf. Sect.2.1.7 and Appendix 2.A). The relevant constraints are: I) the normalization $\sum P(v) = 1$; II) the average energy per molecule, $\langle E \rangle = \langle E_v \rangle + \langle \varepsilon \rangle$, where E_v is the vibrational energy and ε is the translational-rotational energy; III) the average number of vibrational quanta

$$\langle v \rangle = \sum_v vP(v) \tag{2.5.29}$$

which is not modified by the V-V processes. Clearly it is the last constraint which distinguishes between $P_q(v)$ and an ordinary canonical distribution. [It should be borne in mind that (2.5.29) is essentially an approximate requirement valid only for the time and energy intervals where the effect of V-T processes is small (Fig.2.54).] The details of the derivation of $P_q(v)$ through the maximal entropy formalism are provided in Appendix 2.D. The result is

$$P_q(v) = P_q(0) \exp[-E_v/kT-\gamma v] \quad , \tag{2.5.30}$$

where $P_q(0)$ is the population of v=0, T is the translational-rotational temperature, and γ is the Lagrange parameter conjugated to $\langle v \rangle$. [That $P_q(v)$ is indeed a stationary solution of the V-V equation (2.5.28) can be verified by substitution.] The values of T and γ are determined by the initial conditions (Appendix 2.D). For excited initial distributions where the initial value of $\langle v \rangle$ is higher than its final equilibrium value one obtains $\gamma < 0$ and $P_q(v)$ shows population inversion. This is more clearly illustrated by rewriting (2.5.30) as

$$P_q(v) = P_q(0) \exp[-vE_1/kT_q+(vE_1-E_v)/kT] \quad , \tag{2.5.31}$$

where

$$\frac{1}{T_q} = \frac{1}{T} + \frac{k\gamma}{E_1} \tag{2.5.32}$$

is an effective vibrational temperature characterizing the population ratio of the first two levels; $P_q(1)/P_q(0) = \exp(-E_1/kT_q)$. For the low lying levels, where the anharmonicity is still small, $vE_1 \simeq E_v$, and $P_q(v)$ is approximately Boltzmann with the effective temperature T_q. As v increases the positive term in the exponent, $vE_1 - E_v > 0$, becomes dominant and $P_q(v)$ grows rapidly with v. For $\gamma < 0$ this increase is steeper and begins at a lower v the higher is $|\gamma|$ or equivalently T_q/T. Clearly, the real P(v) does not really "blow up" since at the high v region P(v) is "pulled down" by the V-T processes (Fig.2.54). Yet, $P_q(v)$ is a reasonable approximation for the low v region for $t \ll \tau_{V-T}$. At longer times $t \sim \tau_{V-T}$ the relaxation of all the vibrational levels is governed by V-T transfer. Consequently, $<v>$ is no longer a conserved quantity; hence $\gamma \to 0$ and the system approaches the ordinary Boltzmann form.

The effects of anharmonicity are small when the mean vibrational energy per oscillator is low, so that only the low lying (and nearly harmonic) levels are significantly populated. This, quite frequent, situation is encountered for example in the next section where we consider the relaxation of polyatomic molecules following the absorption of a few infrared laser photons. The limiting case of V-V and V-T relaxation of harmonic oscillators is therefore interesting from a physical point of view and we shall conclude this section by reviewing its main characteristics. (For details see [2.167].) Again we assume $\tau_{V-V} \ll \tau_{V-T}$ (now more justifiable since V-V transfer among harmonic oscillators is exactly resonant). As distinguished from the case of anharmonic oscillators the average vibrational energy $<E_v> = \hbar\omega <v>$ is conserved throughout the fast V-V stage of the relaxation. Moreover, it is the only constraint (besides normalization) on the (quasi)equilibrium vibrational distribution $P_q(v)$ established at $\tau_{V-V} < t < \tau_{V-T}$. Hence, $P_q(v)$ is a canonical distribution

$$P_q(v) = P_q(0) \exp[-v\hbar\omega/kT_q] \quad , \tag{2.5.33}$$

where the vibrational temperature T_q is determined by the initial value of the mean vibrational energy

$$<E_v(0)> = \sum_v P_q(v)E_v \quad . \tag{2.5.34}$$

Of course, (2.5.33) could be derived from (2.5.31) as a special case corresponding to the harmonic oscillator level scheme $E_v = vE_1$. The slow V-T relaxation stage carries the oscillators to their final equilibrium value at temperature T. During this process the much faster V-V processes ensure that

$P(v)$ is always a canonical distribution whose temperature goes over monotonically from T_q to T.

2.5.5 Intermode V-V Transfer

Our next example of time scale separation deals with intermode energy transfer in systems involving two (or more) nearly resonant vibrational frequencies $\nu_i \simeq \nu_j$ ($\nu_i = \omega_i/2\pi$). These frequencies may characterize different modes (Sect. 3.3.5) of a polyatomic molecule, e.g., the $\nu_3 \cong 1050\ \mathrm{cm}^{-1}$ and $\nu_6 \cong 1190\ \mathrm{cm}^{-1}$ modes of CH_3F (see below). They may also refer to different molecules; for instance the N_2 vibration $\nu = 2360\ \mathrm{cm}^{-1}$ and the CO_2 antisymmetric stretch $\nu_3 = 2350\ \mathrm{cm}^{-1}$. The efficient V-V transfer between electrically vibrationally excited N_2 and ν_3 is a common pumping mechanism of CO_2 lasers, leading to population inversion in the bands $CO_2(v_1{=}0, v_2{=}0, v_3{=}1) \to CO_2(1,0,0) + h\nu \sim 940\ \mathrm{cm}^{-1}$ and $CO_2(0,0,1) \to CO_2(0,2,0) + h\nu \sim 1040\ \mathrm{cm}^{-1}$ (cf. Sects.3.3.5 and 3.4.4). Due to the large difference between ν_{N_2} and the symmetric stretch ν_1 and bending ν_2 modes of CO_2, these modes are not effectively populated by N_2-CO_2 collisions and thus remain "cold". In other words ν_1 and ν_2 are characterized by lower temperatures than ν_3. Another case where different modes are associated with different temperatures arises when nearly (but not exactly) resonant modes $\nu_i \simeq \nu_j$ are strongly coupled by V-V transfer. Strong coupling, i.e., efficient energy transfer, between systems normally leads to equalization of temperatures. It is thus interesting to find out what are the conditions, or in our terminology — the constraints, which are responsible for the localized energy distributions implied by the existence of specific mode temperatures.

To simplify the discussion let us first consider the case where ν_i and ν_j are vibrational frequencies of two chemical species, say A and B. Suppose further that either one or both species were vibrationally excited but the excitation is not too high so that only the low lying, nearly harmonic levels $E_{v_i} = h\nu_i v_i$ are significantly populated. Based on energy gap considerations (Sect.2.4.2 and Fig.2.37) the dominant V-V pathways in the system are

$$A(v_i) + A(v_i') \rightleftarrows A(v_i \pm 1) + A(v_i' \mp 1) \tag{2.5.35}$$

$$B(v_j) + B(v_j') \rightleftarrows B(v_j \pm 1) + B(v_j' \mp 1) \tag{2.5.36}$$

$$A(v_i) + B(v_j) \rightleftarrows A(v_i \pm 1) + B(v_j \mp 1) \quad , \tag{2.5.37}$$

where the restriction to $\Delta v = \pm 1$ transitions is only for the sake of simplicity. Let $\tau_{V\text{-}V}$ be the time required for these processes to reach equilibrium. The V-T relaxation time characterizing processes like

$$A(v_i) + M \rightleftarrows A(v_i \pm 1) + M \qquad (2.5.38)$$

is expected to be much longer, $\tau_{V-T} \gg \tau_{V-V}$. We now seek the quasiequilibrium distribution appropriate for times $\tau_{V-V} < t < \tau_{V-T}$. This can be derived, like the Treanor distribution, via the maximum entropy procedure (cf. Appendix 2.D). The crucial constraint here is the conservation of $\langle v \rangle$ in all the V-V processes (2.5.35-37) [2.168,169]. However, equipped with the results derived for the previous example, a few shortcuts can be made.

The resonant "intraspecies" processes (2.5.35) redistribute the vibrational energy among the A molecules so as to establish a canonical distribution

$$[A(v_i)]/[A(v_i-1)] = \exp(-h\nu_i/kT_i) \quad , \qquad (2.5.39)$$

where T_i, the vibrational temperature of A, is uniquely determined by $\langle E_{V_i} \rangle$. Similarly, (2.5.36) implies that $B(v_j)$ is characterized by a temperature T_j. In the absence of an energy communication channel between A and B the vibrational temperatures T_i and T_j and the translational-rotational temperature T are all different and determined by the (independent) initial values of $\langle E_{V_i} \rangle$, $\langle E_{V_j} \rangle$, and $\langle \varepsilon \rangle$. However, the existence of (2.5.37) permits energy transfer between A to B. If $\nu_i \neq \nu_j$ this process also couples the vibrational modes to the heat bath which must compensate for the energy mismatch $\Delta E = h\nu_i - h\nu_j$. A relationship between T_i, T_j and T is imposed by the detailed balance (equilibrium) condition for (2.5.37). That is

$$\frac{[A(v_i+1)][B(v_j-1)]}{[A(v_i)][B(v_j)]} = \exp\left(-\frac{h\nu_i - h\nu_j}{kT}\right) \quad , \qquad (2.5.40)$$

where $h\nu_i - h\nu_j = \Delta E$ is the energy mismatch corresponding to the transition $v_i + v_j \rightarrow (v_i+1) + (v_j-1)$ and T is the translational-rotational temperature. Combining (2.5.39) and the analogous expression for $B(v_j)$ with (2.5.40), we find [2.165,167-169]

$$\nu_i/T_i - \nu_j/T_j = (\nu_i - \nu_j)/T \quad . \qquad (2.5.41)$$

Alternatively we could derive this result by combining the expressions (2.5.32) for the vibrational temperatures of molecules A and B. If the dominant intermode coupling pathway is not a simple single quantum resonance

$\nu_i \simeq \nu_j$ but rather $n\nu_i \simeq m\nu_j$, where n and m are positive integers, then instead of (2.5.41) we get

$$m\nu_i/T_i - n\nu_j/T_j = (m\nu_i - n\nu_j)/T \quad . \tag{2.5.42}$$

The proof that this relation also holds for the case that ν_i and ν_j correspond to different vibrational modes of the same polyatomic molecule is straightforward. For nearly exact resonant modes, like the N_2 vibration and the CO_2 antisymmetric stretching vibration, $(\nu_i - \nu_j)/T \simeq 0$ and (2.5.42) implies $T_i \simeq T_j$. In such cases the intermode transfer does not involve energy exchange with the translational-rotational degrees of freedom and $T_i(T_j)$ is fixed by the initial energy content of the vibrational modes. Clearly $T_i \neq T$ since $T_i = T_j = T$ corresponds to complete equilibrium. This conclusion is valid also when $\nu_i \neq \nu_j$ since $T_i = T$ implies $T_j = T$ and vice versa. When $\nu_i \neq \nu_j$, the vibrational temperatures are also different. If more than two vibrational degrees of freedom are coupled by rapid V-V exchange each of them may have a different temperature. We now turn to describe an experiment demonstrating these features for the energy distribution among the vibrational modes of CH_3F [2.168].

The CH_3F molecule has nine normal modes of vibration of which three are doubly degenerate (see Fig.3.25). The corresponding frequencies are $\nu_1 \simeq \nu_4 \simeq$ 2980 cm^{-1}, $\nu_2 \simeq \nu_5 \simeq 1470$ cm^{-1}, $\nu_3 \simeq 1050$ cm^{-1}, and $\nu_6 \simeq 1190$ cm^{-1}; ν_4, ν_5, and ν_6 are the degenerate modes. The ν_3 (C-F stretching) mode was selectively excited [2.168] by exposing a sample cell containing a few Torr of CH_3F molecules to a CO_2 laser pulse [$\sim$3 μs long pulse of the P(32) line of $CO_2(0,0,1) \to CO_2(0,2,0)$]. The time between collisions is of the order of 10^{-7} s $Torr^{-1}$. Thus collisions start taking place during the laser pulse and modify the vibrational populations by inducing both inter- and intramolecular energy transfer. By monitoring the infrared fluorescence of different modes it was found that a steady vibrational distribution was established by V-V processes after about 75 gas kinetic collisions (corresponding to $\tau_{V-V} \sim 7 \cdot 10^{-6}$ s at 1 Torr). Complete equilibrium due to V-T,R transfer was reached much later, after about 15 000 collisions (corresponding to $\tau_{V-T} \sim 1.4 \cdot 10^{-3}$ s at 1 Torr). The quasiequilibrium distribution observed in the time interval $\tau_{V-V} < t < \tau_{V-T}$ indicates a nonstatistical energy partitioning among the vibrational modes and can be characterized by different mode temperature, as implied by (2.5.42) and described in Fig.2.55. The set of dominant intermode V-V pathways responsible for this behavior was identified as [2.168]

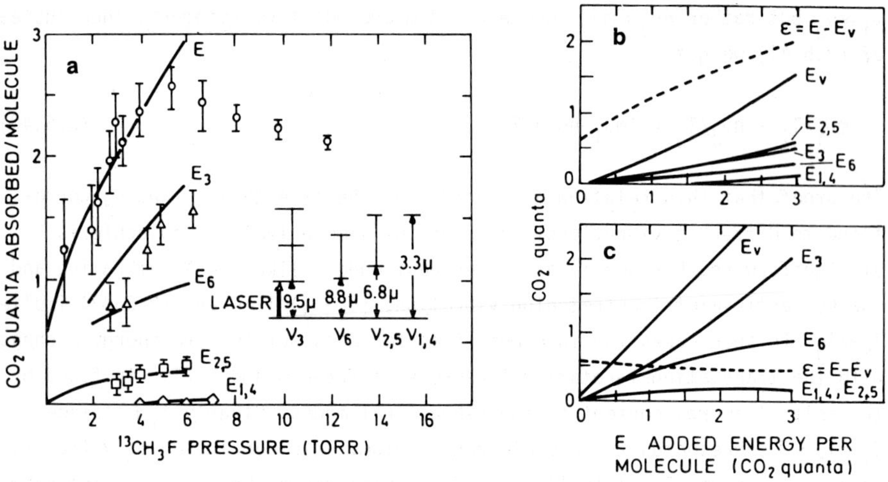

Fig. 2.55a-c. Nonthermal energy partitioning among the vibrational modes of CH_3F following CO_2 laser excitation of the ν_3 mode. (a) Experimental results showing the total energy absorbed E (in units of CO_2 laser quanta, ~3 kcal/mole) and the energies stored in the various modes, E_i, for different CH_3F pressures. The inset shows relevant energy levels. The smooth curves for E_i were computed for each E by applying (2.5.42) to the assumed V-V flow path (2.5.43-45). E_i and T_i are related via the harmonic oscillator expression (2.5.9). The panels on the right contrast the partition of the total energy E, at thermal equilibrium (case b; so-called "Bunzen burner heating"), and after complete V-V equilibrium (case c). Shown are the translational-rotational energy $\varepsilon = E - E_v$ and the energies in the different vibrational modes, E_i, vs E (adapted from [2.168])

$$CH_3F(\nu_3) + CH_3F \rightleftarrows CH_3F(\nu_6) + CH_3F \quad (\Delta E=140\ cm^{-1}) \tag{2.5.43}$$

$$CH_3F(\nu_6) + CH_3F \rightleftarrows CH_3F(\nu_{2,5}) + CH_3F \quad (\Delta E=280\ cm^{-1}) \tag{2.5.44}$$

$$CH_3F(2\nu_{2,5}) + CH_3F \rightleftarrows CH_3F(\nu_{1,4}) + CH_3F \quad (\Delta E=-40\ cm^{-1}) \quad . \tag{2.5.45}$$

The notation, in (2.5.43) for example, means that a ν_3 quantum is converted to a ν_6 quantum at the same molecule or of the collision partner. The much slower V-R,T processes involve large energy gaps, e.g.,

$$CH_3F(\nu_3) + CH_3F \rightarrow CH_3F(0) + CH_3F \quad (\Delta E=-1050\ cm^{-1}) \quad . \tag{2.5.46}$$

By adding (2.5.43) and (2.5.44) we get, in shorthand notation, $\nu_3 \leftrightarrow \nu_2(\nu_5)$. Hence what the three V-V pathways imply is a V-V coupling scheme which can be summarized as $\nu_1(\nu_4) \leftrightarrow 2\nu_2(\nu_5) \leftrightarrow 2\nu_3 \leftrightarrow 2\nu_6$. This yields three independent relations of the type (2.5.42) between the mode temperatures. The fourth relation necessary to determine $T_1(=T_4)$, $T_2(=T_5)$, T_3, and T_6 is provided by the energy conservation condition. The various temperatures determine the average vibrational energies of the corresponding modes, via $\langle E_i \rangle = h\nu_i/[\exp(h\nu_i/kT_i)-1]$. The values of $\langle E_i \rangle$ calculated in this fashion are shown together with the experimental results in Fig.2.55 for different gas pressures and thus different total absorbed laser energies. The figure clearly demonstrates that the V-V processes lead to a substantial localization of the energy in certain modes like the CF stretch (ν_3) and the CH_3 rock (ν_6). On the other hand, the CH stretching (ν_1,ν_4) and bending (ν_2,ν_5) modes are poorly populated. This situation is very different from the case that the same amount of energy is distributed among all degrees of freedom such that all of them have the same temperature $T_i = T$ as in ordinary "Bunsen burner" heating.

The major conclusion of the above analysis is that on short time scales the energy may be localized within certain modes of a polyatomic molecule. However, the conditions on the possible observation of this phenomenon are quite stringent. First the number of independent V-V pathways should not exceed n-1 where n is the number of different mode frequencies; otherwise there are more relations of the kind (2.5.42) than the number of temperatures T_i and the only solution is $T_i = T_j = T$. This situation is likely to occur when there are many similar frequencies $\nu_i \sim \nu_j$ or when the molecule is large or at high levels of excitation, that is, at high densities of states. In this case each mode is strongly coupled to the dense manifold (or "heat bath") comprised by the other modes, implying a common temperature for all degrees of freedom. It should also be remembered that (2.5.42) is based on the harmonic approximation which fails at high excitations. Moreover because of the narrowing energy gaps the V-T processes become comparable and even more efficient (see, e.g., Fig.2.53) than the V-V ones and the assumption $\tau_{V-V} \ll \tau_{V-T}$ is no longer justified.

2.5.6 From Macroscopic Relaxation to Microscopic Information

On the microscopic level the relaxation of a molecular system is characterized by the level populations P(v,t) (we use here v as a general state symbol). A macroscopic characterization of the relaxation is provided by the time dependence of its macroscopic observables

$$\langle A(t) \rangle = \sum_v A_v P(v,t) \quad , \tag{2.5.47}$$

where A_v is the value of the observable A when the system is in state v, e.g., $A_v = E_v$. Macroscopic observables of special interest are the "moments" of the distribution

$$\mu_k(t) = \langle v^k(t) \rangle = \sum_v v^k P(v,t) \quad . \tag{2.5.48}$$

μ_k is called the kth moment. For $k = 0$ we obtain the normalization condition

$$\mu_0(t) = \sum_v P(v,t) = 1 \quad . \tag{2.5.49}$$

In Sect.2.5.1 we have shown that for harmonic oscillators the first moment $\langle v(t) \rangle \propto \langle E_v(t) \rangle$ decays exponentially to its equilibrium value [cf. (2.5.12)]. We shall now show that this behavior will be observed whenever the detailed rate constants satisfy the sum rule [2.154]

$$\rho \sum_{v'} (E_{v'} - E_v) k(v \to v';T) = [\langle E_v(\infty) \rangle - E_v]/\tau \quad . \tag{2.5.50}$$

To see this we multiply the original master equation (2.5.2) by E_v and sum over all v's

$$\begin{aligned} -\frac{d\langle E_v(t) \rangle}{dt} &= -\sum_v E_v \frac{dP(v,t)}{dt} = \rho \sum_{v,v'} P(v,t) E_v k(v \to v';T) \\ &- \rho \sum_{v,v'} P(v',t) E_v k(v' \to v;T) = \rho \sum_v P(v,t) \sum_{v'} (E_v - E_{v'}) k(v \to v';T) \\ &= \sum_v P(v,t)[E_v - \langle E_v(\infty) \rangle]/\tau = [\langle E_v(t) \rangle - \langle E_v(\infty) \rangle]/\tau \quad . \end{aligned} \tag{2.5.51}$$

In obtaining the third equality from the second, we have interchanged v and v' in the second double sum. The fourth equality follows upon use of the sum rule (2.5.50). Thus (2.5.51) proves the sufficiency of the sum rule as a condition for exponential decay. It is not difficult to show that this is also a necessary condition [2.154].

The physical significance of the left-hand side of (2.5.50) is simply the rate of change of vibrational energy due to collisions involving molecules only in the vibrational state v. If $v = 0$, the molecule can only gain energy per collision. If $E_v > kT$, we expect that the vibrational energy decreases upon collisions. Hence, there must be an intermediate value of E_v where it will neither increase nor decrease on collisions. The sum rule (2.5.50) simply identifies this value as $\langle E_v(\infty) \rangle$. Molecules whose vibrational energy exceeds $\langle E_v(\infty)\rangle$ will, on the average (over the final vibrational states), lose energy upon collision and vice versa. In addition the sum rule states that the rate of vibrational energy transfer from (or to) molecules in state v is proportional to E_v.

The sum rule (2.5.50) provides a link between the microscopic (the detailed rate constants) and macroscopic (the mean vibrational energy) descriptions of the system. Given, for example, that $\langle E_v(t)\rangle$ decays exponentially to equilibrium, one can conclude that the sum rule (2.5.50) is valid and hence use it to characterize the detailed rate constants [2.121]. As a special case, say that the vibrational level scheme is harmonic and that only near neighbor transitions ($v'=v\pm1$) are allowed. It then follows from the sum rule that the rate constants are necessarily of the Landau-Teller form $k(v\to v-1;T) = vk(1\to 0;T)$ [2.154]. There is however no reason to limit the application of the sum rule in this fashion. Subject to the assumption that the exponential gap representation (2.4.17) is valid one can use the sum rule (2.5.50) to determine the magnitude of the adiabaticity parameter λ_v.

A simple practical procedure is to regard the initial vibrational energy as a continuous (classical) variable. The sum rule is then of the form

$$\rho\sum_{v'}(E_{v'}-E_v)k(E_v\to E_{v'};T) = [\langle E_v(\infty)\rangle - E_v]/\tau \quad . \tag{2.5.52}$$

Now, if we take the initial vibrational energy to be equal to the macroscopic equilibrium average vibrational energy $E_v = \langle E_v(\infty)\rangle$, the right-hand side of (2.5.52) vanishes. We are thereby left with an implicit equation for λ_v,

$$\sum_{v'}[E_{v'}-\langle E_v(\infty)\rangle]k[\langle E_v(\infty)\rangle\to E_{v'};T] = 0 \quad . \tag{2.5.53}$$

At any given temperature, one can compute $\langle E_v(\infty)\rangle$ and then plot the left-hand side of (2.5.53) as a function of the variable λ_v. The required value of λ_v is the one for which the lhs of (2.5.53) vanishes. This procedure, known as "surprisal synthesis" [2.121], was used to determine the rate constants shown in Figs.2.36 and 2.53. It was similarly used [2.121] to determine the general dependence of the adiabaticity parameter on $u = hc\omega_e/kT$. This yields $\lambda_v \simeq kT/hc\omega_e$ for $u \gg 1$ and $\lambda_v \simeq$ const for $u \ll 1$ (cf. Fig.2.40). These two different ranges correspond to the energy gaps for vibrational and rotational excitation, respectively. We have previously noted that on an energy basis, vibrational and rotational relaxations are similar, both being governed by an exponential gap law rate constants. The remaining differences are due to the different values of the adiabaticity parameter in the large and small u regimes [2.157].

We now turn to examine what can be deduced about $P(v,t)$ from the time evolution of the macroscopic observables $\langle A(t)\rangle$. It was argued above that given enough moments $\mu_k(t)$ or other macroscopic observables one can exactly evaluate $P(v,t)$. We also know that at equilibrium a single constraint $\langle E_v\rangle$ suffices to specify $P(v,\infty) = p(v|T)$ uniquely. This implies that at equilibrium all the other observables correspond to redundant or "noninformative" constraints. For example,

$$\langle E_v^2\rangle = \sum_v E_v^2 p(v|T) \tag{2.5.54}$$

is fully determined once $\langle E_v\rangle$ (or, equivalently, T) is specified. One may thus expect that the number of informative constraints reduces (to one for relaxation in a heat bath) as $t\to\infty$. At equilibrium

$$\langle E_v\rangle = \sum_v E_v p(v|T) = \sum_v E_v \exp(-\beta E_v - \chi_0) \quad , \tag{2.5.55}$$

where $(k\beta)^{-1} = T$ is the heat bath temperature and $\exp(\chi_0) = Q(\beta)$ is the equilibrium partition function. When the distribution is displaced from equilibrium the magnitude of $\langle E_v\rangle = \langle E_v(t)\rangle = \sum E_v P(v,t)$ will no longer be given by (2.5.55). We can however use the actual magnitude of $\langle E_v(t)\rangle$ to introduce a time-dependent β via (2.5.55). This, in turn, defines a time-dependent distribution

$$P(v,t) = \exp[-\beta(t)E_v - \chi_0(t)] \tag{2.5.56}$$

and a time-dependent partition function

$$Q[\beta(t)] = \exp[\chi_0(t)] = \sum_v \exp[-\beta(t)E_v] \quad . \tag{2.5.57}$$

This procedure generates a distribution which is normalized, consistent with the actual, instantaneous, magnitude of the average energy and maximizes the entropy $S = -k\sum P(v,t)\ln[P(v,t)]$ subject to these conditions. It is the distribution appropriate to a system where the nonequilibrium state is fully characterized by one informative constraint $\langle E_v(t)\rangle$. When the functional form (2.5.56) is valid throughout the relaxation, the system is "canonically invariant" [cf. (2.5.18)]. Typically, however, the time-dependent canonical form is obtained only near equilibrium where the details of the initial distribution have been washed out. This behavior is illustrated for the vibrational relaxation of Cl_2 and HF in Fig.2.56.

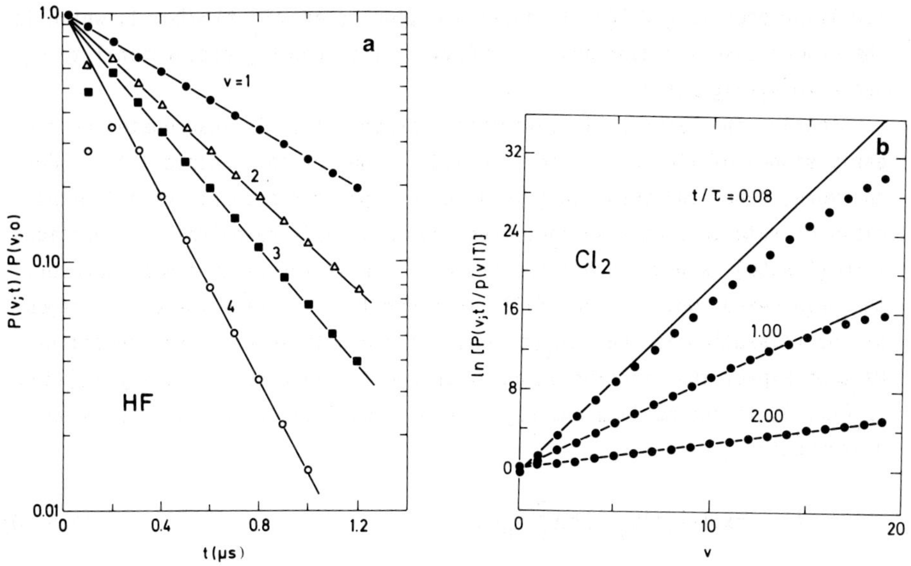

Fig. 2.56a,b. Time evolution of vibrational populations. The dots are detailed solutions of the master equation. HF: computer simulation of experimental results from [2.170]; Cl_2: from [2.171]. τ is the energy relaxation time [cf. (2.5.51)]. The solid lines are the predictions of the canonically invariant form (2.5.56) which corresponds to the single constraint $<E_v(t)>$. Both figures demonstrate the validity of the canonical representations for $t/\tau >> 1$, and even earlier times for the low lying v's. Both the surprisal plot for $[Cl_2(v)]$ and the explicit time evolution of [HF(v)] demonstrate the faster relaxation of higher vibrational levels

Assuming that $<E_v(t)>$ is indeed the only informative constraint on the system, we can derive an equation of motion for P(v,t) which is much simpler than the master equation (2.5.2). First by taking the time derivative of the partition function (2.5.57), we obtain

$$\partial\chi_0/\partial t = \partial \ln(Q)/\partial t = -<E_v(t)>(\partial\beta/\partial t) \quad . \tag{2.5.58}$$

Taking the time derivative of (2.5.56) and using (2.5.58), we find

$$\frac{1}{P(v,t)}\,\frac{\partial P(v,t)}{\partial t} = \frac{\partial \ln[P(v,t)]}{\partial t} = \left(\frac{\partial\beta}{\partial t}\right)[<E_v(t)>-E_v] \quad . \tag{2.5.59}$$

This equation implies the existence of two different classes of states, those with low energy, $E_v < <E_v(t)>$, whose population increases in time (assuming $\partial\beta/\partial t > 0$) and the high lying states, $E_v < <E_v(t)>$, whose population decreases with time. The relaxation rate of level v increases with the deviation of E_v from the mean vibrational energy. Higher vibrational states relax faster than low lying ones (Fig.2.56). In particular, whenever $\langle E_v(t)\rangle/\hbar\omega < 1$, which is the common case for hydrides, $-\partial \ln P(v,t)/\partial t$ is roughly proportional to E_v, hence to v (Fig.2.56).

Clearly, the canonically invariant representation is inadequate for the early stages of the relaxation if P(v,0) is not of the canonical form. Furthermore, canonical invariance is a rather specific property of a limited class of rate constants so that even when P(v,0) is canonical the functional form (2.5.56) is usually not preserved, at least for early times. However, the main reason why (2.5.56) fails in general to describe the early stages of the relaxation is that $\langle E_v(t)\rangle$ is not the only informative constraint. We thus expect that incorporation of an additional constraint, say $\langle E_v^2(t)\rangle = \sum E_v^2 P(v,t)$, in the maximum entropy procedure will yield an extended canonical distribution

$$P(v,t) = \exp[-\beta(t)E_v - \delta(t)E_v^2 - \chi_0(t)] \tag{2.5.60}$$

which provides a more adequate representation of P(v,t). The Lagrange parameters $\beta(t)$, $\delta(t)$, and $\chi_0(t)$ are determined by the three implicit equations $\langle E_v(t)\rangle = \sum E_v P(v,t)$, $\langle E_v^2(t)\rangle = \sum E_v^2 P(v,t)$, and $\sum P(v,t) = 1$, respectively. The equation of motion corresponding to (2.5.60) is [2.172]

$$\frac{\partial P(v,t)}{\partial t} = \left[\left(\frac{\partial\beta}{\partial t}\right)(\langle E_v(t)\rangle - E_v) + \left(\frac{\partial\delta}{\partial t}\right)\left(\langle E_v^2(t)\rangle - E_v^2\right)\right]P(v,t) \quad . \tag{2.5.61}$$

This equation describes the so-called "Fokker-Planck" regime where the distribution function is determined by the magnitudes of the first and second moments of P(v,t). Equation (2.5.61) can be rearranged to the more familiar form of the Fokker-Planck equation. (Fokker-Planck equation is a partial differential equation relating $\partial P(v,t)/\partial t$ to $\partial P(v,t)/\partial v$ and $\partial^2 P(v,t)/\partial v^2$ [2.154].) In general, two or three informative constraints provide a proper description of P(v,t) for most stages of the relaxation (see e.g. Fig.2.56). For some peculiar, e.g., multipeaked, initial distributions additional informative observables might be necessary to describe P(v,t) for $t/\tau < 1$.

2.5.7 Thermodynamics of Molecular Disequilibrium

The prior distribution (Sect.2.1.6 and Appendix 2.A) for molecules in the bulk is the equilibrium distribution which henceforth will be denoted as $P^0(v)$. In analogy to the case of microscopic equilibrium, the deviation of the actual, momentary, distribution P(v,t) from its prior value can be characterized by a surprisal

$$I(v,t) = -\ln[P(v,t)/P^{o}(v)] \quad . \tag{2.5.62}$$

Similarly the average deviation is measured by an entropy deficiency

$$DS[P(v,t)|P^{o}(v)] = \sum_{v} P(v,t)\ln[P(v,t)/P^{o}(v)] = -\langle I(v,t)\rangle \geqq 0 \quad , \tag{2.5.63}$$

where the equality holds only at equilibrium, i.e., when $P(v,t) \equiv P(v,\infty) = P^{o}(v)$ [cf. (2.1.58)].

The major objective of this section is to show that the entropy deficiency $DS[P|P^{o}]$ provides a general molecular definition for the thermodynamic free energy [2.173]. Namely, that it decays monotonically to its equilibrium value in spontaneous relaxation processes and provides an upper bound to the thermodynamic work in controlled processes. To this end let us consider a system of N molecules in volume V and let P(v,t), or in shorthand notation P(v), denote the probability of finding a molecule at time t in state v. As $t\to\infty$, P(v,t) approaches the Boltzmann form $P^{o}(v) = p(v|T) \propto \exp(-E_v/kT)$. The final temperature T depends on the external conditions of the system. Explicitly, we shall consider two cases of common physical interest:

I) Systems coupled to a heat bath, hereafter "isothermal". Here the final temperature is always the bath temperature, regardless of the process carrying the system to equilibrium.

II) Thermally isolated, "adiabatic", systems for which no heat exchange with the surroundings can take place, $Q \equiv 0$. The final temperature T is determined by the final energy content of the system $E^{o} = N\langle E_v\rangle = N\sum p(v|T)E_v$ which depends on the route to equilibrium. In a simple spontaneous relaxation process $E^{o} = E = \text{constant}$. (We use E to denote the initial energy.) On the other hand if the system performs thermodynamic work W, then $E^{o} = E+\Delta E = E - W$.

The entropy deficiency associated with the passage (of a mole) from an initial state P(v) to a final equilibrium state $P^{o}(v) = p(v|T) = \exp(-E_v/kT)/Q$ is (in thermodynamic units, i.e. (2.5.62,3) multiplied by the gas constant R),

$$\begin{aligned} DS[P|P^{o}] &= R\sum_{v} P(v)\ln[P(v)/P^{o}(v)] \\ &= R\sum_{v} P(v)\ln[P(v)] - R\sum_{v} P^{o}(v)\ln[P^{o}(v)] \\ &\quad + R\sum_{v} P^{o}(v)\ln[P^{o}(v)] - R\sum_{v} P(v)\ln[P^{o}(v)] \\ &= -S + S^{o} - R[E^{o}/RT+\ln(Q)] + R[E/RT+\ln(Q)] \\ &= (S^{o}-S) - (E^{o}-E)/T = \Delta S - \Delta E/T \quad , \end{aligned} \tag{2.5.64}$$

where S and E are the initial entropy and energy per mole, and S^o and E^o are the corresponding final values.

For isothermal systems where T is the constant bath temperature (2.5.64) yields

$$DS[P|P^o] = -\Delta A/T \geqq 0 \quad , \tag{2.5.65}$$

where $\Delta A = \Delta E - T\Delta S = A^o - A$ is the molar change in Helmhotz's free energy

$$A = E - TS = \sum P(v)E_v + RT\sum P(v)\ln P(v) \quad . \tag{2.5.66}$$

This expression provides a generalized molecular definition of Helmhotz's function for nonequilibrium molecular systems [2.167,173]. At equilibrium, $P(v) = P^o(v) = \exp(-E_v/kT)/Q$, we regain the familiar statistical expression $A^o = -kT\ln(Q)$.

It should be noticed that T in (2.5.65) is always the constant bath temperature. The appropriate free energy for an isolated system (ΔE=0) is $T\Delta S$ where T is the final equilibrium temperature,

$$DS[P|P^o] = \Delta S \geqq 0 \quad (\Delta E=0) \quad . \tag{2.5.67}$$

Using W and Q to denote the work done and the heat released by the system when its thermodynamic state changes from P(v) to $P^o(v)$ the first law of thermodynamics reads

$$-\Delta E = Q + W \quad . \tag{2.5.68}$$

For isothermal systems $Q = T\Delta S_b$ where ΔS_b is the change in the bath entropy. For adiabatic systems Q = 0. Combining these results with (2.5.64) yields:

For isothermal systems

$$TDS[P|P^o] = -\Delta A = T(\Delta S+\Delta S_b) + W = T\Delta S_T + W \geqq 0 \quad , \tag{2.5.69}$$

where ΔS_T is the change in the total entropy $S_T = S + S_b$.

For adiabatic systems

$$TDS[P|P^o] = T\Delta S + W \geqq 0 \quad . \tag{2.5.70}$$

Since any system plus its surrounding heat bath can be regarded as a thermally isolated system (2.5.69) and (2.5.70) are essentially equivalent.

The maximal work is obtained in a reversible process; $\Delta S_T = 0$ in (2.5.69) or $\Delta S = 0$ in (2.5.70). In both cases we get

$$TDS[P|P^0] = W_{max} \tag{2.5.71}$$

as one should expect from a thermodynamic potential. In the other extreme, i.e., in a completely dissipative process where $W = 0$, the entropy deficiency is the total (system + surroundings) entropy increase

$$DS[P|P^0] = \Delta S_{T,max} \tag{2.5.72}$$

in accordance with (2.5.70) which holds for any isolated system.

To complete the description of $DS[P|P^0]$ as a free energy function we shall now show that its decrease to its equilibrium value in a spontaneous relaxation process is monotonic. That is [2.174],

$$dDS[P|P^0]/dt \leqq 0 \quad , \tag{2.5.73}$$

where the equality holds only at equilibrium. We shall explicitly prove (2.5.73) for a linear-isothermal relaxation process of the type described by (2.5.2). However by using the proper P^0 and the proper detailed balance relationship the proof can be extended to other types of relaxation processes [2.164].

Using the detailed balance relation, $P^0(v)k(v \to v') = P^0(v')k(v' \to v)$, the master equation (2.5.2) can be written as

$$\frac{dP(v)}{dt} = \rho \sum_{v'} k(v' \to v)P^0(v')\left[\frac{P(v')}{P^0(v')} - \frac{P(v)}{P^0(v)}\right] \quad . \tag{2.5.74}$$

Taking the time derivative of $DS[P|P^0]$, noting that $\sum dP(v) = d\sum P(v) = 0$, and using (2.5.74), we obtain

$$\begin{aligned}\frac{dDS[P|P^0]}{dt} &= R\sum_{v} \frac{dP(v)}{dt} \ln\left[\frac{P(v)}{P^0(v)}\right] \\ &= R\rho \sum_{v,v'} k(v' \to v)P^0(v')\left[\frac{P(v')}{P^0(v')} - \frac{P(v)}{P^0(v)}\right]\ln\left[\frac{P(v)}{P^0(v)}\right] \quad .\end{aligned} \tag{2.5.75}$$

The sum is invariant to interchanging v and v'. Writing it again with v and v' replacing each other, adding the result to (2.5.75), and using again the detailed balance relation, we find

$$\frac{dDS[P|P^0]}{dt} = -\frac{R}{2}\sum_{v,v'} P^0(v')k(v'\to v)\left[\frac{P(v')}{P^0(v')} - \frac{P(v)}{P^0(v)}\right]$$

$$\times\left\{\ln\left[\frac{P(v')}{P^0(v')}\right] - \ln\left[\frac{P(v)}{P^0(v)}\right]\right\} \leqq 0 \quad . \tag{2.5.76}$$

The inequality which completes the proof of (2.5.73) follows from the fact that $\ln(x)$ is a monotonically increasing function of x; hence $(x-y)[\ln(x)-\ln(y)] \geqq 0$. For isothermal systems (2.5.76) implies $-dA/dt \geqq 0$. For adiabatic ones $dS/dt \geqq 0$.

2.5.8 Laser Thermodynamics

Laser radiation, being extremely monochromatic, unidirectional, and phase coherent represents a highly ordered system. Quantum mechanically, the laser photons are all in the same quantum state [in the limit of single mode operation (cf. Sect.3.2.3)]. The entropy, $S = -k\sum P(n)\ln[P(n)]$, of a system that all its constituent particles are in a single quantum state m, $P(n) = \delta_{n,m}$ is zero. In thermodynamic terms, the highest efficiency of converting internal energy E to useful work W is obtained when the accompanying entropy increase is minimal. It follows that laser radiation is a form of thermodynamic work [2.175]. This is to be contrasted with spontaneous emission, which in the extreme case of complete spatial, energetic, and phase randomness is a form of heat. Equipped with these basic notions, one can anticipate the detailed discussions in Chaps.3 and 4 and employ the thermodynamic approach developed in this chapter to gain some important qualitative insights into the mechanisms of laser operation.

We have shown in Chap.1 that population inversion $N_f/g_f > N_i/g_i$ is a necessary condition for laser action on the transition $f\to i$; N_f and N_i are the populations of the g_f- and g_i-fold degenerate energy levels E_f and E_i, respectively; $E_f > E_i$. As the simplest application of the thermodynamic approach we shall first show that the requirement for population inversion can be derived from basic thermodynamic considerations [2.175]. To simplify the analysis consider a two (nondegenerate) level system so that $N = N_i + N_f = \text{const}$ or equivalently $P(i) + P(f) = 1$, where $P(i) = N_i/N$ and $P(f) = N_f/N$. Also, we shall set $E_i = 0$ so that $E_f = \varepsilon = h\nu$ is the energy of the photon emitted in the transition $f\to i$.

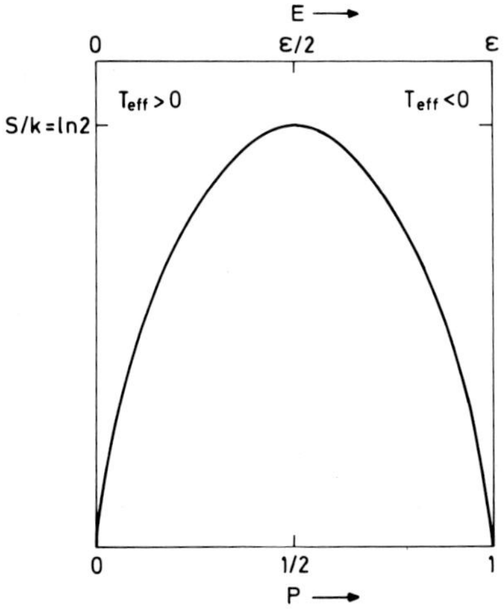

Fig. 2.57. The entropy S vs the upper level population P or the energy E in a two-level system. The region of population inversion $P > \frac{1}{2}$ is characterized by a negative temperature. The entropy shown here is in thermodynamic units. It is obtained from the dimensionless form [cf.(2.1.43) and Fig.2.6] upon multiplying by the Boltzmann's constant k

The average energy E and entropy S per molecule in the two-level system are

$$E = P\varepsilon = Ph\nu \tag{2.5.77}$$

$$S = -k[P \ln(P)+(1-P)\ln(1-P)] \quad , \tag{2.5.78}$$

where $P = P(f)$ and $1-P = P(i)$. These equations imply a simple relationship between E and S as shown in Fig.2.57 (see also Fig.2.6). The entropy is minimal, $S = 0$, when all molecules are in one state (P=0 or P=1) and reaches a maximum $S = k \ln(2)$ when the two levels are equally probable, $P = \frac{1}{2}$. Population inversion $P > \frac{1}{2}$ is often described as a state characterized by a negative effective temperature. This follows from the extension of the ordinary thermodynamic relation $\partial S/\partial E = 1/T$ to systems with a finite number of energy levels. Applying this definition to the two-level system described by (2.5.77,78) and noting $\partial S/\partial E = (1/\varepsilon)\partial S/\partial P$, we easily get the Boltzmann-like relationship

$$P/(1-P) = \exp[-\varepsilon/kT_{eff}] \quad . \tag{2.5.79}$$

Hence $P > \frac{1}{2}$ implies $T_{eff} < 0$. Although the concept of negative temperatures is often instructive it is important to remember that realistic physical systems constituted by atoms or molecules are always associated with an infinite spectrum of levels. For such systems the equilibrium temperature defined by $\partial S/\partial E = 1/T$ is always positive. It is only for a part of the system (e.g., two or more levels) that S may decrease when E increases (Fig.2.57) (a similar S vs E dependence characterizes spin $\pm\frac{1}{2}$ systems [2.54]).

Consider now the changes in the thermodynamic functions when a small amount of molecules, $dN_f = -dN_i = NdP < 0$, is transferred from level f to level i in a radiative process. The first law of thermodynamics requires $dE = NdP = -h\nu dP = -dE_{rad}$, where E_{rad} is the energy of the radiation field. The second law of thermodynamics implies $dS + dS_{rad} \geqq 0$, where dS is the entropy change of the molecular system (2.5.78), while dS_{rad} is the entropy of the emitted photons. Laser photons carry no entropy; hence $dS_{rad} = 0$ and, using (2.5.78), we find

$$dS = -k \ln[P/(1-P)]dP = dE/T_{eff} \geqq 0 \quad . \tag{2.5.80}$$

This necessary lasing condition which follows from basic thermodynamic considerations is nothing else but the familiar requirement for population inversion since for $dP < 0$ (emission) (2.5.80) implies $P > \frac{1}{2}$ (or equivalently $T_{eff} < 0$). Similarly (2.5.80) leads to the conclusion that absorption $dP > 0$ is only possible when $P < \frac{1}{2}$. For noninverted populations $P < \frac{1}{2}$ we find that emission $dP < 0$ is associated with $dS < 0$. To satisfy the second law of thermodynamics $dS + dS_{rad} > 0$ the photons emitted must carry the excess entropy, i.e., $dS_{rad} \geqq -dS \geqq 0$. This is why spontaneous emission (fluorescence) as opposed to laser radiation is entropy rich.

Our derivation of the necessary lasing condition $N_f > N_i$ considered only the pair of states connected by the radiative transition. However, no real laser consists of two isolated levels. Moreover, one can show, either from kinetic or thermodynamic [2.175] considerations, that an initially noninverted two-level system cannot be pumped to a state of population inversion by any thermal source. In all laser systems at least three levels are involved in the pumping process (cf. Sect.3.2). In chemical lasers, especially those based on vibrotational transitions, the number of strongly coupled levels is large and the kinetic scheme is often very complex. In Sects.1.1 and 1.2 we have mentioned the major kinetic processes in these lasers and a detailed kinetic analysis will follow in Chap.4. The kinetic descriptions can often be complemented by thermodynamic ones. In the rest of this section we shall present one example of the thermodynamic approach: the derivation and interpretation of the lasing conditions in rotationally equilibrated chemical lasers [2.175,176].

The allowed vibrotational transitions of diatomic molecules are of the P branch type, $v,J \rightarrow v-1,J+1$, or R branch type, $v,J \rightarrow v-1,J-1$. The corresponding energies of the emitted photons [in the RRHO scheme; cf. (1.2.8,9)] are $h\nu_P = \hbar\omega - 2B(J+1)$ and $h\nu_R = \hbar\omega + 2BJ$. At high buffer gas pressures one can apply

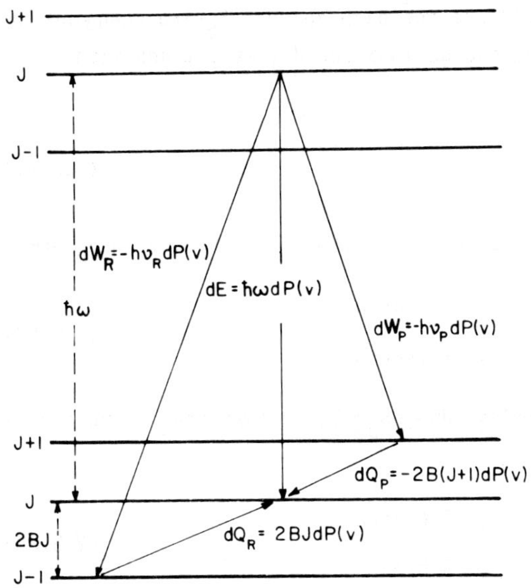

Fig. 2.58. Energy balance in P and R branch transitions in vibrotational chemical lasers at rotational equilibrium. The net energy change dE associated with the passage of molecules from v to v-1 is the sum of the thermodynamic work dW which appears as laser radiation and the heat transfer to the bath dQ necessary to maintain rotational equilibrium. Note that $dP(v) < 0$. Since the laser radiation carries no entropy the lasing condition follows from the requirement (the second law) that the sum of entropy changes corresponding to dE and dQ (dS and dS_b) must be positive (see text)

the assumption of rotational equilibrium according to which $N_{v,J} = N_v p(J|v) = NP(v)p(J|T)$ where P(v) is the vibrational distribution and $p(J|T)$ is the equilibrium rotational distribution [cf. (1.2.13) and Sect.2.5.3 and note that in the RRHO picture $p(J|v,T) = p(J|T)$]. Since the Boltzmann shapes of the rotational distributions are preserved during the lasing only P(v) is subject to variations. Suppose now that a small fraction of molecules, $dN_v = NdP(v) = -NdP(v-1) < 0$, is transferred from the upper vibrational manifold v to the lower manifold v-1, due to lasing on either the P line $v,J \rightarrow v-1,J+1$ or the R line $v,J \rightarrow v-1,J-1$. The net change in the energy of the lasing molecules is, for both cases, $dE = \hbar\omega dP(v) = -\hbar\omega dP(v-1)$, per molecule (Fig.2.58). However, the energy (thermodynamic work, dW) carried by the emitted photons is not dE. If lasing took place on the P branch transition, $dW_P = -h\nu_P dP(v) = -[\hbar\omega - 2B(J+1)] \times dP(v)$. Similarly, $dW_R = -h\nu_R dP(v) = -(\hbar\omega + 2BJ)dP(v)$. The differences $dQ_P = -dE - dW_P = -2B(J+1)dP(v) > 0$ and $dQ_R = -dE - dW_R = 2B(J+1)dP(v) < 0$ correspond to the heat exchanges with the heat bath which are necessary to maintain the rotational equilibrium. Since the laser radiation does not carry entropy only the vibrational entropy $S = -k\sum P(v)\ln[P(v)]$ and the bath entropy S_b are changing in the processes described above. The change in S for a $v \to v-1$ transition, $[dP(v)=-dP(v-1)]$, is

$$dS = -k \ln[P(v)/P(v-1)]dP(v) \tag{2.5.81}$$

and is the same for both P and R branch transitions. The changes in S_b are

$$dS_b = dQ/T = \begin{cases} -[2B(J+1)/T]dP(v) & , \text{ P branch} \\ (2BJ/T)dP(v) & , \text{ R branch} \end{cases} . \tag{2.5.82}$$

From the second law of thermodynamics, $dS + dS_b \geqq 0$, follows the lasing condition

$$(kT/2B)\ln[P(v-1)/P(v)] \begin{cases} \leqq J+1 & , \text{ P branch} \\ \geqq -J & , \text{ R branch} \end{cases} . \tag{2.5.83}$$

This shows that lasing on R branch lines can only occur under conditions of "complete inversion" $P(v) > P(v-1)$, it is restricted to low J's, and is favored at high temperatures. Much less restrictive is the lasing condition for P branch transitions. Here lasing can take place even under partial inversion, $P(v) < P(v-1)$, provided that J is sufficiently high. Also, low temperatures are favorable. Equation (2.5.83) provides a qualitative explanation to two important experimental observations (Sects.4.2 and 4.4). First, that rotationally equilibrated lasers are efficient due to the possibility of extracting internal energy even from partially inverted populations. Second, that lasing in these systems occurs preferably on P branch lines. The thermodynamic origin of these phenomena is rather simple: Lasing at partial inversion, $P(v)/P(v-1) < 1$, reduces the entropy of the lasing molecules [cf. (2.5.81)]. In the case of P branch emission this is compensated for by positive entropy increase of the bath. This compensation is not possible when the emission occurs on the R branch as the bath must supply (rather than absorb) some of the emitted photon energy, $(h\nu_R > \hbar\omega)$ (Fig.2.58).

Finally, it should be noted that the lasing conditions (2.5.83) could more easily be derived by direct substitution of $N_{v,J} = NP(v)p(J|T)$ into $N_{v,J}/g_J > N_{v-1,J\pm1}/g_{J\pm1}$. Clearly, however, the purpose of our relatively lengthy derivation was to illustrate the thermodynamic aspects of this result.

Appendix 2.A. The Prior Distribution

The concept of the prior distribution is based on the grouping property of the entropy, which can be derived as follows. Consider the distribution P(n) over quantum states. Let these states be grouped so that each state is a member of one (and only one) group. The state n is thus uniquely specified by n = v,m, where v denotes the group and m is the label of n within v. The probability of the state n can therefore be resolved as

$$P(n) = P(v,m) = P(v)P(m|v) \quad , \tag{2.A.1}$$

where P(v) is the fraction of states in the group v and P(m|v) is the fraction of the states in the group v which have the label n,

$$\sum_m{}' P(m|v) = 1 \quad , \tag{2.A.2}$$

where the prime on the summation in (2.A.2) indicates a restriction to those states which belong to the group v. Since every state belongs to only one group, one can rewrite the entropy [cf. (2.1.43)] as follows:

$$\begin{aligned} S &= -\sum_n P(n)\ln[P(n)] = -\sum_{vm}{}' P(v)P(m|v)\ln[P(m|v)P(v)] \\ &= -\sum_v P(v)\ln[P(v)] + \sum_v P(v)\left\{-\sum_m{}' P(m|v)\ln[P(m|v)]\right\} \\ &= -\sum_v P(v)\ln[P(v)/g(v)] \quad . \end{aligned} \tag{2.A.3}$$

Here ln[g(v)],

$$\ln[g(v)] = -\sum_m{}' P(m|v)\ln[P(m|v)] \tag{2.A.4}$$

is the entropy of the distribution of quantum states within the group v. If all the states within the group are equiprobable, i.e., $P(m|v) = 1/N_v$ where N_v is the degeneracy (number of states) of v, then the entropy $\ln[g(v)]=\ln(N_v)$ so that $g(v) = N_v$ is the degeneracy in the usual sense. If P(m|v) is not uniform then g(v) is the effective number of states. We shall find it worthwhile to use both these interpretations depending on the particular physical situation. For isolated binary collision processes at a given total energy, where in the absence of constraints all quantum states are equiprobable, we shall take P(m|v) to be uniform so that g(v) is the actual number of quantum states.

(The details of the counting are discussed below.) When we come to discuss macroscopic systems where only the mean energy is known, then quantum states are distributed according to the canonical distribution, $P(n) \propto \exp(-E_n/RT)$; the prior distribution $P^o(v) = g(v)/\sum g(v)$ is, itself, a canonical distribution.

The prior distribution for the final states of a binary collision (at the total energy E to E+dE) clearly depends on the particular grouping of states which has been carried out. We start with a very detailed situation: I) Products in a definite internal quantum state. Given the total energy E and the internal state m, the relative translational energy $E_T = E - E_m$ is known. When the (vector) momentum for the relative motion is in the range $\underline{p}$ to $\underline{p}+d\underline{p}$, the number of quantum states (per unit volume) is $d\underline{p}/h^3$, where h is Planck's constant. Since $d\underline{p} = p^2 dp d\Omega$, where $d\Omega$ is the solid angle element spanned by $d\underline{p}$, and for $E_T = p^2/2\mu$ $dE_T = (p/\mu)dp$ we obtain the number of states (per unit volume) when the translational energy is in the range E_T, $E_T + dE_T$ as[9]

$$4\pi p^2 dp/h^3 = 4\pi\mu p dE_T/h^3 = 4\pi\mu(2\mu E_T)^{\frac{1}{2}} dE_T/h^3 \quad . \tag{2.A.5}$$

By introducing $A_T = \mu^{3/2}/2^{1/2}\pi^2 h^3$, we can thus write the density of translational states as

$$\rho_T(E_T) = A_T E_T^{\frac{1}{2}} \quad . \tag{2.A.6}$$

From this result the density of states for II) Products in any group v and total energy in the range E, E+dE is

$$\rho(v;E) = \sum_m{}' \rho_T(E-E_m) = A_T \sum_m{}' (E-E_m)^{\frac{1}{2}} \quad . \tag{2.A.7}$$

In (2.A.7), E_m is the energy of the internal state m, $E = E_m + E_T$, and the summation is restricted to those states m in the group v.

The prior distribution for products in the group v, $P^o(v;E)$ or in shorthand notation $P^o(v)$, is proportional to their statistical (microcanonical) weight, namely

$$P^o(v) = \rho(v;E)/\sum_v \rho(v;E) = \rho(v;E)/\rho(E) \quad , \tag{2.A.8}$$

---

9 The expression (2.A.5,6) for $\rho_T(E_T)$ can also be derived by counting the states of a particle μ in a three-dimensional box [2.54]. This yields $N(E_T) = (2/3)A_T E_T^{3/2}$ for the number of states (per unit volume) with energy less than or equal to E_T. Hence $\rho_T(E_T) = dN/dE_T = A_T E_T^{\frac{1}{2}}$.

where $\rho(E)$ is the total density of states, i.e., $\rho(E)dE$ is the total number of products states within E, E+dE.

Once the group of states v is specified the evaluation of $P^o(v)$ reduces to a state counting problem. Let us illustrate this procedure through several examples. A simple case which was extensively discussed in Sect.2.2 is that of an exoergic atom-diatom exchange reaction, $A+BC \rightarrow AB+C$. The total energy available to the products is $E = E_{Re} - \Delta E_0 = E_T + E_m$, where E_{Re} is the sum of translational and internal energies of the reactants and ΔE_0 is the exoergicity ($\Delta E_0<0$ for exoergic reactions). The internal product state m is fully specified by $m = \underline{q}_{AB}, \underline{q}_C, v, J, m_J$, ($m_J=-J,-J+1,\ldots,0,J-1,J$), which denote, respectively, the electronic quantum numbers of AB and of C, the vibrational, rotational, and magnetic (azimuthal) quantum numbers of AB. If the total energy E does not allow electronic excitation of either AB or C (i.e., $\underline{q}_{AB}$ and $\underline{q}_C$ are the ground electronic states) the internal state is specified by $m = v,J,m_J$. In most experiments it is not possible to measure the distribution over the 2J+1 degenerate magnetic states m_J. (The degeneracy is removed in the presence of a magnetic field.) More commonly as for instance in infrared chemiluminescence measurements (see, e.g., Figs.2.14,15) the best resolution that can be achieved involves specification of v and J. This corresponds to dividing the final product states into groups[10] v,J; each group v,J includes all the translational and magnetic quantum states with energy $E-E_{v,J}$. The number of these states (per unit energy interval) is $(2J+1)\rho_T(E-E_{v,J})$. Formally we can obtain this result from (2.A.7) by setting v,J for v and summing over all states $m = v,J,m_J$ consistent with this specification. Clearly in this case the summation involves only m_J and noting that $E_{v,J,mJ} = E_{v,J}$ is independent of m_J, we find

$$\rho(v,J;E) = A_T \sum_{m_J}{}' \rho_T(E-E_{v,J}) = (2J+1)\rho_T(E-E_{v,J}) \quad . \tag{2.A.9}$$

The corresponding prior distribution is simply [cf.(2.2.6)]

$$P^o(v,J) = \rho(v,J;E)/\rho(E) = (2J+1)\rho_T(E-E_{v,J})/\sum_{v,J}(2J+1)\rho_T(E-E_{v,J}) \quad , \tag{2.A.10}$$

where the summation is restricted to $E_{v,J} \leqq E$. A less detailed group specification involves only the vibrational state v. Now the sum in (2.A.7) should be carried out over both J and m_J. The result, using obvious notation, is

10 Note that we use the same symbol v for the group label and the vibrational quantum number

$$P^0(v) = \sum_J P^0(v,J) = \rho(v;E)/\rho(E) \quad . \qquad (2.A.11)$$

Given the vibrotational energy levels of the diatomic molecule, $E_{v,J}$, one can easily compute $P^0(v,J)$ or $P^0(v)$ using (2.A.6) and (2.A.9-11). In many cases one can derive rather accurate closed-form expressions. For instance, in the vibrating rotor (VR) model (Sect.3.3) $E_{v,J} = E_v + E_J(v) = E_v + B_v J(J+1)$, where B_v is the rotational constant for molecules in vibrational level v. $P^0(v,J)$ is obtained by substituting this expression for $E_{v,J}$ in (2.A.10). An approximate expression for $P^0(v)$ is obtained if $\rho(v;E)$ is evaluated by replacing the sum over J in (2.A.10) by an integral, that is,

$$\rho(v;E) = A_T \int dJ(2J+1)[E-E_v-B_vJ(J+1)]^{1/2} = \left(\frac{A_T}{B_v}\right)\int_0^{E-E_0} dx[E-E_v-x]^{1/2} = A_v(E-E_v)^{3/2} , \qquad (2.A.12)$$

where we have used $x = B_vJ(J+1)$, $dx = B_v(2J+1)dJ$, and $A_v = (2A_T/3B_v)$. Replacing summation over J by integration is legitimate when the rotational energy spacings $2B_vJ \ll E-E_v$. This corresponds to a classical description of the rotational motion and we thus refer to (2.A.12) as the 'classical VR model'. In this approximation the prior distribution is

$$P^0(v) = B_v^{-1}(E-E_v)^{3/2} / \sum_v B_v^{-1}(E-E_v)^{3/2} \qquad (2.A.13)$$

(cf. Fig.2.8). The last expression is simple enough to calculate the prior vibrational distribution using realistic models (e.g., Morse oscillator) for E_v. An even simpler expression can be obtained in the classical RRHO approximation where we set $B_v = B$, $E_v = \hbar\omega v$ and evaluate the sum by integration over v. This yields the expression (2.2.7) (Fig.2.8).

Suppose now that the energy released in the exoergic reaction $A + BC \rightarrow AB + C$ allows electronic excitation of, say, the molecular product and, as in (2.2.30) only the electronic energy level q of AB is detected. That is, neither the g_q degenerate electronic states nor the vibrational rotational and translational states are resolved. In this case the group v in (2.A.7) is identified as q and the summation is over all the electronic, translational vibrational, and rotational quantum numbers with energy $E-E_q$; this yields

$$\begin{aligned} \rho(q;E) &= A_T g_q \sum_{v,J} (2J+1)[E-E_q-E_v(q)-E_J(v,q)]^{1/2} \\ &\simeq A_v g_q \sum_v [E-E_q-E_v(q)]^{3/2}/B_{vq} \qquad \text{(VR)} \\ &\simeq A_q g_q (E-E_q)^{5/2} \qquad \text{(RRHO)} \quad , \qquad (2.A.14) \end{aligned}$$

where $A_q = (4A_T/15B_q\hbar\omega_q)$. In deriving the second equality we have used, as in (2.A.12), the classical VR model with $A_v = 2A_T/3B_{vq}$ where B_{vq} is the rotational constant for level v,q. The third equality follows from the RRHO approximation where we set $B_{vq} = B_q$, $E_v(q) = \hbar\omega_q v$ and replace summation over v by integration over $E_v(q)$; $0 \leq E_v(q) \leq E - E_q$. The prior electronic energy distribution is simply $P^o(q) = \rho(q;E)/\rho(E)$ with $\rho(E) = \sum_q \rho(q;E)$. The electronic branching ratios (Table 2.2) are defined as $\Gamma_q^o = P^o(q)/P^o(q^*) = \rho(q;E)/\rho(q^*;E)$ where q^* is a reference electronic level.

The basic expression (2.A.7) is the key for evaluating the prior product distributions for any collision which takes place at a given total energy E and yields two species R_1 and R_2 as products. [This includes nonreactive (inelastic, $\Delta E_0=0$) and unimolecular dissociation processes as special cases.] R_i may stand for atoms or diatomic or polyatomic molecules. The internal quantum state of R_i is fully specified by $\underline{q}_i$, $\underline{v}_i$, and $\underline{J}_i$, where $\underline{q}_i$ denotes the collection of quantum numbers which determine the electronic state, $\underline{v}_i$ is the set of vibrational quantum numbers [$3N_i-6$ for nonlinear molecules, $3N_i-5$ for linear molecules where N_i is the number of atoms of R_i (cf. Sect. 3.3.5)], and $\underline{J}_i$ is the set of rotational quantum numbers (2 for linear, i.e., J,m_J, and 3 for nonlinear molecules). To simplify the analysis we shall assume that both products appear in the ground electronic state. Furthermore, the most detailed resolution of final states that we shall consider refers to energy sharing. Namely, what is the prior probability, $P^o(E_V^1,E_V^2,E_R^1,E_R^2;E)dE$, for finding the species R_1 with total vibrational energy E_V^1, total rotational energy E_R^1, and species R_2 with E_V^2 and E_R^2 when the total available energy lies in the small interval E, E+dE. By energy conservation the relative translational energy is $E_T = E - (E_V^1+E_V^2+E_R^1+E_R^2)$. Note also that as long as E_V^i and E_J^i are treated as discrete quantities the energy allowance dE is due to the continuous translational energy; $dE = dE_T$. From (2.A.7) or by simply extending (2.A.9), $P^o(E_V^1,E_V^2,E_R^1,E_R^2)$ is proportional to

$$\rho(E_V^1,E_V^2,E_R^1,E_R^2;E) = g_V^1 g_V^2 g_R^1 g_R^2 \rho_T(E_T) \quad , \tag{2.A.15}$$

where g_V^i and g_R^i are the vibrational and rotational degeneracies associated with E_V^i and E_R^i. As a test of consistency it is easily verified that for atom-diatom collisions where $g_V^i = 1$, $g_R^1 = 1$, $g_R^2 = 2J+1$, (2.A.15), reduces to (2.A.9). For polyatomic molecules the machinery of state counting required for determining the various degeneracy factors may be quite complicated [2.100]. A considerable and usually justified simplification is achieved by treating the rotational motions classically. This corresponds to treating E_R^i as a continuous variable and replacing the degeneracy factor g_R^i by $\rho_R^i(E_R^i)dE_R^i$, where $\rho_R^i(E_R^i)$ is the density of rotational states. In this classical representation $P^0(E_V^1,E_V^2,E_R^1,E_R^2)dE_R^1dE_R^2$ is the prior probability of finding the products with vibrational energies E_V^i and rotational energies in the range $E_R^i, E_R^i+dE_R^i$. This prior distribution is proportional to the density of states

$$\rho(E_V^1,E_V^2,E_R^1,E_R^2;E) = g_V^1 g_V^2 \rho_R^1(E_{R1})\rho_R^2(E_R^2)\rho_T(E_T) \quad . \tag{2.A.16}$$

[Note the different dimensions of (2.A.16) and (2.A.15). In (2.A.15), the units are energy^{-1}; in (2.A.16) energy^{-3}.] There are several procedures to derive $\rho_R(E_R)$ [2.100]; we shall suffice in quoting the result

$$\rho_R(E_R) = A_R E_R^{r/2-1} \quad , \tag{2.A.17}$$

where r is the number of rotational degrees of freedom (r=2 and 3 for linear and nonlinear molecules, respectively). The coefficient $A_R = a(I_\alpha I_\beta I_\gamma)^{\frac{1}{2}}\hbar^3$ for r=3 and $A_R = bI/\hbar^2$ for r=2 where I is the moment of inertia corresponding to the principal axis α of the nonlinear rotor (top) and $I = \mu R_e^2$ is the moment of inertia of the linear molecule; a and b depend on the symmetry of the molecule. For diatomic molecules b=2, r=2 and thus $\rho_R = 2I/\hbar^2 = 1/B$ is energy independent; $B = \hbar^2/2I$ is the rotational constant. This last result could easily be derived from the level scheme of a quantal rigid rotor: $E_R = E_J = BJ(J+1)$ and J is $g_J = (2J+1)$-fold degenerate. The average number of states per unit energy interval is then $\simeq(\frac{1}{2})(g_{J-1}+g_J)/(E_J-E_{J-1}) = 1/B = \rho_R$.

Comparison of (2.A.6) and (2.A.17) reveals that both translational and rotational densities of states are of the form $\rho \propto E^{f/2-1}$ where f is the number of degrees of freedom. On the other hand, for classical harmonic vibrations $\rho_V(E_V) \propto E^{s-1}$ where s is the number of vibrational degrees of freedom; see, e.g., the classical RRK expression (2.3.11) and below. (For a single harmonic oscillator s=1 we simply have $\rho_V=1/\hbar\omega$ where $\hbar\omega$ is the level spacing.) These results are special cases of the general rule which states that for a system described by a Hamiltonian containing n square terms (in the momenta p_i or the coordinates q_i) the density of states is $\rho \propto E^{n-1}$. Thus one-dimensional translations and rotations which involve only one kinetic energy term p_i^2 contribute n=1. On the other hand, a one-dimensional oscillator involves both kinetic and potential energy terms and thus contribute n=2 to the densit[y] of states.

When one or both of the collision products R_i is a polyatomic molecule the vibrational degeneracies g_V^i are essentially continuous and rapidly increasing functions of E_V^i. In such cases it is convenient to replace g_V^i by $\rho_V^i(E_V^i)dE_V^i$. Thus $P^0(E_V^1,E_V^2,E_R^1,E_R^2)$ is now

$$P^0 \propto \rho(E_V^1,E_V^2,E_R^1,E_R^2;E) = \rho_V^1(E_V^1)\rho_V^2(E_V^2)\rho_R^1(E_R^1)\rho_R^2(E_R^2)\rho_T(E_T) \quad . \tag{2.A.18}$$

If we are not interested in the detailed energy distribution between E_V^1,E_V^2, E_R^1,E_R^2 and E_T but only in the partitioning of the energy among E_V^1,E_V^2 and $\varepsilon = E_R^1 + E_R^2 + E_T = E - E_V^1 - E_V^2$ then

$$\begin{aligned} P^0(E_V^1,E_V^2;E) \propto \rho(E_V^1,E_V^2;E) &= \iint dE_R^1 dE_R^2 \rho(E_V^1,E_V^2,E_R^1,E_R^2;E) \\ &= \rho_V^1(E_V^1)\rho_V^2(E_V^2)\rho_{TR}(\varepsilon) \quad , \end{aligned} \tag{2.A.19}$$

where

$$\rho_{TR}(\varepsilon) = A_{TR}\varepsilon^{\alpha/2-1} \tag{2.A.20}$$

is the translational-rotational density of states; $\alpha = r_1 + r_2 + t$ is the sum of translational and rotational degrees of freedom. In deriving (2.A.20) we have used (2.A.6) and (2.A.17), and integrated over the range $0 \leqq E_R^1+E_R^2 \leqq E-E_V^1-E_V^2$. Note that when $R_1 + R_2 = A + BC$, (2.A.20) yields exactly (2.A.12) since $\varepsilon = E - E_V$ and $A_{TR} = A_V$.

The simplest approximation for $\rho_V(E_V)$ is the classical RRK expression (2.3.11) which corresponds to s degenerate harmonic oscillators, $\rho_V \propto E_V^{s-1}$. For the special case that R_1 is an atom and R_2 is a nonlinear polyatomic molecule we have $s_1 = 1$, $s_2 = s$ and $\alpha = 6$; hence $P^0(E_V;E) \propto \varepsilon^2(E-\varepsilon)^{s-1}$ which was used in Sect.2.4.5 to describe the energy distribution in $Ar + SF_6$ collisions [cf. (2.4.25) and Fig.2.42].

The derivation of (2.3.11), $\rho_V(E_V) = (E_V/h\nu)^{s-1}/(s-1)!h\nu$, was based on counting the number of ways to divide $n = E_V/h\nu$ quanta among s identical and independent harmonic oscillators. This expression can easily be extended to the case where not all oscillators have the same frequency. Using combinatorics to count the number of ways of distributing $E_V = \sum v_i h\nu_i$ among s oscillators, one obtains [2.100] that at the classical (high energy, $E_V \gg h\nu_i$) limit

$$\rho_V(E_V) = E_V^{s-1}/(s-1)!\prod_{i=1}^{s} h\nu_i \quad , \tag{2.A.21}$$

where ν_i is the frequency of oscillator i. For the special case of s degenerate oscillators, $\nu_i = \nu$, (2.A.21) reduces to (2.3.11). The classical expression

(2.A.21) gives too low values at low energies (compared to exact counting of quantum states). This deficiency is partly removed by replacing E_V on the rhs of (2.A.21) by $E_V + E_Z$ where $E_Z = \frac{1}{2}\sum h\nu_i$ is the quantal zero point energy. This modification, which corresponds to adding E to the amount of energy that may be distributed among the oscillators, is called the semiclassical approximation [2.177]. (Note however that E_V is the vibrational energy above the ground vibrational level.) A more general extension of (2.A.21) is

$$\rho_V(E_V) = (E_V + aE_Z)^{s-1}/(s-1)!\prod_{i=1}^{s} h\nu_i \quad , \tag{2.A.22}$$

where a is an adjustable parameter. An extensively used empirical expression relating a to E, E_Z, and the ν_i's has been suggested by Whitten and Rabinovitch [2.178]. It should be kept in mind, however, that the accuracy of (2.A.22) is measured by comparing it to direct counting of harmonic oscillator states. At high excitations the effects of anharmonicity become important and must be taken into account. For a detailed discussion of these and other factors that influence $\rho_V(E_V)$ see, for example, [2.100].

So far in this appendix we have considered prior distributions at a well-defined total energy E. These provide a satisfactory representation for highly exoergic processes where ΔE, the uncertainty in E, is mainly due to the width of the reactant translational-rotational distribution and thus $\Delta E \ll \langle E \rangle \equiv E$. For endoergic or thermoneutral processes this assumption is no longer justified and one has to properly average over E [cf. Sect.2.4.1, in particular (2.4.5)]. For a given E and any specifications v and v' of the internal states of the reactants and the products we always take

$$P^0(v';E) \propto k^0(v \rightarrow v';E) = R\rho(v';E) \quad . \tag{2.A.23}$$

Assuming that the proportionality constant R is independent of E, the E-averaged prior rate constant is

$$\langle k^0(v \rightarrow v';E) \rangle = R\int dEP(E)\rho(v';E) \quad , \tag{2.A.24}$$

where P(E) is the total energy distribution. The actual evaluation of (2.A.23) depends on the nature of the collision (unimolecular, bimolecular, inelastic, atom-diatom, atom-polyatom, etc.) and the specification of v and v'. One example, corresponding to vibrational energy transfer in $Ar + SF_6$ collisions, was mentioned in Sect.2.4.5 [cf. (2.4.26) and Fig.2.42]. Let us consider another common example: A reactive atom-diatom collision where v and v' denote the initial and final vibrational states, i.e., $A+BC(v)\rightarrow AB(v')+C$; the translational-rotational (T-R) distributions of the reactants are Boltzmann at temperature T. Measuring energies from the ground vibrational state of the products we have $E=E_v+\varepsilon-\Delta E_0=E_{v'}+\varepsilon'$, where $\varepsilon=E_R+E_T$ and $\varepsilon'=E_R'+E_T'$ are the T-R energies of the reactants and products and ΔE_0 is the exoergicity. Treating E_R and thus ε as classical variables the thermal reactant T-R distribution is $p(\varepsilon|T)=\rho_{TR}(\varepsilon)\exp(-\varepsilon/kT)/Q_{TR}$, where $\rho_{TR}(\varepsilon)=A_{TR}\varepsilon^{\alpha/2-1}=A_{TR}\varepsilon^{3/2}=A_v(E-E_v+\Delta E_0)^{3/2}$ [cf. (2.A.20)]. In this case averaging over E means averaging over ε with the weighting function $p(\varepsilon|T)$. Noting in addition that $k^0(v\rightarrow v';E)\propto(E-E_{v'})^{3/2}=\varepsilon^{3/2}=(E_v-E_{v'}-\Delta E_0+\varepsilon)^{3/2}$, we get

$$k^0(v\rightarrow v';T) = \langle k^0(v\rightarrow v';E)\rangle \propto \int_{\varepsilon_{min}}^{\infty} d\varepsilon(2kT\Delta_{vv'}+\varepsilon)^{3/2}\varepsilon^{3/2}\exp(-\varepsilon/kT) \quad , \tag{2.A.25}$$

where $\Delta_{vv'}=(E_v-E_{v'}-\Delta E_0)/2kT$ is the quantity defined in (2.4.6). For the lower integration limit we have $\varepsilon_{min}=0$ for exoergic processes ($\Delta_{vv'}>0$) and $\varepsilon_{min}=2kT\Delta_{vv'}$ for endoergic processes ($\Delta_{vv'}<0$). The behavior of (2.A.25) was depicted in Fig.2.33. Explicit expressions for other types of collisions can be found elsewhere [2.11,35,125].

Appendix 2.B. Practical Surprisal Analysis

Surprisal analysis seeks to find m constraints A_r such that an observed distribution P(v) can be well represented by

$$P(v) \simeq P^{ME}(v) = P^0(v)\exp\left[-\lambda_0-\sum_{r=1}^{m}\lambda_r A_r(v)\right] \quad . \tag{2.B.1}$$

We show that the optimal fit obtains when the magnitude of the m+1 (Lagrange) parameters is chosen by the maximal entropy prescription: that the expectation value of A_r computed for the distribution $P^{ME}(v)$ equals the observed mean value, for $r=1,\dots,m$

$$\sum_v P^{ME}(v)A_r(v) = \sum_v P(v)A_r(v) \equiv \langle A_r\rangle \quad . \tag{2.B.2}$$

Of course, a perfect fit is always possible when m+1 equals the number of states in the distribution. We are concerned however with the case where m is definitely below this limit.

As a convergence criterion we shall employ [2.35], [cf. (2.1.58],

$$DS[P|P^{ME}] = \sum_v P(v)\left\{\ln[P(v)/P^o(v)]+\lambda_0+\sum_{r=1}^{m} \lambda_r A_r(v)\right\} \tag{2.B.3}$$

which is the weighted difference between the surprisal of the actual results $-\ln[P(v)/P^o(v)]$ and of the assumed functional form for the surprisal. In seeking the best fit of the form

$$-\ln[P(v)/P^o(v)] \simeq \lambda_0 + \sum_{r=1}^{m} \lambda_r A_r(v) \tag{2.B.4}$$

for the surprisal, a larger weight is assigned to the deviations of the more probable points. $DS[P|P^{ME}]$ is nonnegative and equals zero if and only if $P(v) = P^{ME}(v)$. In practice, it is only required to match $P(v)$ to within the experimental error bars. One can show that this is roughly equivalent to making $DS[P|P^{ME}]$ smaller than $(n-m-1)\sigma^2/2$ where n is the number of states and σ^2 is the standard deviation [2.42], $n\sigma^2 \simeq \sum_v P(v)[\delta P(v)/P(v)]^2$. Here $\delta P(v)/P(v)$ is the fractional error in the probability $P(v)$.

To locate the minimal value of $DS[P^{ME}|P]$ we vary the m parameters $\lambda_1,\ldots,\lambda_m$. λ_0 is a function of these parameters given by (2.1.56) which implies that

$$\frac{\partial\lambda_0}{\partial\lambda_r} = -\sum_v P^{ME}(v)A_r(v) \quad . \tag{2.B.5}$$

Hence, the extremum is at

$$0 = \partial DS[P|P^{ME}]/\partial\lambda_r = \sum_v P(v)A_r(v) - \sum_v P^{ME}(v)A_r(v) \quad , \qquad \text{Q.E.D.} \tag{2.B.6}$$

One can also show [2.39] that the extremum is indeed a minimum. If a given number of constraints does not suffice to account for the data then adding another constraint will either improve the fit or not change it. In the latter case the added constraint is noninformative. The proof is again based on (2.1.58). $DS[P|P^o]$ is unchanged by the inclusion of additional constraints. $DS[P^{ME}|P^o]$ will either increase or remain at the same value. The reason is that adding a constraint serves to further restrict the set of distributions from which P^{ME} is selected. Upon seeking the distribution of maximal entropy among the more limited set we can either retain the previous maximum (so that the restriction is not informative) or obtain a new and necessarily lower maximum.

Appendix 2.C. Statistical Models, Prior Distributions, and Collision Dynamics

The purpose of this appendix is to provide a collision theoretic intepretation of the prior distribution [2.179]. As part of the discussion we shall also derive the RRK form (2.3.16) for the rate of unimolecular dissociation. To keep the discussion simple, classical mechanics will be employed. Quantal statistical approximations and transition state theory are discussed in more detail elsewhere [2.24,179]. The discussion will also be limited to collisions at a given total energy.

Our first task is the computation of P(n), the fraction of reactive collision, at the total energy E, for reagents in the internal state n. [The argument is analogous to that leading to (2.1.28) except that here we use a microcanonical rather than a canonical ensemble of reactants.] Consider reagents in the initial internal state n with an impact parameter b [2.1] in the range b, b+db. Choose N(n,b)db initial conditions and solve the classical equations of motion. Let the number of reactive collisions be $N_r(n,b)db$. The cross section for reactive collisions for reagents in the internal state n is then given by [2.1]

$$\sigma_r(n) = 2\pi \int_0^\infty b[N_r(n,b)/N(n,b)]db \quad . \tag{2.C.1}$$

Fortunately, integrating (2.C.1) is easy, provided that some care is taken. Choose the distribution of initial conditions such that

$$N(n,b) = [2\pi b/\pi B^2]N(n) \quad . \tag{2.C.2}$$

Here B is a large value of the impact parameter such that for $b > B$ reaction is not possible. N(n) is the total number of trajectories for reagents in the state n,

$$N(n) = \int_0^B N(n,b)db \quad . \tag{2.C.3}$$

Using (2.C.2) in (2.C.1)

$$\sigma_r(n) = \pi B^2 \int_0^B N_r(n,b)db/N(n) = \pi B^2 N_r(n)/N(n) \quad . \tag{2.C.4}$$

Here $N_r(n)$ is the total number of reactive trajectories for reagents in the state n. The reaction rate constant at a given energy $k(n\rightarrow;E)$ is

$$k(n\rightarrow;E) = \underline{v}_n \sigma_r(n) = \pi B^2 \underline{v}_n N_r(n)/N(n) \quad . \tag{2.C.5}$$

Here $\underline{v}_n$ is the initial relative velocity, $\mu \underline{v}_n^2/2 = E - E_n$. The rate constant k(E) for a microcanonical ensemble is, by the canon,

$$\vec{k}(E) = \sum_n [\rho(n;E)/\rho(E)] k(n\rightarrow;E) \quad , \tag{2.C.6}$$

where the term in the square brackets is the fraction of reagents which are in the internal state n, for the microcanonical ensemble. Summation in (2.C.6) can again be carried out if we choose

$$N(n) = [\rho(n;E)/\rho(E)] N \quad , \tag{2.C.7}$$

where N is the total number of trajectories that are used, $N = \sum_n N(n)$. Using (2.C.7) and (2.C.5) in (2.C.6)

$$\vec{k}(E) = \pi B^2 \sum_n \underline{v}_n N_r(n)/N = \pi B^2 < \underline{v}_r > N_r/N \quad . \tag{2.C.8}$$

Here $N_r = \sum_n N_r(n)$ is the total number of reactive trajectories and $< \underline{v}_r >$ is the average initial velocity of those trajectories that result in a reaction,

$$< \underline{v}_r > = \sum_n \underline{v}_n N_r(n) / \sum_n N_r(n) \quad . \tag{2.C.9}$$

From (2.C.8), (2.C.7), and (2.C.5) [cf. (2.1.28)]

$$P(n) = [\rho(n,E)/\rho(E)] k(n\rightarrow;E)/\vec{k}(E) = \underline{v}_n N_r(n)/ < \underline{v}_r > N_r \quad . \tag{2.C.10}$$

The density of translational states $\rho(n,E)$ is proportional to $\underline{v}_n$ [cf. (2.A.5) and (2.A.6)]. Hence any deviance of P(n) from $P^0(n)$ where $P^0(n) \propto \underline{v}_n$ shows that $N_r(n)/N_r$ is n dependent. As used in the text and elsewhere [2.33-35] selectivity of energy consumption means that not all initial quantum states are equally reactive [i.e., that $N_r(n)$ is n dependent]. The rate constant in the prior limit can be computed from (2.C.5). $N_r^0(n)$ is n independent while $N(n) \propto \rho(n;E) \propto \underline{v}_n$. Hence $k^0(n\rightarrow;E)$ is the same for all initial quantum states and

$$P^0(n) = \underline{v}_n / \sum_n \underline{v}_n = \rho(n;E)/\rho(E) \quad . \tag{2.C.11}$$

It is however possible to define the absence of selectivity by the condition that $[N_r(n)/N_r] = [N(n)/N]$. Then the reference distribution is given by

$$P^S(n) = \underline{v}_n N(n)/\langle \underline{v} \rangle N = \underline{v}_n N(n)/\sum_n \underline{v}_n N(n) \quad . \tag{2.C.12}$$

We have used the superscript "S" for this reference for it is the one adopted in the statistical theories based on the scattering matrix [2.24,180,181]. In practice, such theories need to specify B [whereas $P^0(n)=\rho(n;E)/\rho(E)$ is independent of B]. For most reasonable choices for B, $P^S(n)$ yields results which are quite close to $P^0(n)$.

Transition state theory [2.180,182,183] seeks to avoid the computation of classical trajectories required to evaluate $\vec{k}(E)$ via (2.C.8). It does so by introducing a hypersurface, separating reactants, and products, which all reactive trajectories must cross. It then computes the flux for all trajectories which cross the surface in the direction from reactants to products. This is clearly an upper bound to the required flux since at most all those trajectories originate from the reactants and will proceed to form products. The evaluation of the flux is particularly easy if the motion perpendicular to the surface is separable, $E = E_T + E_m$, where E_T is the translational energy for the perpendicular motion. Then

$$\vec{k}(E) \leqq k^{TST}(E) \equiv N^{\ddagger}(E)/h\rho(E) \quad , \tag{2.C.13}$$

where $N^{\ddagger}(E)$ is the number of states not counting the perpendicular motion.

The considerations that led to (2.C.8) can also be employed to compute the rate for unimolecular dissociation of an energy-rich polyatomic molecule (cf. Sect.2.3.5). Let r now designate those trajectories that lead to the formation of the energy-rich species. When the excitation energy is fully equilibrated the rate of dissociation $\overleftarrow{k}(E)$ is related to the rate of formation $\vec{k}(E)$ by detailed balance [2.24,25]

$$\rho_r(E)\overleftarrow{k}(E) = \rho(E)\vec{k}(E) \quad . \tag{2.C.14}$$

Here $\rho_r(E)$ is the density of states of the energy-rich intermediate at the energy E. If in addition to the assumption that the excitation energy is randomly distributed we also make the transition state approximation for $\vec{k}(E)$, (2.C.13), we obtain for the dissociation rate

$$\overleftarrow{k}(E) = N^{\ddagger}(E)/h\rho_r(E) \quad . \tag{2.C.15}$$

In the RRK approximation, where the energy-rich intermediate is regarded as a collection of s degenerate oscillators, $\rho_r(E) = E^{s-1}/(h\nu)^s$ while $N^{\ddagger}(E) = [(E-E_0)/h\nu]^{s-1}$ leading to (2.3.16).

Appendix 2.D. Derivation of the Treanor Distribution

The Treanor distribution, $P_q(v)$ [cf. (2.5.30)], was defined in Sect.2.5.4 as the equilibrium distribution in a (hypothetical) system governed only by V-V processes. Thus, $P_q(v)$ is the maximal entropy distribution subject to the constraints implied by those processes. Explicitly, let us consider a system of N diatomic molecules, e.g., CO, in which energy can be exchanged among the molecules but not with the surroundings. The entropy (per molecule) is given by[11] [cf. (2.A.3)]

$$S = -\int d\varepsilon \sum_v P(\varepsilon,v)\ln[P(\varepsilon,v)/\rho_{TR}(\varepsilon)] \quad . \tag{2.D.1}$$

There are three constraints on $P(\varepsilon,v)$:

I) The normalization

$$\int d\varepsilon \sum_v P(\varepsilon,v) = \int P(\varepsilon)d\varepsilon = \sum_v P(v) = 1 \quad . \tag{2.D.2}$$

II) The total energy of the N molecules is conserved. Hence the average energy per molecule is constant

$$\langle E \rangle = \int d\varepsilon \sum_v P(\varepsilon,v)(\varepsilon+E_v) = \int d\varepsilon P(\varepsilon)\varepsilon + \sum_v P(v)E_v = \langle \varepsilon \rangle + \langle E_v \rangle = \text{const.} \tag{2.D.3}$$

Note that because of the vibrational anharmonicity $\langle \varepsilon \rangle$ and $\langle E_v \rangle$ are not conserved quantities; only their sum is conserved.

III) The total number of vibrational quanta in the system is a conserved quantity as long as only V-V processes modify P(v). Thus

$$\langle v \rangle = \sum_v P(v)v = \int d\varepsilon \sum_v P(\varepsilon,v)v = 1 \quad . \tag{2.D.4}$$

11 We employ here the group symbol "v" of (2.A.3) to denote the vibrational level v and all the translational-rotational states within the energy interval $\varepsilon,\varepsilon+d\varepsilon$. Thus for P(v) of (2.A.3) we set $P(\varepsilon,v)d\varepsilon$ which is the probability of finding a molecule in vibrational level v and T-R energy within $\varepsilon,\varepsilon+d\varepsilon$ and for the degeneracy factor g(v) we set $\rho_{TR}(\varepsilon)d\varepsilon$ [cf. (2.A.20)].

Maximization of S subject to these three constraints yields

$$P_q(\varepsilon,v) = \rho_{TR}(\varepsilon)\ \exp[-\beta(\varepsilon+E_v)-\gamma v]/Q_q(\beta,\gamma)\quad , \tag{2.D.5}$$

where $\beta = 1/kT$ and γ are the Lagrange multipliers conjugated to $\langle E \rangle$ and $\langle v \rangle$, respectively, and $Q_q(\beta,\gamma)$ is the normalizing constant (partition function). By summing $P_q(\varepsilon,v)$ over v or integrating over ε we obtain, respectively, the T-R and vibrational distributions corresponding to maximal S', namely

$$P_q(\varepsilon) = \rho_{TR}(\varepsilon)\ \exp(-\beta\varepsilon)/Q_{TR}(\beta) \tag{2.D.6}$$

$$P(v) = P_q(v) = \exp[-\beta E_v-\gamma v]/Q_q'(\beta,\gamma)\quad , \tag{2.D.7}$$

where $Q_q'(\beta,\gamma)Q_{TR}(\beta) = Q_q(\beta,\gamma)$ of (2.D.5). $P_q(\varepsilon)$ is an ordinary Boltzmann T-R distribution. $P_q(v)$ is the Treanor distribution (2.5.30). The values of the Lagrange parameters β and γ are uniquely determined by the initial values of $\langle E \rangle$ and $\langle v \rangle$.

A few remarks should conclude our derivation: a) It is somewhat lengthy but not very difficult to show that $P_q(\varepsilon,v)$; hence $P_q(\varepsilon)$ and $P_q(v)$ have exactly the same forms as above when the N molecules are coupled to a heat bath. In this case $T = (k\beta)^{-1}$ is the heat bath temperature. b) In the case of harmonic oscillators the V-V processes are exactly resonant so that both $\langle E_v \rangle$ and $\langle \varepsilon \rangle$ are separately conserved but then $\langle v \rangle = \langle E_v \rangle/\hbar\omega = \text{const.}$ is not an independent constraint. Thus, instead of the two independent constraints $\langle \varepsilon \rangle + \langle E_v \rangle = \text{const.}$ and $\langle v \rangle = \text{const.}$ for anharmonic oscillators, harmonic oscillators are governed by the two constraints $\langle E_v \rangle = \text{const.}$ and $\langle \varepsilon \rangle = \text{const.}$ In this case both $P(\varepsilon)$ and $P(v)$ are Boltzmann but with different temperatures, T and T_q, respectively [cf. (2.5.33)]. c) The Treanor distribution can also be derived [2.165] as a first-order stationary solution of the master equation via a perturbative procedure known as the Chapman-Enskog expansion. As in Sect.2.5.3, $P_q(v)$ appears as an approximate, quasiequilibrium, solution of the vibrational master equation (2.5.28) for times $\tau_{V-V} < t \ll \tau_{V-T}$.

3. Photons, Molecules, and Lasers

To make full use of the unique properties of lasers, one needs an understanding of the fundamental principles which govern their operation. In the first two sections of this chapter we provide an introductory discussion, which should be sufficient for most of the applications treated elsewhere in the book. Extensive representations of the subject can be found in various textbooks, to which the reader is referred [3.1-7]. Chapters 4 and 5 require working knowledge of spectroscopic concepts. An overview of the topics important for chemical lasers and laser chemistry is provided in Sect.3.3. Standard monographs should be consulted for detailed and systematic discussions [3.8-14]. A description of some of the more important laser sources used by chemists is given in the last section. In view of the rapid technological developments in the field, quoted specifications should not be considered as a state of the art. Better performances of existing lasers and novel transitions are announced virtually by the day!

3.1 Interaction of Molecules with Radiation

The introductory description of the interaction between light and matter presented in Chap.1 applies to radiation sources with a broad and continuous ("white") spectral distribution. The description must be modified for highly monochromatic sources like lasers. In this section we present a somewhat more quantitative picture of the interaction between molecules and radiation, especially laser radiation. We shall follow the semiclassical approach in which the molecules are treated quantum mechanically and the radiation field classically. While emphasizing the physical significance of the assumptions and the results we shall skip some of the mathematical details. These can be found in a variety of quantum mechanics and laser physics textbooks [3.1-7].

3.1.1 The Golden Rule

Consider a molecule that at $t=0$ is exposed to an electromagnetic field containing a frequency component ν that is nearly equal to the transition frequency between a pair of levels i and f; $h\nu \approx h\nu_{fi} = E_f - E_i > 0$ (Fig.3.1).

Assuming that no other pair of levels is in near resonance with the radiation field, we consider the system as a two-level problem. For simplicity, we shall also assume that the levels are nondegenerate.

The time evolution of the system can be described by the nonstationary Schrödinger equation

$$(H_0+V)\psi = i\hbar \frac{\partial\psi}{\partial t} \quad , \tag{3.1.1}$$

where H_0 is the molecular Hamiltonian (without the field) and V is the interaction potential of the molecule with the field. If the wavelength of the radiation is large compared to the molecular size then

$$V = -\underline{\mu} \cdot \underline{E} \tag{3.1.2}$$

$$\underline{E} = \underline{E}_0 \cos 2\pi\nu t \quad , \tag{3.1.3}$$

where $\underline{\mu}$ is the dipole moment (vector) operator and $\underline{E}$ is the electric field vector.

The wave function ψ can be expressed as a linear combination of the unperturbed wave functions ϕ_i with time-dependent coefficients $C_i(t)$. For a two-level system

$$\psi(\underline{r},t) = C_i(t)\phi_i(\underline{r})\, e^{-iE_it/\hbar} + C_f(t)\phi_f(\underline{r})\, e^{-iE_ft/\hbar} \tag{3.1.4}$$

$$H_0\phi_i = E_i\phi_i \quad , \quad H_0\phi_f = E_f\phi_f \tag{3.1.5}$$

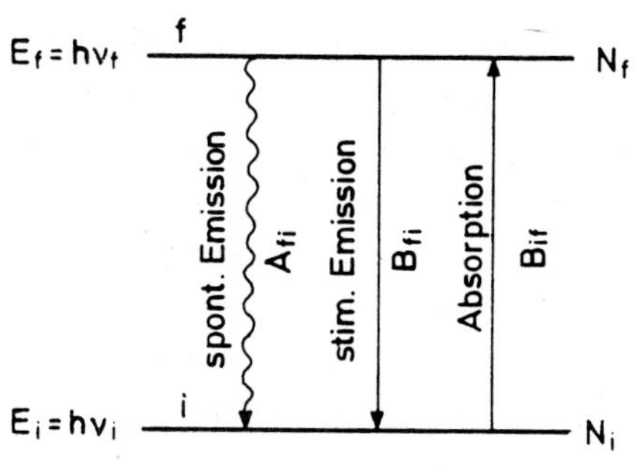

Fig. 3.1. Radiative processes in a two-level system

$$|C_i(t)|^2 + |C_f(t)|^2 = 1 \quad . \tag{3.1.6}$$

The problem can be solved exactly for any magnitude of V. The exact solution due to Rabi is of interest for strong fields and short interaction times which may lead to the so-called coherence effects (Sect.3.1.3). Yet, in most cases of interest, the interaction potential is small and an approximate perturbative procedure can be employed. The most important result of the approximate solution is Fermi's golden rule which states that the rate of transitions between the levels is independent of time.

The probability of finding the system in state i at time t is $|C_i(t)|^2$. For the initial condition that the field is turned on at $t=0$ and $C_i(0)=1$, $C_f(0)=0$, the quantity $P_{if}(t) \equiv |C_f(t)|^2$ is the transition probability from i to f due to interaction of duration t. In other words, $P_{if}(t)$ is the light absorption probability.

The result of first-order perturbation theory for this quantity is

$$P_{if}(t) = |C_f(t)|^2 = \Omega^2 t^2 (\sin x/x)^2 \quad , \tag{3.1.7}$$

where

$$x = \pi(\nu - \nu_{fi})t \tag{3.1.8}$$

$$\Omega = |\underline{\mu}_{if} \cdot \underline{E}_0|/2\hbar \tag{3.1.9}$$

$$\underline{\mu}_{if} = \langle \phi_i | \underline{\mu} | \phi_f \rangle \quad . \tag{3.1.10}$$

Thus the transition probability depends on the "detuning frequency" $\nu - \nu_{fi}$, the "Rabi frequency" $2\Omega = \mu E_0/\hbar$ (Sect.3.1.3), and on the matrix element of the electric dipole moment operator $\underline{\mu}_{if}$ (the "transition dipole"). In many cases $\underline{\mu}_{if} = 0$ and consequently $P_{if} = 0$. The transition is then termed "forbidden". For instance, if ϕ_i and ϕ_f are electronic wave functions which can be expressed as products of spatial and spin functions $\phi_i = u_i(\underline{r})\sigma_i$, $\phi_f = u_f(\underline{r})\sigma_f$ and the spin functions correspond to different multiplicities, $\langle \sigma_i | \sigma_f \rangle = 0$, then $\underline{\mu}_{if} = \langle \phi_i | \underline{\mu} | \phi_f \rangle = \langle u_i | \underline{\mu} | u_f \rangle \langle \sigma_i | \sigma_f \rangle = 0$. In practice, $\underline{\mu}_{if}$ is calculated for approximate molecular wave functions, so that the resulting selection rules are not rigorously obeyed. An important example is the spin-orbit interaction that results in the observation of transitions between zero-order states belonging to different spin multiplicities. The forbidden nature of such transitions is reflected by a small transition moment. As we shall see, such transitions are of considerable importance in lasers.

Being the result of first-order perturbation theory (3.1.7) is valid only if $P_{if}(t) << 1$. The function $f(x) = (\sin x/x)^2$ reaches its maximum at $x = 0$, $f(x=0) = 1$, and obtains (rapidly decaying) secondary maxima at $x = \pm k\pi/2$, $k = 1,2,3\dots$. Hence $P_{if}(t) << 1$ is ensured when $\Omega t << 1$, i.e., for interaction times which are much shorter than the inverse of Rabi frequency $t << \Omega^{-1}$.

The width of the central peak of $f(x) = (\sin x/x)^2$ is $\delta x \simeq 1$ so that $\delta\nu \simeq 1/\pi t$. Consequently the area of the central peak of P_{if} (as function of $\nu - \nu_{if}$) is $\Omega^2 t^2 \delta\nu \simeq \Omega^2 t$. Hence if the interaction between the radiation and molecules occurs over a continuous frequency interval $\Delta\nu >> \delta\nu \simeq (\pi t)^{-1}$ the integrated (over $\nu - \nu_{if}$) transition probability is proportional to the interaction time, $P_{if}(t) \propto t$, and the transition rate is independent of time $\dot{P}_{if}$ = const. A spread in $\nu - \nu_{fi}$ is always the case and results either from a continuous broad distribution of the field frequencies ν or broadening of ν_{fi}. This broadening is usually expressed by a line shape function $g(\nu_{fi}, \nu_0)$ of the molecular transition line ν_{fi}, schematically depicted in Fig.3.2 and further discussed in the next section. To derive the expression for the transition probability in these cases, note that for $\Delta\nu >> \delta\nu \simeq (\pi t)^{-1}$ we can write[1]

$$W_{if} = \dot{P}_{if} = \Omega^2 \delta(\nu - \nu_{fi}) \quad . \tag{3.1.11}$$

The result (3.1.11), where Ω [cf. (3.1.9)] is the coupling matrix element, is known as the golden rule (which applies to a variety of problems besides radiation-matter interactions; see Sect.2.2). Of course (3.1.11) becomes a working expression only after integrating over a frequency interval. The net transition rate is obtained by integrating over ν and ν_{fi} with appropriate weighting functions. To integrate over ν we first transform each component of the electric field according to the classical electromagnetic relation

$$< E_x(\nu)^2 > = E_{x,0}(\nu)^2 < \cos^2 2\pi\nu t > = \frac{1}{2} E_{x,0}(\nu)^2 = 4\pi\rho_x(\nu)d\nu \quad . \tag{3.1.12}$$

Here $\underline{E}_0(\nu)$ is the field amplitude[1a] at frequency ν and $\rho_x(\nu)$ is the x component of the spectral radiation density (energy per unit volume and unit frequency) at frequency ν. For isotropic radiation where $\rho_x(\nu) = \rho_y(\nu) = \rho_z(\nu) = (1/3)\rho(\nu)$, or for molecules in the gas phase where $<\mu_x^2> = <\mu_y^2> = <\mu_z^2> = (1/3)<\underline{\mu}^2>$ and $<\mu_x\mu_y> = 0$ the Ω^2 term in (3.1.11) is replaced by [cf.(3.1.9)]

1 The limit of $[\sin \pi yt/\pi y]^2$ for $\pi yt \to \infty$ is $t\delta(y)$ where $\delta(y)$ is the Dirac delta function.

1a The averaging in (3.1.12) is over the time.

$$\Omega^2 = \frac{2\pi}{3\hbar^2} |\underline{\mu}_{if}|^2 \rho(\nu) d\nu \quad . \tag{3.1.13}$$

The integration over ν_{fi} should be weighted by the normalized line shape function $g(\nu_{fi},\nu_0)$,

$$\int g(\nu_{fi},\nu_0) d\nu_{fi} = 1 \quad . \tag{3.1.14}$$

The line shape function, to be discussed in more detail below, represents the relative probability that the molecule will absorb or emit monochromatic radiation of frequency ν_{fi}. ν_0 is the line center frequency. Using (3.1.9) and (3.1.13) and integrating over ν_{fi} the transition rate is given by

$$\begin{aligned} W_{if} = \dot{P}_{if} &= \frac{2\pi}{3\hbar^2} |\underline{\mu}_{if}|^2 \int g(\nu_{fi},\nu_0)\rho(\nu)\delta(\nu-\nu_{fi}) d\nu d\nu_{fi} \\ &= B_{if} \int \rho(\nu) g(\nu,\nu_0) d\nu \quad , \end{aligned} \tag{3.1.15}$$

where

$$B_{if} = \frac{2\pi}{3\hbar^2} |\underline{\mu}_{if}|^2 \tag{3.1.16}$$

is the Einstein coefficient for absorption, $i \to f (E_i < E_f)$, cf. Sect.1.1.

In the case that levels i and f are degenerate (3.1.16) should be modified into

$$B_{if} = \frac{2\pi}{3\hbar^2} g_f |\underline{\mu}_{if}|^2 \quad , \tag{3.1.17}$$

where g_f is the degeneracy of the final level and $|\underline{\mu}_{if}|^2$ is the degeneracy averaged transition moment. That is, $|\underline{\mu}_{if}|^2$ is the average value of the (square of the) transition dipole from one of the g_i degenerate states of level i to one of the degenerate states of level f. Hence $|\underline{\mu}_{if}|^2 = |\underline{\mu}_{fi}|^2$ is symmetric with respect to interchanging i and f. On the other hand, B_{if} which determines the average transition probability from *one* of the g_i initial states into *all* g_f states is an asymmetric quantity. Explicitly, by solving (3.1.1-5) for the initial condition $C_i(0) = 0$, $C_f(0) = 1$, one finds that W_{fi} is related to the Einstein coefficient for stimulated emission B_{fi} exactly as in (3.1.15) but with i and f interchanged. Similarly, B_{fi} is related to $|\underline{\mu}_{fi}|^2$ by (3.1.17) with i and f interchanged. Since $|\underline{\mu}_{if}|^2 = |\underline{\mu}_{fi}|^2$, we find

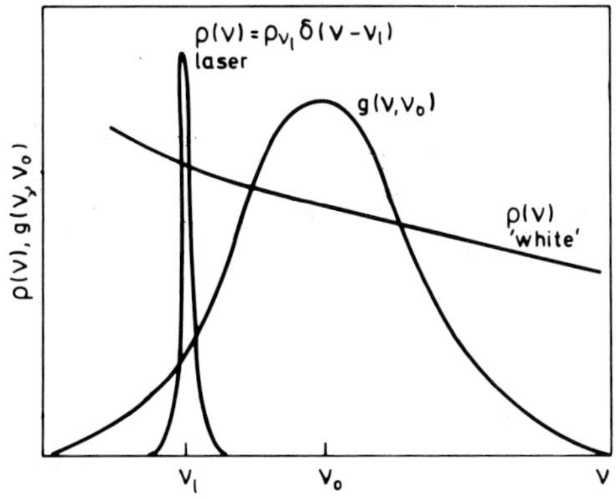

Fig. 3.2. The line shape function $g(\nu,\nu_0)$ and two types of radiation density $\rho(\nu)$. In the case of "white" spectral distribution ρ is much broader than g and the absorption is proportional to $\rho(\nu_0)$. In the case of a monochromatic, e.g., laser, radiation ρ is much narrower than g and the absorption is proportional to $g(\nu_\ell,\nu_0)$. The absorption is thus determined by the "weaker" function of the integrand in (3.1.15)

$$g_i B_{if} = g_f B_{fi} \quad , \tag{3.1.18}$$

which was derived in Sect.1.1 from thermodynamic considerations.

We now consider two important limiting cases: 1) A radiation field with broad ("white") spectral distribution, such as the black body radiation. 2) The radiation source is highly monochromatic, that is, a laser (see Fig. 3.2). In the first case $\rho(\nu)$ is much broader than $g(\nu,\nu_0)$ so that in (3.1.15) $\rho(\nu_0)$ can be taken out of the integral [formally, equivalent to setting $g(\nu,\nu_0) = \delta(\nu-\nu_0)$],

$$W_{if} = W_{if}(\nu_0) = B_{if}\rho(\nu_0)\int g(\nu,\nu_0)d\nu = B_{if}\rho(\nu_0) \quad . \tag{3.1.19}$$

This is the expression used in Sect.1.1.

In the case of laser radiation $\rho(\nu)$ is much narrower than the linewidth and can be expressed as[2]

$$\rho(\nu) = \rho_{\nu_\ell}\delta(\nu-\nu_\ell) \quad , \tag{3.1.20}$$

where ρ_{ν_ℓ} is the energy density at the laser frequency ν_ℓ. From (3.1.20) and (3.1.15) we find

$$W_{if} = W_{if}(\nu_\ell,\nu_0) = B_{if}g(\nu_\ell,\nu_0)\rho_{\nu_\ell} \quad . \tag{3.1.21}$$

---

2 Note that ρ_{ν_ℓ} and $\rho(\nu)$ bear different units. While $\rho(\nu)$ is the energy per unit volume and unit frequency ($erg \cdot s \cdot cm^{-3}$), ρ_{ν_ℓ} is the radiation energy per unit volume ($erg \cdot cm^{-3}$).

Note that both golden rule expressions for the transition rate (3.1.19) and (3.1.21) have the form (2.2.12), namely, a product of a "dynamical factor" and a "density of state factor". The dynamical factors, as in (2.2.13), are proportional to the square of the interaction potential $|V_{if}|^2 \propto |\underline{\mu}\cdot\underline{E}_0|^2 \propto \Omega^2$.

If there are N_i molecules per unit volume, then the rate of change of N_i due to absorption of monochromatic radiation of frequency ν is

$$-\left(\frac{dN_i}{dt}\right)^{st}_{i\to f} = W_{if}N_i = B_{if}g(\nu,\nu_0)\rho_\nu N_i \quad , \tag{3.1.22}$$

where the "st" symbol reminds one that absorption is a stimulated process. This equation is very similar to the equation describing the rate of change of N_i resulting from bimolecular collisions. The analogy is completed if the process of stimulated radiation (absorption in the present case) is regarded as a bimolecular collision between the molecule (say, A) in state i and a photon of energy $h\nu$ giving as a product a molecule in state f

$$A(i) + h\nu \rightarrow A(f) \quad . \tag{3.1.23}$$

Defining a photon density by $\phi_\nu = \rho_\nu/h\nu$, we can rewrite (3.1.22) in the form of a "bimolecular" rate equation

$$-\left(\frac{dN_i}{dt}\right)^{st}_{i\to f} = k_{if}(\nu)\phi_\nu N_i = c\sigma_{if}(\nu)\phi_\nu N_i \quad , \tag{3.1.24}$$

where

$$k_{if}(\nu) = c\sigma_{if}(\nu) = B_{if}h\nu g(\nu,\nu_0) \tag{3.1.25}$$

can be interpreted as the rate constant (cm^3/particle·s) of the process (3.1.23). As for collisional processes, the ratio between the rate constant and the relative velocity (in this case the speed of light c) defines the absorption cross section $\sigma_{if}(\nu)$, (cm^2).

In analogy to (3.1.23) the process of stimulated emission can be represented by the "bimolecular reaction"

$$A(f) + h\nu \rightarrow A(i) + 2h\nu \quad . \tag{3.1.26}$$

Correspondingly

$$-\left(\frac{dN_f}{dt}\right)^{st}_{f\to i} = k_{fi}(\nu)\phi_\nu N_f = c\sigma_{fi}(\nu)\phi_\nu N_f \quad , \tag{3.1.27}$$

where the rate constant $k_{fi}(\nu)$ and the cross section $\sigma_{fi}(\nu)$ for stimulated emission are given by (3.1.25) with i and f interchanged. Using the detailed balance relation (3.1.18), we obtain

$$g_f k_{fi}(\nu) = g_i k_{if}(\nu) \tag{3.1.28}$$

and, since the speed of light is constant

$$g_f \sigma_{fi}(\nu) = g_i \sigma_{if}(\nu) \quad . \tag{3.1.29}$$

Spontaneous emission from the upper level f is a "unimolecular" process

$$A(f) \rightarrow A(i) + h\nu \quad . \tag{3.1.30}$$

However, the semiclassical approach adopted for describing stimulated emission cannot account for this process and one must employ either the fully quantum mechanical [3.1,6] or the phenomenological description of Chap.1. Both descriptions yield the well-known relation

$$A_{fi} = (8\pi h\nu^3/c^3)B_{fi} = (64\pi^4\nu^3/3hc^3)g_i|\underline{\mu}_{fi}|^2 \quad . \tag{3.1.31}$$

It is important to remember that A_{fi} is the integrated rate of spontaneous emission over the entire width of the emission line. The rate of spontaneous emission of photons in the frequency range $\nu - \nu + d\nu$ is

$$-\left(\frac{dN_f}{dt}\right)^{sp}_{f\rightarrow i} = N_f A_{fi} g(\nu,\nu_0)d\nu \quad . \tag{3.1.32}$$

The total emission rate is obtained by integrating over ν. Since $\int g(\nu,\nu_0)d\nu = 1$, this yields

$$-\left(\frac{dN_f}{dt}\right)^{sp}_{f\rightarrow i} = N_f A_{fi} = \frac{N_f}{\tau_{sp}} \quad , \tag{3.1.33}$$

where $\tau_{sp} = 1/A_{fi}$ is called the spontaneous, or natural, lifetime of level f. Using (3.1.25) and (3.1.31), we obtain

$$\sigma_{fi}(\nu) = (c^2/8\pi\nu^2)g(\nu,\nu_0)A_{fi} = (c^2/8\pi\nu^2)g(\nu,\nu_0)/\tau_{sp} \quad . \tag{3.1.34}$$

Equations (3.1.24,27,32) which determine the rate of change of N_f and N_i are called the (population) rate equations. These equations provide the principal tool for describing the time evolution of laser systems. As explained in Chap.1 and will be elaborated in Sect.3.2, the photons emitted from lasers are generated by stimulated rather than spontaneous emission. The ever-present spontaneous emission at the laser frequency is, for all practical purposes, negligible compared to the stimulated process, and can be ignored in the kinetic scheme, except for the early buildup of the oscillations (Sect. 3.2.4). In general the populations are affected by nonradiative as well as radiative processes. The net (time) rate of change of the upper or lower level populations by stimulated emission and absorption only is

$$-\left(\frac{dN_f}{dt}\right)^{st} = \left(\frac{dN_i}{dt}\right)^{st} = \left(\frac{dN_i}{dt}\right)_{i\to f}^{st} + \left(\frac{dN_i}{dt}\right)_{f\to i}^{st}$$

$$= k_{fi}(\nu)[N_f-(g_f/g_i)N_i]\phi_\nu \quad , \tag{3.1.35}$$

where we have used $(dN_i/dt)^{st}_{f\to i} = -(dN_f/dt)^{st}_{f\to i}$ and (3.1.24,27,28). Equivalent forms of (3.1.35) are obtained by using the radiation density $\rho_\nu = h\nu\phi_\nu$ (energy/cm^3) or the intensity $I_\nu = c\rho_\nu = ch\nu\phi_\nu$ (erg/cm^{2}s).

Since the passage of a molecule from level f to level i is accompanied by the emission of a photon $h\nu$, (3.1.35) implies

$$\left(\frac{d\phi_\nu}{dt}\right)^{st} = k_{fi}(\nu)[N_f-(g_f/g_i)N_i]\phi_\nu \quad . \tag{3.1.36}$$

Hence, amplification $d\phi_\nu/dt > 0$ takes place only when the populations are inverted, $\Delta N = N_f - (g_f/g_i)N_i > 0$.

The rate equations are based on the existence of time-independent transition rates as implied by the golden rule expressions (3.1.11,19,21). It should be remembered that these expressions are valid when the interaction period t is such that

$$\Omega << 1/t << \Delta\nu \quad , \tag{3.1.37}$$

where Ω is the Rabi frequency and $\Delta\nu$ is the frequency range over which the interaction between the radiation field and the molecules occurs; in the case of laser radiation $\Delta\nu$ is the spectral linewidth [cf. (3.1.21) and Sect.3.1.2]. The inequality $1 << t\Delta\nu$ allows one to replace the oscillating factor in (3.1.7) by the delta function as in (3.1.11). The inequality $\Omega t << 1$ ensures the validity of first-order perturbation theory. At higher field intensities I_ν, ($I_\nu = ch\nu\phi_\nu = c\rho_\nu \propto \Omega^2 \propto |\underline{\mu}_{if}\cdot\underline{E}_0|^2$) the first-order perturbation approximation may fail and the rates of stimulated transitions are no longer linear in I_ν [see, e.g., (3.1.21)]. In particular, such conditions may be realized when the molecular system interacts with the highly monochromatic and intense radiation of lasers. There is a very wide range of nonlinear phenomena which may be observed when the molecular system interacts with one or few radiation fields [3.15-18], and which are of great interest in laser spectroscopy and reaction dynamics of chemical systems. Among these phenomena (the detailed description of which is beyond the scope of this book) are the coherent anti-Stokes Raman scattering (CARS), widely used as a diagnostic technique (Sect. 5.5), and the simultaneous absorption of more than one photon which will be mentioned in the context of multiphoton processes in Sect.5.4. The major difference between the linear and the nonlinear regimes of light-matter interaction are the different intensity dependencies of the transition rates and the different selection rules for the radiative transitions. For instance, when a molecular system with levels E_f, E_i separated by the energy gap $h\nu_{fi} = E_f - E_i$ interacts with a strong laser field of frequency $\nu \simeq \nu_{fi}/2$ (or with two laser beams with frequencies ν_1 and ν_2 which satisfy $\nu_1 + \nu_2 \simeq \nu_{fi}$) a simultaneous absorption of two photons may take place. The rate of absorption will be proportional to I_ν^2 (or $I_{\nu_1} I_{\nu_2}$). The selection rule for the two-photon transition is no longer governed by $\langle \phi_i | \underline{\mu} | \phi_f \rangle$ (but rather by $\langle \phi_i | \underline{\mu} | \phi_n \rangle \langle \phi_n | \underline{\mu} | \phi_f \rangle$). Consequently, transitions between states (i,f) of similar parity which are forbidden in a single-photon absorption become allowed for two-photon absorption and vice versa. This opens up new possibilities in laser spectroscopy and laser-induced chemistry.

3.1.2 The Line Shape Function

The line shape function $g(\nu,\nu_0)$ accounts for the fact that transitions between levels i and f can be induced by radiation fields with frequencies different from the line center frequency $\nu_0 = (E_f - E_i)/h$. The same function describes the spectral distribution (intensity vs frequency) of the radiation emitted in the spontaneous transition $f \to i$.

The shape and width of the line shape function are determined by the "broadening mechanism". There are two basic types of broadening mechanisms — "homogeneous" and "inhomogeneous". Homogeneous broadening results from processes which assign to each individual molecule a finite probability to absorb or emit radiation over a continuous range of frequencies ν, around ν_0. The corresponding probability distribution function is called the homogeneous line shape function $g_H(\nu,\nu_0)$. Inhomogeneous broadening is a consequence of the fact that different molecules may have different line center frequencies

ν_0. A major inhomogeneous broadening mechanism in the gas phase is due to the Doppler effect, i.e., to the dependence of the frequency emitted by a "monochromatic", moving light source on its relative velocity with respect to the observer [cf. (3.1.40)].

A qualitative insight into the nature of broadening effects can be gained through the uncertainty principle between time and frequency, $\Delta\nu\Delta t \sim 1$. Suppose, for simplicity, that the lower level i is the ground state. If we identify Δt with the spontaneous lifetime of the higher level f, $\Delta t = \tau_{sp} = 1/A_{fi}$, then the so-called "natural linewidth" is $\Delta\nu_{sp} \sim 1/\tau_{sp} = A_{fi}$. For vibrational transitions in diatomic molecules such as HF or HCl, where $\tau_{sp} \sim 10^{-2}$ s, we find $\Delta\nu_{sp} \sim 10^{2}$ s^{-1} ($\sim 10^{-8}$ cm^{-1}) which is extremely narrow compared to the transition frequency $\nu_f - \nu_i \sim 10^{14}$ s^{-1} ($\sim 10^{3}$ cm^{-1}). On the other hand, where electronic transitions for which $\nu_f - \nu_i \sim 10^{15}$ s^{-1} ($\sim 10^{4}$ cm^{-1}) are concerned, the spontaneous lifetimes may be very short, $\tau_{sp} \sim 10^{-9}$s; $\Delta\nu_{sp} \sim 10^{9}$ s^{-1} ($\sim 10^{-2}$ cm^{-1}) is higher.

There are additional decay routes besides spontaneous emission which shorten the lifetime of level f and thereby contribute to the broadening. For instance, if the upper level f is depleted by absorption to a higher level g, $f \rightarrow g$, the effective lifetime of molecules in f will be shortened thus increasing $\Delta\nu$ of the $i \leftrightarrow f$ line. This effect is manifested for example during sequential multiphoton absorption (Sect.5.4). More commonly however the level populations are modified by nonradiative mechanisms. The major nonradiative lifetime broadening mechanisms of small molecules in the gas phase are inelastic collisions. In Sect.1.2 we estimated the time between successive collisions of an HF molecule diluted in 50 Torr Ar as $\tau_{coll} \simeq 10^{-9}$ s. Since the probability for inelastic (R-T) transfer in HF-Ar collisions is $\sim 10^{-1}$ (see Fig.1.9), the nonradiative lifetime is $\tau_{nr} \sim 10^{-8}$ s. Hence, at this pressure $\Delta\nu_{nr} \sim 1/\tau_{nr} \sim 10^{8}$ s^{-1} ($\sim 10^{-3}$ cm^{-1}) is considerably larger than the natural linewidth. In general, the overall lifetime broadening is $\Delta\nu \approx \tau^{-1} = \tau_{sp}^{-1} + \tau_{nr}^{-1} \approx \Delta\nu_{sp} + \Delta\nu_{nr}$.

Radiationless transitions between different electronic manifolds of a polyatomic molecule (such as internal conversion $S_1 \rightarrow S_0$ or intersystem crossing $S_1 \rightarrow T_1$, Sect.2.4.9) as well as intramolecular vibrational energy transfer (and dephasing, Sects.2.4.9 and 5.3.3) from one vibrational mode to the dense manifold of the other vibrational modes present additional decay routes, and thus broadening mechanisms of molecular energy levels. Typical electronic radiationless processes are associated with $\Delta\nu \sim 1/\tau_{nr} \sim 10^{8}$ s^{-1} and vibrational energy transfer (at high excitations) with $\Delta\nu \sim 10^{12}$ s^{-1}.

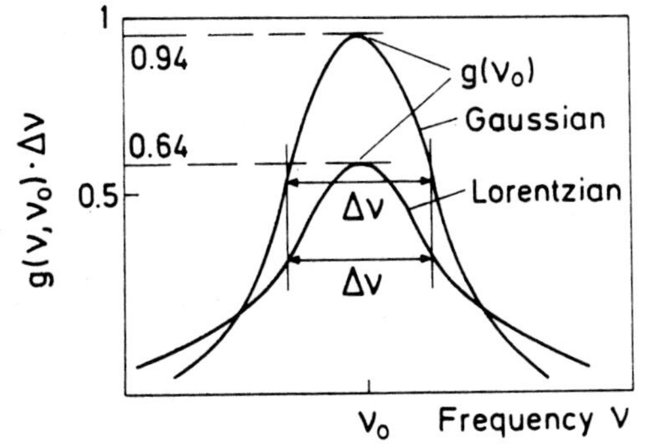

Fig. 3.3. Frequency dependence of Gaussian and Lorentzian line shape functions with the same $\Delta\nu$, the width at half maximum

Lifetime broadening, either radiative or nonradiative, is quite often only a minor broadening effect. A major contribution to the homogeneous linewidth comes from elastic collisions. These collisions do not alter the level populations but destroy the phase relation between the molecules and the interacting radiation field. Based on the classical electron model or via a quantal-phenomenological approach[3], it can be shown that the linewidth due to the elastic, or more generally "dephasing", processes is $\Delta\nu_{coll} \sim 1/\tau_{coll} = \omega^{-1}$ where ω is the collision frequency. Hence for our example of HF diluted in 50 Torr Ar, $\Delta\nu_{coll} \sim 10^9$ s^{-1} ($\sim 10^{-2}$ cm^{-1}) which is larger than $\Delta\nu_{nr}$.

The homogeneous line shape function has a Lorentzian form [3.1-3] (Fig.3.3)

$$g_H(\nu,\nu_0) = \frac{2}{\pi\Delta\nu_H} \frac{1}{1+4[(\nu-\nu_0)/\Delta\nu_H]^2} \quad , \tag{3.1.38}$$

where $\Delta\nu_H$ is the total homogeneous linewidth. The maximum of g_H obtains at line center $\nu = \nu_0$,

$$g_H(\nu_0,\nu_0) = g_H(\nu_0) = 2/\pi\Delta\nu_H = 0.64/\Delta\nu_H \quad . \tag{3.1.39}$$

When molecules interact with very intense radiation fields, the homogeneous linewidth is further broadened. This effect, called power broadening, will be considered in the next section.

---

3 In the quantum mechanical description the effect of dephasing collisions is introduced via a phenomenological dephasing time T_2, describing the relaxation of the nondiagonal density matrix elements (which account for phase correlations between the molecules, cf. Sect.2.3.7). In analogy to the theory of magnetic resonance T_2 is also called the transverse relaxation time. The energy, or level population, relaxation time T_1 is introduced into the equation of motion for the diagonal matrix elements $T_1 \sim \tau_{nr}$. In the so-called "impact approximation" $T_2 \sim \tau_{coll} = \Delta\nu_{coll}^{-1}$.

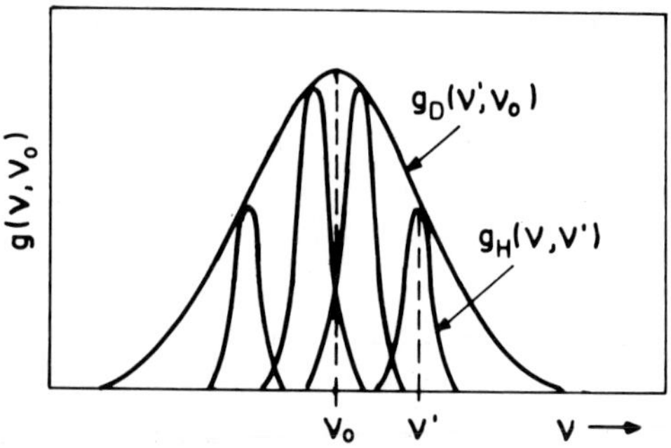

Fig. 3.4. A superposition of homogeneous line shapes with Gaussian weights. Here $\Delta\nu_D > \Delta\nu_H$. The absorption or emission probability at frequency ν is proportional to $\int d\nu' g_H(\nu,\nu') g_D(\nu',\nu_0)$

The line center frequency, now denoted as ν', of a molecule moving with velocity v_x along the x axis is Doppler shifted by an amount $\nu' - \nu_0 = (v_x/c)\nu_0$ with respect to ν_0 which we now use for the frequency of molecules with $v_x = 0$. Since at thermal translational equilibrium the probability distribution function of v_x is Gaussian (Maxwellian), so is the distribution of $\nu' - \nu_0$. This latter probability distribution function is the inhomogeneous line shape function with the Gaussian form (Fig.3.4)

$$g_D(\nu',\nu_0) = \left[\frac{4\ln(2)}{\pi}\right]^{\frac{1}{2}} \frac{1}{\Delta\nu_D} \exp\left[-4\ln(2)\left(\frac{\nu'-\nu_0}{\Delta\nu_D}\right)^2\right] \tag{3.1.40}$$

with the Doppler width

$$\Delta\nu_D = [8kT\ln(2)/mc^2]^{\frac{1}{2}}\nu_0 \quad . \tag{3.1.41}$$

At line center

$$g_D(\nu',\nu_0) = [4\ln(2)/\pi]^{\frac{1}{2}}\Delta\nu_D^{-1} = 0.94/\Delta\nu_D \quad . \tag{3.1.42}$$

Example: For HF at T = 300 K the Doppler width of, say, the v = 1→0 transition is $\Delta\nu_D \simeq 10^{-2}$ cm^{-1}. Thus, comparing with the above results when the pressure is ~1 Torr, $\Delta\nu_D >> \Delta\nu_H$. When the pressure is raised to ~100 Torr, $\Delta\nu_D \sim \Delta\nu_H$. Note that $\Delta\nu_D \propto \nu_0$. Hence, $\Delta\nu_D$ is very high for electronic transitions and very small for pure rotational transitions.

The net response of a spectral line to radiation at frequency ν is obtained by superimposing the Lorentzian profiles $g_H(\nu,\nu')$ according to the statistical distribution of line center frequencies ν' (Fig.3.4). Since the statistical weight (i.e., the fraction) of molecules with central frequencies ν' is $g_D(\nu',\nu_0)$, we find

$$g(\nu,\nu_0) = \int g_H(\nu,\nu')g_D(\nu',\nu_0)d\nu' \quad . \tag{3.1.43}$$

It can easily be verified that in the two extreme cases $\Delta\nu_D \gg \Delta\nu_H$ and $\Delta\nu_D \ll \Delta\nu_H$ we obtain, as expected, $g(\nu,\nu_0) = g_D(\nu,\nu_0)$ and $g(\nu,\nu_0) = g_H(\nu,\nu_0)$, respectively.

When $\Delta\nu_D \sim \Delta\nu_H$, $g(\nu,\nu_0)$ is given by the integral (3.1.43) which can be expressed in terms of the error function [erf(x)] (see, e.g., [3.19]). The width of the line shape function, or, as it is also called, the Voigt profile, is roughly $\Delta\nu_D + \Delta\nu_H$. Since for both purely Doppler and purely Lorentz broadened lines we can write [see (3.1.39,42)]

$$g(\nu_0) = g_0/\Delta\nu \quad , \tag{3.1.44}$$

where $g_{0,H} = 2/\pi = 0.64$ and $g_{0,D} = [4\ln(2)/\pi]^{\frac{1}{2}} = 0.94$, the cross section for stimulated emission at line center $\nu = \nu_0$ can be expressed as

$$\sigma_{fi}(\nu_0) = (c^2/8\pi\nu_0^2)(g_0/\Delta\nu)A_{fi} \quad . \tag{3.1.45}$$

As a concrete example, consider again a vibrational transition of a diatomic molecule like HF or HCl where $\nu_0 \sim 10^{14}\ s^{-1}$, $A_{fi} \sim 10^2\ s^{-1}$, and $\Delta\nu \sim 10^8\ s^{-1}$. Hence, $\sigma_{fi} \sim 10^{-16}\ cm^2$ or $k_{fi} \sim 10^{-6}\ cm^3 molec^{-1} s^{-1}$. These values are very high in comparison to collisional rate constants, a typical value for a collisional bimolecular rate constant at room temperature being $10^{-10}\ cm^3 molec^{-1} s^{-1}$. The enormous rate constants associated with stimulated emission, due basically to the speed of light, are of key importance in chemical lasers, as they often dominate over the slower, collisional, processes (Sect.4.4).

3.1.3 Coherent Interaction

Consider a molecule interacting with a very short laser pulse of duration $t = t_p \ll \Delta\nu_H^{-1}$, contrary to the requirement in (3.1.37). Since $\Delta\nu_H \simeq 1/T_2 \simeq 1/\tau_{coll}$, where T_2 is the dephasing time and τ_{coll} is the time between successive collisions, the interaction between the molecule and the field cannot be disturbed by dephasing collisions. In other words, the motion in phase of the field and the induced dipole moment is not interrupted during the short interaction time. This type of interaction is called coherent. Hence, coherent effects are expected when the pressure is low, the interaction period is short, and (as will be argued below) when the field is strong.

The rate equation approach is appropriate when the interaction is incoherent, i.e., when dephasing events wash out the definite phase relations between the molecules and the field. In the following we shall examine the condition for coherence effects and thus delineate the operational regime of chemical lasers if such effects are to be avoided. For detailed descriptions of coherent phenomena (such as self-induced transparency, superradiance, or photon echo) see, for example, [3.1,6].

It was shown by Rabi in 1937 (see, e.g., [3.6]) that for the two-level problem the coefficients C_i and C_f in (3.1.4) can be evaluated exactly. Rabi's ("flopping") formula[4] for absorption, i.e., for $C_i(0)=1$, $C_f(0)=0$, is

$$P_{if}(t) = |C_f(t)|^2 = \Omega^2 \frac{\sin^2\lambda t}{\lambda^2} \quad , \tag{3.1.46}$$

where

$$\lambda = [\pi^2(\nu-\nu_{fi})^2+\Omega^2]^{\frac{1}{2}} \quad . \tag{3.1.47}$$

As a test of consistency, note that for weak fields, $\Omega\to 0$, we recover the first-order result (3.1.7).

The probability of finding the system in the upper state oscillates with a frequency $\lambda/2$ where λ, defined by (3.1.47), is dependent on both the field strength Ω and the detuning gap $\nu-\nu_{fi}$. The amplitude of these oscillations, i.e., the maximal value of $|C_f|^2$ as a function of these parameters, is

$$A = |C_f|^2_{max} = \Omega^2/[\pi^2(\nu-\nu_{fi})^2+\Omega^2] \quad . \tag{3.1.48}$$

Figures 3.5 and 3.6 show A and $|C_f(t)|^2$ as functions of the detuning frequency. Equation (3.1.48) and Fig.3.5 indicate that due to the Ω^2 term in the denominator the spectral response of the system is broadened as the field strength increases. This effect is known as "power broadening". Like collisional broadening it is a homogeneous effect, represented by a Lorentzian profile with full width at half maximum, $\Delta\nu=\Omega/\pi$, (3.1.46)[5]. As in Sect.3.1.2 the uncertainty relation provides a qualitative explanation for this broadening mechanism: Since the field drives the system to oscillate between the

4 Rabi derived his formula for the probability of spin flipping due to applied RF magnetic field on spin 1/2 atoms in a Stern-Gerlach apparatus.

5 It can be shown [3.1] that the homogeneous line shape function due to the combined effect of dephasing and power broadening is given by (3.1.38) with $\Delta\nu_H \simeq \Delta\nu_{dephasing} + \Delta\nu_{power}$.

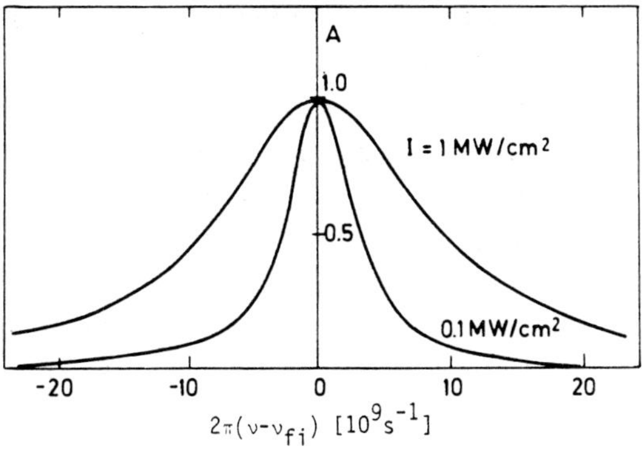

Fig. 3.5. The amplitude of oscillation vs detuning frequency for two different radiation intensities $I(\mathrm{W/cm^2})$. $I = cE_0^2/8\pi \propto (\Omega/\mu)^2$, ($\mu$=0.1 Debye). Increase in the intensity leads to broadening of the spectral response (cf. Fig.1.20)

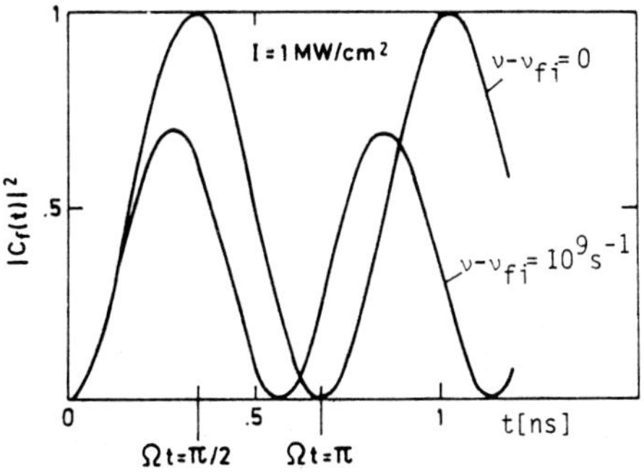

Fig. 3.6. Probability of finding the system in the upper state as function of time for two different frequencies. For $\nu = \nu_{fi}$ the system is totally excited when $(\mu E_0/\hbar)t = 2\Omega t = \pi$ and totally unexcited when $2\Omega t = 2\pi$

levels with frequency of the order of Ω the level lifetime is uncertain by an amount $\Delta t \simeq 1/\Delta\nu = \pi/\Omega$.

An interesting special case is that of exact resonance, $\nu = \nu_{fi}$. Here

$$P_{if}(t) = |C_f(t)|^2_{\nu=\nu} = \sin^2\Omega t \tag{3.1.49}$$

and the probability of finding the system in the upper state oscillates between 0 and 1 (see Figs.3.5 and 3.6). These values are reached when $(\mu E_0/\hbar)t = 2\Omega t = 0,\ 2\pi,\ 4\pi$, and $\pi,\ 3\pi,\ 5\pi, \ldots$, respectively. Hence, after a laser pulse with duration $t_p = \pi/2\Omega = \pi(\hbar/\pi E_0)$ has passed through the system, the molecules are left totally excited. Such a pulse is called a "π pulse". On the other hand, a pulse characterized by $t_p = \pi/\Omega$ leaves the system totally unexcited, a phenomenon known as "self-induced transparency".

It is interesting to compare the conditions for observing such coherent effects with the requirements (3.1.37) for the validity of the golden rule. In order that $|C_f(t)|^2$ will reach the value 1 it is necessary that the pulse duration t_p and the field strength as expressed via Ω satisfy $\Omega t_p > 1$. As mentioned above, coherent interaction requires $t_p << T_2$. Hence, the field should be intense enough to satisfy $\Omega >> 1/T_2$. (Thus, power broadening should exceed the dephasing broadening $\Omega/\pi = \Delta\nu_{power} >> \Delta\nu_{dephasing} = 1/\pi T_2$.) Hence the condition for coherent interaction $\Omega >> 1/t = 1/t_p >> 1/T_2$ is the opposite of the condition for the validity of the golden rule (3.1.37).

Our discussion so far was limited to a "collision-free" two-level system interacting with a purely monochromatic field. The effect of an elastic dephasing collision is to erase the phase history of the interaction with the field and let the system evolve from a new initial condition[6]. If $t >> T_2$, there are many dephasing events during the interaction period and one has to average over a sequence of varying initial conditions, characterized by a random distribution of the phases γ_i, γ_f. A detailed analysis shows that in this approximation the level and photon populations are governed by rate equations with field-dependent cross sections. In the limit of very intense fields $\Omega t >> 1$ the coefficients $|C_i|^2$, $|C_f|^2$ approach a stationary value depending on the detuning frequency $\nu - \nu_{fi}$. ($|C_i|^2 = |C_f|^2 = 1/2$ when $\nu = \nu_{fi}$.) In this case we say that the transition is saturated by the field. In the opposite limit of weak fields $\Omega t << 1$, one obtains rate equations of the form (3.1.35) and (3.1.36) which we derived through the golden rule.

The vast majority of chemical laser experiments have been carried out at relatively high pressures where coherent effects cannot be observed and the absorption and emission phenomena can be accounted for in the rate equation approximation. Based on our previous estimate of $T_2 \simeq \tau_{coll} \simeq 10^{-8}$ s at $P = 1$ Torr and the requirement $t_p << T_2$, we can expect that for total pressures below ~ 0.1 Torr coherent effects may start contributing for pulse duration $t_p < 0.1$ μs. Power broadening and possibly some coherent excitation are believed to be important in the multiphoton dissociation of polyatomic molecules by an intense infrared laser (see Sect.5.4 and Fig.1.20).

6 Elastic collisions do not affect the populations $|C_i|^2$ and $|C_f|^2$ but change the phases γ_i and γ_f of the coefficients, $C_f = |C_f| \exp(i\gamma_f)$. After many dephasing collisions $t >> T_2$ the average phase difference $\gamma_i - \gamma_f$ is zero. T_2 is the characteristic time for relaxation of the correlation $C_i^* C_f = |C_i C_f| \times \exp[i(\gamma_i - \gamma_f)]$, i.e., for $t >> T_2$, the average of the nondiagonal density matrix elements $C_i^* C_f = 0$. Inelastic collisions change both but are relatively improbable and can be neglected for $T_2 << t << T_1$, where T_1 is the energy relaxation time (see also Sect.2.3.7).

3.2 Essential Physics of Lasers

This section outlines briefly some of the basic characteristics of laser operation and laser radiation. The discussion of these topics which will be rather qualitative and often only sketchy represents the threshold requirement for the interpretation of the operational characteristics of lasers. Whenever a more detailed description is called for the reader should consult one of the many excellent textbooks on laser physics and quantum electronics [3.1-7].

3.2.1 The Gain Coefficient

The basic requirement for light amplification by stimulated emission of radiation is population inversion. The device in which laser radiation is generated is called a laser oscillator. However, the amplification principle is most easily introduced by considering the operation of a laser amplifier. An amplifier is a volume containing the active medium which may be molecules (atoms, ions) in the gas, liquid, or solid phase. The molecules are pumped to a state of population inversion with respect to one (or more) of their radiative transitions so that a signal of radiation with frequency corresponding to this transition will be amplified while passing through the medium.

Suppose that the active transition is between levels E_f and E_i, $E_f > E_i$, with the corresponding degeneracies g_f and g_i. Without specifying the nature of the pumping mechanism let us assume that the molecules have been pumped to a constant inversion level $\Delta N = N_f - (g_f/g_i)N_i$, independent of the spatial location in the amplifier. Let us assume further that the radiation signal entering the amplifier tube at $z=0$ is a short monochromatic pulse with frequency $\nu \simeq \nu_0 = (E_f - E_i)/h$, where ν_0 is the line center frequency of the active transition. The radiation density (erg/cm^3) of the signal which passes through the amplifying medium with velocity c will be denoted by $\rho_\nu(z)$, or $\rho_\nu(t)$ where $z = ct$ is the axial location of the pulse at instant t.

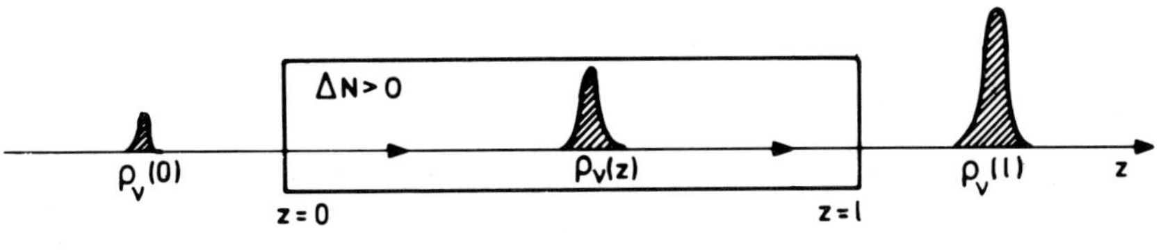

Fig. 3.7. A schematic illustration of a laser amplifier. The small input radiation signal is amplified while travelling in the cell containing the inverted medium

The radiation density is related to the intensity $I_\nu(z)$ and the photon density $\phi_\nu(z)$ through $I_\nu(z) = c\rho_\nu(z) = c\phi_\nu(z)h\nu$. Since no amplification takes place at $z < 0$ or $z > \ell$ we have $\rho_\nu(z<0) = \rho_\nu(z=0) = \rho_\nu(0)$ and $\rho_\nu(z>\ell) = \rho_\nu(z=\ell) = \rho_\nu(\ell)$.

At any point $z = ct$ along its direction of propagation the radiation signal interacts with a new, inverted, population of molecules. Neglecting spontaneous emission and relaxation processes (which can modify the uniform inversion density along the amplifier) the change in the energy density of the signal is given by

$$\frac{d\rho_\nu(z)}{dz} = \sigma_{fi}(\nu)\Delta N\rho_\nu(z) = \gamma(\nu)\rho_\nu(z) \quad , \tag{3.2.1}$$

where we have used $z = ct$, $d\rho_\nu(z)/dz = (1/c)d\rho_\nu(t)/dt$, and (3.1.36). $\Delta N = N_f - (g_f/g_i)N_i$ is the population inversion and [cf. (3.1.25)]

$$\gamma(\nu) = \sigma_{fi}(\nu)\Delta N = g(\nu,\nu_0)B_{fi}(h\nu/c)\Delta N \tag{3.2.2}$$

is the "small signal gain coefficient". This quantity measures the relative amplification per unit length. Note that (3.2.1) is also valid when $\Delta N < 0$, leading to a net absorption rather than emission amplification. Thus, $-\gamma(\nu)$ which can also be expressed as $-\gamma(\nu) = \sigma_{if}[N_i - (g_i/g_f)N_f] = \varepsilon(\nu)$ is the familiar expression for the absorption coefficient.

The additional "small signal" associated with $\gamma(\nu)$ indicates that (3.2.1) is only valid when $\rho_\nu(z)$ is small enough to ensure that the population inversion is not considerably modified by the interaction with the radiation signal or, alternatively, that all components of the pulse sample the same population inversion. If the radiation density is high the front edge of the pulse may deplete the upper level population to such an extent that its latter components interact with a considerably reduced population inversion and thus are amplified to a lesser degree. This phenomenon leads to effective temporal narrowing of the pulse which cannot be accounted for by the simple equation (3.2.1). In the limit of very large radiation densities the amplification process "saturates", that is, the population inversion is completely drained by the signal so that behind the pulse $\Delta N = 0$. In this case $\rho_\nu(z)$ grows linearly with z rather than exponentially as implied by (3.2.1).

Integration of (3.2.1) yields

$$\rho_\nu(z) = \rho_\nu(0)\,\exp[\gamma(\nu)z] \quad . \tag{3.2.3}$$

Hence, the overall amplification ratio, or the total gain per pass, is

$$G(\ell) = \rho_\nu(\ell)/\rho_\nu(0) = \exp[\gamma(\nu)\ell] \quad . \tag{3.2.4}$$

Laser amplifiers can also be used to amplify continuous, rather than pulsed, signals. It is clear however that in this case the amplifying medium should be continuously pumped and relaxation processes which tend to reduce the population inversion must be taken into account. It is also clear that effective amplification (of either pulsed or continuous signals) requires that ν, the frequency of the signal entering the amplifier tube, should nearly or exactly match the inverted transition frequency ν_0. Hence the most reasonable and common experimental use of amplifiers is in combination with laser oscillators which generate radiation signals by stimulated emission on the same radiative transition utilized in the amplifier (cf. Sects.3.4 and 4.3). Laser oscillators are the subject of the next section.

3.2.2 Laser Oscillators

Laser amplifiers are driven by external signals. Laser oscillators are the generators of these highly monochromatic and unidirectional radiation signals. The oscillator consists of an open optical resonator containing the active laser material and a pumping device to bring this material into a state of population inversion. It can thus be regarded as an amplifier bounded by two mirrors which are aligned perpendicular to the long (laser) axis[7] (Fig.3.8).

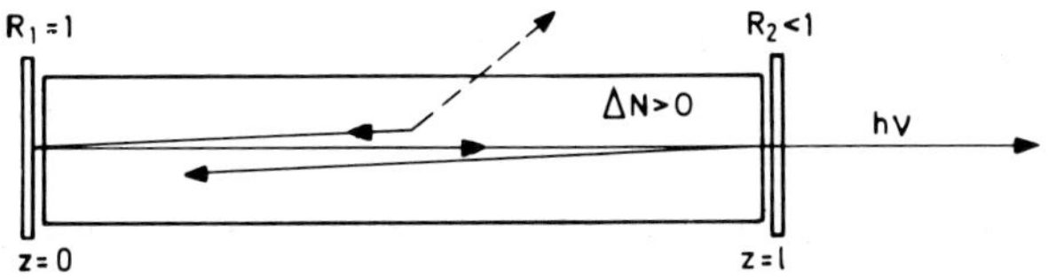

Fig. 3.8. Schematic illustration of a laser oscillator with two plane parallel reflectors. A part of the intense radiation field generated in the cavity by stimulated emission is coupled out as a narrow beam through the partially transmitting mirror. Off-axial radiation leaks out through the side walls

7 In Fig.3.8 as well as throughout this section we shall assume that the length of the active laser material ℓ is equal to the distance between the mirrors L. This is not always the case in real laser devices, as will be demonstrated by various examples in Chaps.4 and 5. We make the assumption $\ell = L$ to keep the formal expressions simple. Their extension to the case $\ell < L$ is usually straightforward.

The mirrors which may be flat or curved comprise the open optical cavity. One (or both) of the mirrors is partially transmitting at the wavelength of the laser radiation. The buildup of laser oscillations is triggered by spontaneous emission from the upper level of the (inverted) transition. The emitted photons induce stimulated emission of photons from other molecules which stimulate further transitions and so on. The photons travelling along or very close to the axial direction will be reflected by the mirrors and thus continue to induce additional transitions. Those which move in off-axial directions will leak out from the optical cavity through the transparent side walls, either immediately or after a few reflections. In the case of a resonator with two plane parallel mirrors the processes just described generate an intense wave of radiation bouncing back and forth between the end reflectors and (almost) exactly along the laser axis. Such a standing wave pattern is called the lowest transverse electromagnetic mode of the resonator and is designated by TEM_{00}. For reasons to be clarified in the next section, most of the experimental arrangements involve at least one curved, rather than two plane parallel, reflectors. In these cases as well, the lowest transverse mode is the most, though not the only, stable wave pattern.

A part of the radiation travelling back and forth along the resonator axis is coupled out through one of the mirrors as a highly collimated and monochromatic signal propagating with an almost constant phase across its wave front. (These special characteristics of laser radiation will be discussed in some more detail in the next section.) The output coupling reduces the radiation density within the optical cavity. Diffraction effects which lead to divergence and leakage of radiation through the side walls, absorption, and scattering by the mirrors and within the laser medium introduce additional, unwanted, loss mechanisms for the radiation in the cavity. Thus, in order to start or sustain laser oscillations the gain of photons by stimulated emission should overcome the loss of radiation by both the useful output coupling and the dissipative loss processes. Equivalently, the net relative gain of photons in one full round trip between the mirrors should be equal to or larger than 1. This requirement can be expressed in the form

$$R_1R_2G(2\ell)\exp(-2\chi\ell) = R_1R_2\exp\left\{2[\gamma(\nu)-\chi]\ell\right\} \geqq 1 \quad . \tag{3.2.5}$$

The factor $G(2\ell)=\exp[2\gamma(\nu)\ell]$ is the total relative gain by stimulated emission per round trip [cf. (3.2.4)]. R_1 and R_2 are the reflectivities of the mirrors (R is the fraction of photons reflected by a mirror). Hence R_1R_2 is the fraction of photons remaining in the cavity after the two reflections.

The factor $\exp(-2\chi\ell)$ represents the fraction of radiation which is not lost by diffraction, absorption, and the other useless processes. [Assuming that these losses are independent of the axial location and proportional to the radiation density, i.e., $d\rho/dz = -\chi\rho$, the relative loss per round trip is $\exp(-2\chi\ell)$ and χ can be interpreted as the loss per unit length]. The inequality (3.2.5) can be rewritten in the form

$$\gamma(\nu) \geqq \gamma_t(\nu) \equiv \chi + (1/2\ell)\ln(1/R_1R_2), \tag{3.2.6}$$

where $\gamma_t(\nu)$ is the threshold value of the gain coefficient. The last equation reflects the requirement that the (differential) gain per unit length should exceed the average (differential) loss per unit length. Another common statement of this requirement is obtained by transforming from length to time units. Using c to denote the speed of light in the laser medium, we can rewrite (3.2.5) as

$$\alpha(\nu) \equiv c\gamma(\nu) \geqq c\gamma_t(\nu) = \alpha_t(\nu) = 1/\tau_c \tag{3.2.7}$$

with

$$\frac{1}{\tau_c} = \chi c + \frac{c}{2\ell}\ln\left(\frac{1}{R_1R_2}\right) \quad . \tag{3.2.8}$$

$\alpha(\nu)$ is usually called the gain coefficient per unit time. $\alpha_t(\nu)$ is its threshold value. τ_c is called the photon lifetime in the cavity. The origin of this name becomes clear if we note that in a "cold" or "passive" cavity which is simply an optical cavity without the amplifying medium (hence $\gamma(\nu)\equiv 0$), the reduction in the radiation density during the time interval of one full round trip, $T = (2\ell/c)$, is

$$\rho(t+T)/\rho(t) = R_1R_2\exp(-2\chi\ell) = \exp(-T/\tau_c) \quad . \tag{3.2.9}$$

This equation implies that $1/\tau_c$ is the number of cycles (round trips) passing until the radiation energy density decays to 1/e of its initial value. If ρ changes only slightly during the time interval T then $\exp(-T/\tau_c) \simeq 1 - T/\tau_c$; hence $\Delta\rho/T = [\rho(t+T)-\rho(T)]/T \simeq -\rho(t)/\tau_c$ which upon replacing $\Delta\rho/T$ by $d\rho/dt$ becomes

$$\frac{d\rho}{dt} = -\frac{\rho}{\tau_c} \quad . \tag{3.2.10}$$

The decay lifetime of the radiation energy τ_c is related to another common characteristic of resonators known as the quality factor Q. The quality factor is defined as the ratio between the energy stored in the resonator $E_{st} = \rho V$ (V is the volume) and the power dissipated, $P = -dE_{st}/dt = -Vd\rho/dt$, times the angular frequency $\omega = 2\pi\nu$ of the radiation. Thus, using (3.2.10), we find

$$Q = (E_{st}/P)\omega = 2\pi\nu\tau_c \quad . \tag{3.2.11}$$

The photon lifetime provides a measure for the mean residence time of the radiation in the passive cavity. Hence, using the uncertainty relation between time and frequency, one can obtain a simple estimate for the spectral width $\Delta\nu_c$ of this radiation. Namely,

$$\Delta\nu_c \simeq 1/2\pi\tau_c = \nu/Q \quad . \tag{3.2.12}$$

The gain coefficient is proportional to the population inversion $\gamma(\nu) = \sigma(\nu)\Delta N$. Thus, the threshold condition (3.2.7) can be expressed as

$$\Delta N_t = 1/c\sigma(\nu)\tau_c \tag{3.2.13}$$

$$= \frac{8\pi\nu^2}{c^3 g(\nu,\nu_0)} \frac{\tau_{sp}}{\tau_c} \quad , \tag{3.2.14}$$

where $\Delta N_t = \gamma_t(\nu)/\sigma(\nu)$ is the threshold value of the population inversion. The second equality has been obtained by using (3.1.34). If lasing takes place at the line center frequency $\nu = \nu_0$, we can use $g(\nu_0,\nu_0) = g_0/\Delta\nu$ where $\Delta\nu$ is the linewidth of the lasing transition [cf. (3.1.44)]. Hence

$$\Delta N_t = \frac{8\pi\nu_0^2\Delta\nu\tau_{sp}}{c^3 g_0\tau_c} \quad . \tag{3.2.15}$$

This equality, for the minimal inversion required to sustain oscillation, as well as all its equivalent forms above, is known as the Schawlow-Townes threshold condition.

In order that lasing oscillations start to occur it is necessary that $\Delta N \geq \Delta N_t$. Steady-state operation requires $\Delta N = \Delta N_t$. Clearly, to maintain this constant value of population inversion the transition must be continuously pumped. In Sect.3.2.4 we shall show that the minimal pumping rate required for steady-state operation is proportional to $\Delta N_t/\tau_2$ where τ_2 is the overall relaxation time of the upper level of the lasing transition. If nonradiative relaxation is slower than spontaneous radiative relaxation, then $\tau_2 \simeq \tau_{sp}$ and from (3.2.15) it is clear that the minimal pumping rate is proportional to $\nu_0^2 \Delta\nu / \tau_c$. This implies that the pumping requirements become more demanding for transitions with broad linewidth and high frequency and for cavities with low quality factors. This is why infrared lasers are generally more easily operated than visible and ultraviolet lasers.

Finally, in order to gain some idea about the inversion densities involved in typical laser experiments, let us consider two typical examples: a chemical laser and a dye laser. (These laser types are discussed in detail in Chap.4 and Sect.3.4.3, respectively.) For simplicity, we assume that output coupling is the only loss mechanism. Thus for an $\ell = 100$ cm long cavity with one perfectly reflecting mirror, $R_1 = 1$, and a 90% reflecting output coupling mirror, $R_2 = 0.9$, we find $\tau_c \sim 10^{-7}$ s [use (3.2.8) with $\chi=0$]. The spontaneous lifetimes of infrared vibrotational transitions in molecules such as HF or HCl are of the order $\tau_{sp} \sim 10^{-2}$ s and ν is typically 10^{13}-10^{14} s^{-1}. The linewidth in the few Torr range is $\Delta\nu \sim 10^8$ s^{-1}. Since $g_0 \sim 1$ we find, using (3.2.15), $\Delta N_t \sim 10^{11}$ molec/cm^3. For comparison, the number densities corresponding to pressures in the few Torr region are 10^{16}-10^{17} molec/cm^3. Thus, although typical chemical lasers such as those utilizing HF or HCl molecules involve usually 10 to 50 active transitions, the pumping reactions can easily generate inversion densities of 10^{13}-10^{15} molec/cm^3 per transition, which is far above the threshold inversion required. [It should be noted however that very soon after the attainment of threshold the inversion density is clamped to its threshold value (Sect.3.2.4).]

In a dye laser, using the same cavity parameters, $\tau_c = 10^{-7}$ s, $\nu = 6 \times 10^{14} s^{-1}$ (λ=5000 Å), and $\Delta\nu = 3 \times 10^{13}$ s^{-1}, $\tau_{sp} = 10^{-9}$ s (an allowed transition is involved). One has to add a refraction index correction to (3.2.15) by multiplying by n_0^3, $n_0 \simeq 1.5$ for most solvents. ΔN_t becomes again $\sim 10^{11}$ molec/cm^3. Note that in this case the expected increase in ΔN_t due to the high frequency and broader linewidth of the dye laser is compensated by the decrease in τ_{sp}. The operational difficulty referred to above is reflected in larger pumping rates required for dye lasers compared to HX lasers. Dye concentrations normally used in laser systems are 10^{15}-10^{18} molec/cm^3, and threshold population is usually obtained by optical pumping with either a fixed frequency laser (N_2, Ar^+, Nd^{3+}: YAG doubled) or with an intense flashlamp.

3.2.3 Laser Radiation and Modes

Thermal sources such as ordinary lamps radiate into all directions, over a wide spectral range, and with random phase relations between the radiating atoms. On the other hand, laser radiation is highly collimated, highly monochromatic, and the phase relations between the various components of the emitted signal are nearly constant in time and space. The differences between thermal and laser radiation stem from a combination of two factors. First, the stimulated emission responsible for the generation of laser radiation ensures that the phases of the stimulating and emitted photons are equal, as opposed to the random character of the spontaneous emission from thermal sources. Second, the repeated oscillations in the laser resonator reshape the radiation field into a highly monochromatic and unidirectional wave pattern. To understand how these properties are achieved we need some acquaintance with the concept of cavity modes.

Modes in a Resonant Cavity

The laser resonator with the two mirrors perpendicular to its long axis is an open optical cavity. The analysis of the possible field structures in such cavities is usually rather complicated and it proves instructive to consider first the properties of radiation fields in closed cavities. The simplest closed optical cavity is a rectangular box of volume V bounded by plane parallel reflecting walls. To retain the similarity with laser cavities let us assume that ℓ_z, the linear longitudinal dimension of the box, is considerably larger than the transverse dimensions ℓ_x and ℓ_y. The electromagnetic field present in any given enclosure can be expressed as a weighted (Fourier) sum of the allowed standing waves in this enclosure, which are the stationary solutions of the appropriate wave equation. The standing waves in the rectangular box of volume $V = \ell_x\ell_y\ell_z$ are of the form $\sin(k_x x)\,\sin(k_y y)\,\sin(k_z z)$ with

$$k_x = n\pi/\ell_x \quad , \quad k_y = m\pi/\ell_y \quad , \quad k_z = q\pi/\ell_z \quad , \tag{3.2.16}$$

where $n,m,q = 0,1,2,\ldots$. k_x, k_y, and k_z are the magnitudes of the x, y, and z components of the wave vector $\underline{k}$. The magnitude of the wave vector $k = |\underline{k}|$ is related to the frequency ν, or the wavelength λ, of the wave by the dispersion relation

$$k^2 = k_x^2 + k_y^2 + k_z^2 = 4\pi^2/\lambda^2 = \nu^2/c^2 \quad . \tag{3.2.17}$$

Each triad n,m,q defines a cavity mode, or more precisely a doubly-degenerate mode since a wave characterized by these numbers can oscillate in two independent directions of polarization. Each mode defines a standing wave oscillating in the direction specified by the relative magnitudes of the wave vector components, (3.2.16). The frequency of the wave is determined by (3.2.17). Since $\ell_z \gg \ell_x, \ell_y$ it is common to distinguish between the longitudinal (axial) mode number q and the transverse mode numbers n,m.

The spectral composition of an electromagnetic field enclosed in the cavity is determined by the relative contributions of the different modes. In the case of radiation fields in the UV, visible, and infrared regions and macroscopic volumes $\ell_x, \ell_y, \ell_z \gg \lambda$, there will be many possible modes within any small frequency interval $\nu, \nu+d\nu$. This means, for example, that the radiation field resulting from spontaneous emission in an atomic or molecular transition with spectral linewidth $\Delta\nu$ will usually contain a multitude of cavity modes. It is thus common and convenient to denote the number of modes in the frequency interval $\nu, \nu+d\nu$ as $p(\nu)d\nu$ where $p(\nu)$, the mode density, is the number of modes per unit frequency. $p(\nu)$ can be determined as follows: From (3.2.16) we find $\Delta k_x \Delta k_y \Delta k_z = (\pi^3/V)\Delta n \Delta m \Delta q$; hence the number of modes per unit "volume" $(\Delta k_x \Delta k_y \Delta k_z = 1)$ in the wave vector space is $2\Delta n \Delta m \Delta q = 2V/\pi^3$ (the factor 2 is due to the two independent polarizations). Now, (3.2.16) describes an equation of a sphere in the $\underline{k}$ space. The volume of such a sphere with radius k is $(4/3)\pi k^3 = (4/3)\pi(2\pi\nu/c)^3$. Since k_x, k_y, k_z are positive, only one octant of this sphere contains cavity modes. Multiplying this volume $(1/8)(4/3)\pi(2\pi\nu/c)^3$ by the number of modes per unit volume $2V/\pi^3$, we find that $\Omega(\nu)$, the total number of modes whose frequency is less than or equal to ν, is given by $\Omega(\nu) = (8/3)\pi V\nu^3/c^3$. The number of modes in the frequency interval $\nu, \nu+d\nu$ is $p(\nu)d\nu = \Omega(\nu+d\nu) - \Omega(\nu)$. Hence

$$p(\nu) = \frac{d\Omega(\nu)}{d\nu} = \frac{8\pi\nu^2}{c^3} V \quad . \tag{3.2.18}$$

Although we have derived this expression for the special case of a rectangular box it can be shown to be valid for any arbitrary shape of the volume, as long as $\lambda^3 \ll V$.

In the quantum theory of radiation the quantum state of a photon is specified by its mode and its direction of polarization. If the walls of the cavity are in thermal equilibrium or the cavity contains matter in thermal equilibrium at temperature T, then the average number of photons per quantum state (mode and polarization) is known from Bose-Einstein statistics to be (see, e.g., [2.54])

$$\phi(\nu) = [\exp(h\nu/kT)-1]^{-1} \quad . \tag{3.2.19}$$

where ν is the frequency of the mode. The average radiation energy per unit volume and unit frequency interval in a cavity at thermal equilibrium is thus

$$\rho(\nu) = p(\nu)\phi(\nu)h\nu/V = \frac{8\pi h\nu^3}{c^3}\frac{1}{\exp(h\nu/kT)-1} \quad , \tag{3.2.20}$$

which is Planck's formula for black body radiation [cf. (1.1.13)].

The number of cavity modes $p(\nu)\Delta\nu$ within the linewidth $\Delta\nu$ of an infrared or a visible transition is typically very large ($p(\nu)\Delta\nu \sim 10^8$ for the He-Ne laser transition, for $V=1\ cm^3$. Other examples will be given below). These modes cover densely the spectral linewidth and correspond to wave vectors pointing at all possible (4π) directions. On the other hand laser radiation propagates in a single well-defined direction and its spectral width $\Delta\nu_\ell$ is much narrower than the transition linewidth $\Delta\nu$. This is due to the fact that laser cavities support oscillations only on very few modes. These are the modes with the lowest, $n=0$, $m=0$, or the few lowest, transverse components which correspond to oscillations parallel or very nearly parallel to the laser axis direction. The practical route for eliminating the high order transverse modes is simply to replace the closed cavity by the open one. To be precise, our description of the closed cavity modes cannot be simply translated to the case of open cavities. The modes of such cavities are determined, like those of closed ones, by the stationary solutions of the wave equations. The field patterns corresponding to these modes depend critically on the geometrical parameters of the resonator, such as the distance between the mirrors, as well as on their size, shape, and curvature. Except for a few special cases the stationary wave patterns are found by numerical solutions of the wave equation with the appropriate geometrical boundary conditions [3.1-3]. We shall not elaborate on this point and limit ourselves to a qualitative but instructive description based on the terminology and notation of modes in closed cavities.

The simplest case to describe is that of a plane wave cavity (Fig.3.8). Here radiation with nonzero transverse momentum components $k_x, k_y \neq 0$ will escape from the resonator after very few reflections. Stable oscillations can only occur in the lowest transverse mode for which $k_x = k_y = 0$, or equivalently, $n=m=0$. Of course, the axial component $k_z \neq 0$; hence $q \neq 0$. The modes with $n=0$, $m=0$ are commonly denoted by TEM_{00}. The number of possible axial modes q within the transition linewidth is simply $\Delta\nu/\Delta\nu_q$, where $\Delta\nu_q$ is the frequency difference between two neighboring axial modes. The standing wave condition

(3.2.16) implies that the wavelengths of these modes must satisfy $q(\lambda_q/2) = \ell_z = \ell$, where ℓ is the distance between the mirrors. Hence

$$\Delta\nu_q = \nu_q - \nu_{q-1} = c/\lambda_q - c/\lambda_{q-1} = c/2\ell \quad . \tag{3.2.21}$$

Thus the frequency separation between the axial modes is a constant, inversely proportional to the length of the resonator. Typically for infrared gas lasers $\ell \sim 100$ cm, the speed of light equals its vacuum value $3 \cdot 10^{10}$ cm^3/s so that $\Delta\nu_q \sim 10^8$ s^{-1} which is of the same order as $\Delta\nu_D \sim 10^8$ s^{-1}. Hence only one or few axial modes fall within the transition linewidth. In the case of transitions in the visible region where $\Delta\nu \sim 10^9$-10^{13} s^{-1} the number of modes is larger but remains very much below the corresponding number $p(\nu)\Delta\nu$ in closed cavities.

The Gain Profile

The number of modes lying within the linewidth $\Delta\nu$ of the emission line provides only a rough estimate for the number of modes which actually oscillate. The number of these active modes is determined by the threshold condition (3.2.6) and the broadening mechanism. To simplify the discussion let us assume (as is often the case) that only the lowest transverse mode TEM_{00} can reach the threshold of oscillations so that we are only interested in determining what is the number and the identity of the active axial modes. The threshold condition (3.2.6) implies that for lasing on the ν_q mode to be possible $\gamma(\nu_q) \geqq \gamma_t$. The threshold gain coefficient γ_t can be safely taken as constant over the frequency range of interest which is the transition linewidth $\Delta\nu$. Figure 3.9 displays schematically the gain coefficient of an inhomogeneously broadened line $\gamma(\nu)$ and the level of (useful and useless) radiation losses γ_t. Only three of the axial modes which fall within $\Delta\nu = \Delta\nu_D$ have sufficient gain to overcome the losses and thus reach threshold and oscillate. Figure 3.9 also illustrates that the spectral width of an oscillating mode $\Delta\nu_\ell$ is narrower than the corresponding width $\Delta\nu_c$ in the passive cavity. The higher the value of $\gamma(\nu_q)$ of a lasing mode, the higher is the power and the narrower is the width $\Delta\nu_\ell$ of the radiation emitted through this mode (see Sect.3.2.4), as reflected schematically in the figure.

The intense radiation field established after the attainment of threshold tends to reduce the gain coefficient $\gamma(\nu_q)$ down to the threshold value γ_t (see Sect.3.2.4). This phenomenon, known as "gain saturation", reflects one of the most significant differences between homogeneously and inhomogeneously

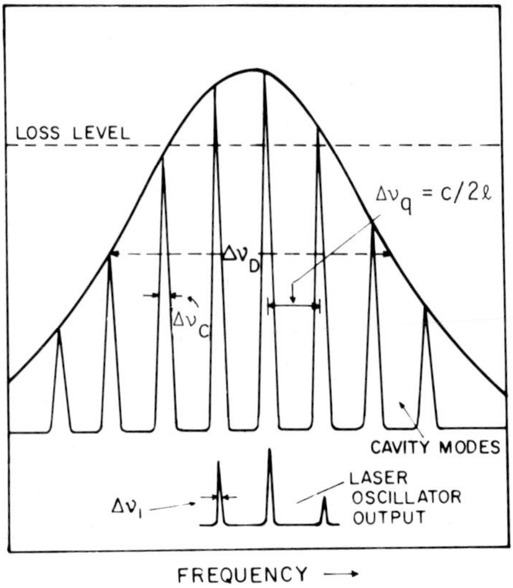

Fig. 3.9. Gain profile of an inhomogeneously broadened line. $\Delta\nu_D$ is the width of a Doppler broadened gain curve; $\Delta\nu_q = c/2\ell$ is the spacing between two adjacent modes. Only three modes are above the oscillation threshold and appear as laser output. The spectral width of the laser $\Delta\nu_\ell$ is much smaller than the width of the modes in the passive cavity $\Delta\nu_c$

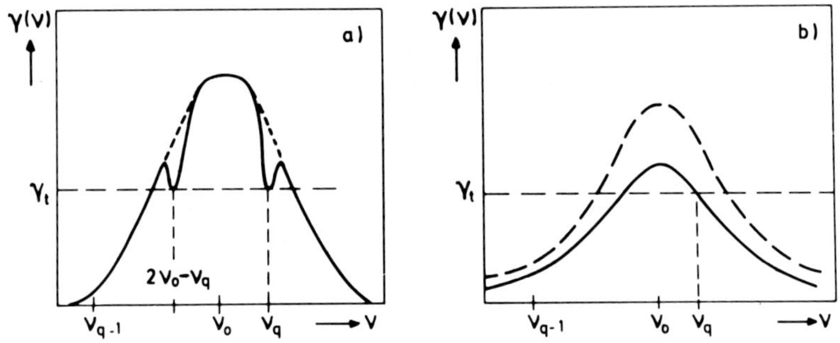

Fig. 3.10. Schematic illustration of gain saturation in inhomogeneously (a) and homogeneously (b) broadened transitions. The dashed curves are the unsaturated (zero field) gain profiles while the solid curves are the saturated gains, γ_t is the gain threshold, and only one cavity mode, ν_q, has a higher gain. In the case of homogeneous broadening this leads to uniform attenuation of the gain profile, while hole burning can occur in inhomogeneous broadening. In the case of Doppler broadening an off-center mode $\nu_q \neq \nu_0$ implies the existence of two holes

broadened lines. Figuratively, the high radiation density tends to "burn a hole" in the gain profile $\gamma(\nu)$ around $\nu = \nu_q$. Yet a hole may be burnt only in the case of inhomogeneous broadening. To explain this behavior consider the two limiting cases of purely inhomogeneous, i.e., Doppler, and purely homogeneous broadened lines in gas lasers. In the Doppler case only a small fraction of the gas molecules, those which can radiate at $\nu \simeq \nu_q$, participate in the lasing process; the population of the others neither contributes nor is affected by the stimulated emission. Hence $\gamma(\nu_q) \rightarrow \gamma_t$ whereas $\gamma(\nu \neq \nu_q)$ is the same as the unsaturated (zero field) gain (Fig.3.10a). It can be shown that the hole in the gain profile has a Lorentzian shape with width $\Delta\nu_{hole}$ which at high radiation densities $[\rho(\nu)]$ is proportional to $[\Delta\nu_H \rho(\nu_q)]^{\frac{1}{2}}$ [3.1], where $\Delta\nu_H$ is the homogeneous linewidth.

In general, the number of holes in an inhomogeneous gain profile equals the number of modes above threshold (see Fig.3.9). In the case of Doppler broadening each off-center $\nu_q \neq \nu_0$ mode implies the possible appearance of two holes symmetrically located around the line center frequency, one hole at $\nu = \nu_q$ and an "image hole" at $\nu = \nu_0 - (\nu_q - \nu_0) = 2\nu_0 - \nu_q$ (Fig.3.10a). This is due to the fact that the radiation field at the mode frequency ν_q interacts with two classes of molecules. When moving in the +z direction it is amplified by molecules with velocity component v_z around $v_z = v_0[1+(\nu_q - \nu_0)/\nu_0]$. On its way back, in the -z direction, it interacts with molecules with velocity $v_0[1-(\nu_q - \nu_0)/\nu_0]$ (cf. Sect.3.1.2). Since the Doppler gain profile $\gamma(\nu)$ is a simple reflection of the velocity distribution two classes of molecules interact with the intracavity radiation field, giving rise to two holes in $\nu = \nu_0 \pm (\nu_q - \nu_0)$. In the case that $\nu_q = \nu_0$ only one class of molecules, i.e., those with $v_z = 0$, contributes to the amplification process. Hence, in the vicinity of ν_0 the net amplification rate decreases as $\nu_q \rightarrow \nu_0$, reaching a local minimum at $\nu_q = \nu_0$, a phenomenon known as the "Lamb dip".

Different behavior is displayed in the case of homogeneously broadened transitions. Here all molecules can emit over the entire gain profile. Therefore when threshold is first reached at a frequency ν_q, corresponding to the mode with the highest gain, all molecules contribute to the stimulated emission process and the gain coefficient saturates homogeneously, as illustrated in Fig.3.10b. Since the lasing reduces the gain coefficient of the highest gain mode to its threshold value, and the line profile preserves its shape, it is clear that the gain coefficients of all the other modes are reduced below threshold and therefore cannot lase. Thus, while inhomogeneously broadened transitions can oscillate on several axial modes, homogeneously broadened transitions imply single-mode operation. It should be mentioned however that for certain resonator geometries multimode operation can take place even in the case of homogeneously broadened lines. This is because the field patterns of different transverse modes sample out different spatial regions of the laser cavity (see below). On the other hand, multimode operation may result in effectively homogeneous depletion of inhomogeneous lines. For instance, in high gain chemical lasers operating at low pressures (a few Torrs) multimode operation can take place and the Doppler profiles of the emitting lines are drained homogeneously. In chemical and other gas lasers there is another important mechanism which often masks hole burning effects although the lasers operate at very low pressures and the lines are predominantly Doppler broadened. These are the momentum transfer (elastic) collisions which, besides contributing to collisional broadening, tend to maintain the Maxwell-Boltzmann shape of the velocity distribution. Consequently, there is a constant flux of molecules into the frequency region in the Doppler profile depleted by the lasing process so as to fill the hole which the stimulated emission tends to drill.

Although the oscillations in the lowest order transverse mode are parallel to the principal laser axis some divergence and consequently losses of radiation are unavoidable owing to diffraction. These losses can be estimated through the (linear) far field diffraction angles θ_x and θ_y. For a parallel beam reflected from a mirror with linear dimensions ℓ_x and ℓ_y we have[8] $\theta_x \simeq \lambda/\ell_x$ and $\theta_y \simeq \lambda/\ell_y$. For infrared and visible lasers where $\lambda/\ell_x << 1$ the diffraction losses per one, or few traversals of the laser resonator can be kept quite small. The intensity distributions corresponding to higher order transverse modes obtain [(n+1)(m+1)] maxima at off-axial locations and their diffraction losses become higher as n and m increase. However, in contrast to the TEM_{00} which peaks along the principal laser axis, the higher modes may be amplified by molecules located far from the axis. The relative losses of the various transverse modes depend on the resonator geometry. By proper choice of the mirrors' curvatures and their distance it is possible to control the gain and the losses of the higher order modes.

In general, plane parallel resonators involve relatively high losses. In addition, such resonators are rather unstable in the sense that any small misalignment of the reflectors causes a loss of radiation from the cavity. (More rigorously, plane wave resonators are on the borderline of the stability diagram [3.1-7].) Therefore in most of the experimental arrangements at least one reflector is a concave spherical mirror. Divergence of the radiation in the cavity is sometimes desirable, e.g., when extraction of energy from all parts of the active laser medium is preferred over beam directionality. In such cases unstable laser resonators comprising one or two convex mirrors can be advantageous.

---

8 In addition to physical optics derivations θ_x can be estimated through the uncertainty principle. The uncertainty in the x component of the momentum $p_x = \hbar k_x$ of photons travelling along the z axis with momentum $p = p_z = \hbar k = h\nu/c$ is given by $\Delta p_x \Delta x \simeq h$, where Δx is the position uncertainty. For photons reflected or emitted from a surface with a linear x dimension ℓ_x we have $\Delta \ell_x = \ell_x$. Hence the diffraction angle is $\theta_x = \Delta p_x/p \simeq h/p\Delta x = h/p\ell_x = \lambda/\ell_x$.

Beam Quality and Brightness

Lasers are often described as sources of coherent radiation. A rigorous definition of the properties of coherent radiation is beyond our scope. For the present purpose, the somewhat loose but common interpretation that the coherence of laser signals is reflected by their very narrow spectral bandwidth, their very small spatial divergence, and the definite phase relations across their wave fronts will suffice. Equipped with some basic knowledge about the properties of the radiation within the laser resonator, we now turn to a somewhat more specific discussion of the cohrent nature of the radiation coupled out through the mirrors as the laser signal.

The stationarity of the field patterns corresponding to the various resonator modes means a constant amplitude and phase of the radiation field everywhere in the laser cavity, including in particular at every point on the mirrors. Consequently the signal emerging through the partly transmitting mirror will have the same, stationary, phase distribution as the radiation incident on the internal side of the mirror (except for fluctuations due to absorption and scattering by the matter composing the mirror). The signal then continues to propagate with the same phase distribution across its wave front. Detailed analysis of the field distribution of the TEM_{00} reveals that the transverse intensity distribution has a Gaussian form. That is, the intensity profile at any axial location z is of the form $\exp[-2(x^2+y^2)/w(z)^2]$, where w(z), the "spot size", measures the linear width of the Gaussian beam. The z dependence of the spot size varies from one resonator to another. For example, if both mirrors have the same radius of curvature, w(z) reaches a minimal value $w_0 = w(z_0)$, usually referred to as the "waist" of the beam, at the midpoint z_0 between the mirrors. In the very common arrangement of half-symmetric resonators, of the type shown in Fig.3.11, the waist is in the plane mirror which usually

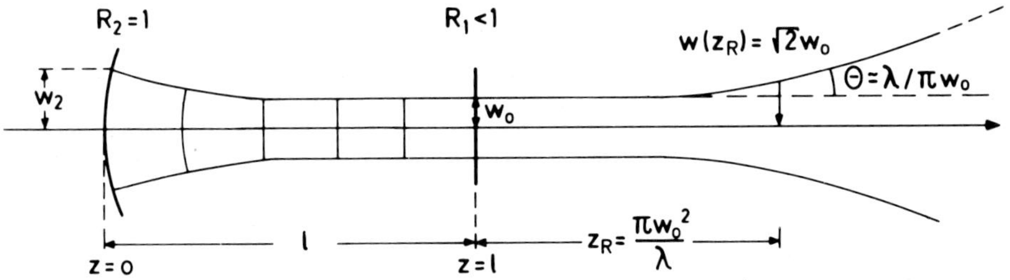

Fig. 3.11. Beam geometry in a half-symmetric resonator. The laser beam propagates with very little divergence over a range z_R and then diverges with the far field diffraction angle θ. (The divergence is greatly exaggerated in the figure)

serves as the output coupler. For the special case where ℓ equals half the radius of curvature of the spherical mirror the spot sizes at the mirrors are $w_1 = w_0 = (\ell\lambda/\pi)^{\frac{1}{2}}$ and $w_2 = 2(\ell\lambda/\pi)^{\frac{1}{2}} = 2w_0$. The plane mirror now serves as a source of plane wave radiation with surface area $\pi(w_0/2)^2$. Diffraction analysis shows that the beam originating from this mirror propagates with essentially no divergence over the (Rayleigh) range $z_R = \pi w_0^2/\lambda$ and then starts to diverge with a far field angle $\theta \simeq \lambda/w_0$, corresponding to a solid angle of divergence $\Omega \simeq \lambda^2/w_0^2$. These connections between z_R and θ and the waist size w_0 are valid also for resonators in which the waist occurs somewhere between the mirrors. Combining the expressions for w_0 and θ, we find $\theta \simeq (\lambda/\ell)^{\frac{1}{2}}$. Using a dye laser as an example, we have $\lambda \sim 5\times10^{-5}$ cm, $\ell \sim 50$ cm, yielding $\theta = 10^{-3}$ rad. Indeed laser beam divergence is typically in the milliradian range, meaning that the beam can propagate over large distances without appreciable intensity decrease.

We conclude this section by comparing the properties of laser and thermal radiation. One direct way of demonstrating the difference between these two kinds of radiation is to compare the average number of photons per mode $\phi(\nu)$. In the thermal case $\phi(\nu) \sim 1$-10 for infrared frequencies and $\phi(\nu) \ll 1$ for visible light, even for very hot source with $T = 10^3$-10^4 K [cf. (3.2.19)]. On the other hand, the photon density in a given mode of a weak laser with a power level of about 1 mW can easily exceed 10^7 photons/cm^3. Thus there is no way of filtering and focusing thermal radiation into the spectral and spatial width corresponding to one or even very many modes of laser light. A more illustrative comparison between laser and ordinary radiation is provided by their respective "brightness". The brightness of a radiation source is defined by

$$B_0(\nu) = P/\Delta A \Delta\Omega \Delta\nu \quad , \tag{3.2.22}$$

where P is the power radiated from a surface area ΔA of the source into a solid angle $\Delta\Omega$ around the normal to the surface and within the frequency range $\Delta\nu$. In the case of thermal sources $B_0(\nu) = \alpha c \rho(\nu)/4\pi$ where $\rho(\nu)$ is the black body radiation density and $\alpha < 1$ is the emissivity of the surface. Even for very hot surfaces, say $T = 10^3$-10^4 K, the radiation density in the infrared region ($\nu \sim 10^{13}$ s^{-1}) is $\rho(\nu) \sim 10^{-17}$ erg·s/cm^3; hence $B_0(\nu) \sim 10^{-10}$ erg/cm^2. For comparison, in ordinary infrared gas lasers $P \sim 10^{-3}$ Watt, and $\Delta\nu = \Delta\nu_\ell \sim 10^6$ s^{-1}. The effective aperture and the far field diffraction angle are related via $\lambda^2/\Delta A$; hence $B_0(\nu) = P/\Delta A \Delta\Omega \Delta\nu \simeq P/\Delta\nu_\ell \lambda^2$ is of the order of 10^{-3} erg/cm^2 which is far brighter than any (even very hot) thermal source. For visible radiation the brightness ratio can be much larger.

3.2.4 The Laser Rate Equations

The laser rate equations describe the time evolution of the molecular level populations and photon densities in the laser cavity. These are first-order kinetic equations similar to the master equations describing the approach to equilibrium of nonequilibrium chemical systems (Sect.2.5). The only difference between ordinary master equations and the laser rate equations is that the latter incorporate the effects of stimulated and spontaneous radiation as in (3.1.35,36). In the derivation of these equations from the quantum mechanical equations of motion, we have averaged over the phases of the radiating molecules. Consequently, the laser rate equations contain no information about phase relations between the molecules and the radiation field. However, the information about the time dependence of photon and molecular populations provided by the rate equations suffices to determine many important characteristics of laser operation, for example, the spectral and temporal properties of the laser output power, the pumping requirements, and the effects of various collisional and radiative processes on the laser performance. In multilevel systems, like chemical lasers, the rate equation formalism constitutes the only practical approach for studying their dynamical behavior, as shown in detail in Sect.4.4. Similarly, the rate equation approach is the principal tool for interpreting multiphoton activation experiments (Sect.5.4).

To simplify the description of the basic properties of the rate equations and their solutions let us consider again a laser system with one active lasing transition $f \rightarrow i$ (Fig.3.12). We shall also assume that the laser operates on a single mode. Clearly the frequency of this mode ν must be within the spectral linewidth of the transition, that is, $|\nu-\nu_0| < \Delta\nu/2$, where $\nu_0 = E_f - E_i$

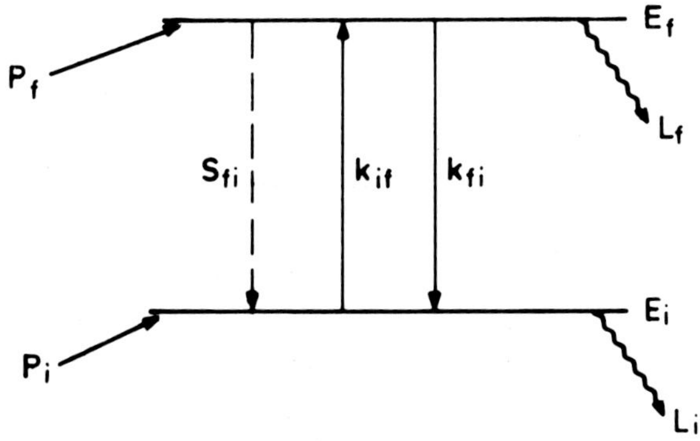

Fig. 3.12. Scheme of the processes governing the rate of change of level and photon populations (see text)

is the line center frequency. Without specifying the nature of the pumping and relaxation processes, we shall denote the pumping rate (molec/s·cm^3) of levels i and f by P_i and P_f, and their overall relaxation rates (molec/s·cm^3) by L_i and L_f.

The rate equation for the photon density $\phi = \phi(\nu)$ in the active mode is

$$\frac{d\phi}{dt} = k_{fi}\Delta N\phi - \frac{\phi}{\tau_c} - S_{fi} \quad , \tag{3.2.23}$$

where $k_{fi} = k_{fi}(\nu) = c\sigma_{fi}(\nu)$ is the rate constant for stimulated emission and $\Delta N = N_f - (g_f/g_i)N_i$ is the population inversion [cf. (3.1.36)]. The second term accounts for the loss of photons from the cavity [cf. (3.2.10) and recall $\rho=\phi h\nu$]. The last term represents the contribution of spontaneous emission to the photon density in the oscillating mode.

To derive S_{fi} we note that the rate of spontaneous emission into all modes within a frequency interval $d\nu$ around the laser mode frequency ν is $N_fA_{fi}g(\nu,\nu_0)d\nu$. The number of modes in this interval, according to (3.2.18), is $p(\nu)d\nu = (8\pi\nu^2/c^3)Vd\nu$. Hence,

$$VS_{fi} = \frac{N_fA_{fi}g(\nu,\nu_0)d\nu}{(8\pi\nu^2/c^3)d\nu} = N_fg(\nu,\nu_0)h\nu B_{fi} = k_{fi}N_f \quad , \tag{3.2.24}$$

where we have used (3.1.31) for the relation between A_{fi} and B_{fi} and (3.1.25) for k_{fi}. Remembering that $k_{fi}N_f\phi$ is the rate of stimulated emission into the active mode, the last equation implies that for $\phi = 1/V$, i.e., if there is one photon in the cavity, the rates of spontaneous and stimulated emission per mode are equal. Above the lasing threshold $\phi >> 1$ and the spontaneous emission term is thus negligible. Combining (3.2.23) and (3.2.24), we get

$$\frac{d\phi}{dt} = k_{fi}N_f(\phi + \frac{1}{V}) - k_{if}N_i - \frac{\phi}{\tau_c} \quad , \tag{3.2.25}$$

where $k_{if} = (g_f/g_i)k_{fi}$.

The rate equations for the upper and lower level populations are

$$\frac{dN_f}{dt} = P_f - k_{fi}\Delta N\phi - L_f \tag{3.2.26}$$

$$\frac{dN_i}{dt} = P_i + k_{fi}\Delta N\phi - L_i \quad , \tag{3.2.27}$$

where the effects of spontaneous emission on the level populations have been incorporated into the relaxation terms. It should be noted that S_{fi} is only a very small fraction of the overall rate of spontaneous emission from level f, N_fA_{fi}. In infrared lasers where $A_{fi} \sim 10\text{-}10^2\ s^{-1}$, N_fA_{fi} is generally negligible compared to the rate of nonradiative processes. However in the visible region where typically $A_{fi} \sim 10^7\text{-}10^9\ s^{-1}$, N_fA_{fi} may be the major contribution to L_f (hence also a significant, negative, contribution to L_i).

Equations (3.2.25-27) are coupled nonlinear equations which except for few special cases must be solved numerically. We shall conclude this section with the description of two such cases, which are instructive for understanding the pumping requirements and output characteristics in laser systems. The first case is that of steady-state oscillations characteristic to all the continuous wave lasers. The second is that of short pulse generation by Q switching.

Steady-State Solutions

The mathematical description of the steady-state solutions of the rate equations can be greatly simplified and yet reveal the essential physics of the problem by assuming that the population of the lower level is always negligible, $N_i = 0$. This assumption implies that the depletion rate L_i is much larger than the rates of the processes which populate the lower level. This condition is valid for example in some four-level laser systems, such as the Nd^{3+} laser (Fig.3.13). The relaxation rate of the upper level will be represented as N_f/τ_f, where τ_f is the "lifetime" of this level. Let us also simplify the notation by using k instead of k_{fi}, and P instead of P_f. Since $N_i = 0$ and consequently $dN_i/dt = 0$, we are left with only two rate equations, which at steady state, $d\phi/dt = 0$, $dN_f/dt = 0$, are of the form

$$\frac{d\phi}{dt} = kN_f\left(\phi + \frac{1}{V}\right) - \frac{\phi}{\tau_c} = 0 \tag{3.2.28}$$

$$\frac{dN_f}{dt} = P - kN_f\phi - \frac{N_f}{\tau_f} = 0 \quad . \tag{3.2.29}$$

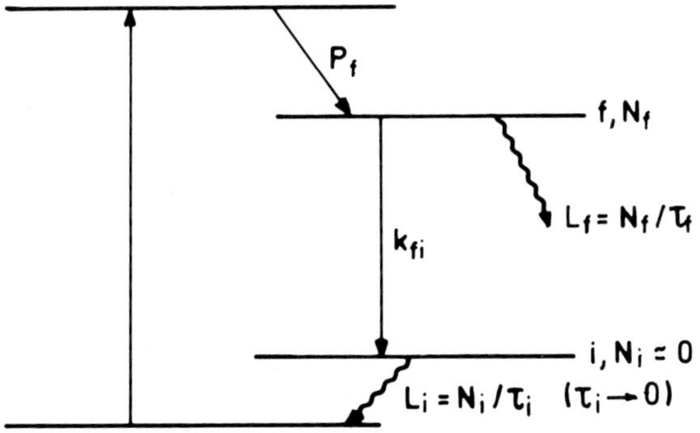

Fig. 3.13. Kinetic scheme of a four-level system

From the first equation we find

$$N_f = \frac{1}{k\tau_c}\frac{\phi}{\phi+1/V} = N_{f,t}\frac{\phi}{\phi+1/V} \tag{3.2.30}$$

or

$$\phi = \frac{N_f/V}{N_{f,t}-N_f} \quad , \tag{3.2.31}$$

where we have used the expression $1/k\tau_c = \Delta N_t = N_{f,t}$ for the threshold inversion. From the second equation we get

$$\phi = \frac{P}{kN_f} - \frac{1}{k\tau_f} \quad . \tag{3.2.32}$$

During the lasing process the photon density in the cavity is very large, $\phi \gg 1$, and (3.2.30) reveals that N_f approaches very nearly (though never actually reaches) its threshold value $N_{f,t}$. Alternatively, from (3.2.31) it follows that when $N_f \to N_{f,t}$ the photon density ϕ can reach exceedingly high values. The absolute value of N_f and hence ϕ depends on the pumping rate. To evaluate this dependence we note that above threshold, that is, when $\phi \gg 1$ or $(N_{f,t}-N_f)/N_f \ll 1$, we can replace in (3.2.32) N_f by $N_{f,t} = 1/k\tau_c$ to get

$$\frac{\phi}{\tau_c} = P - \frac{N_{f,t}}{\tau_f} = P - P_t \quad , \tag{3.2.33}$$

where $P_t = N_{f,t}/\tau_f$ is called the threshold pumping rate. The last equation is simply an energy balance relation demonstrating that the rate (per unit volume) at which photons leave the cavity is equal to the difference between the pumping rate P and the relaxation rate $L_f = N_{f,t}/\tau_f = P_t$. This relation is sometimes called the "gain-equal-loss" (or "on-threshold") condition. This condition is exact for continuous wave (cw) lasers but it also provides a good approximation for the energy balance of pulsed lasers at post-threshold times (see Sect.4.4).

The total rate at which radiation energy leaves a laser cavity of volume V is $h\nu\phi V/\tau_c$. Since part of this energy is lost by diffraction, absorption, and other dissipative processes, the useful laser power is

$$\mathcal{P} = f\frac{h\nu}{\tau_c}V\phi = fh\nu V(P-P_t) \quad , \tag{3.2.34}$$

where f is the fraction of useful output coupling; $f = 1$ if $\chi = 0$ in (3.2.8); more generally $f = (c\tau_c/2\ell)\ln(1/R_1R_2)$.

Two remarks should be made before proceeding to the next example.

1) In the last section we have mentioned that the spectral width of the laser $\Delta\nu_\ell$ is inversely proportional to the output power. A qualitative explanation based on the uncertainty principle and (3.2.28) can be given as follows: In the absence of stimulated emission, as in passive cavities, $\Delta\nu = \Delta\nu_c = 1/2\pi\tau_c$. In the active laser cavity the stimulated emission which continuously adds photons to the cavity increases the effective lifetime of the photons by decreasing their effective leakage rate from $1/\tau_c$ to $1/\tau = 1/\tau_c - kN_{f,t}$. Setting $N_f = N_{f,t}$ in (3.2.28), we find $1/\tau = kN_{f,t}/\phi V$. Hence $\Delta\nu_\ell \simeq 1/2\pi\tau = kN_{f,t}/2\pi\phi V = 1/2\pi\tau_c\phi V$, or using (3.2.34) $\Delta\nu_\ell \sim fh\nu/\tau_c^2 P$. Note that this width is due to the spontaneous emission term kN_f, which represents the random addition of photons to the coherent wave generated by stimulated emission. This can be regarded as a dephasing process associated with the dephasing time τ.

2) The steady-state solutions are exact for laser systems operating under constant pumping rate. They remain a very good approximation when the pumping and consequently the stimulated emission and relaxation rates vary slowly in time. A detailed analysis shows that for such cases the inversion ΔN is clamped to its threshold value ΔN_t and the laser power can be derived from the gain-equal-loss condition (3.2.34), with a time-dependent pumping rate $P(t)$ replacing the constant pumping rate P. Thus, instead of (3.2.34), we have

$$P(t) = f\,\frac{h\nu}{\tau_c}\,V\phi(t) = fh\nu V[P(t) - N_{f,t}/\tau_f] \quad . \tag{3.2.35}$$

This approximation fails when $P(t)$ or other rate processes vary rapidly in time. For example, the attainment of threshold following a sudden switch-on of the pumping mechanism in cw lasers is accompanied by a transient period of several rapid oscillations of the population inversion and the photon density around their steady-state values. The amplitude of these oscillations, known as "relaxation oscillations", decays rapidly to zero and the inversion and photon density stabilize at their steady-state values. Another example where (3.2.35) does not hold is when the photon lifetime τ_c, or equivalently, the quality factor Q, is suddenly increased. This technique, which is known as "Q switching", is commonly used for the generation of short and intense laser pulses and provides our next example of the rate equation approach.

Q Switching

The principle of the technique is to keep the quality factor on a low value during the pumping period and then suddenly increase its value and thereby induce a sharp buildup of photon density and laser power. As long as Q is kept low the photons leak out rapidly from the cavity. Consequently, ϕ is small and laser oscillations cannot take place. Meanwhile a very high inversion density ΔN is built up by the pumping process without being depleted by stimulated emission. A mathematical interpretation of this phenomenon is provided by the relation $\Delta N_t = 1/k\tau_c = \nu/kQ$. If Q is small, the threshold inversion is large. In particular as $Q \to 0$ the lasing condition $\Delta N \geqq \Delta N_t$ implies that a very high inversion density need be reached before lasing can take place. If after ΔN has reached a very high value Q is suddenly increased, then $\Delta N \gg \Delta N_t = \nu/kQ$ which gives rise to a rapid generation of photons, manifested by a short and intense laser pulse. There are several experimental techniques of Q switching. One of them is to use a rotating mirror. During the pumping period the mirror is aligned parallel to the laser axis so that R and thus Q are effectively zero. When the mirror is rotated to be perpendicular to the axis, Q reaches a finite value and lasing starts. Another method, known as passive Q switching, uses a saturable absorber inside the laser cavity. Such materials absorb strongly the laser radiation at low power levels and become transparent (bleach) at high power levels due to saturation of the absorbing transition. A common example is the use of SF_6 to ("passively") Q switch CO_2 lasers.

The stimulated emission starting immediately after Q is switched to its high value is so intense that one can ignore the effects of all other rate processes on the populations of the levels connected by the lasing transition. Thus, throughout the pulse $N_i + N_f = N_{i,0} + N_{f,0} = N$, where N is the sum of the lower and upper level populations $N_{i,0}$ and $N_{f,0}$ at the beginning of the pulse, $t = 0$. Also, at $t = 0$, $\Delta N_0 = N_{f,0} - (g_f/g_i)N_{i,0} \gg \Delta N_t = 1/k_{fi}\tau_c$, where τ_c is the photon lifetime during the pulse. Neglecting the pumping, spontaneous emission, and relaxation terms and noting that $N = \text{constant}$, so that $dN/dt = 0$, the three rate equations (3.2.25-27) can be reduced into the two independent equations

$$\frac{d\phi}{d\tau} = \left(\frac{\Delta N}{\Delta N_t} - 1\right)\phi \qquad (3.2.36)$$

$$\frac{d\Delta N}{d\tau} = -2\,\frac{\Delta N}{\Delta N_t}\,\phi \quad , \qquad (3.2.37)$$

where $\Delta N_t = 1/k_{fi}\tau_c$ and $\tau = t/\tau_c$. The time dependence of ΔN and ϕ can be determined by numerical integration of these coupled nonlinear equations (Fig.3.14). However, an analytic expression for the relation between ΔN and ϕ can be obtained if we divide (3.2.36) by (3.2.37). This yields

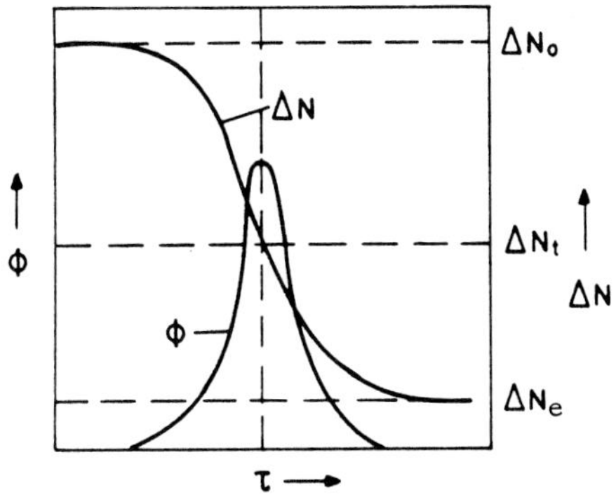

Fig. 3.14. Time dependence of the inversion ΔN and photon density ϕ in a Q switched system (after [3.5])

$$\frac{d\phi}{d\Delta N} = \frac{1}{2}\left(\frac{\Delta N_t}{\Delta N} - 1\right) \quad . \tag{3.2.38}$$

After integration, we obtain

$$\phi = \phi_0 + \frac{1}{2}\left[\Delta N_t \ln\left(\frac{\Delta N}{\Delta N_0}\right) - (\Delta N - \Delta N_0)\right] \quad , \tag{3.2.39}$$

where ϕ_0 and ΔN_0 are the photon density and the population inversion at $t = 0$, respectively; clearly $\phi_0 \simeq 0$. Neglecting radiation losses other than output coupling, the instantaneous laser power $P = h\nu V\phi/\tau_c$ is

$$P = \frac{h\nu}{2\tau_c} V\left[\Delta N_t \ln\left(\frac{\Delta N}{\Delta N_0}\right) - (\Delta N - \Delta N_0)\right] \quad . \tag{3.2.40}$$

Differentiation of P with respect to ΔN shows that the peak power P_{max} is obtained when $\Delta N = \Delta N_t$ (Fig.3.14). Since the initial inversion is much larger than the threshold inversion, $\Delta N_0 \gg \Delta N_t$, the expression for P_{max} becomes [set $\Delta N = \Delta N_t$ in (3.2.39) and note that $x \ln(x) \to 0$ when $x = \Delta N_t/\Delta N_0 \to 0$]

$$P_{max} = \frac{h\nu}{2\tau_c} V\Delta N_0 \quad . \tag{3.2.41}$$

The duration of the pulse t_p approaches τ_c. This can be shown, somewhat indirectly, as follows: At the end of the laser pulse $\phi = \phi_e = 0$. Setting this value, as well as $\Delta N = \Delta N_e$, and $\phi_0 = 0$ in (3.2.39), we find

$$\Delta N_e/\Delta N_0 = \exp[(\Delta N_e - \Delta N_0)/\Delta N_t] \quad . \tag{3.2.42}$$

From this implicit relation we find that $\Delta N_0/\Delta N_t \gg 1$ implies, as expected, $\Delta N_e \ll \Delta N_0$. The total number of photons emitted during the pulse Φ is equal to the number of molecules which fell from the upper to the lower level, that is, $\Phi = V(N_{f,0} - N_{f,e})$. Since $N_{f,0} + N_{i,0} = N_{f,e} + N_{i,e} = N$ we get $\Phi = V(\Delta N_0 - \Delta N_e)/2 \simeq V\Delta N_0/2$. Hence, the total pulse energy is

$$E = \Phi h\nu = V\Delta N_0 h\nu/2 \quad . \tag{3.2.43}$$

An appropriate definition for the laser pulse duration t_p is $E = P_{max} t_p$. Combining (3.2.41) and (3.2.43), we thus find that for $\Delta N_0 >> \Delta N_t$, $t_p \simeq \tau_c$.

3.3 Survey of Atomic and Molecular Spectroscopy

The rate of absorption and emission of radiation is determined by the magnitude of the transition dipole μ_{if}. In this section we briefly review the selection rules governing the transition probabilities in atoms and molecules. Our purpose is to introduce the terminology and concepts used extensively in chemical lasers and laser chemistry research (Chaps.4 and 5).

3.3.1 Atomic Spectra

The frequencies of spectral lines observed in atomic spectra obey the Rydberg-Ritz combination principle

$$\nu_{\ell m} = T_m - T_\ell \quad . \tag{3.3.1}$$

For every atom there exists a set of numbers T_i such that *all* the transitions associated with the atom may be obtained from (3.3.1). The numbers T_i are called term numbers. They are usually expressed in cm^{-1}, the corresponding energies (in ergs) being hcT_i. A given electronic state of a polyelectronic atom is sometimes (somewhat loosely) called a term and is represented by a term symbol of the general form

$$n^{(2S+1)}L_J \quad ,$$

where n is the principal quantum number of the highest occupied orbital, L is the total electronic orbital momentum, S is the total spin angular momentum, and J is the total angular momentum, given by $\underline{J} = \underline{L} + \underline{S}$. The possible values of J are

$$J = L+S, L+S-1, \ldots, (L-S) \quad .$$

This description of the term is valid in the Russel-Saunders coupling case: interactions between the orbital angular momenta of individual electrons are stronger than spin-orbit coupling. Thus L may be obtained by summing up the contributions of the separate electrons: $L = \sum \ell_i$. Terms for which $\sum \ell_i$ is even are called even terms; the others are odd terms. Similarly, S is obtained by summing over individual spin values $S = \sum s_i$. In general, one tries to define a term using "good" quantum numbers, i.e., such that are associated with operators that very nearly commute with the total Hamiltonian. Selection rules only hold to the extent that the quantum numbers define a true stationary state. Since in practice this condition is never rigorously obeyed, all selection rules are only approximate (cf. Sect.3.1).

The symbols for L = 0,1,2,3 are S,P,D,F, respectively. The quantity 2S+1 is called the term multiplicity. Thus the term symbol for a ground state sodium atom is $3^2S_{1/2}$, showing that the highest occupied orbital is the third one, total spin is ½, total electronic orbital momentum is zero, and total angular momentum is ½. Similarly, the ground state of cesium is $6^2S_{1/2}$.

Transition intensities are determined by selection rules, as follows (for electric dipole transitions):

$$\Delta L = 0,\pm1 \qquad \Delta J = 0,\pm1 \qquad \Delta S = 0$$

$$\text{but} \quad J = 0 \not\leftrightarrow J = 0$$

$$L = 0 \not\leftrightarrow L = 0$$

and by LaPorte's rule: odd terms combine only with even terms. The symbols $\leftrightarrow$ and $\not\leftrightarrow$ designate allowed and forbidden transitions, respectively.

The angular momenta discussed so far determine the coarse and fine structures of the spectra. Interaction with the nuclear spin I leads to hyperfine structure, by removing the degeneracy of the fine structure components. Further splitting may be brought about by external magnetic and electric fields. In particular, for ions embedded in crystals or glasses, large shifts are caused by local variations in crystal field parameters. These finer details are sometimes of crucial importance for laser operation.

An example for an efficient laser operating on a LaPorte forbidden transition is the iodine laser, emitting at 1.315 μm (Sect.4.3.1). The upper and lower states are fine structure components of the 5^2P term, derived from a $5p^5$ configuration. The energy level diagram along with hyperfine splitting

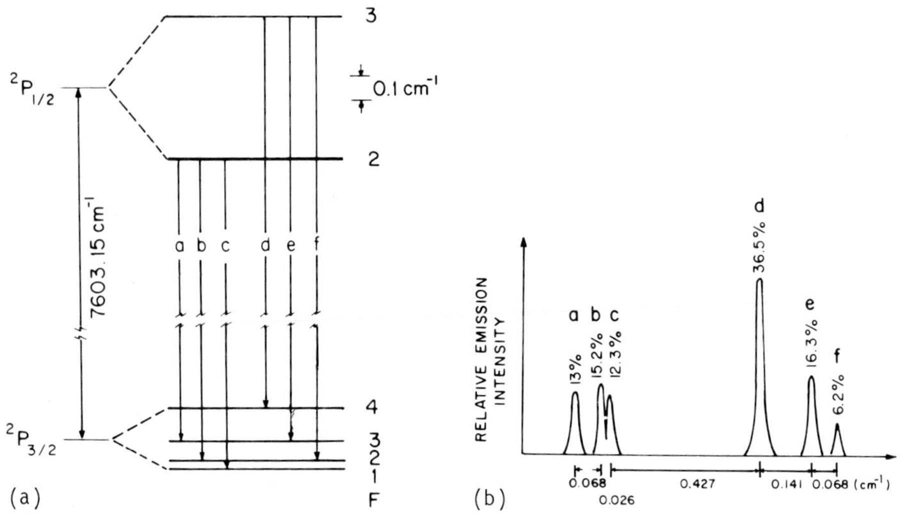

Fig. 3.15. (a) Energy level diagram of ^{127}I, showing the hyperfine splitting due to interaction with the nuclear spin. (b) The resulting spectrum. Energy units are cm^{-1}. This spectrum was obtained in emission; the excited iodine atom was prepared by photodissociation of C_3F_7I. Note that this is the spontaneous emission spectrum; the relative intensities reflect the degeneracy of the transitions (see also Sect.4.3.1). (Adapted from [3.20])

due to interaction with the nuclear spin $I = 5/2$ for ^{127}I is shown in Fig.3.15. The vector sum of I and J is designated as F. It can assume the values $F = I + J$, $I + J - 1, \ldots |I - J|$. The splitting between F levels is usually very small compared to that caused by spin-orbit coupling. In the case of iodine the upper $^2P_{1/2}$ level is split into two (F=2 and 3) while the lower $^2P_{3/2}$ level is split into four (F=1,2,3 and 4) sublevels. The selection rules are $\Delta F = 0, \pm 1$, but $F = 0 \nleftrightarrow F = 0$, leading to a splitting of the $^2P_{1/2}$-$^2P_{3/2}$ transition into six hyperfine components with a total energy spread of 0.73 cm^{-1} (Fig.3.15). This spread is comparable to pressure broadening (Sect.3.1.2) at relatively low pressures. Under normal operating conditions, the lines are merged (see Fig.4.12).

This example serves to illustrate an important property of forbidden transitions: the Einstein A and B coefficients of such transitions are small. Consequently, the upper state is long lived, or metastable. In the absence of deactivating collisions, it will last for a relatively long period of time, and act as a trap, or sink, for the activation process. The energy is thus stored in the upper state and may be released, by proper triggering, as a powerful stimulated emission pulse.

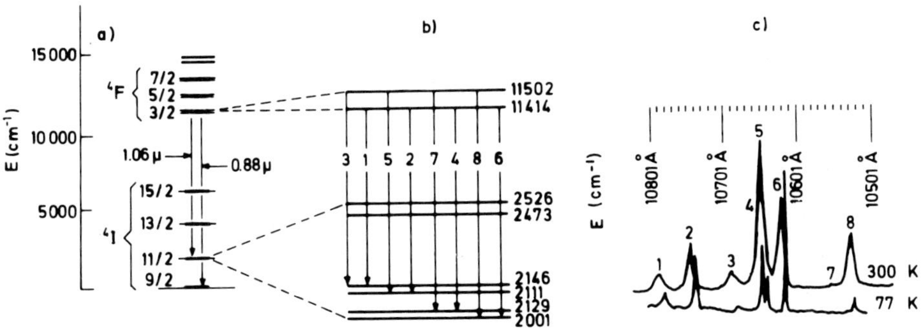

Fig. 3.16. (a) Schematic energy level diagram of Nd^{3+}, showing some of the low lying levels. The strongest transitions in emission originate from the $^4F_{3/2}$ level, and appear at about 1.06 and 0.88μ. (b) Detailed energy level diagram, showing the hyperfine splitting of the $^4F_{3/2}$ and $^4I_{11/2}$ levels. The figure applies to Nd^{3+} embedded in yttrium aluminum garnet (YAG) crystal, widely used as a laser material. (c) Observed fluorescence spectrum of Nd^{3+}: YAG at 300 and 77 K. All eight possible transitions are observed at room temperature. Cooling reduces the intensity of lines 3,5,7, and 8, originating from the higher sublevel of $^4F_{3/2}$ (adapted from [3.21])

Another forbidden transition is employed in the well-known Nd^{3+}:YAG and glass lasers. An energy level diagram is shown in Fig.3.16. The transition is an intra-f-shell one, and thus LaPorte forbidden. As in the case of the iodine laser, the spontaneous lifetime is very long, leading to good storage capacity.

3.3.2 Molecular Spectra

In molecules, the spherical symmetry characteristic of atoms no longer exists. In addition, one has to consider the internal degrees of freedom for the relative motion of the nuclei. The description of molecular spectra is usually based on the assumption that the total energy E may be written as the sum of independent contributions

$$E = E_E + E_N \quad , \tag{3.3.2}$$

where E_E is the energy associated with electronic motion, and E_N the nuclear energy. The physical basis for this separation was first given by Born and Oppenheimer in 1927 [3.8-14] and is commonly referred to as the Born-Oppenheimer (BO) approximation. The total Hamiltonian operator of the molecule may be written as

$$H = T_E + T_N + V_{EN} + V_{EE} \quad , \tag{3.3.3}$$

where $T_E = (-\hbar^2/2m_e)\sum_i \nabla_i^2$ is the kinetic energy operator for all the electrons, $T_N = -\sum_I(\hbar^2/2M_I)\nabla_I^2$ is the kinetic energy operator of all the nuclei, $V_{EN} = -\sum_{I,i} Z_I e^2/r_{Ii}$ represents electrostatic interactions between the electrons and the nuclei, and $V_{NN} = \sum_{I>J} Z_I Z_J e^2/R_{IJ}$ and $V_{EE} = \sum_{i>j} e^2/r_{ij}$ likewise represents electrostatic interactions between nuclei and between electrons, respectively. In the BO approximation we assume that the motion of the nuclei is much slower than that of the electrons. Thus, we may obtain an approximate solution for the electronic motion by letting $T_N = 0$ and V_{NN} = a constant. This leads to a modified Hamiltonian operator H_E and an "electronic" Schrödinger equation

$$H_E\phi_E = E_E\phi_E \quad , \quad \text{with} \quad H_E = T_E + V_{EN} + V_{EE} \quad . \tag{3.3.4}$$

ϕ_E is a function of the electronic coordinates r_i and parametrically also of the nuclear coordinates R_I. One can solve (3.3.4) for each set of nuclear coordinates R_I, and obtain an electronic energy $E_E(R_I)$. For every set of fixed nuclear coordinates one obtains a number of values for E_E^q, corresponding to various electronic excitation degrees q. These values are usually well separated, and on gradually varying R_I, one can obtain a set of smoothly varying curves $E_E^q(R_I)$. These curves are referred to as molecular electronic energy potential curves, or surfaces. These potential energy surfaces are now used in writing the nuclear Hamiltonian. The Schrödinger equation for the nuclear motion is thus

$$(T_N+V_{NN}+E_E)\chi_N = E_N\chi_N \quad . \tag{3.3.5}$$

The total wave function is given in the BO approximation by

$$\psi = \phi_E\chi_N \tag{3.3.6}$$

and the total energy by

$$E = E_E + E_N \quad . \tag{3.3.7}$$

Using (3.3.5-7), one obtains the following equation for the full Hamiltonian:

$$H\psi = H\phi_E\chi_N = (E_E+E_N)\phi_E\chi_N \quad . \tag{3.3.8}$$

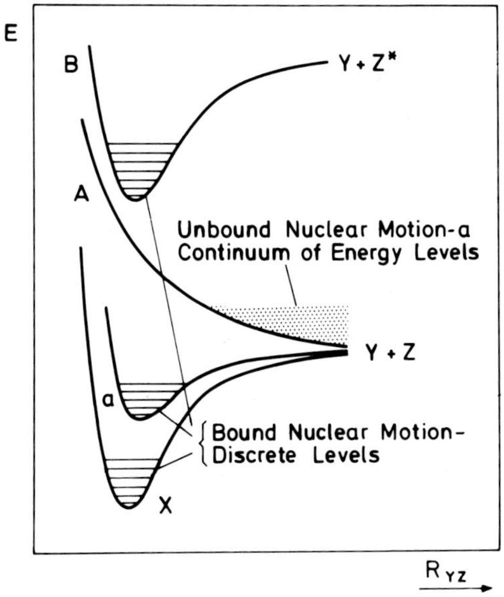

Fig. 3.17. Schematic energy level diagram of a diatomic molecule

Inspection shows that in (3.3.8) one neglects terms such as $\partial^2\phi_E/\partial R^2$ and $(\partial\phi_E/\partial R)(\partial\chi_N/\partial R)$. An important special case where these terms may not be neglected is when the electronic potential curves approach each other. This situation is referred to as curve crossing (Sects.1.3.3,2.4.7) and the BO approximation may at best be used as a starting point for a perturbation analysis. A graphical representation of BO energy levels for a diatomic molecule YZ is shown in Fig.3.17. Since there is only one internuclear distance, an energy level diagram may be drawn using a two-dimensional surface. The curves designated by X, a, A, and B are obtained by solving the electronic Schrödinger equation (3.3.4). The ground state is commonly designated as X, excited states with the same multiplicity as the ground state are designated A,B,C,... in ascending order of energy, and excited states of different multiplicity by a,b,c,... . At very large internuclear distances, the atoms are essentially separated, and may be described by the proper atomic configurations. Symmetry considerations limit the number of molecular states that may be correlated with a given pair of atomic states (the Wigner-Whitmer rules [3.8]). Two types of potential curves may be distinguished. Those that reach a minimum energy only at infinite separation (e.g., curve A in Fig.3.17) are termed repulsive potentials and correspond to an unbound state. Others have a minimum at some intermediate internuclear separation and can support bound states. The ground state is not necessarily bound. In an important group of laser systems — the rare gas excimers — the ground state is repulsive (see Sect.3.4.2).

The electronic energy curves shown in Fig.3.17 serve as potential energy surfaces for the nuclear motion along the coordinate R_{YZ}. According to the laws of quantum mechanics, such a motion is quantized for bound states, giving rise to a discrete set of energy levels. These are schematically shown in the figure by the horizontal lines. The motion associated with these states is the molecular vibration. Repulsive curves are associated with a continuum of states, resulting from the free translational motion of the atoms. A further nuclear motion is rotation around the center of mass of the molecule. It can be shown [3.8-14] that rotation and vibration are independent to a first approximation, so that the total internal energy of the molecule may be represented by the sum

$$E = E_E + E_V + E_R \quad .$$

Experimentally, it is found that the separation between electronic levels is typically 10^4-10^5 cm^{-1}, between vibrational levels $10^2 \sim 10^3$ cm^{-1}, and between rotational levels 0.1-10 cm^{-1}. Thus one can distinguish three major parts in a molecular spectrum: the visible-UV spectrum relates to electronic transitions, the infrared spectrum to vibrational transitions, and the far infrared (microwave range) to pure rotational transitions.

3.3.3 Electronic Spectra of Diatomic Molecules

The dipole moment operator of a molecule may be expressed as a sum

$$\underline{\mu} = \underline{\mu}_E + \underline{\mu}_N \tag{3.3.9}$$

with $\underline{\mu}_E$ operating on the electronic wave function only and $\underline{\mu}_N$ on the nuclear wave function only. Using BO states, the transition moment $\underline{\mu}_{fi}$ may be written as [3.8-13]

$$\begin{aligned} \underline{\mu}_{fi} &= \left\langle \phi_E^f \chi_N^f \middle| \underline{\mu} \middle| \phi_E^i \chi_N^i \right\rangle = \left\langle \phi_E^f \chi_N^f \middle| \underline{\mu}_E + \underline{\mu}_N \middle| \phi_E^i \chi_N^i \right\rangle \\ &= \left\langle \phi_E^f \chi_N^f \middle| \underline{\mu}_E \middle| \phi_E^i \chi_N^i \right\rangle + \left\langle \phi_E^f \chi_N^f \middle| \underline{\mu}_N \middle| \phi_E^i \chi_N^i \right\rangle \\ &\simeq \left\langle \phi_E^f \middle| \underline{\mu}_E \middle| \phi_E^i \right\rangle \left\langle \chi_N^f \middle| \chi_N^i \right\rangle + \left\langle \phi_E^f \middle| \phi_E^i \right\rangle \left\langle \chi_N^f \middle| \underline{\mu}_N \middle| \chi_N^i \right\rangle \quad . \end{aligned} \tag{3.3.10}$$

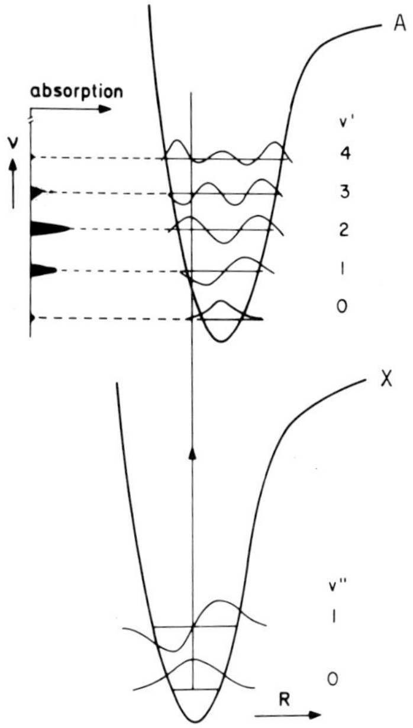

Fig. 3.18. Vibronic structure of an electronic transition according to the Franck-Condon principle. The intensity of the individual vibronic bands, shown on the left-hand side, is determined by the overlap between the v"=0 vibrational eigenfunction of the ground electronic state and the vibrational eigenfunctions of the excited electronic state

In the last line of (3.3.10), $\langle\phi_E^f\chi_N^f|\underline{\mu}_E|\phi_E^i\phi_N^i\rangle$ was approximated as the product of two terms (the Condon approximation). The physical basis for this simplification is that $\langle\phi_E^f|\underline{\mu}_E|\phi_E^i\rangle \equiv R_{fi}$ is usually only weakly dependent of the internuclear distance. This term determines the overall electronic transition probability. The other factor $\langle\chi_N^f|\chi_N^i\rangle$ is an overlap integral between a vibrational state associated with the initial electronic state and the final one. The transition probability is proportional to $|\underline{\mu}_{fi}|^2$ (Sect.3.1.1), leading to $|\langle\chi_N^f|\chi_N^i\rangle|^2$, a term usually called the Franck-Condon (FC) factor. The origin of the FC factor may be understood by reference to Fig.3.18.

Assume that the molecule is in its ground vibrational state. Absorption of light populates the excited electronic state A in one of the allowed vibronic states. During the transition time ($\sim 10^{-16}$ s) the nuclei are essentially stationary (the period of vibration being $\sim 10^{-13}$ s). Thus the transition may be termed vertical, in the sense that A is formed with the same nuclear separation R as that of X prior to the transition. The quantum mechanical probability amplitude for finding the molecule at this R is given by $|\chi_N^f|^2$ and the transition probability is quantitatively given by the FC factors. These factors thus determine the coarse structure of the spectrum, as shown on the left-hand side of Fig.3.18. Note that similar considerations apply also to transitions to repulsive states, except that the spectrum is continuous.

The overall transition probability is determined by $\underline{\mu}_{fi}$ and depends on the selection rules. As in atoms these refer to changes in eigenvalues of operators that (approximately) commute with the molecular Hamiltonian. Such operators are used in the description of term symbols. The symbol is usually written as

$$^{2S+1}\Lambda_{\Omega} \quad .$$

Here, S is the spin quantum number. Λ is the absolute value of the component of the electronic orbital angular momentum along the molecular axis which assumes the values $0, \pm 1, \pm 2, \pm 3, \ldots$. States with $\Lambda = 0,1,2,3,\ldots$ are designated $\Sigma, \Pi, \Delta, \Phi, \ldots$, respectively. All states with $\Lambda \neq 0$ are doubly degenerate. Ω is the absolute value of the component of the total electronic (orbital plus spin) along the molecular axis. Thus CN $A^2\Pi_{3/2}$ denotes the first excited state of CN with the same multiplicity as the ground state. It is a doublet state (S=1/2) with $\Lambda = 1$, $\Omega = 3/2$. States are distinguished as Σ^+ or Σ^- according to whether the electronic wave function does or does not change sign when reflected at any plane passing through both nuclei, respectively. In homonuclear molecules a further property is important: The electronic wave function may change sign

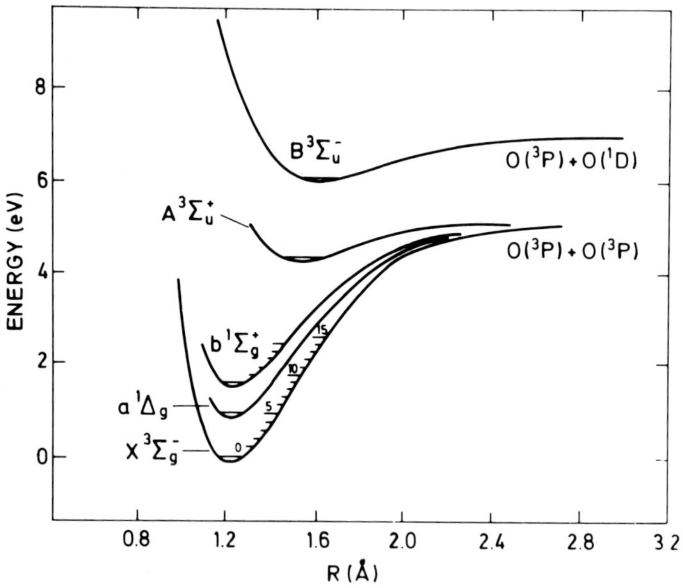

Fig. 3.19. A partial energy level diagram of molecular oxygen, adapted from [3.22]

on reflection through the center (i.e., when the electronic coordinates are replaced by their negatives). A wave function that does not change sign is termed g and one that changes its sign is termed u.

As an example, the potential energy diagram of O_2 is shown in Fig.3.19 with the appropriate term notations.

Selection rules for electric dipole transitions are

$$\Delta S = 0$$

$$\Delta\Lambda = 0,\pm 1$$

$$\Delta\Omega = 0,\pm 1$$

$$\Sigma^+ \leftrightarrow \Sigma^+ \quad , \quad \Sigma^- \leftrightarrow \Sigma^- \quad \text{but} \quad \Sigma^+ \not\leftrightarrow \Sigma^-$$

$$g \leftrightarrow u \quad \text{but} \quad g \not\leftrightarrow g \quad \text{and} \quad u \not\leftrightarrow u \quad .$$

Taking O_2 again as an example, it is seen that transitions to the first three excited states are forbidden (spin forbidden to a and b, symmetry forbidden to A). These states are thus metastable, and are characterized by a long radiative lifetime.

The $^1\Delta_g$ state plays an important role in the pumping mechanism of the I^* chemical laser (Sect.4.1). The first allowed transition is to the B state, and gives rise to the strong atmospheric absorption of O_2 below 2000 Å (Schumann-Runge bands).

The energy level diagram of O_2 may be used to illustrate a common phenomenon in molecular spectroscopy – predissociation. It is noted that discrete vibrational levels of the $B^3\Sigma_u^-$ state are degenerate with the dissociative continuum belonging to lower states. After being excited to the B state, the molecule is energetic enough to dissociate, but in order to do so, a crossing from the bound $^3\Sigma_u^-$ potential curve to one of the repulsive curves has to take place. If such a crossing takes place, the overall effect is that absorption into a discrete manifold results in molecular dissociation in the absence of collisions. This effect is termed predissociation and is quite common in molecular spectroscopy. Evidently, the BO states used to account for the appearance of the spectrum are not eigenstates of the molecular Hamiltonian (cf. Sect.1.3). Predissociation is thus one example of a breakdown of the BO approximation. Experimentally, predissociation is observed by broadening of absorption spectral lines, decrease in fluorescence lifetime and yield, and formation of dissociation products in the absence of collisions.

The phenomenon is of interest to practising photochemists, as it is in a sense a case of "eating the cake and having it too": Spectral lines are often sharp enough to allow highly selective excitation and yet immediate reaction follows, ensuring retention of the selectivity. Thus, predissociation has been successfully used in laser isotope separation experiments (see Sects. 1.3 and 5.3).

3.3.4 Infrared Spectra of Diatomic Molecules

The second term in the last line of (3.3.10) vanishes when $\phi_E^f \neq \phi_E^i$ because of the orthogonality of different electronic states. When $\phi_E^f = \phi_E^i$, $<\phi_E^f|\phi_E^i> = 1$, yielding the dipole moment for transitions within a given electronic state. The intensity of infrared transitions is determined by $|<\chi_N^f|\underline{\mu}_N|\chi_N^i>|^2$. A good approximation for this expression can be obtained by expanding the nuclear dipole moment $\underline{\mu}_N(\underline{R})$ [$=\underline{\mu}_N(R)$ or, in shorthand notation $\mu(R)$, for diatomic molecules] as a power series of $R-R_e$,

$$\mu(R) = \mu(R_e) + \left(\frac{\partial\mu}{\partial R}\right)_{R=R_e}(R-R_e) + \frac{1}{2}\left(\frac{\partial^2\mu}{\partial R^2}\right)_{R=R_e}(R-R_e)^2 + \dots \qquad (3.3.11)$$

and retaining only the first two terms. R_e is the equilibrium internuclear distance. The first term $\mu(R_e)$ is a constant; hence due to the orthogonality of χ_N^i and χ_N^f its contribution to $<\chi_N^f|\mu(R)|\chi_N^i>$ vanishes. The first and major nonzero contribution is thus

$$\left(\frac{\partial\mu}{\partial R}\right)_{R=R_e}<\chi_N^f|(R-R_e)|\chi_N^i> \quad . \qquad (3.3.12)$$

This expression shows that to obtain an infrared transition $(\partial\mu/\partial R)_{R=R_e}$ must differ from zero. As a corollary, infrared transitions cannot appear in the spectrum of homonuclear diatomic molecules where $\mu(R) \equiv 0$ and $\partial\mu/\partial R \equiv 0$. The potential curves of bound electronic states may be approximated by several analytic functions (e.g., the Morse function). For spectroscopic analysis, a most useful one is the series expansion in q, where $q = R - R_e$ is the displacement from the equilibrium position

$$V(q) = V_0 + \left(\frac{\partial V}{\partial q}\right)_{q=0}q + \frac{1}{2!}\left(\frac{\partial^2 V}{\partial q^2}\right)_{q=0}q^2 + \frac{1}{3!}\left(\frac{\partial^3 V}{\partial q^3}\right)_{q=0}q^3 + \dots \quad . \qquad (3.3.13)$$

The first (constant) term may be arbitrarily set to be equal to zero. The second term contains the first derivative at the minimum and thus vanishes. The first nonzero term is of the form $\frac{1}{2}kq^2$, recognizable as the potential energy of a harmonic oscillator. Terms containing higher powers of q may be considered as anharmonic correction terms. They are usually negligible near the equilibrium position but may become dominant at large q's.

A complete description of nuclear motion also includes rotations which are often treated as those of a rigid rotor (RR). The large difference between energies associated with vibrations and rotations is reflected by a similar difference in the typical period of motion (10^{-11} s for rotations, 10^{-13} s for vibrations). Thus, in any given vibrational state the molecule may be excellently approximated as a rigid rotor, albeit with a moment of inertia dependent on the vibrational quantum number.

Once again, energies are expressed in spectroscopic terms, G(v) for vibrations, F(J) for rotations, as follows:

$$E_{v,J} = E_v + E_J = hc[G(v)+F(J)] \quad , \tag{3.3.14}$$

where

$$G(v) = \omega_e(v+\tfrac{1}{2}) \quad ; \quad \omega_e = (2\pi c)^{-1}(k/\mu)^{\frac{1}{2}} \tag{3.3.15}$$

and

$$F(J) = B_e J(J+1) \quad ; \quad B_e = (h/8\pi^2\mu c R_e^2) \quad . \tag{3.3.16}$$

v is the vibrational quantum number, v = 0,1,2,...; J is the rotational quantum number, J = 0,1,2,... . μ is the reduced mass of the molecule, R_e the internu-

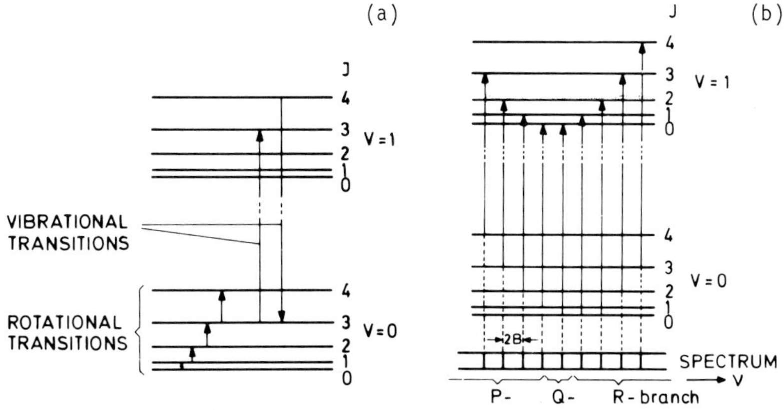

Fig. 3.20a,b. Energy levels and observed transitions in the RRHO approximation. (a) Schematic diagram, showing pure rotational transitions (in absorption) and vibrational transitions in absorption and emission. (b) Details of the v=0 → v=1 absorption transitions, showing the origin of the P, Q, and R branches

clear separation at equilibrium, and k the force constant. The energy level diagram in the RRHO approximation is shown in Fig.3.20a.

Observed transitions occur either between rotational states of a given vibrational state or between two different vibrational states. Selection rules are

$$\Delta J = 0, \pm 1 \quad (\text{for } \Lambda=0,\ \Delta J=\pm 1) \tag{3.3.17}$$

$$\Delta v = \pm 1 \quad . \tag{3.3.18}$$

Thus, the pure rotational spectrum is composed of lines with frequency separation given by

$$B_e J(J+1) - B_e J(J-1) = 2B_e J \quad . \tag{3.3.19}$$

B_e is typically 0.1-10 cm^{-1}, so that rotational spectra are normally observed in the microwave range.

Pure rotational lasing has been observed in several molecules, e.g., HF, within several vibrational levels (Sect.4.1 and Fig.2.41).

Transitions between vibrational states in the HO approximation are subject to the selection rule (3.3.18). The fine structure is determined by (3.3.17). We thus observe sets of lines for which $\Delta J = 1$, $\Delta J = -1$, and for $\Lambda \neq 0$ also $\Delta J = 0$. These are called R, P, and Q branches, respectively. The appearance of the spectrum is shown schematically in Fig.3.20b.

The position of the Q branch (a single line in the RRHO approximation) is termed the band origin. For molecules with $\Lambda = 0$ (e.g., HF, HCl, CO) the Q band is absent and the origin is recognizable by a gap in the sequence of lines. The separation between consecutive lines is $2B_e$, as in the pure rotational transitions.

A spectrum such as the one shown in Fig.3.20b is never encountered in practice. Deviations from the RRHO model must be considered, particularly in the analysis of vibrational spectra. The most important ones are due to an anharmonicity correction, and to changes in the rotational constant due to the vibrational motion.

Anharmonicity is displayed by the fact that overtone transitions are observed for practically all molecules. This deviation from the $\Delta v = \pm 1$ selection rule may be accounted for by adding terms to the series expansion of V(q), (3.3.13). The energy terms are corrected according to

$$G(v) = \omega_e(v+\tfrac{1}{2}) - \omega_e x_e(v+\tfrac{1}{2})^2 + \omega_e y_e(v+\tfrac{1}{2})^3 + \ldots \quad . \tag{3.3.20}$$

Table 3.1. Observed and calculated overtone transitions in HCl

Δv	ν(obs) [cm^{-1}]	ν(calc) [cm^{-1}]
1	2885	2885
2	5668	5668
3	8346	8347
4	10927	10923
5	13396	13396

An example of observed overtone transitions is given in Table 3.1. ν(calc) is found from (3.3.20) with only the first two terms retained; $\omega_e = 2988.7\ cm^{-1}$, $\omega_e x_e = 51.6\ cm^{-1}$.

The intensity of overtone transitions is quite weak, decreasing roughly by a factor of ten on unit increase in v. Nevertheless, these transitions provide a direct means to produce highly excited vibrational states, a fact utilized in recent laser experiments (Sects.1.3 and 5.4).

Coupling between vibrational and rotational motions is manifested by the fact that the separation between fine structure components is not constant. In general, R branch separation decreases on increasing J, while the P branch separation increases. The vibrational period is usually so much shorter than the rotational period that one expects the internuclear distance determining B to be equal to the mean nuclear separation of each vibrational state. This effect leads to a v dependence of B that may be expressed as

$$B_v = B_e - \alpha_e(v+\tfrac{1}{2}) \quad , \text{ with } \quad \alpha_e \ll B_e \quad . \tag{3.3.21}$$

Lower states are usually designated by a double prime (") and upper states by a single prime ('). Thus $B_v'' > B_v'$ and the formulae for the R and P branches become

$$\nu = \nu_0 + (B_v' + B_v'')m + (B_v' - B_v'')m^2 \tag{3.3.22}$$

with $m = J+1$ for the R branch and $m = J$ for the P branch.

As an example, the fine structure of the 0-1 transition of HBr is shown in Fig.3.21.

So far we discussed only factors affecting the separation between lines. As Fig.3.21 shows, absorption intensity varies strongly from line to line. For molecules in thermal equilibrium one can calculate the distribution of vibrational and rotational populations using the Boltzmann factor. The intensity of a transition in absorption is then roughly proportional to the

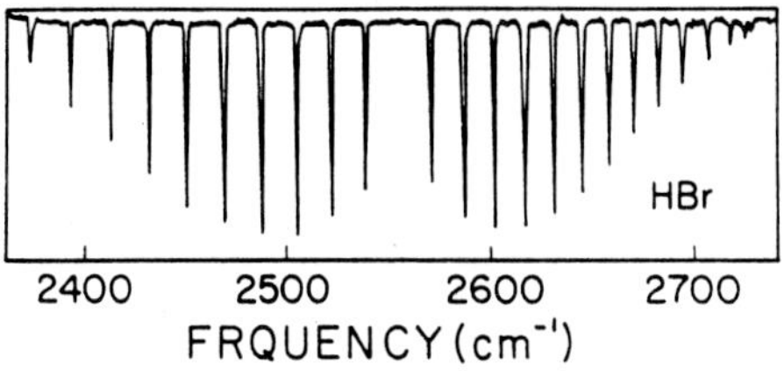

Fig. 3.21. The room temperature absorption spectrum of HBr, showing the P and R branches

population of the level from which absorption takes place. In the case of rotational states, one obtains (in the rigid rotor approximation)

$$N(J) = (N/Q_R)(2J+1)\,\exp[-hcB_e J(J+1)/kT] \quad . \tag{3.3.23}$$

The actual relative intensities are obtained on multiplying (3.3.23) by the transition moment $|<v',J'|\mu|v'',J''>|^2$. The v,J dependence obtained is important for chemical laser operation, and is further discussed in Sect.4.2.

3.3.5 Energy Levels of Polyatomic Molecules

Polyatomic molecules appear in this book in two major contexts: as laser materials and as substrates in laser chemistry experiments. Lasing can occur in the infrared, between rotational or vibrotational levels (e.g., HCN, CO_2, CF_4 lasers), or in the UV-visible, between electronic levels (e.g., dye lasers). Molecular spectroscopy has been very useful in interpreting the nature of the transitions involved in these molecular lasers. As we shall see, the spectroscopic concepts may sometimes be less usefull in interpreting laser chemistry results, particularly when excitation is to high energies (i.e., close to the dissociation energy) as in infrared multiphoton dissociation (cf. Sects.1.3 and 5.4 for a discussion of the quasicontinuum region).

Our starting point is again the general molecular Hamiltonian (3.3.3) and the BO approximation. Consequently, we write the molecular wave functions as a product

$$\psi = \phi_E \chi_N \tag{3.3.24}$$

and the total energy as a sum

$$E = E_E + E_N \quad . \tag{3.3.25}$$

We shall divide the discussion into two separate sections, one dealing with nuclear motion, the other with electronic motion.

Vibrational and Rotational Energies of Polyatomic Molecules

The nuclear configuration of a molecule containing N atoms can be fully described by 3N coordinates. A possible set are the 3N Cartesian coordinates with respect to some reference point in space. This is not a very useful set, though, as bonding in the molecule severely restricts the relative motion of the atoms, compared to a group of independent (unbound) atoms. It can be shown [3.14] that in polyatomic molecules, as in diatomic ones, the total energy related to nuclear motion may be written as the sum

$$E_N = E_T + E_R + E_V \quad .$$

Here E_T is the kinetic energy associated with the translation of the center of mass (CM), E_R the rotational energy of the whole molecule about axes passing through the center of mass, and E_V the energy associated with the relative motion of the nuclei, i.e., vibrational energy. The term "degree of freedom" is often used to describe independent nuclear motions. Thus, an atom has 3 degrees of freedom, and an N atomic molecule has 3N degrees of freedom. The number of degrees of freedom is not changed when atoms get associated in a molecule, but it is useful to reclassify them. We assign three degrees of freedom to the translational motion of the CM, three degrees of freedom to the rotations around the CM (for nonlinear molecules, see below), and are left with 3N-6 degrees of freedom that have to be assigned as vibrations. Each degree of freedom is associated with an energy term. Translation is described by $\frac{1}{2}M\dot{q}^2$ with M the molecular mass and q a Cartesian coordinate. Similarly, rotational energies are given by $\frac{1}{2}I\dot{\theta}^2$, where I is the moment of inertia and θ is the angular displacement. It turns out that one can define a set of 3N-6 internal coordinates Q_i for which the (classical) vibrational energy may be written as

$$E_V = \frac{1}{2}\lambda_i Q_i^2 + \frac{1}{2}\dot{Q}_i^2 \quad . \tag{3.3.26}$$

These coordinates are termed normal modes of vibration, the λ's are constants depending on the interactions between neighboring atoms.

An important special case should be mentioned: In linear molecules the moment of inertia for rotation around the molecule axis is zero, so that only

2 rotational degrees of freedom are available. Consequently, in linear molecules one has 3N-5 independent vibrations. For example, acetylene (C_2H_2) has 7 normal modes, while hydrogen peroxide (H_2O_2) has only 6.

Translational motion of the CM does not affect the spectrum except for broadening effects (Sect.3.1.2). We shall thus concentrate on the energy levels associated with rotations and vibrations. Again, it is useful to start with the RRHO approximation. In fact, the energy terms of a normal mode coordinate [cf. (3.3.26)] is that of a harmonic oscillator with a force constant λ (the reduced mass is incorporated in the coordinate Q). Within the normal mode approximation, one can write the molecular wave function (nuclear part) as

$$\chi_N = \left\{\prod_i \psi_i(Q_i)\right\}\Theta \qquad (3.3.27)$$

with $\psi_i(Q_i)$ the wave function associated with the normal coordinate Q_i and Θ the rotational wave function. The vibrational part of the problem is thus reduced to that of 3N-6 (3N-5 for linear molecules) independent harmonic oscillators. As in the case of diatomic molecules, the harmonic approximation is useful only at small amplitudes of vibration. Anharmonic corrections have to be introduced as soon as vibrations lead to large changes in geometry. These corrections may be of crucial importance for understanding the operation of lasers and even more so for the details of molecular excitation by lasers. In this introductory section we shall discuss the energy level diagrams of some polyatomic molecules within the harmonic approximation.

A general procedure for obtaining the normal modes of a given molecule, called normal mode analysis, is available. A detailed description may be found in [3.14]. The method is mathematically straightforward, but except for molecules with high symmetry, quite lengthy and will not be outlined here. Rather, we shall discuss the results for some illustrative examples that will be used in a later section of the book. It can be shown that each normal mode transforms as one of the symmetry species of the point group to which the molecule belongs. In highly symmetric molecules this leads to the occurrence of degenerate vibrations: two or more independent modes of vibrations with exactly the same frequency. Such degeneracies are removed when finer details of the molecular Hamiltonian are considered, such as interactions between vibrations and rotations, nuclear spin effects, etc. Another important case of degeneracy in polyatomic molecules arises when two different vibrational modes (or overtones) happen to have the same frequency (cf. CO_2 and CH_3F examples below). When states showing the accidental degeneracy also belong to the same symmetry group, a strong interaction takes place between them. The situation is referred to as a Fermi resonance. Fermi resonances are very common, and should always be considered in the analysis of vibrational spectra of polyatomic molecules.

Let us illustrate the concepts introduced by considering the energy levels of three molecules CO_2, CH_3F, and SF_6. The first one is the active species in one of the most widely used molecular lasers. The other two have played dominant roles in laser energy transfer and multiple photon dissociation studies.

CO_2

CO_2 is a linear, symmetric molecule, belonging to the $D_{\infty h}$ point group. With N = 3 there are four normal modes, also called fundamentals. They are shown in Fig.3.22 along with their assignment. It is seen that the ν_2 mode is of type Π_u and thus doubly degenerate. The motions associated with the normal modes are as follows: ν_1 is a symmetric stretch, in which the central carbon atom is stationary and the two oxygen atoms move synchronously toward or away from it. Note that this motion is very much like that of the diatomic molecule O_2, and likewise no change in dipole moment occurs so that an electric dipole transition is forbidden (see below). ν_2 is a bending mode – two independent motions can be visualized: in the plane of the paper and perpendicular to it (see Fig.3.22). ν_3, the asymmetric stretch, is such that the central atom moves towards one of the oxygens, while both O atoms move in

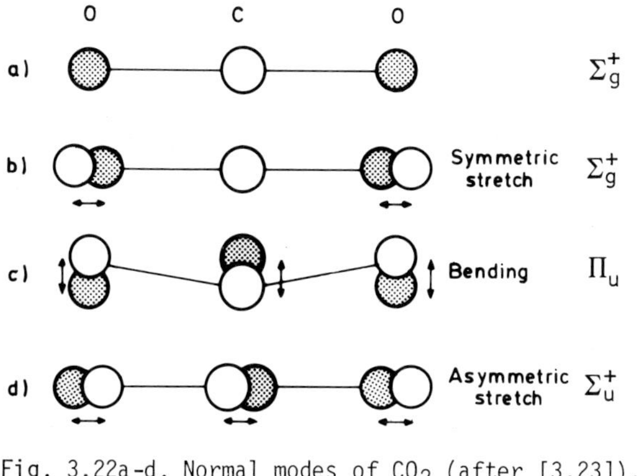

Fig. 3.22a-d. Normal modes of CO_2 (after [3.23]). (a) Equilibrium (vibrationless) position. (b) The symmetric stretch mode ν_1. (c) The doubly degenerate bending mode ν_2. (d) The asymmetric stretch mode ν_3

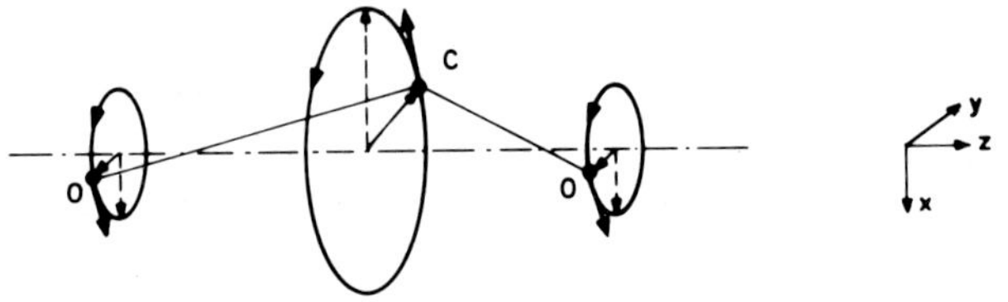

Fig. 3.23. Vibrational angular momentum in CO_2 (adapted from [3.8c]). Two bending modes are shown, along the x axis (dashed arrows) and along the y axis (solid arrows). The resultant motion of each atom is along a circle around the equilibrium position, as shown. The heavy arrows designate the momentary tangent velocity vector associated with each atom

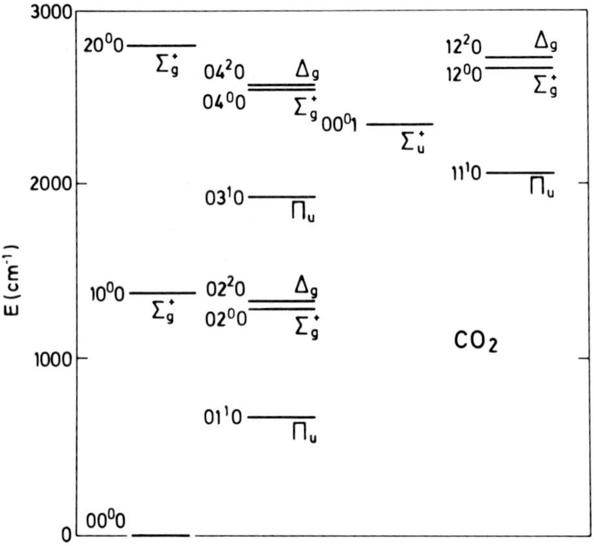

Fig. 3.24. Low lying vibrational energy levels of CO_2 (after [3.8c])

the opposite direction. In this mode one C-O bond contracts, the other extends in each half period. The term symbol commonly used is (v_1, v_2^{ℓ}, v_3) where v_1, v_2, and v_3 denote the quantum numbers of the ν_1, ν_2, and ν_3 frequencies, respectively. ℓ is the quantum number of the vibrational angular momentum. As seen from Fig.3.23, when the two degenerate ν_2 vibrational modes are shifted by a phase difference of 90°, their superposition may be considered as a rotation, and correspondingly is characterized by a rotational quantum number ℓ.

A partial energy level diagram of CO_2 is shown in Fig.3.24. An interesting case of Fermi resonance is observed in CO_2. The vibrational frequencies of the normal modes (fundamentals) are $\nu_1 = 1388\ \mathrm{cm}^{-1}$, $\nu_2 = 667\ \mathrm{cm}^{-1}$, $\nu_3 = 2349\ \mathrm{cm}^{-1}$. It is seen that $2\nu_2$ is nearly degenerate with ν_1. The $2\nu_2$ mode belongs to the $\Pi_u \times \Pi_u$ representation, i.e., two species should arise: Σ_g^+ and Δ_g. Since the 10^00 level is also a Σ_g^+ species, it interacts strongly with the former causing a relatively large splitting between the $02^00(\Sigma_g^+)$ and the $02^20(\Delta_g)$ levels. The CO_2 laser operates on the transitions $00^01 \rightarrow 10^00$ (10.6 μm band) and $00^01 \rightarrow 02^00$ (9.6 μm band) (cf. Sect.3.4.4).

CH_3F

CH_3F belongs to the C_{3v} point group. It has nine fundamental modes, three of which are doubly degenerate (Fig.3.25). As in CO_2, one could use the notation $(v_1v_2v_3v_4v_5v_6)$ to represent energy terms, but usually a shorthand notation is used, e.g. (0,2,0,0,0,3) is denoted as $2\nu_2 + 3\nu_6$. The fundamental frequencies of CH_3F are also shown in Fig.3.25, and a partial energy level diagram in Fig.3.26. It is evident that on top of the real degeneracies (i.e., ν_4, ν_5, and ν_6) there is a large number of accidental near degeneracies. Note that even though the ν_2 and ν_5 frequencies differ by only 3 cm^{-1} the interaction between them is small as they belong to different symmetry species.

We use the example of CH_3F to introduce the concept of local modes (Sects. 1.3.2 and 2.3.7). Inspection of Fig.3.25 shows that ν_3 may be assigned as

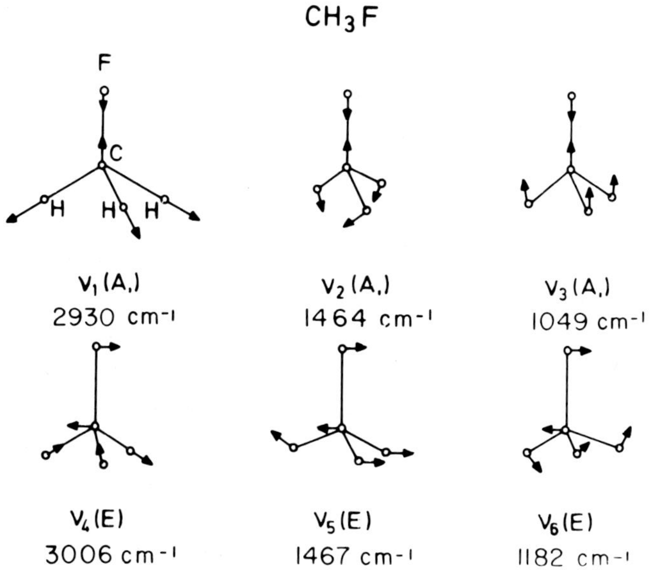

Fig. 3.25. Normal modes of CH_3F. The E-type modes are doubly degenerate. Vibrational frequencies are from [3.24]

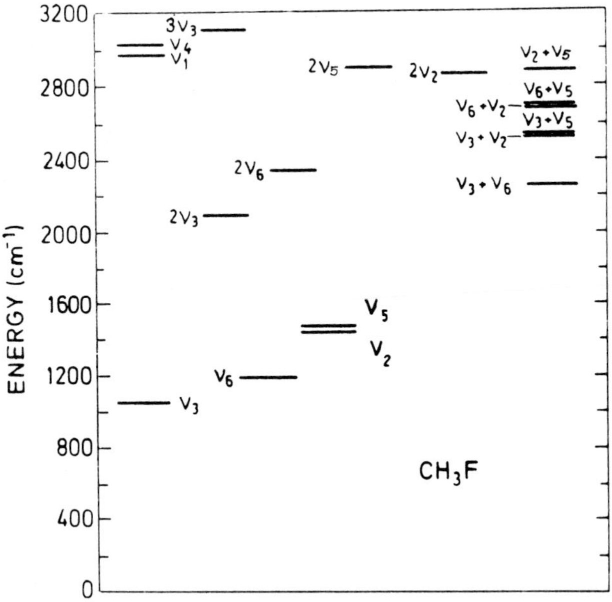

Fig. 3.26. A partial energy level diagram of CH_3F (adapted from [3.25])

Table 3.2. Highest observed frequencies [cm^{-1}] in some C-H containing molecules

CH_3F	3006	CH_2F_2	2967	CHF_3	3035
CH_3Cl	3041	CH_2Cl_2	3048	$CHCl_3$	3033
CH_3Br	3055	CH_2Br_2	3076	$CHBr_3$	3040
CH_3I	3060				
CH_4	3020				
C_2H_4	3272				
C_2H_2	3350				
C_6H_6 (benzene)	3099				

largely due to a C-F stretching motion. It turns out that molecules containing C-F bonds do indeed have fundamental modes with similar frequencies (CH_2F_2 - 1262 cm^{-1}, CHF_3 - 1152 and 1209 cm^{-1}, CF_4 - 904 and 1265 cm^{-1}). It appears that even though the molecules widely differ in their symmetry properties, the frequencies associated with the C-F stretch are rather constant. Similarly, ν_1 and ν_4 may be considered as C-H stretch frequencies (symmetric and asymmetric, respectively). Table 3.2 compares the highest vibrational frequencies of some molecules containing C-H bonds. Again it is obvious that 3000 cm^{-1} may be considered as the frequency of the C-H stretch, the exact value being determined by the actual chemical environment. In fact, this "local mode" description is most useful in the assignment of complex spectra of very large molecules, where normal mode analysis is extremely complicated. Furthermore, the local mode representation is more amenable to chemical interpretation, particularly when discussing reactivity (Fig.1.18).

SF_6

The normal modes of the seven-atomic molecule SF_6 are shown in Fig.3.27. Fifteen modes are expected, but due to the very high symmetry of the molecule, only six fundamental frequencies are encountered. The triply degenerate ν_3 mode coincides with the strong CO_2 laser emission lines (Fig.1.20) and has been extensively utilized in laser studies (cf. Sect.5.4). Note that it is typical of the S-F stretch mode (e.g., SF_5Cl ν=909 cm^{-1}, SF_5CF_3 ν=902 cm^{-1}).

Rotations of Polyatomic Molecules

Linear polyatomic molecules, like diatomic molecules, have only one moment of inertia. Their rotational energy levels are thus similar to those of diatomics. Examples are CO_2, N_2O, C_2H_2.

The treatment of nonlinear molecules is complicated in the general case. An important special case is that of the symmetric top. It has been stated above that three independent rotational degrees of freedom suffice to completely determine the rotational energy of a nonlinear molecule. Consequently one can define three orthogonal axes of rotation, called the principal axes

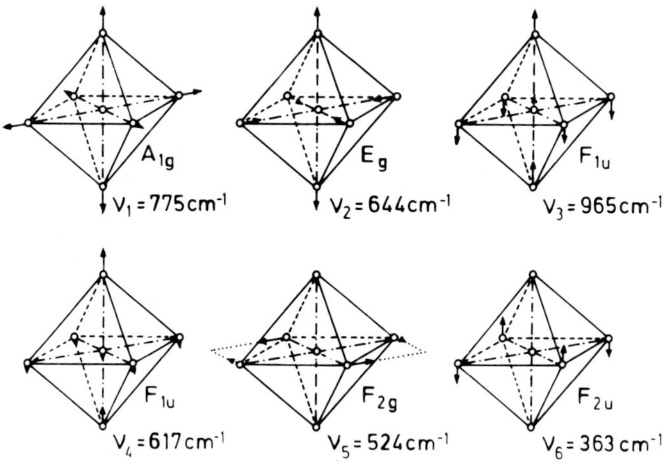

Fig. 3.27. The fifteen normal modes of SF_6. The E_g vibration is doubly degenerate. Each F vibration is triply degenerate

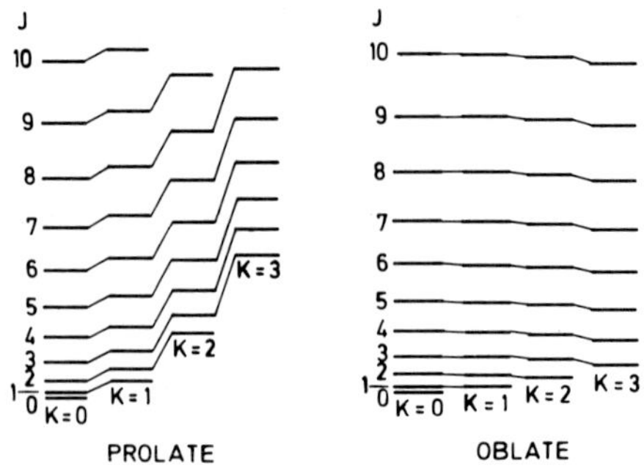

Fig. 3.28. Energy levels of a prolate and an oblate symmetric top

of rotation. Each principal axis is associated with a moment of inertia I so that in general $I_A \neq I_B \neq I_C$. In a symmetric top, $I_A \neq I_B = I_C$. One distinguishes between a prolate top for which $I_A < I_B$, and an oblate top for which $I_A > I_B$. Examples of prolate tops are CH_3F, NH_3, H_2O and of oblate tops benzene, BCl_3, C_2H_4. A special case is that of the spherical top, in which $I_A = I_B = I_C$. This is the case for all molecules belonging to cubic point groups, such as CH_4, SF_6, etc.

The energy levels of a symmetrical top are [3.8c]

$$E(J,K)_{sym.top} = BJ(J+1) + (A-B)K^2 \quad , \tag{3.3.29}$$

where

$$B = \frac{h}{8\pi^2 cI_B} \qquad A = \frac{h}{8\pi^2 cI_A}$$

and K is the component of J along the principal (A) axis

$$K = J, J-1, \ldots, -J+1, -J \quad .$$

Since energy depends on the square of K, all levels with $K > 0$ are doubly degenerate. In nonplanar molecules, inversion may either leave the wave function unchanged, or reverse its sign. This leads to inversion doubling [3.8c]. The splitting of the doublet is usually very small, resulting in an overall four-fold degeneracy of each J level in nonplanar symmetric tops. The energy levels of a symmetric top are shown in Fig.3.28. Note that for a given K, E(J) increases with J for a prolate top, and decreases with J for an oblate top.

In a spherical top, $A = B$, and the energy levels are given by

$$E(J)_{spher.top} = BJ(J+1) \tag{3.3.29}$$

just as in diatomic molecules. The statistical weight of a given level, however, is not $2J+1$ but rather $(2J+1)^2$. The second $2J+1$ factor arises from the fact that all K sublevels of a given J state are degenerate in a spherical top (compare Fig.3.28, where each J,K sublevel is $2J+1$-fold degenerate).

Electronic Energy Levels of Polyatomic Molecules

Given the complexity of the energy level diagram of the electronic ground state, one expects a hopeless situation in electronic transitions. Indeed, a general classification scheme such as exists in atomic and molecular (diatomic) spectroscopy is not available. It is clear however that molecules have to be classified according to their symmetry properties, making group theory instrumental in elucidating visible and UV spectra of molecules. Having laser applications in mind, we shall first discuss a generalized energy level diagram that is frequently used in the discussion of dye lasers and photochemical experiments. This diagram introduces terms and concepts relating to electronic excitation that will be used in subsequent chapters. As with infrared transitions, electronic transitions may be sometimes ascribed to a particular group of atoms in the molecular structure. Such a group is termed a chromophore, if the properties of the transition (oscillator strength, frequency range) are largely determined by the group alone and not by the entire molecule. An important chromophore is the carbonyl group >C=O, present in aldehydes and ketones. We shall discuss it at some length as a typical example of valence electronic transitions. Such transitions conserve the principal quantum number of the atomic orbital to which the electron undergoing the transition belongs. In contrast, Rydberg transitions involve a change in that quantum number, as is often the case in atoms. These are often at higher energies.

It was stated that the BO approximation is applicable also to polyatomic molecules, so that the solution to the electronic Hamiltonian serves as a potential surface for the nuclear motion. Obviously a pictorial representations such as given in Fig.3.17 is a hopeless task. For an N atomic molecule one needs a 3N-5 dimensional surface. Still, it is useful to draw quasidiatomic potential curves as shown in Fig.3.29. The term "internuclear distance" may be thought of as a displacement along a normal coordinate, or a local

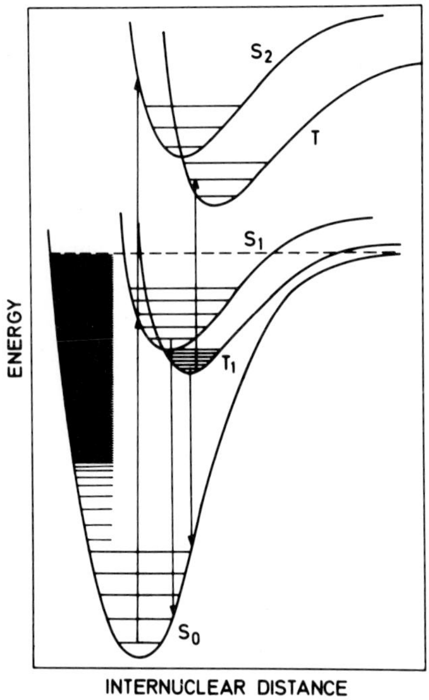

Fig. 3.29. Schematic energy level diagram of a polyatomic molecule in the Born-Oppenheimer, weak spin-orbit coupling approximation. Electronic energy levels are shown as a function of one nuclear coordinate, allowing a two-dimensional representation. In reality, a 3N-5 (3N-4 for linear molecules) hypersurface is required for an N atomic molecule. The dashed line designates the dissociation energy of the ground state (S_0) for motion along the nuclear displacement in question. Note the high density of states of S_0 in the vicinity of low lying vibrational states of S_1 and T_1

mode, and the figure is thus a projection of the 3N-5 dimensional surface along a particular direction. As the figure implies, the Franck-Condon factor is applicable for transitions in polyatomic molecules. One often observes progressions, with separation given by the vibrational frequency of the mode that underwent the largest geometry change. In most stable molecules, the ground electronic state is a singlet, designated as S_0 in Fig.3.29. Excited states are usually separated into two manifolds: singlet states (e.g., S_1, S_2 in the figure) and triplet states (T_1,T_2). This separation assumes that spin-orbit coupling is small compared to Coulombic interactions. By Hund's theorem, the triplet energy is lower than that of the corresponding singlet; hence the shown sequence. Transitions intensities are governed by the selection rule $\Delta S = 0$, by orbital symmetry considerations, and by Franck-Condon factors. Symmetry considerations introduce other selection rules as in the infrared region (cf. carbonyl example in Sect.3.3.7). The following radiative transitions are thus observed in large molecules: strong absorption to S_1, S_2, etc. (oscillator strength, $f \simeq 0.1$-1, $\varepsilon_{max}=10^4$-10^5 lit/mole cm), and much weaker absorption to T_1, $f \simeq 10^{-5}$, appearing at longer wavelengths than the absorption to S_1. In many molecules, light is reemitted in the form of fluorescence ($S_i \rightarrow S_0$ transitions) or phosphorescence ($T_i \rightarrow S_0$ transitions). In contrast with diatomic molecules, resonance fluorescence is not common in large polyatomic molecules, as collision-induced vibrational relaxation is very efficient. Light emission is thus usually observed from the lowest vibrational state, except for very low-pressure conditions. Furthermore, radiationless transitions (Sect.2.4.9) leading to deactivation of higher excited states S_i or $T_i (i>1)$ are very rapid, and emission is normally observed only from the lowest excited state of a given multiplicity, namely S_1 or T_1. Radiative lifetimes are typically 10^{-9}-10^{-7} s for singlets and 10^{-3}-1 s for triplets. The long lifetime of the triplet state makes it much more vulnerable to deactivating collision-induced processes than the singlet state. Consequently, phosphorescence is normally observed only in the low-pressure gas phase or in solids at cryogenic temperatures. A notable exception is biacetyl that phosphoresces in liquid solution at room temperature (Sect.5.3.3).

The advent of lasers allowed careful investigation of selectively excited states of polyatomic molecules. It appears that even under collision-free conditions the emission quantum yield is not unity and in many cases is excitation wavelength dependent. This topic, already discussed in Sect.2.4.9, will be taken up again in Chap.5.

3.3.6 Infrared Spectra of Polyatomic Molecules

Since we shall be concerned mostly with the infrared vibrational spectrum in this book, pure rotational transitions will not be discussed.

Linear Molecules

As with diatomic molecules, transitions are allowed only if $\mu_{if} \neq 0$. Here the transition moment has the form

$$\underline{\mu}_{fi} = \left\langle v_1^f, v_2^f, \ldots v_m^f \middle| \underline{\mu} \middle| v_1^i, v_2^i, \ldots v_m^i \right\rangle \quad . \tag{3.3.30}$$

Selection rules are conveniently derived using group theoretical considerations. It is required that the direct product of the representations of the terms appearing in (3.3.30) will contain the totally symmetric representation. Levels that may be reached by infrared absorption from the ground state are called infrared active. Thus the 10^00 state of CO_2 is infrared inactive while the 01^00 and 00^01 states are infrared active.

In diatomic molecules, the transition moment is always parallel to the molecular axis. In linear polyatomic molecules, one distinguishes between parallel transitions (similar to the diatomic case) and perpendicular transi-

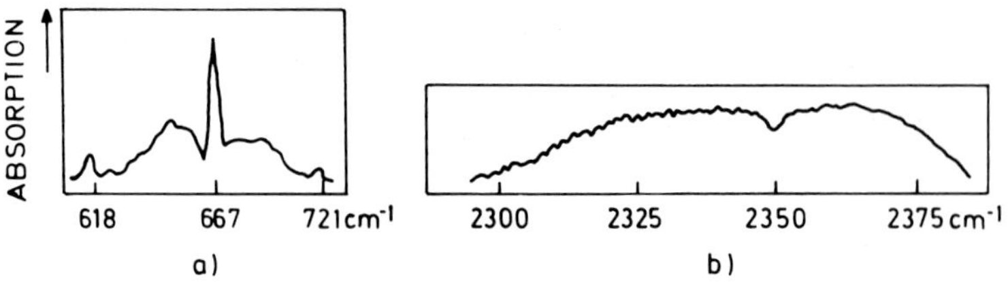

Fig. 3.30a,b. Perpendicular and parallel absorption bands of CO_2. (a) The perpendicular band at 15μm due to the $00^00 \rightarrow 01^00$ transition. (b) The parallel band at 4.3 μm due to the $00^00 \rightarrow 10^00$ transition. Higher resolution shows that only transitions originating from even J values are observed. This effect results from nuclear statistics and is extensively discussed in [3.8], from which the figure is adapted. Note the prominent Q band in the 15μm-band

tions, in which $\underline{\mu}$ is perpendicular to the molecular axis. Thus, the $00^00 \rightarrow 00^01$ transition in CO_2 is parallel, while the $00^01 \rightarrow 01^00$ transition is perpendicular.

The selection rules are $\Delta v = \pm 1$ and $\Delta J = \pm 1$ for parallel transitions, $\Delta v = \pm 1$ and $\Delta J = 0, \pm 1$ for perpendicular transitions. Consequently, the familiar P and R branches are observed in the former while in the latter a Q branch is also present. As an example, two transitions of CO_2 are shown in Fig.3.30.

In addition to the fundamental transitions. one observes overtones and combination transitions. The former arise from anharmonicity, much in the same manner as discussed for diatomic molecules. The latter are due to transitions to "mixed" states, i.e., states in which more than one normal vibration is populated. A further complication is due to the appearance of hot bands (sometimes called difference bands) originating from low lying overtones. This effect considerably complicates the infrared spectrum of molecules with low frequency vibrations, in particular at elevated temperatures.

Symmetric Top Molecules

The terms parallel and perpendicular transitions are retained in the case of symmetric top molecules. They refer to bands in which the transition moment is parallel or perpendicular to the A axis, respectively.

The selection rules for parallel bands are

$$\Delta K = 0 \qquad \Delta J = 0, \pm 1 \qquad \text{if} \qquad K \neq 0 \tag{3.3.31}$$

$$\Delta K = 0 \qquad \Delta J = \pm 1 \qquad \text{if} \qquad K = 0 \ . \tag{3.3.32}$$

The appearance of the resulting spectrum is shown in Fig.3.31.

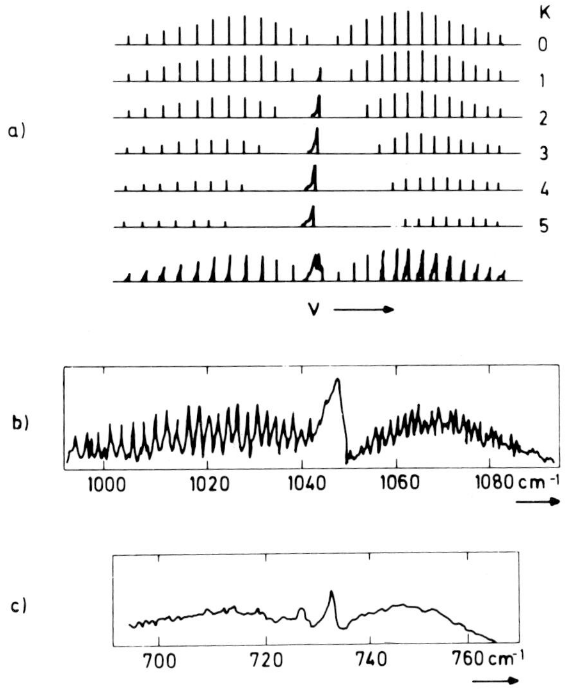

Fig. 3.31a-c. Parallel bands of a symmetric top (adapted from [3.27]). (a) Schematic representation, showing the construction of the observed band from individual K sub-bands. (b) The 9.55 μm band of $CH_3F(\nu_3)$. (c) The 13.6 μm band of $CH_3Cl(\nu_3)$

The selection rules for perpendicular bands are

$$\Delta K = \pm 1 \qquad \Delta J = 0, \pm 1 \quad . \tag{3.3.33}$$

A schematic representation of the spectra is shown in Fig.3.32. It is seen that one usually obtains a sequence of Q branches, on top of a more smeared-out P and R structure. Anharmonicity effects in symmetrical (and spherical) tops may be treated much in the same way as for diatomics. In general, the energy separation in a given mode becomes progressively smaller. Thus, it is usually not possible to use a monochromatic source, such as a laser, to reach high excitation energies: a frequency that is in resonance with the $0 \rightarrow 1$ transition of a given mode, is in general out of resonance with, say, the $2 \rightarrow 3$ transition of the same mode. On the other hand, one should recognize that due to the large number of combination and difference bands, polyatomic molecules display a finite, albeit weak, absorption over a very large frequency range. This is particularly important at elevated temperatures, as many excited vibrational levels may be appreciably populated. In this context, elevated temperatures means $kT > h\nu$, where ν is a vibrational frequency. The re-

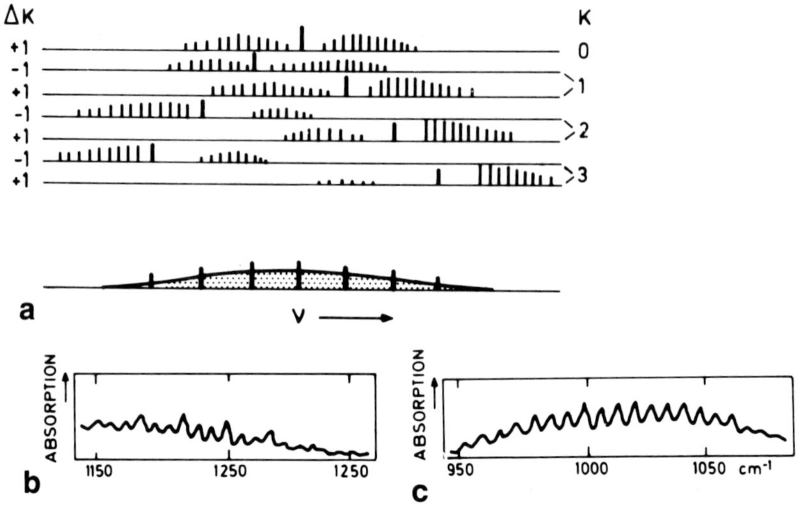

Fig. 3.32a-c. Perpendicular bands of a symmetric top (adapted from [3.27]). (a) Schematic representation. The Q bands are the most prominent feature, and often appear as narrow bands imposed on a seemingly continuous background due to P and R bands. The experimentally observed spectrum is a superposition of many K subbands and is shown at the bottom. The quasicontinuous background is represented by the dotted section. (b) The 8.5 μm band of $CH_3F(\nu_6)$. (c) The 9.8 μm band of $CH_3Cl(\nu_6)$

sulting spectrum is quasicontinuous in appearance and is sometimes referred to as "background absorption". It is of considerable importance in laser experiments, particularly with high-power infrared laser excitation (see Sects. 1.3 and 5.4).

Spherical Top Molecules

The advent of high-resolution infrared lasers, in particular tunable diode lasers, allowed for the first time the realization of truly high-resolution spectra of complicated large molecules. Classical texts [3.8c] did not treat the vibration-rotation spectrum of an octahedral molecule (belonging to the O_h point group) in any detail, in the absence of experimental data. A recently well studied spectrum is that of the ν_3 mode of SF_6 (Fig.3.27). Each $J \neq 0$ level of this three-fold degenerate mode is split due to vibration-rotation interaction (Coriolis splitting). Further splitting is due to the fact that the molecule is not really a rigid rotor. Each J level can split into components identified by their group-transformation properties as A_1A_2, E_1F_1, and F_2 [3.28,29]. In the spectrum, the lines tend to form "clusters", which have been of much help in the analysis. Figure 3.33 shows the results of such an

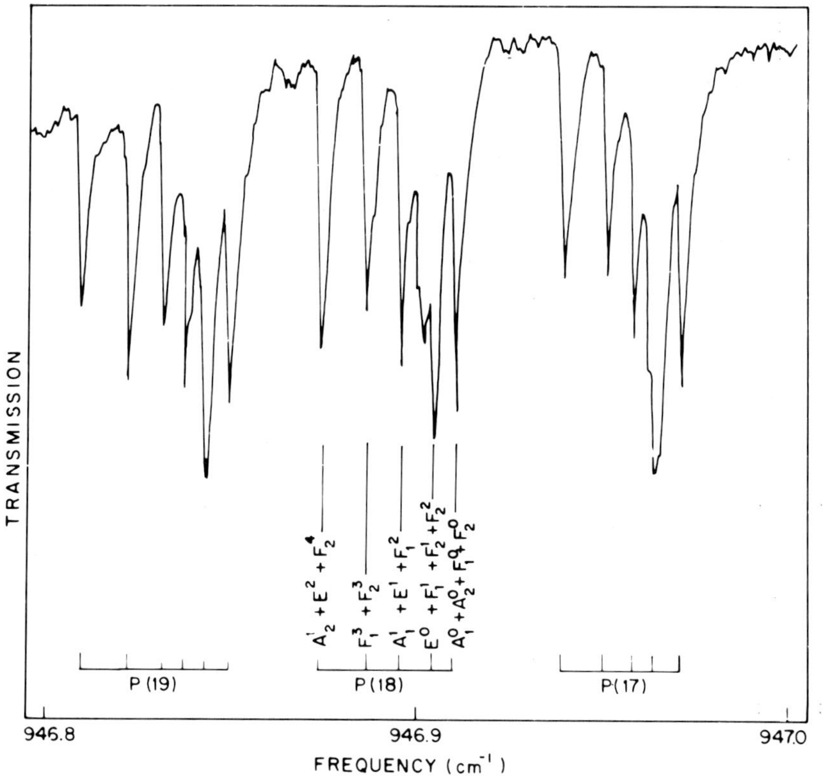

Fig. 3.33. A portion of the $0 \rightarrow 1$ ν_3 absorption transition in SF_6, recorded at 135 K with a resolution of about 10^{-4} cm^{-1}. The calculated Doppler width under these conditions is $\sim 2 \times 10^{-3}$ cm^{-1} [cf. (3.1.41)]. Each P branch component is a manifold of lines, which are seen to be composed of further, unresolved, components (adapted from [3.30]; the notation is from [3.31])

analysis for a small section of the ν_3 absorption band. The figure should be compared with Fig.1.17. For details of the analysis and further theoretical background the reader is referred to the original literature.

3.3.7 The Near Ultraviolet Spectrum of Carbonyl Compounds

As mentioned above, the analysis and interpretation of the electronic spectra of polyatomic molecules is often complicated, and of a highly specific nature. It requires detailed knowledge of the molecular symmetry (which may be different for the two electronic states involved) and must take account of vibrational-electronic interactions, often dominating the spectroscopy of "forbidden" transitions. We shall demonstrate some of the arguments and me-

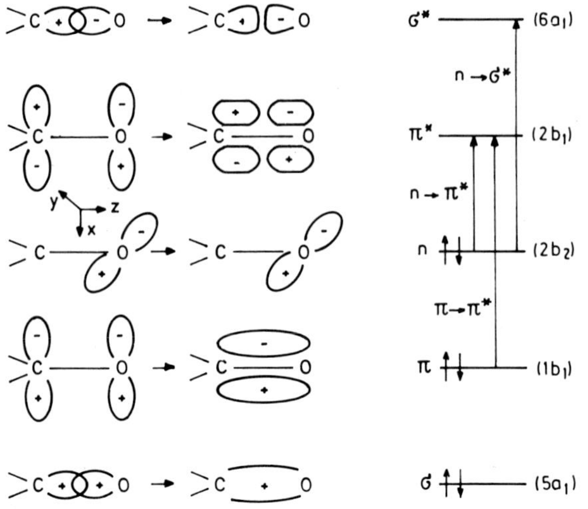

Fig. 3.34. Localized molecular orbitals of the carbonyl group. The symmetry species on the right are those of the C_{2v} group, to which the formaldehyde molecule belongs (adapted from [3.10])

thods used in such analyses for the case of a specific example — the n-π* transition of carbonyl compounds, primarily that of the parent compound, formaldehyde. The discussion is sufficient for the applications discussed in Chap.5, but the original literature should be consulted for more details [3.32]. We shall make use of the results of this section in Sect.5.3.2.

Molecules containing the carbonyl group show a characteristic absorption in the near UV. The transition is rather weak[9] ($f=10^{-4}$-10^{-2}) and for small molecules (e.g., formaldehyde, H_2CO) shows a pronounced vibrational structure. It has been termed an n-π* transition [3.10], a notation derived from the MO description of the bond. Figure 3.34 shows the localized orbitals of a C=O bond. In the ground state the carbon is thought of as having an sp^2 hybridization, with one hybrid orbital forming a σ bond with the p_z orbital of the oxygen atom. A σ* orbital is also formed, leading to a higher energy level. The other two hybrid orbitals are used to form bonds with the substituent groups (H atoms in the case of formaldehyde). The p_x orbitals of the carbon and oxygen atoms form a π bond, completing the double bond, with concurrent formation of an antibonding π* state. Finally, the p_y orbital of the oxygen is not participating in the bond and is termed a nonbonding (n) orbital. In the ground state, the six available electrons (two electrons from

9 f is the oscillator strength defined in footnote 4 on p.9. For a more detailed discussion of this concept see, for example, [3.26]

Table 3.3. Some properties of the lowest $n \rightarrow \pi^*$ transition in selected carbonyl compounds

Compound	Absorption region [Å]	Origin [cm^{-1}]	$\omega''(CO)$ [cm^{-1}]	$\omega'(CO)$ [cm^{-1}]	$r''(CO)$ [Å]	$r'(CO)$ [Å]
H_2CO	3500-2300	28258	1738	1177	1.22	1.32
CH_3CHO	3500-2500	~28700	1743	1125	1.21	1.32
$(CH_3)_2CO$	3300-2200	30924	1710	-	1.24	1.33
$(CHO)_2$	4600-3400	21973	1745	1391		
$(CH_3CO)_2$	4800-3400	21850	1718	-		

the carbon, four electrons from the oxygen) occupy the lowest three orbitals, namely the σ, π, and n orbitals. The configuration may be described as $(\sigma)^2(\pi)^2(n)^2$. The lowest excited states are formed by the following transitions:

$$(\sigma)^2(\pi^2)(n^2) \rightarrow (\sigma)^2(\pi)^2(n)(\pi^*) \quad ,$$

termed a $n \rightarrow \pi^*$ transition, which is normally the transition of lowest energy;

$$(\sigma)^2(\pi)^2(n^2) \rightarrow (\sigma^2)(\pi)(n^2)(\pi^*) \quad ,$$

termed a $\pi \rightarrow \pi^*$ transition; and

$$(\sigma)^2(\pi)^2(n)^2 \rightarrow (\sigma)^2(\pi)^2(n)(\sigma^*) \quad ,$$

termed a $n \rightarrow \sigma^*$ transition.

Table 3.3 gives data on some singlet-singlet $n \rightarrow \pi^*$ transitions in carbonyl compounds. The table shows that the CO bond length increases appreciably on excitation, and the CO stretch frequency decreases. These observations are consistent with weakening of the C-O bond due to promotion of an electron to an antibonding state. In many cases and particularly at high excitation energies, spectral lines are diffuse, due to predissociation.

The forbidden nature of the transition may be explained by reference to the particular case of formaldehyde, the parent compound. The ground state is of C_{2v} symmetry, and assuming that the $(n-\pi^*)$ state belongs to the same symmetry group, the transition would be an $A_1 \rightarrow A_2$ transition, which is forbidden in C_{2v}. It can be vibronically allowed, by out of plane vibrations such as ν_4 (see Fig.3.35 for the normal modes of formaldehyde). In fact, it turns out that the excited state is not planar, but has a pyramidal configu-

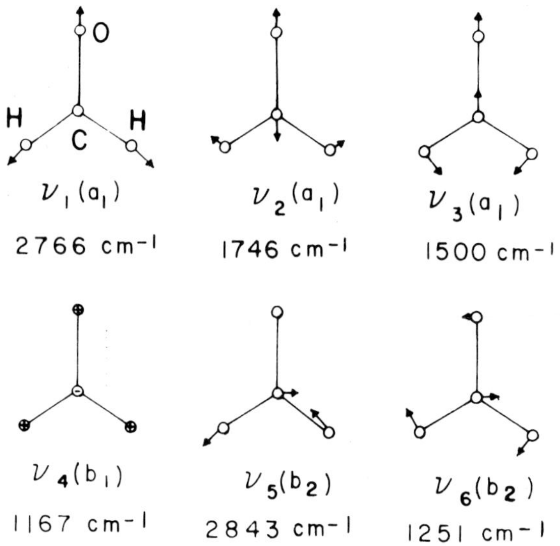

Fig. 3.35. Normal modes of formaldehyde (adapted from [3.33])

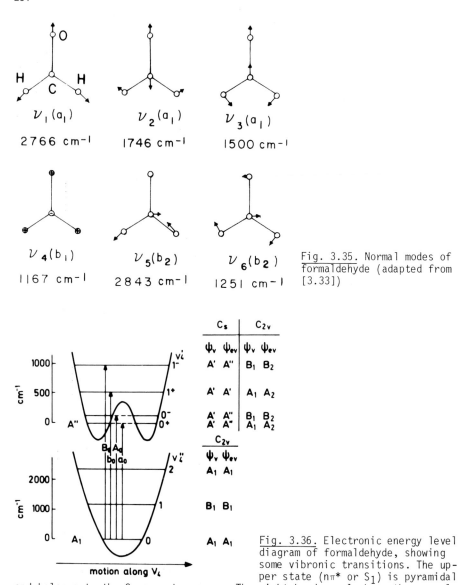

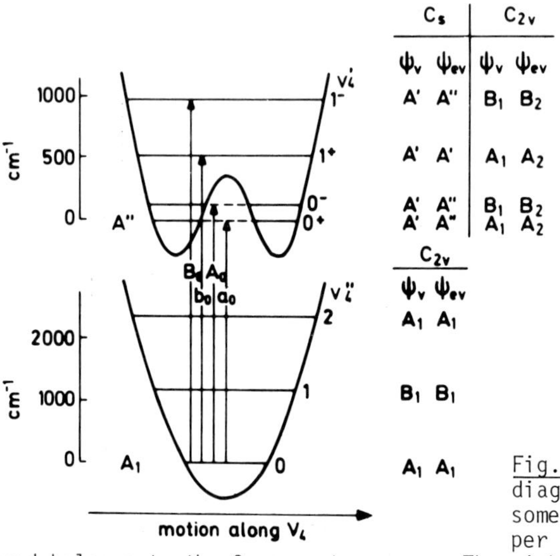

Fig. 3.36. Electronic energy level diagram of formaldehyde, showing some vibronic transitions. The upper state ($n\pi^*$ or S_1) is pyramidal and belongs to the C_s symmetry group. The right-hand panel shows the correlation between the C_s and C_{2v} group elements. Motion along ν_4 is chosen for the nuclear coordinate (compare Fig.3.29) since ν_4 appears (usually in combination with other modes) in most observed bonds. In C_s ν_4 corresponds to "umbrella inversion" motion leading to a double minimum in the potential curve [3.8c]

ration. Two such configurations are possible, and they can be interchanged by moving the carbon atom through the plane defined by the oxygen and two hydrogen atoms. This situation leads to potential curves shown in Fig.3.36,

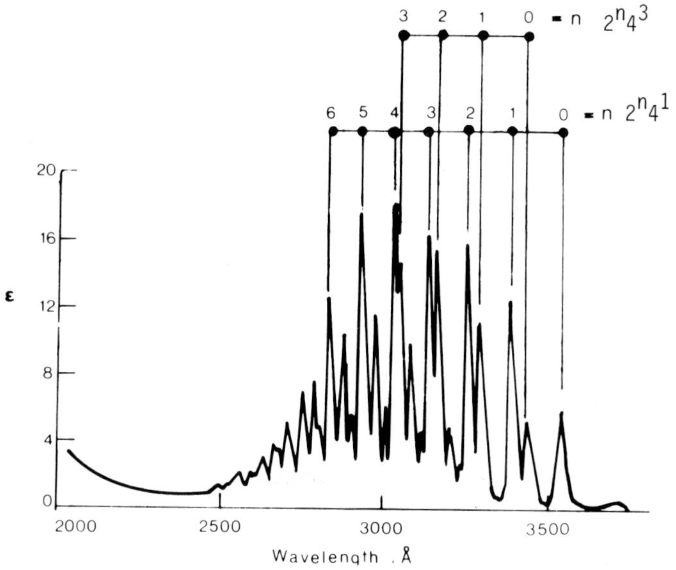

Fig. 3.37. Low-resolution spectrum of the n-π^* absorption transition in formaldehyde. Some vibronic assignments are shown

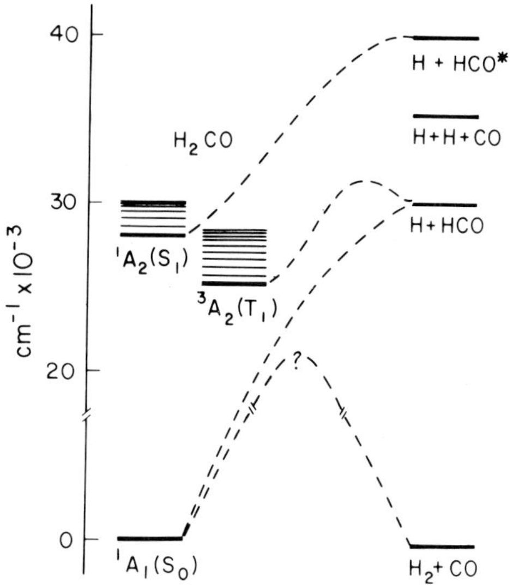

Fig. 3.38. An alternative representation of formaldehyde energy level diagram, showing possible dissociation channels (adapted from [3.34])

i.e., having a double minimum. The excited state strictly belongs to the C_s point group, and the transition should be termed $A_1 \to A''$ transition. In order to determine selection rules, it is useful to consider the deviation from C_{2v} symmetry as a small one, and retain C_{2v} notation for vibrational states of the electronically excited state, as shown in the figure. The observed spectrum (Fig.3.37) shows prominent progressions in ν_2, the carbonyl stretch.

A different representation of the energy level diagram of formaldehyde is shown in Fig.3.38. This figure should be compared with the general scheme for polyatomic molecules (Fig.3.29). With appropriate changes, it may serve for all carbonyl compounds.

S_1 and T_1 are the singlet and triplet n-π* states, respectively. The triplet state is observed in both absorption (3600-3800 Å) and in emission. It is of great importance in the photochemistry of carbonyl compounds (Sect.5.3). Predissociation may be seen to arise from coupling to the continua of $H_2 + CO$ and H + HCO, leading to molecular and radical dissociation channels, respectively.

In aliphatic carbonyl compounds, of the type (RR')CO such as acetone, the forbidden nature of the $n \rightarrow \pi^*$ transition is maintained. There, the molecular dissociation is not encountered, possibly due to a high barrier (shown in Fig.3.16). Predissociation is still important, due to the radical dissociation process

$$RR'CO \rightarrow R + R'CO \quad .$$

In dicarbonyl compounds, such as glyoxal and biacetyl (Sect.5.3.3), the energy level diagram is somewhat more complicated, but the $n \rightarrow \pi^*$ transition is still the lowest lying one. Interaction between the two carbonyl groups leads to lowering of the energy levels so that the transition appears in the blue region of the spectrum (around 4000 Å): the fluorescent and phosphorescent properties of these compounds proved to be a good testing ground for theories on internal energy flow in polyatomic molecules. Lasers played a crucial role in these studies, as detailed in Chap.5.

3.4 Laser Sources

The number of available laser sources is very large, and is growing virtually by the day. We begin this section by enumerating specifications that should be considered in choosing a laser for a particular application (Sect.3.4.1). We than proceed to discuss in some detail three types of lasers that appear to be particularly useful in initiating chemical reactions: excimer lasers (Sect.3.4.2) in the UV range, dye lasers (Sect.3.4.3) in the visible, and CO_2 lasers (Sect.3.4.4) in the infrared. A fourth class of lasers, the chemical lasers, is extensively dealt with in Chap.4.

3.4.1 Laser Specifications

The important wavelength range for practicing chemists is between 0.1 and 30 μm. Figure 3.39 shows that this range is reasonably well covered by known laser transitions. Of the many laser sources shown, only relatively few were extensively used in chemical research, particularly as initiators of chemical reactions. The chosen lasers are characterized by a combination of properties that may be expressed by a set of specifications. We proceed to discuss these specifications separately, but it should be borne in mind that a successful source is one that combines several desired features (such as narrow bandwidth and high power). Frequently, these properties are interdependent, allowing the experimentalist to trade one for another.

Wavelength Range and Tunability

The frequency of the laser radiation is determined by the gain profile which in turn depends on the spectroscopic properties of the active medium, and on cavity parameters (Sect.3.2). Lasers operating on atomic or ionic transitions in gas or solid phases are usually restricted to a few lines only. Tunability, when achievable, is obtained within the rather narrow gain profile (of the order of 0.1-1 cm^{-1}). A larger frequency range is achieved with molecular gas lasers, such as CO_2 and hydrogen halides. These devices can be tuned, usually with gaps, over a few hundred wave number units, by using different vibrota-

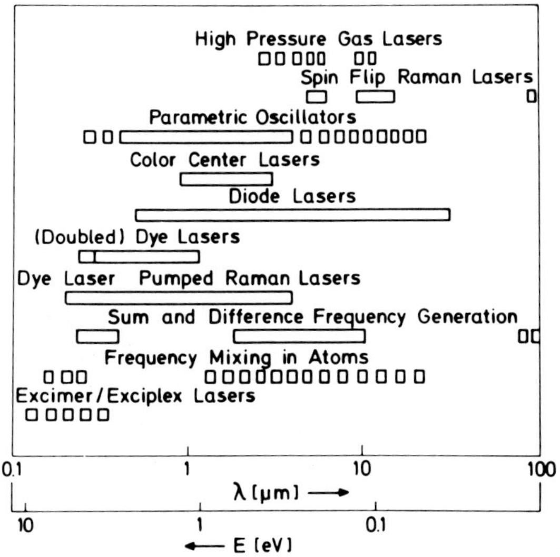

Fig. 3.39. Survey of important laser sources suitable for experiments in chemistry and spectroscopy. (▭) Fully tunable; (▫▫▫) incompletely tunable or fixed frequency lasers. Most of the gaps shown will be closed by further technical development

tional transitions. Increasing the pressure to several tens of atmospheres allows continuous tuning over the high gain portions of the bands. Truly continuous laser sources include dye lasers [3.35], diode lasers [3.36], and lasers based on various nonlinear operations, such as optical parametric oscillators, sum and frequency generation by three- and four-wave mixing [3.37], and Raman spin-flip lasers [3.38].

Of special interest in this book are electric discharge lasers operating on molecular vibrational transitions in the infrared. The more important ones are shown in Fig.3.40.

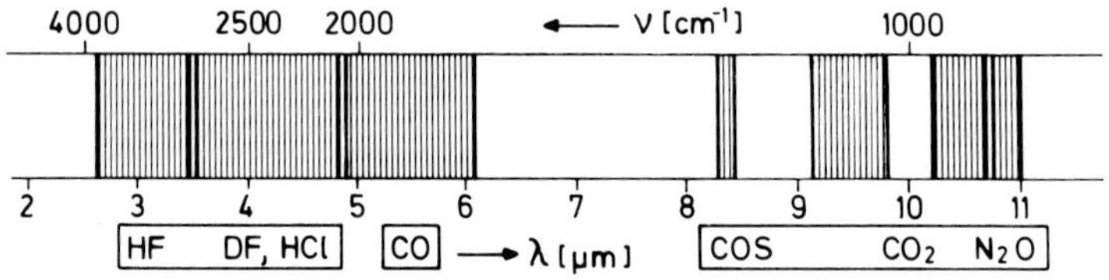

Fig. 3.40. Lasing range of direct discharge-excited molecular infrared gas lasers. Shaded areas show tuning range, but tuning is usually limited to discrete wavelengths within the range. True continuous wavelength scan is possible in high-pressure systems, operated in the transverse excitation mode (see Sect.3.4.4; compare also Fig.3.49)

Power

Lasers are operated in either a continuous wave (cw) fashion or in a pulsed mode. The average power is defined as the energy output E_{out} per unit time, measured over a period much longer than the pulse width or the interval between shots. In the case of a cw laser, the average power coincides with the peak power, defined as the maximum power output, i.e., the power at the moment when $dE_{out}/dt = 0$. The two quantities are usually drastically different for pulsed lasers. A typical example is a CO_2 laser operating at 1 Hz repetition rate with an energy output of 1 J in 10^{-7} s. The peak power is 10 MW, while the average power is only 1 W. High repetition rates are desirable not only for larger conversion efficiencies in laser-induced chemical reactions, but also for signal processing purposes. Therefore peak power should always be quoted at the appropriate repetition rate. In practice, power can often be traded for spectral purity or beam quality (Sect.3.2.3). Thus operating at many longitudinal modes results in a large frequency spread, which in an inhomogeneously broadened system leads to larger power output. Insertion of a dispersing element into the cavity reduces the bandwidth, as fewer modes

Table 3.4. Characteristics of some cw lasers

Laser	Approximate wavelength [μm]	Power (typical) TEM_{00} [W]	Beam divergence [mrad]
Ar^+	0.514	15	0.6
Ar^+ laser pumped dye laser	0.585	0.1	1.5
He-Ne	0.6328	0.005	1
Nd:YAG	1.064	20	2.5
CO_2	10.6/9.4	50	2.2
Diode lasers	IR	0.001	20 [a]
Chemical HF	2.9	10	1
Color center	1.9	0.1	[b]
Spin-flip Raman lasers	6	1	20 [c]

[a] Without external resonator several degrees in one direction, 20 mrad in the other.
[b] Not definitively known.
[c] Can be improved by external resonator.

Table 3.5. Characteristics of some pulsed lasers

Laser	Wavelength [μm]	Pulse length [s]	Pulse energy multimode [J]	Typical rep. rate [pps]
Discharge pumped KrF	0.248	10^{-8}	1	10
Discharge pumped ArF	0.193	10^{-8}	0.2	10
Amplified Nd:YAG	1.064	10^{-8}	1	10
doubled	0.53	10^{-8}	0.5	10
tripled	0.355	10^{-8}	0.3	10
quadrupled	0.265	10^{-8}	0.1	10
Nitrogen laser pumped dye	0.360-0.740	10^{-8}	10^{-3}	10
CO_2-Q switched	10.6	10^{-5}	0.1	100
CO_2 - TEA	10.6	10^{-7}	10	1-100
Discharge pumped HF	2.9	5×10^{-7}	1	10
Discharge pumped CO	5.6	10^{-5}	1	10
Optical parametric oscillators	2	10^{-5}	0.01	100
CF_4	16	10^{-7}	0.10	1

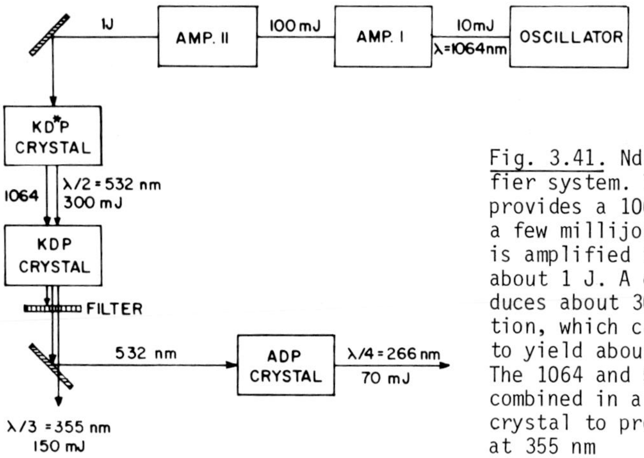

Fig. 3.41. Nd:YAG oscillator amplifier system. The master oscillator provides a 1064 nm TEM_{00} pulse of a few millijoule at 20 ns. The pulse is amplified in a few stages up to about 1 J. A doubling crystal produces about 300 mJ of 532 nm radiation, which can be doubled again to yield about 70 mJ at 266 nm. The 1064 and 532 nm pulses can be combined in a different mixing crystal to produce a ~150 mJ pulse at 355 nm

reach threshold concurrently, leading to lower power output. In the case of transverse modes, one simply reduces the active volume by operating at TEM_{00}, leading to corresponding power attenuation. Consequently, spectral linewidth and beam divergence must be known, when power outputs of different lasers are compared. Tables 3.4 and 3.5 list the properties of some cw and pulsed lasers that are commonly used for chemical applications.

Spectral Purity

Lasers are "monochromatic" sources. Their bandwidth is determined by the dominant broadening mechanism and by the mode structure (Sect.3.2). For example, the wavelength of a gas laser will generally hop within the Doppler profile (cf. Fig.3.9) unless special care is taken to stabilize the frequency. Available techniques allow the operation of a frequency-stabilized laser with a bandwidth of ~1 kHz. Line narrowing is achieved by inserting dispersing elements into the cavity. These include prisms, gratings, birefringent filters, and etalons. It is usually of advantage to operate a frequency-narrowed laser in an oscillator-amplifier arrangement: control of the mode structure (transversal and longitudinal) is much easier at low power levels. Therefore, one uses a small master oscillator as the primary source, and amplifies the high-quality beam in one or more stages. Figure 3.41 shows the schematics of an Nd:YAG (Sect.3.3.1) oscillator amplifier system. See also the discussion of dye lasers below.

Beam Quality

Many applications require tight focusing of the beam. For TEM_{00} lasers (Sect. 3.2), one is limited only by diffraction, namely the beam waist can be made as small as the laser wavelength [3.3,5]. This is to be compared with incoherent sources, where a limit is set by the light source temperature which must be larger than (or equal to) the image temperature. High-quality beams can be easily collimated over long distances, without appreciable loss. Collimation and focusing are best described by two interrelated quantities: the beam waist w_0 and the Rayleigh range z_R (see Fig.3.11). Good beam quality is important in nonlinear applications, such as frequency doubling, parametric oscillations, and CARS (Sect.5.5). In chemical applications, poor beam quality may result in an inhomogeneous distribution of excited species across the laser beam. This is due, for instance, to the occurrence of "hot spots" in the beam. Using conventional sources, excited species concentration is so low that such inhomogeneities do not affect the measured rate. With lasers, interactions between two or more excited species can be important, often making it mandatory to include the beam spatial profile in the kinetic analysis.

Cost and Reliability

Even though the number of known laser sources is very large, only relatively few have found widespread use in laser chemistry. For spectroscopic applications, rather low power levels are often sufficient (10^{-6} W average power). In order to obtain measurable chemical yields, one needs a much higher power. Inasmuch as the primary power source for most practical lasers is electricity, output power depends on electrical input and on conversion efficiency. At present, most lasers are very inefficient converters. The best electricity to light conversion efficiency – obtained with CO_2 lasers – is about 20%. For most other lasers it is much lass – between 0.01 and 0.1%. The operating costs scale with the input power. It has been estimated that in large sizes, capital cost is about $3 per Watt of input power. In the case of UV-visible lasers this translates to $3000 per Watt-year of output power assuming amortization in one year. 1 Watt is a mere 10^{-5} Einstein (moles) of photons per second; even if every photon were used to prepare a product, the cost would be too high for most chemicals. The need for more efficient and reliable lasers is reflected by intense activity, resulting in continuous discovery of new laser systems. An important breakthrough has been achieved with excimer lasers (see below) that operate in the UV with an efficiency of ~1% (cf. also Sect.5.6).

It is evident that a given application is best met by tailoring the laser to its exact requirements. Practically, however, one usually has to do with a limited number of laser sources in a given laboratory. In the following we give a brief discussion of three laser systems that presently appear to be most versatile and flexible, and have found widespread use in chemical laboratories. These are, in order of increasing wavelength, exciplex lasers, dye lasers, and CO_2 lasers.

3.4.2 Exciplex Lasers

An exciplex (*exci*ted com*plex*) is a molecule with a bound excited state and a repulsive (except for the weak van der Waals attraction) ground state. An example are the rare gas monohalides, such as KrF^*. If the complex is formed by two identical atoms or molecules, the term excimer (*exci*ted di*mer*) is commonly used. Examples are the rare gas excimers Xe_2^*, Kr_2^*, etc.

Figure 3.42 shows the emission spectrum of KrF. The strong feature at 248 nm is due to an allowed $^2\Sigma^+ - ^2\Sigma^+$ transition. Since the lower state is practically repulsive at room temperature, KrF^* and similar molecules are expected to be good laser materials. KrF^* is formed in a discharge containing mostly argon, a little krypton, and a trace of fluorine (Fig.3.42; for kinetic details, see below). Placing the reaction mixture in a laser cavity results in strong laser emission at 248 nm [3.39].

Following accepted convention, we include in the discussion species such as $HgCl^*$ and ClF^*, even though their ground state is bound. This is justified by spectroscopic and kinetic as well as technological considerations. The states involved in the lasing transition and the mechanism of the excited state formation are similar to those of the rare gas monohalides. The observed UV transitions covered by this broad definition of exciplex lasers are shown in Fig.3.43.

A unique feature of exciplex lasers is that they may be considered as chemical lasers, since the excited molecules are formed by a chemical reaction or recombination of ions. The term chemical laser, however, is usually reserved for lasers based on reactions involving neutral and unexcited species, in which the ultimate energy source is the exothermicity of the reaction. In most exciplex lasers, the prime source is electrical energy in the form of fast electrons which is converted first to electronic excitation of the reactants.

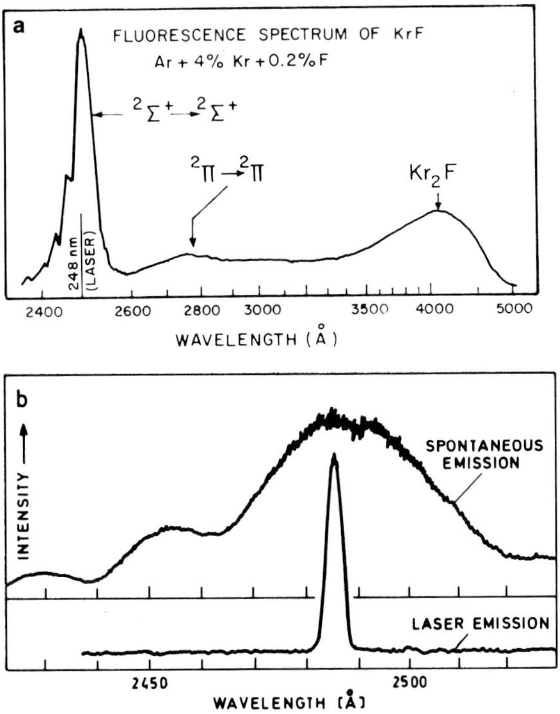

Fig. 3.42. (a) Spontaneous fluorescence spectrum of KrF^* with some contribution on the long wavelength end by Kr_2F^* as obtained in a typical laser gas mixture. (b) The spontaneous emission and laser spectrum of KrF^* at higher resolution than in (a) (adapted from [3.40])

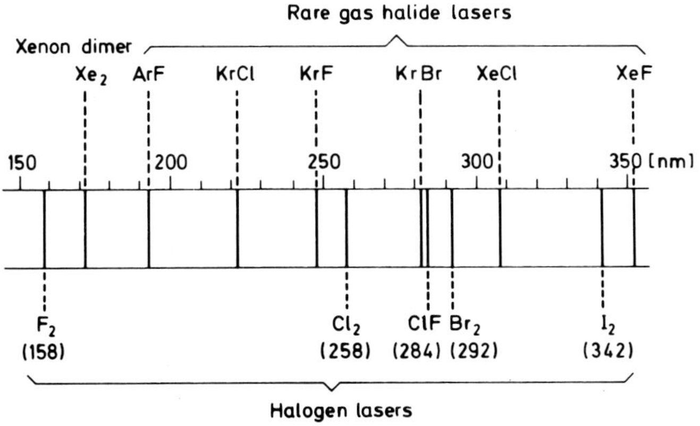

Fig. 3.43. Observed transitions of excimer lasers in the UV range. The lasers are operated in the pulsed mode, pulse width being about 10-40 ns. Energy per pulse is typically 1-100 J with e beam pumping, and 0.01 J with electric discharge pumping

Spectroscopy

The emission spectrum of exciplex molecules is understood in terms of the energy level diagram shown in Fig.3.44.

The ground state correlates with ground state rare gas atom R, and ground state neutral halide atom X. $^2\Sigma$ and $^2\Pi$ molecular states are formed, both being chemically unstable, except for a minimum in the potential energy curve due to weak van der Waals attraction. The well depth increases with atomic number and becomes about 0.15 eV for XeF. For this molecule vibrational structure can be observed in the $X^2\Sigma$ state. Mercury chloride (HgCl) is a stable molecule in the ground state with a well depth of about 1 eV. In this case the lasing transition is to highly excited vibrational states that are rapidly depopulated by collisions. In the heavier halides, spin-orbit splitting is large, and the ground ^2P state is split into $^2P_{3/2}$ and $^2P_{1/2}$ (cf. Sect.3.3.1). Consequently, the $^2\Pi$ state splits into $^2\Pi_{3/2}$ and $^2\Pi_{1/2}$ states, but the $^2\Sigma$ is still the lowest state.

Molecular states correlating with electronically excited atoms ($R+X^*$ or R^*+X) are similar to the ground state in terms of bonding — they are essentially repulsive, having only a shallow minimum due to van der Waals attraction. The bound state, commonly referred to as the B state, correlates with the ion pair $R^+(^2P) + X^-(^1S)$. This ion pair is actually correlated with two molecular terms, a $^2\Sigma(B)$ and a $^2\Pi(C)$, both being stabilized by long-range

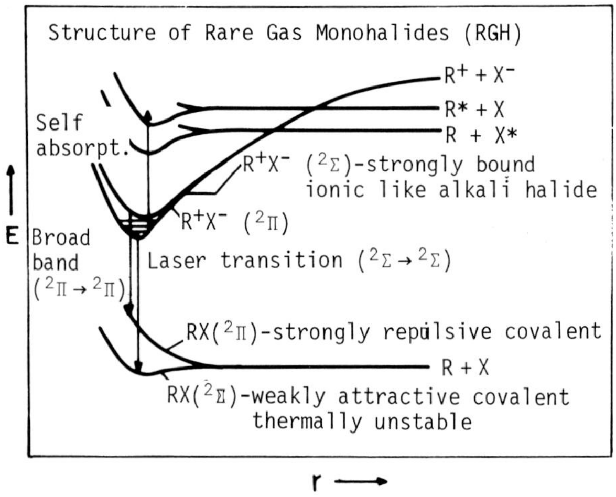

Fig. 3.44. Generalized energy level diagram of rare gas halide molecules. The laser transition is between the bound $R^+X^-(^2\Sigma)$ state and the $RX(^2\Sigma)$ weakly attractive state (adapted from [3.39])

Coulomb attraction. At infinite separation their energy above ground state atoms is given by the ionization potential of the rare gas atom minus the electron affinity of the halide atom. This energy is usually over 10 eV, higher than many electronically excited states of the atoms. Therefore the ionic curves cross excited covalent curves at some intermediate internuclear distance. This crossing provides a route for population of the B state from an electronically excited metastable rare gas atom R^*.

The emission spectra are due to the $^2\Sigma-{}^2\Sigma$ transitions $B \rightarrow X$. They are strongly allowed, with an oscillator strength of 0.1 and a radiative lifetime of a few nanoseconds. In the pressure range usually employed, mainly the vibrational ground state of B is populated. Franck-Condon favored transitions are usually to the repulsive part of the X state potential curve. Consequently the width of the spectrum is strongly dependent on the slope of this curve. As a concrete example, Fig.3.45 shows the energy level diagram of KrF, leading to the emission spectrum given in Fig.3.42. The stimulated emission cross section is very large, $\sim 10^{-16}$ cm^2. From (3.1.25) we find that the stimulated emission rate constant is huge, 3×10^{-6} cm^{3}s^{-1} — easily leading to saturation of the transition unless the lower state population is rapidly removed. The repulsive nature of the lower state is thus seen to be crucial for efficient operation of the laser.

An important spectroscopic consideration in laser systems is parasitic light absorption by intermediate states or by species formed in the active volume during the laser operation. In the present case this includes atomic absorption by X or R, by the B state itself, by atomic metastables, and by

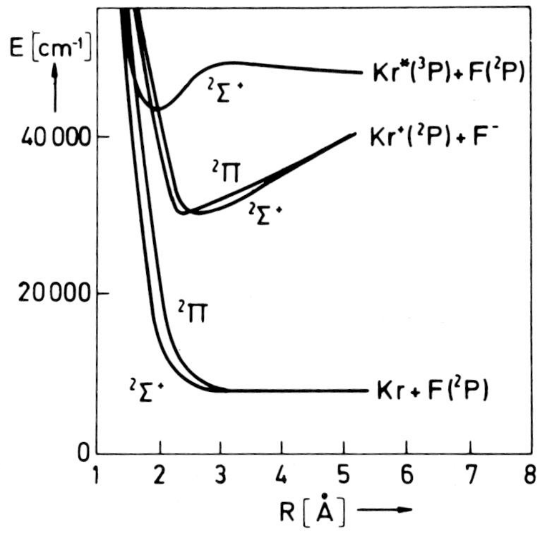

Fig. 3.45. Relevant potential curves for the KrF laser system. The situation pictured here is typical of other rare gas halide lasers, too (cf. Fig.3.44)

other molecules such as rare gas dihalides. These parasitic absorptions are in some cases a serious handicap to power upscaling. Another loss mechanism is dissociation of the excimer during its lifetime, caused by light absorption or by collision. The absence of laser emission from XeI^* is thought to be due to dissociative absorption.

Kinetics and Pumping

The lasing state can be populated by several reactions. Some examples are (using XeF for concreteness)

$$1)\quad XeF_2 \xrightarrow{h\nu} XeF^* + F \tag{3.4.1}$$

$$2)\quad Xe^+ + F^- + M \rightarrow XeF^* + M \tag{3.4.2}$$

$$3)\quad Xe^* + F_2 \rightarrow [Xe^+ + F_2^-] \rightarrow XeF^* + F \quad . \tag{3.4.3}$$

Reaction (3.4.1) is VUV photodissociation, (3.4.2) is a three-body process for which extremely large rate constants were measured ($k \sim 10^{-25}$ $cm^6 molec^{-2} s^{-1}$), and (3.4.3) is a harpoon reaction, proceeding via a charge transfer complex, for which rate constants are also large — 7.3×10^{-10} $cm^3 molec^{-1} s^{-1}$ in the case of $Xe^* + F_2$.

Lasing has been obtained so far only by using fast electrons as a primary source, either from an e beam source, or by electric discharge. The detailed mechanism depends on electron temperature and many other parameters such as mixture composition and pressure. As an example, a KrF laser typically operates with a mixture containing 95% Ar, 5% Kr and 0.2% F_2 or NF_3 (cf. Fig.3.42) In e beam pumped lasers the reaction sequence is believed to be

$$e^- + Ar \rightarrow Ar^+ + 2e^- \tag{3.4.4}$$

$$e^- + F_2 \rightarrow F^- + F \tag{3.4.5}$$

$$Ar^+ + F^- + M \rightarrow ArF^* + M \tag{3.4.6}$$

$$ArF^* + Kr \rightarrow KrF^* + Ar \tag{3.4.7}$$

$$KrF^* \rightarrow KrF + h\nu \text{ (laser)} \quad . \tag{3.4.8}$$

In a discharge pumped laser, the sequence is

$$e^- + Ar \rightarrow Ar^* + e^- \tag{3.4.9}$$

$$Ar^* + F_2 \rightarrow ArF^* + F \qquad (3.4.10)$$

$$ArF^* + Kr \rightarrow KrF^* + Ar \qquad (3.4.11)$$

or

$$e^- + Kr \rightarrow Kr^* + e^- \qquad (3.4.12)$$

$$Kr^* + F_2 \rightarrow KrF^* + F \quad . \qquad (3.4.13)$$

An important loss mechanism is formation of triatomics

$$KrF^* + Ar + M \rightarrow ArKrF^* + M \qquad (3.4.14)$$

$$KrF^* + Kr + M \rightarrow Kr_2F^* + M \quad . \qquad (3.4.15)$$

Rate constants of the order of 10^{-31} $cm^6 molec^{-2} s^{-1}$ were measured for these reactions [3.41]. In the presence of $3 \cdot 10^{19}$ molec cm^{-3} argon, three-body loss occurs in about 10 ns, efficiently competing with spontaneous emission. With a stimulated emission cross section of 10^{-16} cm^2, a photon flux of 10^{25} photons $cm^{-2}s^{-1}$, or about 6 MW/cm^2, is sufficient to overcome collisional deactivation. It is, however, evident that upscaling of these lasers by increasing the pressures is possible only if extremely large light fluences are employed.

Available Lasers

The most powerful and efficient lasers are those pumped by an e beam source. 100 J/pulse was achieved by dumping 2000 J into an Ar-Kr-F_2 mixture. Efficiencies up to 15% (based on energy deposited in the sample) or 1% (wall plug efficiency) were obtained. The necessary instrumentation is, unfortunately, quite expensive and is not likely to be available in most chemistry laboratories.

Discharge pumped lasers yield considerably smaller output energies, of the order of 1 J/pulse for the best mixtures. However, they are commercially available, run reliably at a reasonable repetition rate ($\sim$10 pulses per second), and are easy to maintain. Pulse width is typically 10 to 20 ns, yielding high peak powers. These devices have been used to induce multiphoton dissociation [3.43] and ionization (Sect.5.4.4). Tuning over a limited range by placing a dispersive element in the cavity is possible and has indeed been achieved in several systems, e.g., ArF, KrF, and XeF. Future developments

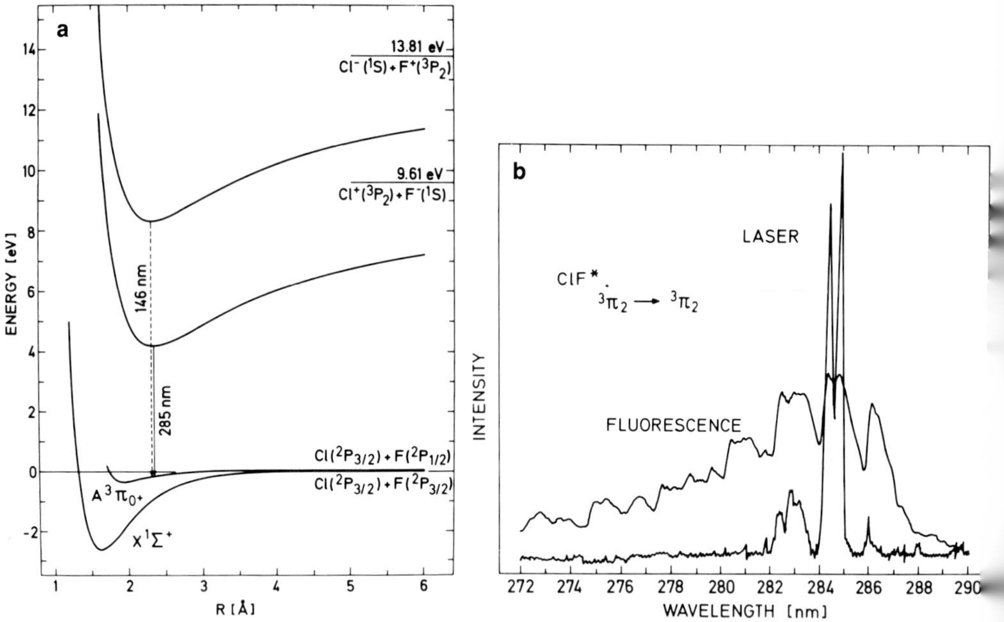

Fig. 3.46a,b. The ClF^* laser. (a) Simplified potential energy diagram. The molecule has a well-defined minimum in the ground electronic state, and cannot be strictly considered as an excimer. However, its energy level diagram is very similar to those shown in Figs.3.44 and 3.45. (b) Fluorescence and laser outputs (adapted from [3.42])

are expected to increase the number of laser systems, as well as to find more tunable systems. As Fig.3.43 shows, the UV range is already covered quite well, making new applications possible. Obvious ones are selective photochemistry, time-resolved VUV dissociation studies, and extension of dye laser range to 300 nm and below.

An example of a recently discovered laser transition is that of ClF^* [3.42]. An e beam pumped mixture of Ne, F_2, and Cl_2 yielded laser emission at 285 nm. The emission spectrum and relevant potential energy curves are shown in Fig.3.46. It is seen to fit well with the general scheme of Fig.3.44. This result may open up a new class of potentially high-power UV lasers.

3.4.3 Dye Lasers

Dyes are polyatomic molecules with an intense absorption band in the range 200 to 1000 nm, arising from an allowed electronic transition. Often the excited state is photostable, making fluorescence the dominant relaxation process. Radiative lifetimes are in the range of 10^{-8} to 10^{-9} s and stimulated emission cross sections can be extremely large (10^{-16} cm^2). When properly pumped, e.g., by a fixed frequency laser or a powerful flashlamp, many of these molecules are found to lase efficiently. Dye lasers are now among the most important laser sources available to chemists, both as primary sources for photoinitiation of reactions and as diagnostic tools. A comprehensive account on their properties and applications is available [3.35]. We shall thus give only a short account of operational principles and describe some of the more recent developments in the field.

Spectroscopy and Excited State Kinetics

The energy level diagram of a typical organic dye molecule is shown in Fig. 3.47 (cf. Fig.3.29). The color is due to singlet-singlet ($S_0 \rightarrow S_n$) transitions. They are rather broad (~ 1000 Å) and essentially continuous. These properties result from the very high density of vibrational states (cf. Chap.2). Excitation to S_n (n>1) is followed by rapid ($k_{IC} \sim 10^{11}$ s^{-1}) internal conversion to

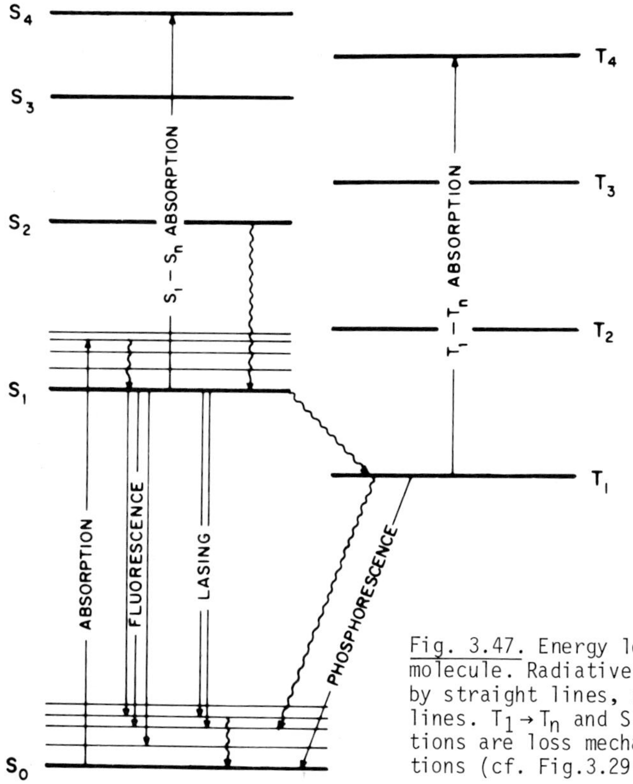

Fig. 3.47. Energy level diagram for a dye molecule. Radiative transitions are shown by straight lines, nonradiative by wavy lines. $T_1 \rightarrow T_n$ and $S_1 \rightarrow S_n$ radiative transitions are loss mechanisms for laser operations (cf. Fig.3.29)

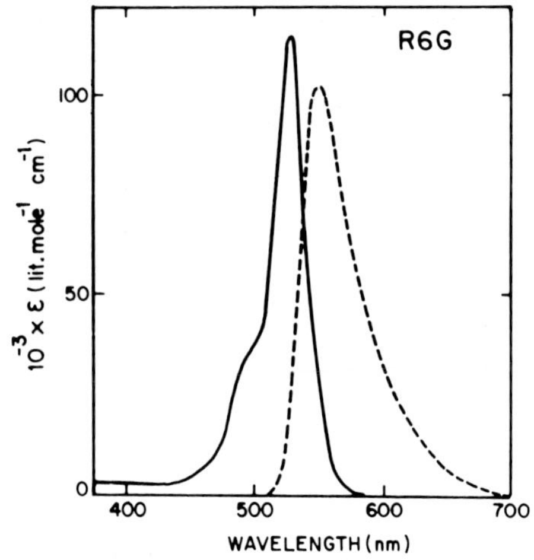

Fig. 3.48. Absorption (——) and emission (---) spectra of rhodamine 6G in ethanol. The transition is very strong in both absorption and emission, as indicated by the large molar absorption coefficients (adapted from [3.35])

S_1 (Sect.2.4.9). Vibrational relaxation in S_1 is also very fast under usual operating conditions – liquid solution or high-pressure gas mixtures. Therefore emission is observed as a rule only from a thermally relaxed population of S_1. The resulting emission is shifted to the red with respect to absorption, and often is a mirror image of the absorption spectrum. This can be understood in terms of displacement of the nuclear coordinate(s) coupled to the electronic transition. Figure 3.48 shows the absorption and emission spectra of rhodamine 6G, one of the most efficient laser dyes. Emission spectra are broad and terminate at various vibrational levels of S_0 (Fig.3.47). These are very rapidly (within 10^{-12} s) depopulated by vibrational relaxation. The situation is similar to that described for the excimer lasers. No appreciable population of the lower state is ever built up, ensuring population inversion even for very fast stimulated emission conditions. The resulting broad emission spectrum makes tuning feasible over a wide range ($\sim$500 Å). Tuning is easily achieved by placing dispersive elements, such as gratings, prisms, or etalons, in the cavity. Strong coupling between different modes ensures minimal optical loss: power is channeled to the highest gain transition, so that intense emission over a narrow bandwidth is obtained. This is an example of homogeneous broadening (Sect.3.2.3). Neglecting photochemical conversions, the lifetime of S_1 is given by $\tau = k^{-1}$, where

$$k = k_f + k_{IC} + k_{ISC} + \sum_n c\sigma_{S_1 \rightarrow S_n} \rho(\nu) \quad . \tag{3.4.16}$$

k_f, k_{IC} and k_{ISC} are rate constants for radiative coupling to S_0, nonradiative internal conversion to S_0, and nonradiative intersystem crossing to T_1, respectively (Sect.2.4.9). The last term gives the depopulation rate of S_1 due to absorption to higher lying singlet states. This term is vanishingly small for low light densities, but becomes important when laser operation is considered. Typical values for efficient laser dyes are: $k_f = 10^8$-10^9 s^{-1}, $k_{IC} < 10^6$ s^{-1}, $k_{ISC} < 10^7$ s^{-1}. As in the case of excimer lasers, parasitic absorption reduces lasing efficiency and may prevent it altogether. $T_1 \rightarrow T_n$ absorption becomes a problem for laser pulses longer than k_{ISC}^{-1}. This was one cause for the belated development of cw dye laser systems. Removal of triplet molecules by suitable quenchers was a key step in the realization of cw dye lasing.

Pumping Arrangements

Dye lasers are usually operated in liquid solutions. Lasing from polymethylmetacrylate (PMMA) solid solutions was reported [3.31], but the efficiency was very low due to heat gradients formed in the material. Vapor phase operation has also been demonstrated by optical [3.44] or electron beam [3.45] pumping. Optical pumping is by far the most important mechanism for achieving population inversion. Pulsed operation is obtained by either flashlamp or fixed frequency laser pumping. Continuous wave operation was thus far obtained only by pumping with a tightly focused gas laser, such as Ar^+. A summary of power levels obtained by various pumping schemes is given in Table 3.6.

Commercially available dye lasers are pumped by Ar^+ lasers, by coaxial or linear flashlamps, and by a variety of powerful pulsed lasers, the most popular being the N_2 laser at 337.1 nm and Nd:YAG based system operating on 530, 353, or 265 nm.

Table 3.6. Power characteristics of some dye lasers

Pumping method	Average power [W]	Pulse duration	Peak power [W]	Pulse energy [J]
Ar^+ laser (10 W)	1-2	cw	-	-
Cu laser	5	20 ns	$5 \cdot 10^4$	10^{-3}
Flashlamp	100	0.5-100 μs	10^6	10
N_2 laser	0.100	10 ns	10^5	10^{-3}
Nd^{3+} laser	1	10 ns	10^7	10^{-1}
Excimer laser	0.100	20 ns	10^6	10^{-2}
Mode-locked Ar^+	0.025	20 ps	10^3	10^{-8}

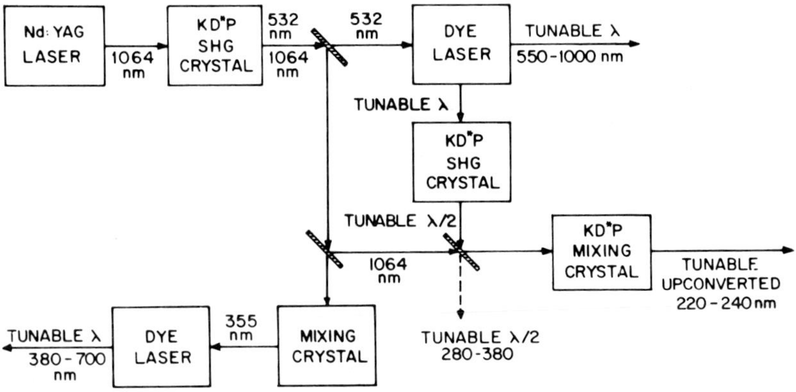

Fig. 3.49 Caption see opposite page

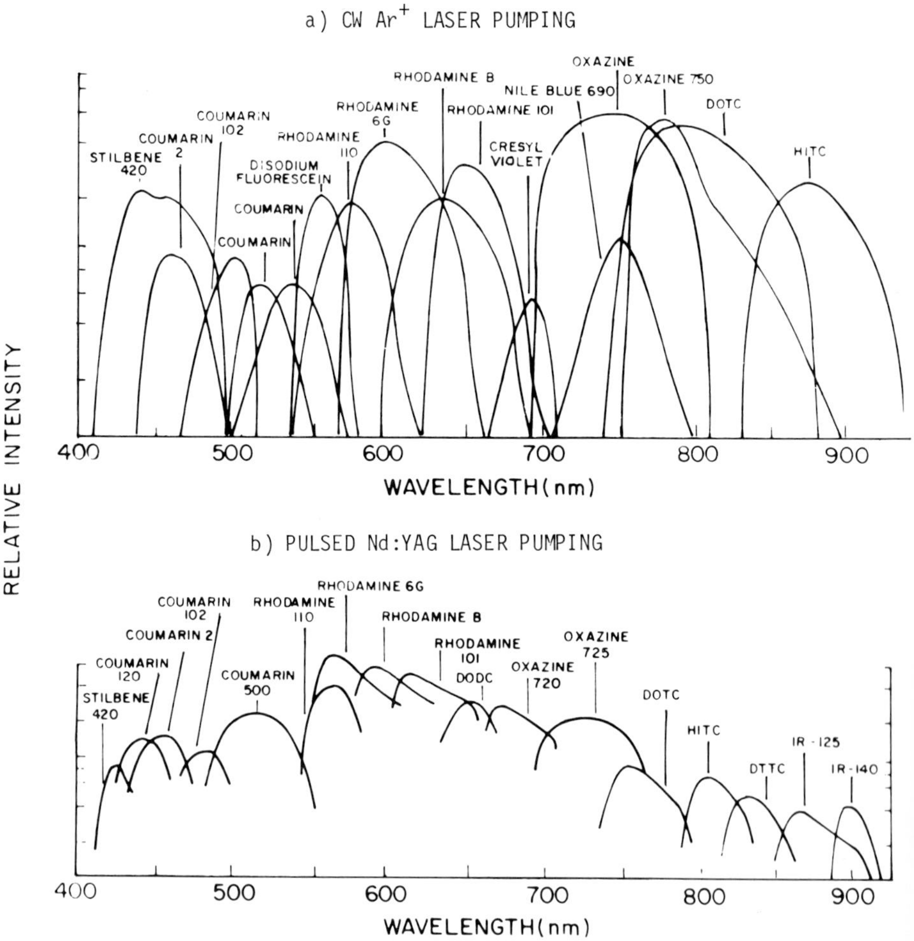

Fig. 3.50 Caption see opposite page

Tuning and Frequency Control

One of the most attractive features of dye lasers is their tuning capability. The transition is homogeneously broadened (cf. Sect.3.1.2), the situation being effectively an extreme case of pressure broadening. The insertion of a frequency selective loss element into the cavity results in considerable narrowing of the spectral width, without excessive loss of power. Thus, the fluence per unit frequency increases. Best results are obtained by using an oscillator amplifier (cf. Fig.3.41) arrangement. Various nonlinear techniques can be used to extend the frequency range of a single laser system to the range 200 to 3000 nm. Figure 3.49 shows a schematic diagram of such a laser system.

The tuning ranges of some commonly used dyes is shown in Fig.3.50.

Applications of Dye Lasers

These are discussed throughout the book. For convenience, some of the more important applications are listed here for quick reference.
Laser chemistry applications:

a) Selective excitation of a given vibrotational level. Pulsed lasers are used to obtain single level lifetimes and bimolecular rate constants. High average power lasers are used to prepare macroscopic amounts of materials (e.g., isotope separation; Sect.5.3.2).
b) Time-resolved photodissociation studies of repulsive transitions (Sect. 5.3.1).
c) Selective photoionization by multiphoton absorption (Sects.1.3.5,5.4.4).
d) Photochemistry of highly vibrationally excited states by overtone absorption (Sects.1.3.2,2.3.7).

Laser diagnostics (Sect.5.5.2):

a) Laser-induced fluorescence of products.
b) Photoacoustic and thermal lensing spectroscopy.
c) Intracavity absorption.
d) Nonlinear monitoring methods: two-photon absorption, CARS.

Fig. 3.49. Dye laser based system, with a large tuning range

Fig. 3.50a,b. Typical tuning curves of some commonly used dye lasers. The dyes are cited by their most frequently used trade names. The pump laser in this case is (a) a cw Ar^+ laser, (b) the second or third harmonic of a Nd:YAG laser

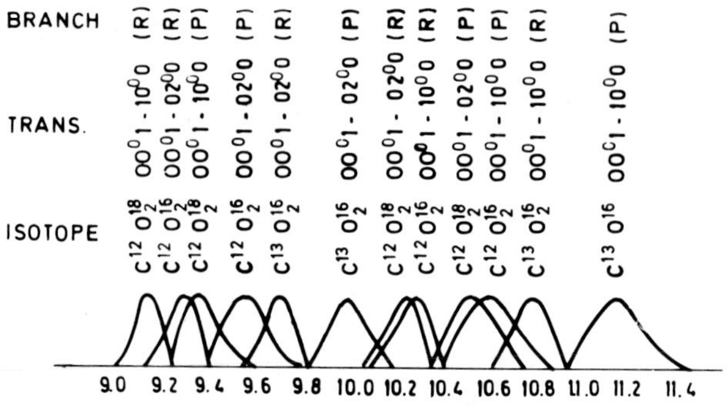

Fig. 3.51. Tuning range of TEA CO_2 laser working with various isotopic CO_2 laser mixtures

3.4.4 CO_2 Lasers

Among the infrared lasers, the pulsed or cw CO_2 lasers have served as the most frequently used tool in laser chemistry studies. This is due to a combination of several reasons. A central one is their ease of operation and high efficiency. Equally important, perhaps, is the fact that they may be tuned over a reasonably wide frequency range (920-1080 cm^{-1}) (see Fig.3.51). A very large number of molecules absorb in this range, and can be conveniently studied using this laser. Finally, in the pulsed mode the laser easily provides 1-10 J in 0.1-1 μs, leading to peak powers of 10^9-10^{10} W/cm^2. These results can be obtained with surprisingly small devices currently priced at $5,000-10,000. Thus, the laser is widely used in multiphoton dissociation studies, one of the most active fields in unimolecular kinetics in recent years (Sect.5.4). For a detailed discussion of CO_2 lasers the reader is referred to [3.46].

Spectroscopy and Kinetics

The vibrational energy level diagram of CO_2 was discussed in some detail in Sect.3.3.5. The laser is usually operated using a gas mixture containing helium, nitrogen, and carbon dioxide, helium being the major component. The need for added gases may be explained by reference to Fig.3.52 (see also Sect.4.2). The most commonly used pumping mechanism is electron impact. Electrons are created by an electrical discharge, and excite vibrational levels of nitrogen and carbon dioxide. The metastable v=1 level of N_2 is a very efficient sink of vibrational excitation due to its long lifetime (~1 s). On collisions with CO_2 molecules, the 00^01 level is preferentially populated, due to the nearly resonant (ΔE=-18 cm^{-1}) process

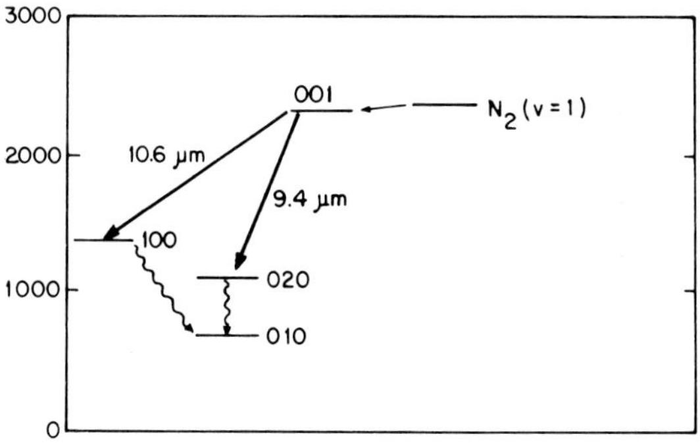

Fig. 3.52. Energy level diagram of the CO_2 laser. Pumping is by energy transfer from the long-lived v=1 level of N_2 to the 00^01 level. Laser transitions are shown by the heavy lines. The lower laser levels are depopulated by collisions (wavy lines) mostly with helium (cf. Fig.3.24)

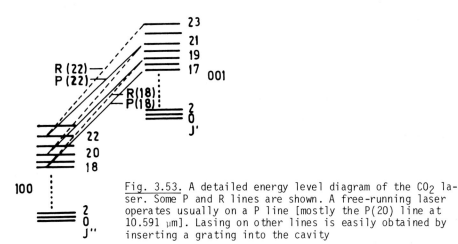

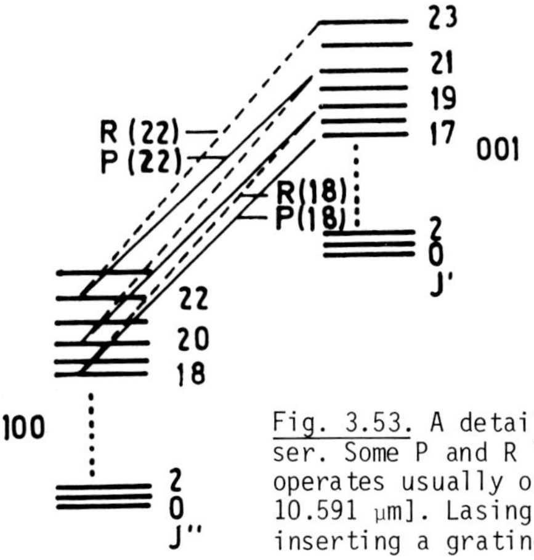

Fig. 3.53. A detailed energy level diagram of the CO_2 laser. Some P and R lines are shown. A free-running laser operates usually on a P line [mostly the P(20) line at 10.591 µm]. Lasing on other lines is easily obtained by inserting a grating into the cavity

$$N_2(v=1) + CO_2(00^00) \rightarrow N_2(v=0) + CO_2(00^01) \quad , \quad k = 1.9\times 10^4\ \mathrm{Torr}^{-1}\mathrm{s}^{-1} \quad . \tag{3.4.17}$$

Lasing occurs on several rotational lines in the R and P branches of the 00^01-10^00 and 00^01-02^00 bands. A detailed transition diagram is given in Fig.3.53.

The lower laser levels must be depopulated by collisions. The best collision partner is one that efficiently depopulates the 10^00 and 02^00 levels, but does not quench the 00^01 level. Helium gas turns out to be an excellent choice. The lower levels are better relaxed by some molecules, such as H_2 and H_2O, but these molecules are also effective in relaxing the upper laser level. This fact and the convenience of using a rare gas atom (no complications due to chemical reactions) made helium the major component in virtually

Table 3.7. Some relaxation rate constants $[(P\tau)^{-1}$ in units of $Torr^{-1}s^{-1}]$ relevant to CO_2 laser operation (adapted from [3.46])

Process: $M + CO_2(v_1,v_2^{\ell},v_3)$

Level	01^00	00^01	$10^00/02^00$
M = CO_2	194	350	2.2×10^3
H_2	6.5×10^4	3.8×10^3	3.3×10^4
H_2O	4.5×10^5	2.4×10^4	1.2×10^6
He	3.3×10^3	85	4.7×10^3
N_2	650	106	26

every CO_2 laser gas mixture. Some collisional rate constants relevant to CO_2 laser operation are listed in Table 3.7.

Operation and Properties of CO_2 Lasers

Two major laser types are commonly used, both pumped by electric discharge. cw lasers are pumped longitudinally and operated at low pressures (total pressure about 20 Torr). Power output is about 50 W per meter of active discharge. The laser is easily Q switched (cf. Sect.3.2.4) either passively or actively, to yield 10 μs pulses with peak power of about 10^4 W.

Transversely excited lasers are commonly operated at atmospheric pressure. These TEA (transversely excited atmospheric pressure) lasers typically produce 1-10 J pulses with a peak power of 1-10 MW. The pulse length depends on the discharge, but also on nitrogen gas content. The rate constant for 00^01 level population, (3.4.17), is $\sim2\times10^4$ $Torr^{-1}$ s^{-1}, yielding a typical transfer time of 1 μs at normal operating conditions. Thus even with a 100 ns discharge pulse width, laser action persists for about 1 μs (see Fig.3.54). This "tail" can be a nuisance for some experiments. It can be eliminated by reducing the partial pressure of nitrogen, albeit with considerable reduction of output energy.

As was discussed in detail in Sects.3.2.3,4 the spectral width of a laser is usually determined by the frequency dependence of the optical gain which in turn is broad in the case of effective collision broadening of the transition at high pressure. Consequently, the output of a cw laser contains much

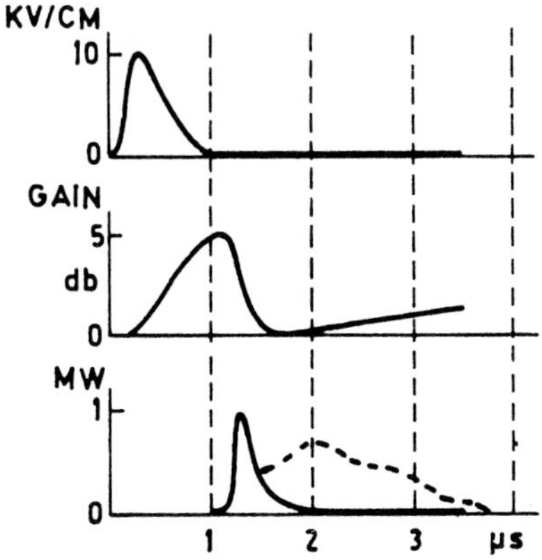

Fig. 3.54. Temporal evolution of gain (*middle curve*) and laser output (*lower curves*) compared to voltage across the discharge electrodes (*upper curve*). The laser pulse is seen to consist of a short spike only in the absence of nitrogen (——), to which a relatively long "tail" is added in the presence of nitrogen in the gas mixtures (---)

fewer modes – and can even be stabilized to only one oscillating mode – than the TEA laser output which typically has several GHz ($\Delta\nu \sim 0.1$ cm^{-1}) bandwidth. This also leads to spiking in the temporal pulse shape due to mode hopping. A narrower frequency spread is sometimes desired, for instance to excite a single molecular transition at a low pressure. This can be achieved, for example, by inserting cells containing low pressure gas mixtures into the resonator. These cw pumped gain cells may be regarded as narrowband frequency filters. Alternatively or additionally injection locking of the TEA laser by an external cw CO_2 laser can be used. In this way up to 5 J laser energy in a single longitudinal mode of roughly 15 MHz bandwidth has been demonstrated [3.47]. The range of CO_2 lines and the tunability within each line for TEA pressure conditions is shown in Fig.3.51. Obviously at very high pressures ($P>10$ atm) complete line overlap and continuous tuning are conceivable. This, however, requires a considerable technical effort and the use of electron beam laser designs.

Applications

CO_2 laser chemical applications are extensively discussed in Chap.5. A summary of the more important ones is provided for quick reference:

a) Multiphoton unimolecular dissociation.
b) Isomarization (by cw or pulsed lasers).
c) Bimolecular reactions of vibrationally excited molecules
d) Vibrational energy transfer studies.

Other relevant applications include: pumping of other infrared lasers; a notable example is the CF_4 laser operating at 16 μm, and monitoring trace quantities by LIF or by LIDAR techniques.

4. Chemical Lasers

Chemical lasers are defined as lasers in which population inversion results from a chemical reaction. This broad definition refers to a variety of pumping processes involving, for example, exoergic reactive collisions, photolytic or electron impact-induced bond rupture, and collisional energy transfer from one chemical species to another.

Chemical lasers have matured from a laboratory curiosity, as they were regarded some ten years ago, to an important tool for studying reaction dynamics and to high-power radiation sources with widespread technological applications. This chapter reviews the principal aspects of chemical laser research. Section 4.1 presents a survey of pumping mechanisms. The basic lasing conditions (of vibrotational chemical lasers) are discussed in Sect.4.2. Some of the widely used experimental designs are described in Sect.4.3. The kinetics of collisional and radiative processes are considered in Sect.4.4. Finally, in Sect.4.5, we list some scientific and technical applications of chemical lasers.

4.1 Survey of Chemical Lasers

The possibility that chemical reactions such as the H_2/Cl_2 explosion may yield sufficient vibrationally excited molecules to reach laser threshold was proposed already in 1961 [4.1]. Yet the first chemical laser system was experimentally demonstrated only in 1964 and involved an electronic transition [4.2]. This was the atomic iodine photodissociation laser where excited iodine atoms $I^*(5\,^2P_{1/2})$ are produced by flash photolysis of RI ($R{=}CH_3, CF_3$, etc.) and lasing occurs on the $I^*(5\,^2P_{1/2}) \rightarrow I(5\,^2P_{3/2})$ transition; $\lambda = 1.315\,\mu m$ (see Sect.4.3.1). In the following year the HCl laser based on the $H + Cl_2 \rightarrow HCl(v) + Cl$ reaction (the second step in the H_2/Cl_2 explosion chain) was demonstrated [4.3]. (Hereafter, in view of the extensive literature, we shall mostly refer to general references such as [4.4-8], where the original studies are cited.) The first descriptions of pulsed HF, DF chemical lasers [e.g., $F+H_2 \rightarrow HF(v)+H$] were published in 1966/1967. At about the same time other photodissociation lasers (NO,CN) as well as the first chemical CO laser were reported. From then on the number of chemical lasers rapidly increased. The study of elimination type reactions (radical combination, insertion, addition) was pioneered and photoelimination lasers were developed and studied in detail. The fact that collisional energy transfer (from the excited reaction products to a molecule which then is capable of lasing) can be used favorably was first demonstrated for DF/CO_2 and HCl/CO_2 in 1968/69. These and other schemes are summarized in Fig.4.1 and Table 4.1. All these chemical laser systems still involved only infrared radiation in the pulsed mode of operation.

Starting about in 1967 and accelerating since 1970 technological interest in chemical lasers began to grow. An important step in this direction was the discovery of cw chemical laser operation (the HF laser based on the $F+H_2 \rightarrow HF+H$ reaction) in 1969. The following years also saw progress in high-energy pulsed chemical lasers. Beginning in 1970 the high power potential of the photochemical iodine laser (Sect.4.3.1) was explored. High pulse energies for the HF laser were reported by various laboratories, the record so far being a multikilojoule HF laser demonstration (Sect.4.3.2).

In subsequent years the direction of chemical laser research changed somewhat as laser action in the visible received increased attention. Some of the concepts investigated involved radiative association [4.9], dissociation processes where selection rules demand excited product formation [4.10], or electronic chemiluminescence in metal atom oxidation (halogenation) reactions [4.11]. None of these approaches has been very successful up to now. Another group of lasers, however, became known and reduced the need for new electronic chemical lasers from the users' point of view. In 1975 laser emission from KrF^* formed in the reaction of an excited krypton atom with fluorine compounds was demonstrated [4.8]. (This new class of rare gas excimer and exciplex lasers has been discussed in Sect.3.4.1 with regard to its operation and physical chemistry.) Stimulated emission on pure rotational transitions extends the spectral range of chemical lasers to the far infrared and microwave regimes. This type of lasing which requires rotational nonequilibrium is now receiving increased attention (cf. Sect.2.4.4 and below). Another most notable recent addition to the family of chemical lasers is the "purely chemical" cw iodine laser pumped by resonant energy transfer from electronically excited molecular oxygen (see below).

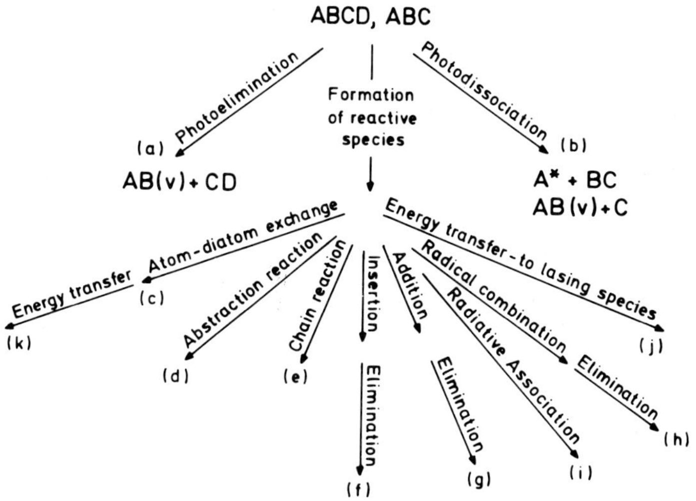

Fig. 4.1. Phenomenological view of reaction steps which compose a chemical laser system. Specific examples of the categories (a)-(k) are given in Table 4.1

As far as the experimental practice is concerned it should be noted that even in cases where the lasing species is a primary product of a chemical reaction no example has become known where this would be a spontaneous one-step reaction. Instead it is always necessary to prepare a (usually highly reactive) reagent in a prereaction. This preceding chemical reaction may involve combustion or spontaneous decomposition, photolysis, electron impact, or shock heating. The important consequences which this condition has for the experimental realization of chemical lasers will be discussed in Sect.4.3. The various schemes employed are shown in Fig.4.1, depicting the different steps from the preparation of the reagents to the pumping process. Table 4.1 serves to exemplify some typical cases. No claim for completeness is made and the interested reader is referred to [4.4-8] for more detailed lists of chemical laser systems.

The first two categories in Table 4.1 list formation of excited species following photolytic bond cleavage. The first group involves β-elimination of hydrogen halides. (Complications arise from the fact that the excited states from which the dissociation proceeds are usually not well known and the role of radiationless processes is not clear.) The number of parent molecules studied is very large and includes, for example, CH_2=CHF, cis and trans CHF=CHF, CH_2=CCl_2, CH_2=CDCl, CH_2=CHBr, CH_2=$CFCH_3$, CHCl=$C(CH_3)_2$, CH_3CCl_3, and $CH_2ClCHCl_2$.

Table 4.1. Types of chemical lasers (the indices a-k correlate with Fig.4.1)

Type of reaction	Typical example(s) [4.4-8]	
a) Photoelimination	C_2H_3Cl	$\rightarrow HCl(v) + C_2H_2$
b) Photodissociation	C_3F_7I	$\rightarrow I^* + C_3F_7$
	$NOCl$	$\rightarrow NO(v) + Cl$
	$(CN)_2$	$\rightarrow CN(v) + CN$
	CH_3NC	$\rightarrow CN^* + CH_3$
	IBr	$\rightarrow Br^* + I$
	$HFCO$	$\rightarrow HF(v) + CO$
	Cs_2	$\rightarrow Cs^* + Cs$
c) Atom-diatom exchange	$F + H_2$	$\rightarrow HF(v) + H$
	$Kr^* + F_2$	$\rightarrow KrF^* + F$
	$Cl + HI$	$\rightarrow HCl(v) + I$
	$O + CH$	$\rightarrow CO(v) + H$
	$CN + H_2$	$\rightarrow HCN(v) + H$
	$Al + F_2$	$\rightarrow AlF(v) + F$
d) Abstraction	$F + CH_4$	$\rightarrow HF(v) + CH_3$
	$Ar + NF_3$	$\rightarrow ArF^* + NF_2$
	$H + ClN_3$	$\rightarrow HCl(v) + N_3$
	$H + O_3$	$\rightarrow OH(v) + O_2$
	$Xe_2^* + N_2O$	$\rightarrow XeO^* + N_2 + Xe$
e) Chain reaction	$H + ClF_3$	$\rightarrow HF(v) + Cl + 2F$
	$\{F + H_2$	$\rightarrow HF(v) + H$
	$\{H + F_2$	$\rightarrow HF(v) + F$
f) Insertion-elimination	$O^* + CHF_3$	$\rightarrow HF(v) + F_2CO$
g) Addition-elimination	$NF + C_2H_4$	$\rightarrow HF(v) + CH_3CN$
h) Radical combination-elimination	$CF_3 + CH_3$	$\rightarrow HF(v) + C_2H_2F_2$
	$CH_3 + NF_2$	$\rightarrow 2HF(v) + HCN$
i) Radiative association	$O^* + Ar$	$\rightarrow (OAr)^* \rightarrow O + Ar + h\nu$
j) Energy transfer to lasing species	$N_2(v) + CO_2$	$\rightarrow CO_2(v) + N_2$
	$O_2^* + I$	$\rightarrow I^* + O_2$
k) Energy transfer following atom-diatom exchange	$DF(v) + CO_2$	$\rightarrow CO_2(v) + DF$

The second category – photodissociation lasers – concerns lasing action following photolytic decomposition of repulsive electronic states. As opposed to photoelimination either only one bond or two bonds located geminally on the same atom are broken in this pumping process. C_3F_7I is just one of the various alkyl iodides that are used in the iodine photochemical lasers (Sect. 4.3.1). CN lasers can be generated not only in $(CN)_2$ photodissociation but have been demonstrated also in the photolysis of CF_3CN, C_2F_5CN, HCN, and CrCN. Besides Cs_2, Rb_2 may also be used to produce alkali atom photodissociation lasers. The atomic bromine laser is the light-homologue analog to the iodine laser and can generally be pumped under similar conditions.

The next two groups of lasers in Table 4.1 involve atom-diatom exchange or abstraction reactions. The list of examples given is by no means complete and at least 25 different hydrogen compounds can be used only for the F + RH(D) reactions. Among them are H_2, HX (X=Cl,Br,I), C_nH_{2n+2}, CH_nX_{4-n}, B_2H_6, S_iH_4, and AsH_3. Also the group of HCl lasers is a large one and similarly the CO laser family. The latter can be formed in many reactions of ground state [$O(^3P)$+CH, CH_2, C_2H_2, CF, CN, CS, CSe] or excited state [$O(^1D)$+C_3O_2, CN] oxygen atom reactions. Triatomic chemical lasers have been produced, besides the case shown (HCN), only in $O_2 + CH \rightarrow CO_2 + H$, yielding a chemical CO_2 laser. A special comment concerns the group of rare gas halide and other excimer and exciplex systems. Almost any combination of rare gas-rare gas, rare gas-halogen, and halogen-halogen systems has been made to give off stimulated emission [4.8]. Similarly, the $Al + F_2$ reaction is a representative of a very large class of pumping reactions between a metal atom (e.g., Li, Mg, Ti, Cu, Pt, U) and an oxidizer (e.g., F_2, O_2, NF_3) [4.11].

As far as the chain reactions listed under e) in Table 4.1 are concerned the assignment to this group is often questionable. Generally speaking many of the reactions under c) and d) can be one step in a chain reaction. The cases shown under e), however, are of a special nature insofar as they either involve chain branching or reactions where all the steps in the chain produce population inversion. For the HF laser this is discussed in somewhat greater detail in Sect.4.3.2.

The next three groups [f),g),h)] refer to chemical pumping in the course of unstable (hot) product formation. Insertion, addition to a double bond or radical combination lead to elimination of the lasing species. Alkyl radicals, hydrogen, and oxygen (^1D or ^3P) atoms have been used to start the pump sequence. Most lasers of this kind work either with HF or with HCl.

The remainder of Table 4.1 contains chemical transfer lasers, the term referring to transfer from either a primary reactive species or from a hydrogen halide laser molecule. The iodine chemical transfer laser is one of the latest additions to this chemical laser family. Its pumping reaction is the following [4.12]

$$O_2(^1\Delta) + I(^2P_{3/2}) \rightarrow O_2(^3\Sigma) + I^*(^2P_{1/2}) \qquad (4.1.1)$$

$$K = \frac{[I^*(^2P_{1/2})][O_2(^3\Sigma)]}{[I(^2P_{3/2})][O_2^*(^1\Delta)]} = 2.9 \ (300\ K) \quad .$$

As the necessary excited oxygen can be chemically generated in a precurser reaction between molecular chlorine and alkaline hydrogen peroxide (with high yield), this laser presents an impressive example for the efficiency of chemical pumping.

Several reactions listed in Table 4.1 show not only vibrational or electronic but also pure rotational laser action. This is true for some of the very early HF and HCl lasers produced in electric discharges in mixtures of fluorine (chlorine) and hydrogen compounds [4.13] and has lateron been found particularly in HF elimination lasers [4.14]. Rotational population inversion is not necessarily connected with the primary pumping process but could be created also in V-R relaxation (cf. Sects.2.4.4, 2.4.6 and Fig.2.41), by J-dependent rotational relaxation [4.15], or by selective depletion of certain rotational levels due to concomitant vibrotational laser emission [4.16]. Here also one new addition to the rotational chemical laser group deserves special mention as it is not covered in the literature reviews quoted above [4.4-8]. The observation of OH (OD) rotational laser action was made when the far infrared water laser was operated under conditions of rare gas dilution [4.17]

$$O(^1D) + H_2 \rightarrow OH(v,J) + H \quad , \quad -\Delta H = 15400\ cm^{-1} \quad . \qquad (4.1.2)$$

The vibrational excitation of the OH molecules can be converted into high rotational excitation of lower vibrational levels by V-R transfer in OH-rare gas collisions, e.g., $OH(v,J) + Ar \rightarrow OH(v-1, J+\Delta J) + Ar$. (Recall that V-R transfer is especially efficient for hydrides, Sect.2.4.4.) The slow R-T relaxation of highly excited rotational levels (cf. Sects.2.5.2 and 4.4.3) helps maintain the population inversion in the high J regime. The requirements for efficient V-R transfer on the one hand and slow R-T relaxation on the other are best met for hydrides. Indeed, pure rotational lasing was observed for molecules like HF [4.13,14,16] (see Fig.2.41), OH [4.17], and NH [4.18].

From the large and diversified list of Table 4.1 only very few examples have been explored in any detail. These comprise the hydrogen halide lasers, the iodine photodissociation laser, the rare gas halide lasers, the chemical CO, and the CO_2 energy transfer lasers. It should be obvious after this short discussion that chemical laser research still contains much unexplored territory for reaction dynamics studies and for subsequent new laser developments. We shall present in the following sections some typical results and show some design principles and construction details. To prepare ourselves for this a summary of the lasing conditions in chemical lasers is appropriate.

4.2 Lasing Conditions in Chemical Lasers

One of the major problems in the description of chemical lasers is the complexity arising from the large number of active laser lines and the multitude of various rate processes which influence the populations of the levels connected by radiative transitions (Fig.4.2). The spectral, temporal, and power characteristics of chemical lasers are thus determined by an intricate interplay between the kinetic processes on the one hand and the optical and mechanical parameters of the laser resonator on the other.

The mutual influence of all these factors on the laser performance can be evaluated, at least in principle, by solving the rate equations governing the time evolution of the level populations and photon densities in the laser cavity. The information provided by this approach as well as the difficulties associated with its application to specific systems will be described in detail in Sect.4.4. In spite of the great complexity of chemical laser systems, considerable insight into the mechanisms governing their operation can be gained from rather simple considerations involving the hierarchy of time scales of different rate processes as described in Chaps.1 and 2 (expecially Sect.2.5) and the basic lasing conditions discussed in Chap.3 (Sect.3.2 in particular).

To pose the questions that a preliminary analysis can resolve let us consider again the output pattern of the $F + H_2 \rightarrow HF + H$ chemical laser [4.19] which has been briefly discussed in Chap.1 (cf. Fig.1.6). (We shall repeatedly refer to this classical example throughout this chapter.) The same results are shown in somewhat different form in Fig.4.3. The fluorine atoms in this system are generated by flash photolysis of CF_3I molecules and a buffer gas is added to ensure fast rotational relaxation. (The initial reagent mix-

Fig. 4.3 ►

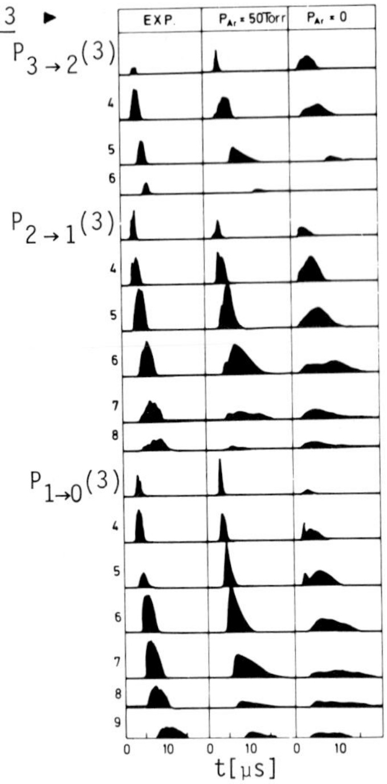

Fig. 4.2

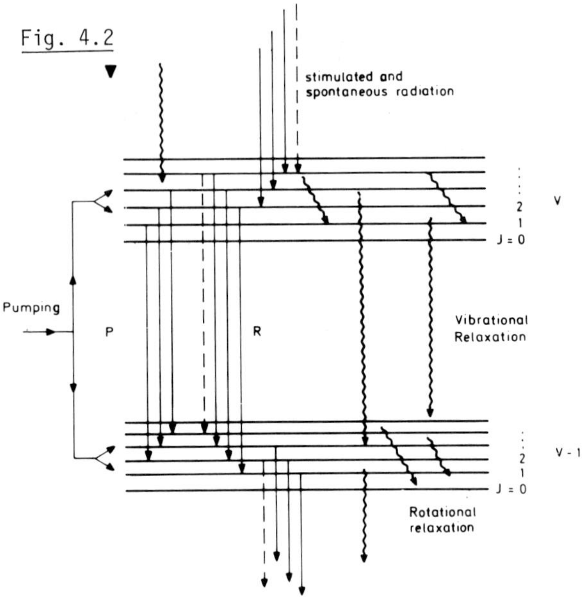

Fig. 4.2. A schematic description of the rate processes which modify the level populations in a vibrotational chemical laser. (Absorption and collisional excitation processes are not indicated for the sake of simplicity)

Fig. 4.3. Laser output patterns of the flash-initiated $F+H_2 \to HF(v,J)+H$ laser. *Left column:* experimental results (from [4.19]) for the initial mixture $CF_3I:H_2:Ar=1:1:50$ Torr. *Center column:* computed results for the same initial conditions; obtained by solving the laser rate equations (adapted from [4.20]). *Right column:* computed results for the same initial reagent mixture but without the inert gas, corresponding to slower rotational relaxation. (See also Fig.1.6)

ture is $CF_3I : H_2 : Ar = 1 : 1 : 50$ Torr and the total HF pressure at the end of the pumping period is of the order of 0.1 Torr.) The chemical energy released in the pumping reaction, $E \simeq 35$ kcal/mole, is sufficient for populating the four lowest levels, $v = 0,1,2,3$, of the HF molecules. The nascent vibrational populations are $N_0:N_1:N_2:N_3 = 0.06:0.34:1.00:0.60$ [4.19] (see also Fig.2.14) as reflected by the areas under the rotational profiles within each vibrational manifold in Fig.4.4. This figure also shows the vibrotational populations after rotational relaxation has been completed but before vibrational relaxation has started. Examination of Fig.4.3 reveals that lasing transitions take place not only in the $v=1 \to 0$ and $v=2 \to 1$ bands for which $N_v > N_{v-1}$ but also in the $v=3 \to 2$ band even though $N_3 < N_2$. This implies that "complete

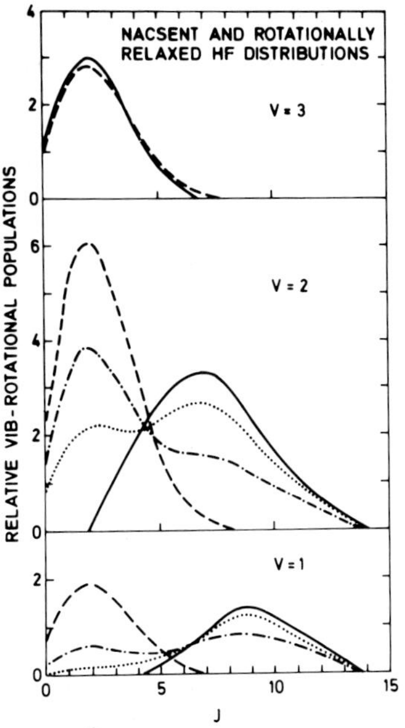

◄ Fig. 4.4

Fig. 4.5
▼

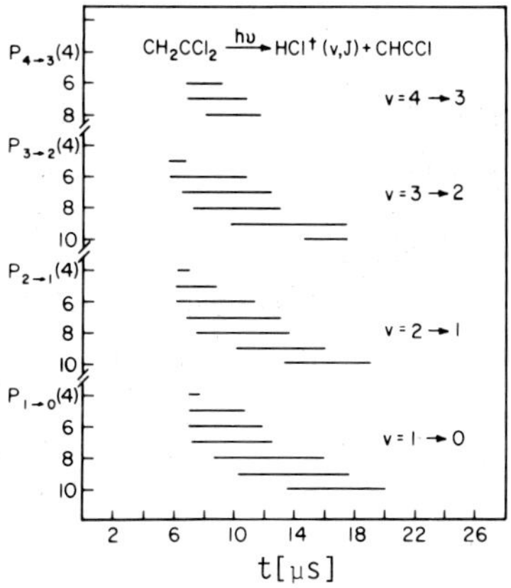

Fig. 4.4. The vibrotational state distribution of the nascent products in the $F+H\rightarrow HF(v,J)+H$ reaction (solid curve) and after complete rotational relaxation (dashed curve). The vibrational populations are proportional to the areas under the rotational distributions. The nascent population of v=0 is small and therefore not shown. Also shown are two partly relaxed rotational distributions (dotted and dash-dotted lines). Rotational relaxation is completed before the attainment of lasing threshold. (For the sources of experimental data of the nascent distributions see Fig.1.6). The partly relaxed distributions obtain by solving the rotational master equation (cf. Sect. 2.5.2) (adapted from [4.20])

Fig. 4.5. Schematic display of the durations of laser pulses for some HCl $P_{v\rightarrow v-1}(J) = v,J-1\rightarrow v-1,J$ lines of the 1,1 dichloroethylene photoelimination laser: $CH_2{=}CCl_2 \xrightarrow{h\nu} CH{\equiv}CCl + HCl(v,J)$. Note the J-shift in lasing in the various vibrational bands. (Adapted from [4.21])

inversion" $N_v > N_{v-1}$ is not a necessary lasing condition for chemical lasers and that "partial inversion" $N_v < N_{v-1}$ is sufficient for supporting the lasing in the $v\rightarrow v-1$ band on some J lines. In fact, as we shall see below, most of the laser energy is extracted from partially inverted populations. Another obvious trend displayed in Fig.4.3 is the gradual upward shifting in the J number of the lasing transition in a given vibrational band. This behavior is not unique to the HF system; it is one of the common characteristics of output spectra from lasers operating under conditions which ensure fast rotational relaxation. For example, Fig.4.5 shows that similar features are observed in the output spectrum of the photoelimination HCl laser:

$CH_2{=}CCl_2 + h\nu \rightarrow CH{=}CCl + HCl$. Among the other characteristics of rotationally equilibrated laser systems are the complete absence of R branch transitions and their high efficiency (compared to rotationally nonequilibrated lasers). Some of these features have been qualitatively explained on thermodynamic grounds in Sect.2.5.8. In this section we shall provide a more detailed, but still rather qualitative, description of these phenomena based on kinetic and gain properties. This will be followed by a quantitative (rate equation) description of chemical laser kinetics in Sect.4.4.

The spectroscopically allowed vibrotational transitions of heteronuclear diatomic molecules are of the P branch type or the R branch type (Sect.3.3.4). The former, $v,J \rightarrow v-1,J+1$, are commonly denoted as $P_{v\rightarrow v-1}(J\rightarrow J+1)$ or in short $P_{v\rightarrow v-1}(J+1)$ and the latter, $v,J \rightarrow v-1,J-1$, or $R_{v\rightarrow v-1}(J\rightarrow J-1)$ or $R_{v\rightarrow v-1}(J-1)$[1]. The lasing threshold condition for the $v,J \rightarrow v-1,J'=J\pm1$ line reads [cf. (3.2.7)]

$$\alpha_{v,J}^{v-1,J'} = c\sigma_{v,J}^{v-1,J'}\Delta N_{v,J}^{v-1,J'} \geqq 1/\tau_c \quad , \tag{4.2.1}$$

where α, σ, and ΔN denote the gain coefficient, the cross section for stimulated emission, and the population inversion of the transition, respectively. τ_c, the photon lifetime, is assumed to be the same for all transitions. The *sufficient* lasing condition (4.2.1) implies the *necessary*, population inversion, condition

$$\Delta N_{v,J}^{v-1,J'} = N_{v,J} - (g_J/g_{J'})N_{v-1,J'} > 0 \quad , \tag{4.2.2}$$

where $g_J = 2J+1$ and $g_{J'} = 2J'+1$ are the degeneracies of the upper and lower levels, respectively. Most of the characteristics of the laser spectrum described above can be explained qualitatively on the basis of the less informative relation (4.2.2). Let us therefore begin our analysis with this relation and then elaborate on the more detailed requirement (4.2.1).

In Sect.1.2 we have shown that the large excess of Ar in the CF_3I/H_2 system ensures rapid rotational equilibration and nearly no vibrational deactivation of the HF molecules at all prethreshold times. Thus when lasing just starts to take place we can write

$$N_{v,J} = N_v p(J|v;T) = N_v g_J \exp[-E_J(v)/kT]/Q_R(v) \quad , \tag{4.2.3}$$

---

1 Note that as for absorption, the rotational quantum number is one unit smaller in the upper vibrational state in P branch transitions and one unit larger in R branch transitions.

where $E_J(v)$ and $Q_R(v)$ are the rotational energy and the rotational partition function of molecules in the vth manifold, respectively. $N_v = \sum_J N_{v,J}$ is the total number of lasing molecules in level v. In the following we shall assume that (4.2.3) holds not only up to threshold but also throughout the lasing process. This assumption implies that rotational relaxation is instantaneous not only with respect to the other collisional processes but also with respect to stimulated emission which, during the lasing period, is a very fast process. Substitution of (4.2.3) into (4.2.2) yields

$$N_v/N_{v-1} > \exp\left\{[E_J(v)-E_{J'}(v-1)]/kT\right\}Q_R(v)/Q_R(v-1) \quad . \tag{4.2.4}$$

In the rigid rotor approximation where $E_J(v) = E_J = BJ(J+1)$ and $Q_R(v) = Q_R$ are independent of v, we get

$$\frac{N_v}{N_{v-1}} > \begin{cases} \exp[-2B(J+1)/kT] & \text{(P branch)} \quad (4.2.5a) \\ \exp[2BJ/kT] & \text{(R branch)} \quad , \quad (4.2.5b) \end{cases}$$

where both inequalities refer to transitions originating at the same upper level, i.e., $v,J \rightarrow v-1,J+1$ and $v,J \rightarrow v-1,J-1$, respectively. These results were derived and interpreted from a thermodynamic point of view in Sect.2.5.8.

The inversion characteristics of P and R branch transitions are illustrated in Fig.4.6. The figure shows the effective (degeneracy averaged) populations in two adjacent vibrational manifolds. The necessary lasing condition (4.2.2) for $v,J \rightarrow v-1,J'=J\pm1$ transitions requires $N_{v,J}/g_J > N_{v-1,J'}/g_{J'}$. From either Fig.4.6 or the explicit inequalities (4.2.4) or (4.2.5) we see that lasing in the R branch requires complete inversion, namely $N_v/N_{v-1} > 1$, but even then only the low lying J lines are inverted. On the other hand under complete inversion all P branch lines fulfill the population inversion condition. Moreover, even at partial inversion, i.e. when $N_v/N_{v-1} < 1$, there is always a range of sufficiently high J lines for which the line inversion $\Delta N_{v,J}^{v-1,J+1} > 0$. This can be more clearly seen by rewriting (4.2.5a) as

$$J+1 > (kT/2B)\ln(N_{v-1}/N_v) \qquad (v,J \rightarrow v-1,J+1) \quad . \tag{4.2.6}$$

Two additional general differences between P and R branch transitions are

I) The inversion on a P line $v,J \rightarrow v-1,J+1$ is always higher than that on the R line originating at the same level, $v,J \rightarrow v-1,J-1$.

II) Decreasing the temperature T favors the lasing in the P branch. The opposite is true for the R branch.

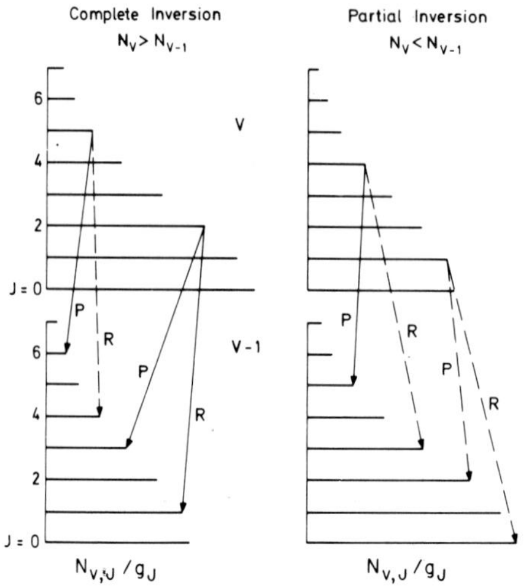

Fig. 4.6. Degeneracy-averaged populations $N_{v,J}/g_J = N_v \exp[-BJ(J+1)/kT]$. Solid arrows indicate inverted transitions, broken arrows noninverted transitions. ($B/kT=0.05$, corresponding to HCl at room temperature. $N_v/N_{v-1}=1.4$ and 0.7 for the cases of complete and partial inversion, respectively)

The cross section for stimulated emission at the line center frequencies $\nu = (E_{v,J}-E_{v-1,J'})/h$ of the emission line is given by

$$\sigma_{v,J}^{v-1,J'} = \left(\frac{g_0}{c\Delta\nu}\right)h\nu B_{v,J}^{v-1,J'} = \left(\frac{g_0 c^2}{8\pi\nu^2\Delta\nu}\right)A_{v,J}^{v-1,J'} = \frac{8\pi^3 g_0\nu}{3hc\Delta\nu}\, g_J|\mu_{v,J}^{v-1,J'}|^2 \quad , \tag{4.2.7}$$

where we have used (3.1.31) and (3.1.45); B, A, μ, and $\Delta\nu$ are the Einstein coefficient for stimulated emission, the Einstein coefficient for spontaneous emission, the dipole moment, and the linewidth of the transition, respectively. The transition dipole is commonly expressed in the form

$$g_J g_{J'}|\mu_{v,J}^{v-1,J'}|^2 = |M_{v,v-1}|^2 S_{J'} F_{v,J}^{v-1,J'} \quad . \tag{4.2.8}$$

Here $|M_{v,v-1}|^2$ and $S_{J'}$ are the vibrational and rotational components of the transition dipole. In the harmonic oscillator approximation $|M_{v,v-1}|^2$ is proportional to v; $|M_{v,v-1}|^2 = v|M_{1,0}|^2$. The rotational factor is $S_{J'} = J' = J+1$ and $S_{J'} = J'+1 = J$ for P and R branch transitions, respectively. The last factor accounts for the coupling between the vibrational and rotational motions. In the RRHO and other approximations which neglect this coupling $F_{v,J}^{v-1,J'} = 1$. Detailed calculations [4.22,23] show that to a first approximation the vibration-rotation interaction factor is independent of v and varies linearly with J. In the P branch $F_{v,J}^{v-1,J+1} \simeq 1 + 8(B_e/\omega_e)(J+1)$ and in the R branch $F_{v,J}^{v-1,J-1} \simeq 1 - 8(B_e/\omega_e)J$.

In the case of pure Doppler broadening $\Delta\nu = \Delta\nu_D$ is proportional to ν; $\Delta\nu_D/\nu = [8kT\ln(2)/mc^2]^{\frac{1}{2}}$, and the v,J dependency of $\sigma_{v,J}^{v-1,J'}$ results only from

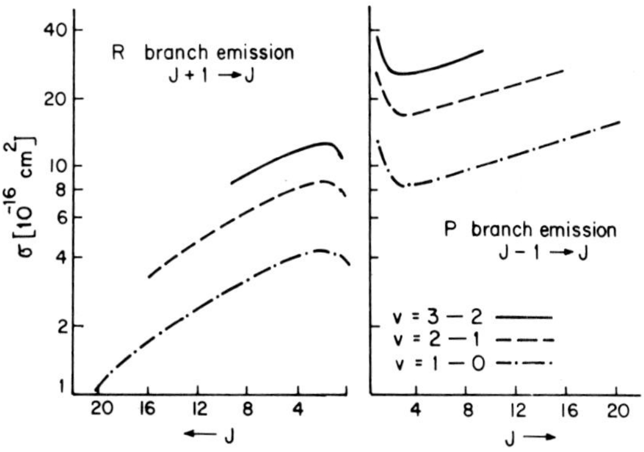

Fig. 4.7. Cross sections for stimulated emission of HF molecules at 300 K, with $\Delta\nu = \Delta\nu_D$ (adapted from [4.24])

the transition dipole. In the case of collisional or combined collision-Doppler broadening $\Delta\nu/\nu$ introduces an additional v,J dependency. As a specific example we show in Fig.4.7 the Doppler broadened cross sections for HF at 300 K. The Doppler effect is the dominant broadening mechanism for this molecule up to pressures of the order of ~10 Torr.

Combining (4.2.1,2,7),

$$\alpha_{v,J}^{v-1,J'} = \left(\frac{8\pi^3 g_0 \nu}{3hc\Delta\nu}\right)|M_{v,v-1}|^2 S_{J'} F_{v,J}^{v-1,J'} \left(\frac{N_{v,J}}{g_J} - \frac{N_{v-1,J'}}{g_{J'}}\right) \quad . \tag{4.2.9}$$

This general expression applies to any vibrotational distribution $N_{v,J}$. In the special, but common case of rotational equilibrium $N_{v,J}$ is given by (4.2.3). As a specific example demonstrating the dependence of the gain coefficient on the vibrational populations, N_v, the temperature T and the J number consider the Doppler broadened ($\Delta\nu/\nu \propto T^{\frac{1}{2}}$) P branch transitions in the v=2→1 band. Using the vibrating rotor level scheme $E_J(v) = B_v J(J+1)$ and the high-temperature ($B_v/kT \ll 1$) approximation for the partition function, $Q_R(v) = kT/B_v$, we obtain

$$\alpha_{v,J}^{v-1,J'} = CS_{J'} F_{v,J}^{v-1,J'} \left(\frac{N_{v-1}}{T^{3/2}}\right)\left\{\left(\frac{N_v}{N_{v-1}}\right) B_v \exp[-B_v J(J+1)/kT] - B_{v-1} \exp[-B_{v-1} J'(J'+1)/kT]\right\} \tag{4.2.10}$$

which is often referred to as the Patel gain equation [4.25]. The constant C includes universal constants and the vibrational matrix element. The gain equation (4.2.10) has served as the basis for determining vibrational energy

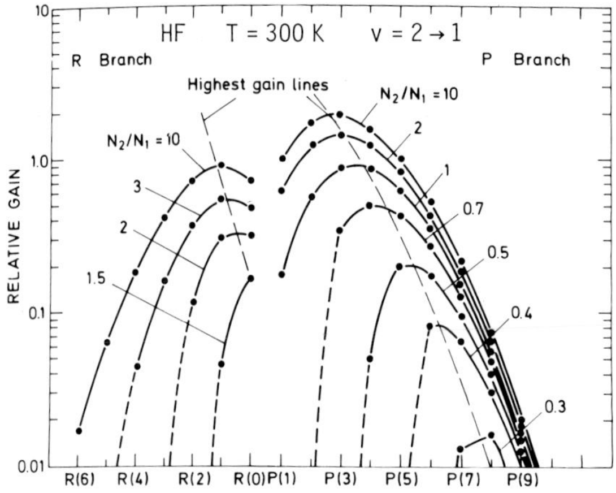

Fig. 4.8. Gain coefficients for the P and R branch lines of the $v=2\to1$ band of HF for different values of N_2/N_1 (relative scale). P(J') and R(J') correspond to $v=2,J\to1,J'=J\pm1$

partitioning among reaction products in many chemical laser experiments (see Sect.4.5).

The gain coefficients for the R and P branch transitions in the $v=2\to1$ band of HF (for the case of Doppler broadening) are shown in Fig.4.8. We note that the highest gain transition in the P branch is shifted towards higher J values as the vibrational population ratio N_v/N_{v-1} decreases, as could be anticipated from (4.2.6). This is not surprising since the J dependence of the gain coefficient is dominated by the population inversion and not by the cross section (cf. Figs.4.6-8).

We are now ready for a more detailed analysis of the output patterns shown in Fig.4.3 (or Fig.1.6). The lasing process is triggered when the total concentration of HF molecules generated by the pumping reaction is large enough so that the population inversion on one of the transitions exceeds its threshold value (4.2.1). The experimental observation (Fig.4.3) is that the first transition reaching threshold is $P_{2\to1}(3)$. This occurs at $t\simeq3$ μs when the rotational distribution is already thermal and the vibrational distribution is still equal to the nascent distribution (Fig.4.4). Thus, $N_{v,J}$ is given by (4.2.3) and one can use (4.2.10) to calculate the gain coefficients. Using the known values of the nascent vibrational populations in the $F+H_2\to HF(v)+H$ reaction (Fig.4.4) and (4.2.10), we find that $P_{2\to1}(3)$ has indeed the highest gain, hence it is the first lasing transition. (Figure 4.8 confirms that for the nascent vibrational populations ratio $N_2/N_1\simeq3$, $P_{2\to1}(3)$ is the highest gain transition in the $v=2\to1$ band.)

The start of oscillations on the first highest gain transition, $P_{2\to1}(3)$, is accompanied by several phenomena which are typical for lasing at rotational equilibrium: The intense stimulated emission tends to "burn a hole" in the Boltzmann profile of the upper, v=2, vibrational manifold at J = 2 and simultaneously to "build a hump" at v=1, J = 3. However, if, as assumed, rotational relaxation is instantaneous no hole burning effects can occur since the fast transfer of molecules between neighboring rotational levels will immediately level out any displacement from equilibrium which the radiation tends to create. Instead, the population of the upper vibrational manifold will be "funnelled" into the lower one through the highest gain transition without spoiling the Boltzmann shape of the rotational distributions. Now, since the stimulated emission clamps the population inversion of the active transition to its threshold value (Sect.3.2) it suppresses the possibility of laser oscillations on neighboring lines in the same band, the gain of which is by definition lower than that of the lasing line.

This picture of funnelling through the single lasing line is analogous to that of single-mode operation in the case of homogeneously broadened spectral transitions. The highest gain transition is equivalent to the highest gain mode, the Boltzmann profile is equivalent to the homogeneous (normally Lorentzian) line shape, and the rotational relaxation is the analogue of the collisional broadening mechanism (see Sects.3.1.2 and 3.2.3).

The transfer of population from v to v-1 reduces the ratio N_v/N_{v-1}. Consequently the highest gain transition is shifted to the next higher J value as implied by Fig.4.8 and confirmed by Fig.4.3 (see also Fig.1.6). This process of gradual J shifting continues until at a certain moment the absolute value of the population inversion on the highest gain transition, though positive, is too low to fulfill the lasing condition (4.2.1) and the lasing process terminates. For the $F + H_2$ system in Fig.4.3 lasing in $v=2\to1$ ceases after $P_{2\to1}(9)$ (not shown in Fig.4.3) has finished lasing. Hence, from Fig. 4.8 the final vibrational population ratio is $N_2/N_1 \simeq 0.3$, implying that most of the laser radiation has been extracted from partially inverted populations. A quantitative account of this fact will be given in Sect.4.4.3.

Our preliminary analysis of the lasing mechanism in chemical lasers utilizing vibrotational transitions has disregarded several important kinetic factors; the effect of radiative cascades from one band to another, the modification of level populations by vibrational deactivation, the output power dependence on the time profile of the initiation-pumping mechanism, and the effects of rotational nonequilibrium. All these points will be treated in detail in Sect.4.4. The discussion will, however, be continued first in the next section by describing some experimental elements of chemical lasers. It is felt that this is appropriate at this point in order to make clear what can be measured in chemical laser experiments and how these experiments look like in the laboratory.

4.3 Operation

This section is intended to show how experimental measures can be used to control the output of chemical lasers. Some experimental design criteria will be outlined for three lasers, namely the iodine photodissociation laser, the hydrogen fluoride laser (pulsed and cw), and the CO laser. They also serve to exemplify the two main laser initiation techniques, flash photolysis and electrical discharge. The iodine laser will be our first example as no difficulties arising from a multilevel laser scheme involving strong level coupling (both by radiation and collisions) are encountered here. Besides, this laser has been developed to very high energies and powers and its rather advanced technology is well suited to demonstrate the characteristic features of pulsed photolysis-initiated chemical lasers.

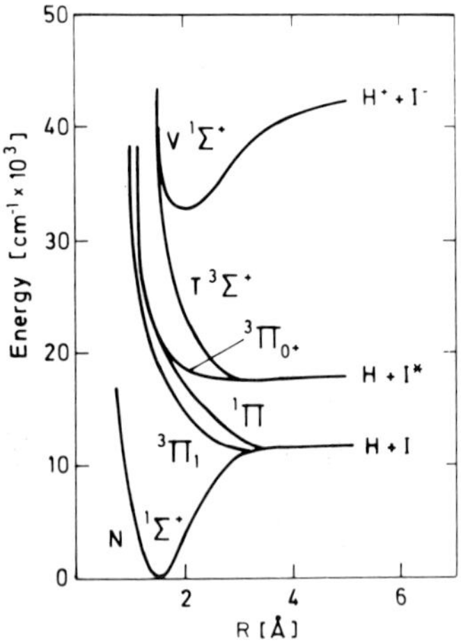

◄ Fig. 4.9

Fig. 4.10 ▼

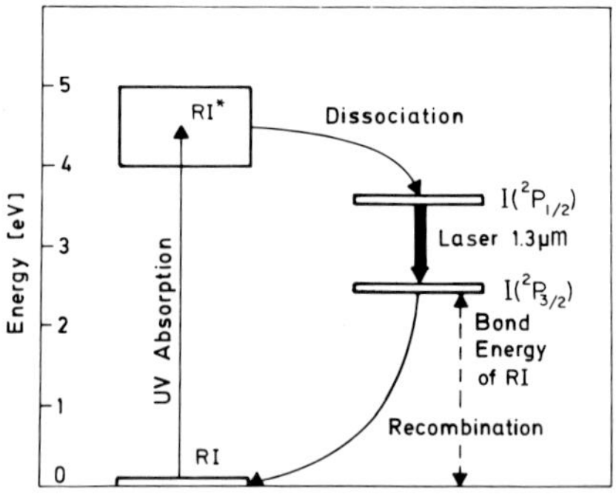

Fig. 4.9. Molecular energy levels of HI, assumed to be representative also for RI; $I^* = I(5\ ^2P_{1/2})$ (adapted from [4.26]; see also Fig.1.23)

Fig. 4.10. Four-level (ground and excited states of RI, excited and ground states of I) scheme of iodine laser operation (approximate energy scale)

4.3.1 Flash Photolysis: The Iodine Laser as a Model Case

The basic excitation scheme of the iodine photochemical laser utilizes an organic coumpound RI whose energy level diagram is similar to that shown in Fig.4.9. Perfluoroisopropyl iodide, i-C_3F_7I, is so far the best substrate, but other organic iodides have also been used.

Absorption of UV radiation from a flashlamp in the range of 35000 to 45000 cm^{-1} excites the molecule to a dissociative upper state. The iodine molecule is formed almost exclusively in the upper state of the ground state multiplet $5^2P_{1/2}$, (I^*),

$$\text{i-}C_3F_7I \xrightarrow{h\nu} \text{i-}C_3F_7 + I^* \quad . \tag{4.3.1}$$

In this process a UV photon of an average energy of ~4.5 eV is converted into an infrared photon of ~1 eV energy. In the case of i-C_3F_7I the dissociation fragments C_3F_7 and iodine, the latter after releasing its excess energy in the form of laser radiation

$$I(5^2P_{1/2}) \rightarrow I(5^2P_{3/2}) + h\nu \quad (\lambda=1.351\ \mu m) \quad , \tag{4.3.2}$$

recombine with high efficiency (see below) to regenerate the parent compound. The laser is thus characterized by a cyclic four-level scheme according to Fig.4.10.

Besides the reaction pathways shown in Fig.4.10 there is a variety of secondary processes which influence the iodine atom and C_3F_7 radical concentrations, during and after the flashlamp pulse. These involve collisional deactivation of I^*, recombination of R + R and R + I, radical molecule reactions, and three-body recombination of iodine atoms. The corresponding rates for the i-C_3F_7I dissociation system are not very well known but for the similar case of CF_3I photolysis sufficiently precise rate data are available to make a reliable kinetic modelling study possible. Table 4.2 shows some of the rate constants and Fig.4.11 the calculated time developments of species concentrations. It can be concluded here that on the time scale of the laser operation, which for reasons given below is ~10 μs, no collisional concentration changes for I^* are to be expected. In the case of CF_3I the radical-radical recombination to form C_2F_6 is favored over the radical-atom recombination. Experiment shows that this is different for i-C_3F_7I where the parent reformation accounts for >95% of the C_3F_7 removal, thus yielding a nearly closed cycle as shown in Fig.4.10.

Table 4.2. Reactions and rate constants used in iodine laser kinetic modelling (adapted from [4.27])

Reactants	Products	$R = CF_3$ $k[(cm^3/molec)^n s^{-1}]$[a]	$R = C_3F_7$	Type
$I^* + RI$	$I + RI$	5.4E-17	8.0E-16	Two-body deactivation
$I^* + R$	$I + R$	3.7E-18		
$I^* + R_2$	$I + R_2$	4.7E-16		
$I^* + O_2$	$I + O_2$	8.6E-12	b	
$I^* + I_2$	$I + I_2$	(1.3E-14) $e^{1650/T}$ 3.2E-12 300 K		
$R + R$	R_2	1.5E-11	1.0E-11	Two-body recombination
$I + R$	RI	5.0E-11		
$I^* + R$	RI	3.0E-12		
$R + RI$	$R_2 + I$	3.0E-16		Radical-molecule reaction
$I^* + RI$	$I_2 + R$	2.5E-19		
$R + I_2$	$RI + I$	4.0E-12		
$I + RI$	$I_2 + R$	1.6E-23		
$2I^* + M$	$I_2 + M$	4.3E-38		Three-body recombination[c]
$2I^* + RI$	$I_2 + RI$	1.0E-37		
$2I^* + I_2$	$2I_2$	1.3E-33		
$2I^* + He$	$I_2 + He$	2.8E-38		
$2I + M$	$I_2 + M$	4.3E-34		Three-body recombination
$2I + RI$	$I_2 + RI$	(7.0E-34) $e^{1600/T}$ 1.5E-31 300 K	(2.1E-33) $e^{1600/T}$ 4.5E-31 300 K	
$2I + He$	$I_2 + He$	(8.27E-29)$T^{-1.7}$ 4.6E-33 300 K		
$2I + I_2$	$2I_2$	(1.1E-15)$T^{-5.9}$ 2.9E-30 300 K		
$2I + O_2$	$I_2 + O_2$	3.7E-32		
$2I + R_2$	$I_2 + R_2$	5.8E-32		
$I^* + I + M$	$I_2 + M$	4.3E-34		Three-body recombination
$I^* + I + RI$	$I_2 + RI$	4.3E-34	1.6E-33	
$I^* + I + He$	$I_2 + He$	9.4E-34	2.2E-33	
$I^* + I + I_2$	$2I_2$	4.3E-32		

[a] Unless otherwise indicated, k values are for 300 K; [b] Blanks indicate use of the rate listed in the $R = CF_3$ [illegible] M stands for all species not explicitly indicated.

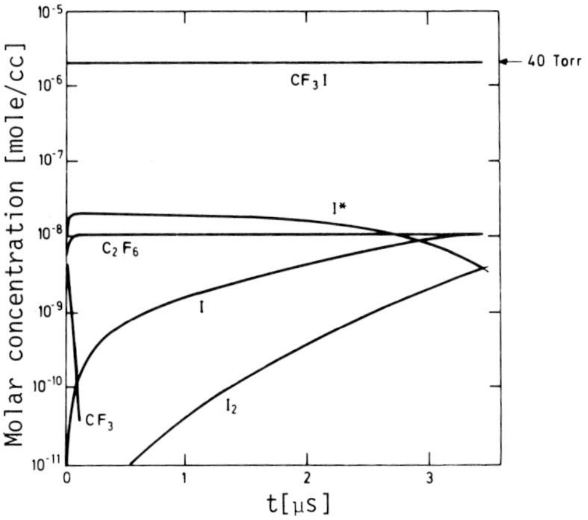

Fig. 4.11. Calculated species concentrations as function of time in an iodine laser (adapted from [4.27])

The photolysis produces iodine atoms in the lowest lying excited state ($5^2P_{1/2}$). The transition to the ground state ($5^2P_{3/2}$) is a magnetic dipole transition. The radiative lifetime is still a matter of some discussion; we shall adopt here the value of ~130 ms. Knowledge of the total radiative lifetime is insufficient to calculate the cross section of stimulated emission and design the laser accordingly since hyperfine splitting of the transition has to be taken into account. The degeneracy of the upper and lower laser levels is removed by the magnetic dipole and the electric quadrupole moments of the ^{127}I nucleus, which has nuclear spin 5/2. The angular momentum of the upper level can be parallel or antiparallel to the nuclear spin. The total angular momentum then obtains as $F = 5/2 + 1/2 = 3$ and $F = 5/2 - 1/2 = 2$. In the lower level there are four hyperfine levels, $F = 5/2 + 3/2,\ 5/2 + 1/2,\ 5/2 - 1/2,\ 5/2 - 3/2 = 4,\ 3,\ 2,\ 1$ (cf. Fig.3.15). Each hyperfine level F is (2F+1)-fold degenerate. Hence the total degeneracy of the upper level ($5^2P_{1/2}$) is $(2\cdot3+1) + (2\cdot2+1) = 12$ and statistically 7/12 of the excited iodine atoms generated in the photolysis appear with $F = 3$ and 5/12 with $F = 2$. Similarly the total degeneracy of the lower level, $5\ ^2P_{3/2}$, is $(2\cdot4+1) + (2\cdot3+1) + (2\cdot2+1) + (2\cdot1+1) = 24$. The selection rule $\Delta F = F_u - F_\ell = 0, \pm1$ implies six transitions (F_u and F_ℓ denote the F values of the upper and lower levels of the transition, respectively). These are $F_u \to F_\ell = 3\to4,\ 3\to3,\ 3\to2$, and $2\to3,\ 2\to2,\ 2\to1$. Due to the larger splitting between the hyperfine levels $F_u = 2$ and 3 of the upper state the six transitions form two groups with three neighboring transitions originating in the $F_u = 3$ level and the other three in the $F_u = 2$ level. The total frequency spread of the six transitions is 0.7 cm^{-1} ($\simeq 2\cdot10^{10}\ s^{-1}$=20 GHz) (see Fig.3.15 and 4.12).

Table 4.3. Einstein coefficients and maximal stimulated emission cross sections of the iodine laser lines (after [4.29])

Transition						
$F_u \to F_\ell$	$3 \to 4$	$3 \to 3$	$3 \to 2$	$2 \to 3$	$2 \to 2$	$2 \to 1$
$\sigma_{u \to \ell}(\nu_0),(10^{-18}\mathrm{cm}^2)$	6.0	2.4	0.66	2.67	3.3	2.55
$A_{u \to \ell}(s^{-1})$	5.0	2.1	0.6	2.4	3.0	2.3

The cross sections for stimulated emission of the six iodine laser lines were measured in several laboratories [4.28] for different conditions. For instance, Table 4.3 lists the maximal, i.e., the line center (cf. Sect.3.1.2) cross sections, $\sigma_{u \to \ell}(\nu_0)$, measured in a system of 20 Torr C_3F_7I [4.29]. The table also gives the Einstein A coefficients of the transitions assuming that all of them have the same linewidth $\Delta\nu$. Since ν_0 is practically the same for all lines ($\nu_0 \simeq 7600$ cm^{-1} is much larger than the frequency range 0.7 cm^{-1} of the six lines), this assumption implies $A_{u \to \ell} \propto \sigma_{u \to \ell}$ [cf. (3.1.45)]. Let $N^* = [I^*]$ denote the overall iodine concentration in the upper electronic manifold, $5\,{}^2P_{1/2}$. Assuming that the upper hyperfine levels are populated according to their statistical weight, $[g_3/(g_2+g_3)]N^* = (7/12)N^*$ and $[g_2/(g_2+g_3)]N^* = (5/12)N^*$ atoms populate the sublevels F_u=3 and F_u=2, respectively; g_3=7 and g_2=5 are the respective degeneracies. The net spontaneous emission rate $I^* \to I + h\nu$ is then given by N^*A, where

$$A = [g_3/(g_2+g_3)][(A_{3\to4}+A_{3\to3}+A_{3\to2}) + [g_2/(g_2+g_3)](A_{2\to3}+A_{2\to2}+A_{2\to1}) \quad (4.3.3)$$

is the effective Einstein A coefficient which according to the "canon" (cf. Sect.2.1.2) is obtained by averaging the individual A's over the initial states and summing over the final states. Using Table 4.3 one finds $A = 7.7\,s^{-1}$ A relation similar to (4.3.3) defines the net stimulated cross section $\sigma(\nu)$ at any frequency ν, namely

$$\sigma(\nu) = [g_3/(g_2+g_3)][\sigma_{3\to4}(\nu)+\sigma_{3\to3}(\nu)+\sigma_{3\to2}(\nu)] + [g_2/(g_2+g_3)][\sigma_{2\to3}(\nu)+\sigma_{2\to2}(\nu)+\sigma_{2\to1}(\nu)] \quad . \quad (4.3.4)$$

The frequency dependence of $\sigma(\nu)$ and thus of the gain depends on the linewidths and the extent of overlap between the individual transitions. The linewidths can be conveniently controlled by varying the pressure and/or the

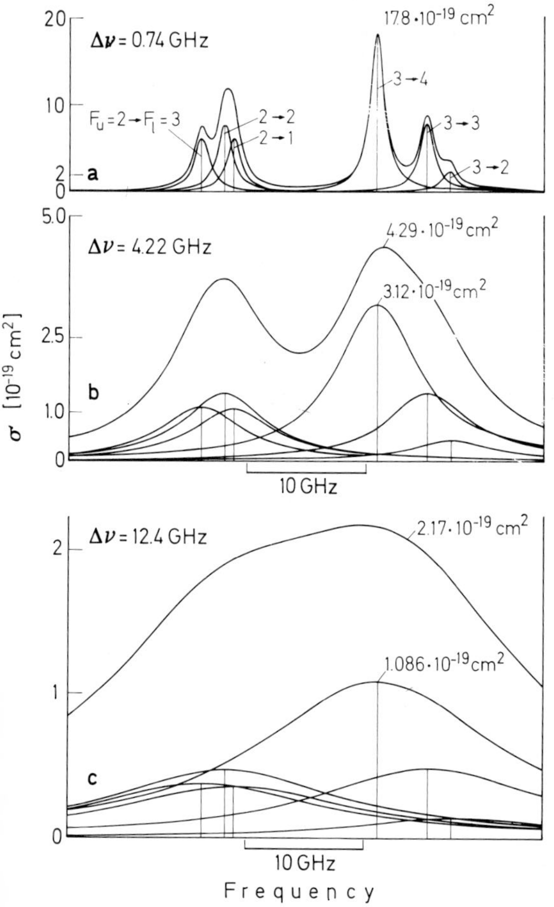

Fig. 4.12a-c. Calculated spectra of the iodine laser transitions for various linewidths. (a) 20 Torr of C_3F_7I; (b) 50 Torr of C_3F_7I + 150 Torr of CO_2; (c) 50 Torr of C_3F_7I + 700 Torr of CO_2. The linewidth $\Delta\nu$ corresponding to each gas mixture is assumed to be the same for all $F_u \rightarrow F_\ell$ lines. The upper curve in every panel represents $\sigma(\nu)$ and is obtained by superimposing the individual cross sections according to (4.3.4) (adapted from [4.30]) (see Fig.3.15 for additional details)

gas mixture in the laser cavity. This possibility of controlling the frequency and intensity of the stimulated emission presents one of the great advantages of the iodine laser. Under conditions where collision broadening dominates over Doppler broadening [typically at $P > 20$ Torr (cf. Sect.3.1.2)], we have $\sigma = a/P$ ($\sigma \propto 1/\Delta\nu$, $\Delta\nu \propto P$ [cf. (3.1.29)]. The proportionality factor a, known as the pressure broadening coefficient, is for many gases in the range of ~10 MHz/Torr. Using such data for mixtures of C_3F_7I and CO_2 the spectra shown in Fig.4.12 can be calculated.

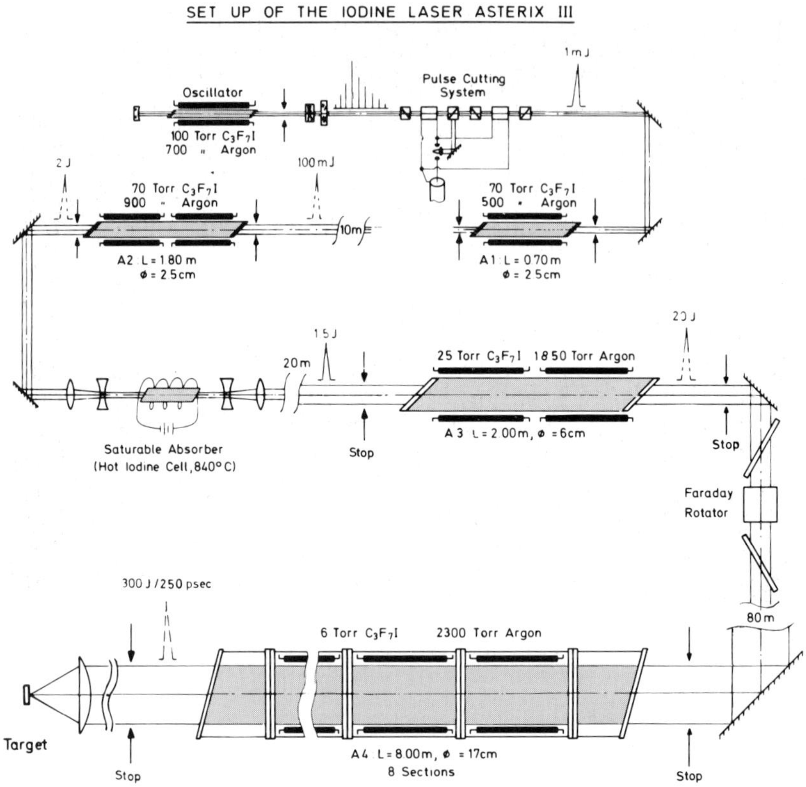

Fig. 4.13a. Schematic design of the photochemical iodine laser Asterix III. The laser radiation originates from a small mode-locked oscillator which delivers a train of mJ pulses. One pulse out of this pulse train is switched out by an optical gate and then subsequently amplified in four stages to 100mJ, 2J, 20J and 300J. Geometrical dimensions are given for the various components. Optical isolation after the 2nd and 3rd amplifier is provided by a saturable absorber and a Faraday rotator. The pulse power of this system is roughly 1 Terawatt, the total light path 150 m. The design elements shown here are more or less typical for all high-power lasers in use today although other laser systems may require a larger number of more sophisticated components (e.g., Nd:glass lasers) or different materials in the beam path (e.g., CO_2 lasers). (Figure adapted from [4.32])

The efficiency of extracting the energy stored in the iodine atoms depends on the coupling between the lines constituting the gain profiles shown in Fig.4.12. [Recall that the gain $\gamma(\nu)=\sigma(\nu)\Delta N$; in the iodine laser $\Delta N \simeq N^*$.] If efficient collisional population transfer among the upper sublevels and among the lower sublevels was taking place then the gain profile could be homogeneously drained at any frequency with a rate proportional to $\gamma(\nu)$. (This picture is completely analogous to the "homogeneously" broadened vibrotational tran-

Fig. 4.13b. Photograph of last amplifier stage of the high-power iodine laser system schematically shown in (a)

sitions in the case of fast rotational relaxation, as described in the previous section.) However, for typical iodine amplifier laser conditions (see below) and pulse durations in the ns range only the lower sublevels are effectively coupled [4.31]. The relaxation between the upper, and relatively distant, sublevels F_u=2 and 3 is too slow. Thus, for pulse durations in the sub-ns range the six transitions fall into two sets, corresponding to F_u=2 and F_u=3, each of which forms an essentially homogeneously broadened line comprising one upper sublevel and the three strongly coupled lower sublevels. This situation is quite clearly reflected in Fig.4.12b. For full energy extraction an emission pulse should therefore contain frequencies of both groups of lines. If this can be achieved up to 2/3 of the initial population inversion ΔN_0 can be extracted as laser energy, $\eta_{ex} \leq (2/3)\Delta N_0$. [This is because $\Delta N = N_u - (g_u/g_\ell)N_\ell$ with the effective degeneracies g_u=12 and g_ℓ=24.] If on the other hand only the $F_u=3 \rightarrow F_\ell=4$ line — the line with the highest σ (see Fig. 4.12) — is active $\eta_{ex} \leq 0.46\ \Delta N_0$.

The iodine photochemical laser has been developed to a very high-power radiation source. Together with the Nd:glass and CO_2 lasers it is now regarded as one of the best candidates for laser-induced initiation of thermonuclear fusion reactions [4.31]. Some devices, such as the **Asterix III** shown in Fig.4.13, are capable of delivering ~1kJ pulses of sub-ns duration with nearly diffraction limited beam quality (cf. Sect.3.2.3) [4.32,33].

The pulse duration t_p and the spectral width of the laser output $\Delta\nu$ are related by the uncertainty principle $\Delta\nu t_p \simeq 1$. For some high-power applications, (e.g., nuclear fusion) a short pulse duration rather than a narrow spectral bandwidth is the desired feature. Hence, for efficient energy extraction from inverted atomic populations by a $<10^{-9}$ s pulse, the (homogeneous) atomic linewidth should preferably be $\Delta\nu > 10^9$ Hz. Figure 4.12 reveals that for the iodine laser transition this requirement can easily be met by adjusting the foreign gas pressure. (For the techniques of short pulse generation see, e.g., [4.32].)

The common experimental arrangements for generating high-energy pulses incorporate a laser oscillator and several stages of laser amplifiers (Sect.3.4), as shown in Fig.4.13. For efficient energy extraction from the gain medium in the amplifiers the inversion density should be as high as possible. As noted in Sect.3.2.1 this ensures that the incoming pulse rapidly reaches the saturation limit. In this limit the pulse is intense enough that in the course of its propagation it reduces ΔN to zero and adds the energy stored in the inverted populations to the laser beam. On the other hand, at very high inversion densities small amounts of spontaneously emitted or scattered photons can trigger out uncontrolled oscillations which may discharge the amplifier prior to the arrival of the laser pulse. These "self" or "parasitic" oscillations start when ΔN exceeds some threshold value ΔN_t, or equivalently when the integrated (small signal) gain $G = \exp(\sigma \Delta N \ell)$ exceeds G_t; ℓ is the length of the amplifier and σ is the cross section for stimulated emission (cf. Sect.3.2.1). The possible onset of parasitic oscillations is the major constraint on ΔN and hence on the maximal energy which can be stored in the amplifier E_{st}. [The energy has to be stored in the amplifier for <15 μs [4.32]. On this time scale collisional quenching is a minor loss mechanism (cf. Table 4.2 and Fig.4.11).] Using A to denote the cross sectional area of the amplifier and $h\nu$ the transition energy, we find

$$E_{st}/A = \ell \Delta N_t h\nu = [h\nu \ln(G_t)]/\sigma \quad (\mathrm{J/cm^2}) \quad , \tag{4.3.5}$$

For typical iodine laser amplifiers incorporated in a multicomponent system, as in Fig.4.13, $G_t \sim 10^2$. Thus at least three amplifier stages are required to amplify a 1mJ pulse delivered by the oscillator into a final output pulse of ~1 kJ (Fig.4.13). Pockel cells, Faraday rotators, or saturable absorbers are commonly used to prevent parasitic optical coupling between adjacent amplifie

[4.32]. (The saturable absorber is a hot[2] iodine cell which absorbs weak radiation, but becomes transparent at high intensities and thus allows the passage of the incoming pulse.)

High radiant fluxes can cause damage to windows, coatings, and other amplifier components. The damage threshold energy density (E_d/A) is of the order of a few J/cm^2. This is another factor which limits the storable energy and has to be taken into account in designing the amplifiers. The length of the amplifier, ℓ, is thus limited by the requirement that the output energy E_{out} should not exceed E_d, namely,

$$E_{out}/A = \eta_{ex}E_{st}/A = \eta_{ex}h\nu\ell\Delta N_t < E_d/A \quad , \tag{4.3.6}$$

where η_{ex} is the extraction efficiency. Typically for the iodine amplifiers, $0.3 < \eta_{ex} \leqq 0.5$ [4.32]. This is due to the fact that the lasing occurs only on the $F=3\rightarrow 4$ lines (Fig.4.12), and that the saturation limit for pulse amplification is not always achieved.

For high beam quality the inversion profile across the amplifier cross section has to be as uniform as possible. This is favored at low pressures of the laser active material RI. On the other hand, such low pressures imply low ΔN values. Consequently, some degree of nonuniformity must be tolerable [$\Delta(\Delta N)<10\%$]. It was found empirically (for i-C_3F_7I lasers) that for good beam quality on the one hand and sufficient population inversion on the other, the RI pressure P_{RI} and the amplifier diameter d should be related via

$$P_{RI}\cdot d = 170 \text{ (Torr cm)} \quad . \tag{4.3.7}$$

The high-power laser arrangement described in Fig.4.13 was designed according to the conditions implied by (4.3.5-7).

As a last point in the description of the iodine photochemical laser we shall consider its overall ("wall plug") efficiency η. This efficiency is defined as the ratio between the laser pulse energy and the electrical energy stored in the capacitor banks. η can be expressed as a product of three factors, $\eta = \eta_p\eta_q\eta_{ex}$:

η_p This is the efficiency of converting the electrical energy stored in the capacitor bank into useful photolysis light. In other words, this is the fraction of electrical energy which is converted to UV photons which reach the RI molecules and in addition have the right frequency to be absorbed by these molecules. Taking into account coupling losses between the capa-

2 Heated in order to form I atoms from I_2 molecules.

citor bank and the flash lamps, the spectral distribution of the flash-light, the transmission of this light to the active laser volume, the incomplete absorption, and the incomplete utilization of the pumped molecules it is found [4.32] that $\eta_p \sim 0.05$.

η_p This is the quantum efficiency associated with the conversion of the UV input photons into IR output photons (λ=1.315 μm) (cf. Fig.4.10). Thus, $\eta_q = (h\nu_{emitted}/h\nu_{absorbed}) \simeq 0.2$.

η_{ex} This is the efficiency of extracting the energy stored in the amplifier media into radiant flux. As mentioned above $\eta_{ex} \sim 0.4$.

Combining the above figures, we find $\eta \simeq 0.4\%$.

As mentioned at the beginning of this section the iodine laser was used here as the first model example because its physical chemistry and spectroscopy are relatively well understood and its technical parameters are known in more detail than those of vibrotational chemical lasers. Our next model example will be the hydrogen fluoride chemical laser (and hydrogen halide lasers in general), where both the kinetic scheme and the experimental arrangements are considerably more complex.

4.3.2 Hydrogen Halide Chemical Lasers

Of the hydrogen halide lasers — HF, HCl, HBr, and their deuterated analogues — HF has received the most attention from laser users, partly because of the high degree of population inversion obtainable and partly because of the possibility of self-propagation of the reaction by a chain or even a branched chain mechanism. The advantage of a chain reaction laser is that the external trigger would only be needed to start the reaction, yielding a potentially high operation efficiency. The two reaction steps of interest are

$$F + H_2 \rightarrow HF(v) + H \quad , \quad \Delta H \simeq -32 \text{ kcal/mole} \tag{4.3.8}$$

$$H + F_2 \rightarrow HF(v) + F \quad , \quad \Delta H \simeq -98 \text{ kcal/mole} \quad . \tag{4.3.9}$$

The HF vibrational levels which can be populated by the energies released in the "cold" reaction (4.3.8), and the "hot" reaction (4.3.9) are indicated in Fig.4.14.

The various attempts to design HF(DF) and other hydrogen halide chemical lasers can be classified into four main categories: 1) *pulsed lasers* using only the cold reaction, 2) *pulsed lasers* with contribution from the hot reaction, 3) *continuous wave* (cw) *lasers*, 4) *chemical transfer lasers* (CTL). In each design scheme there are three kinds of technical problems to solve, namely,

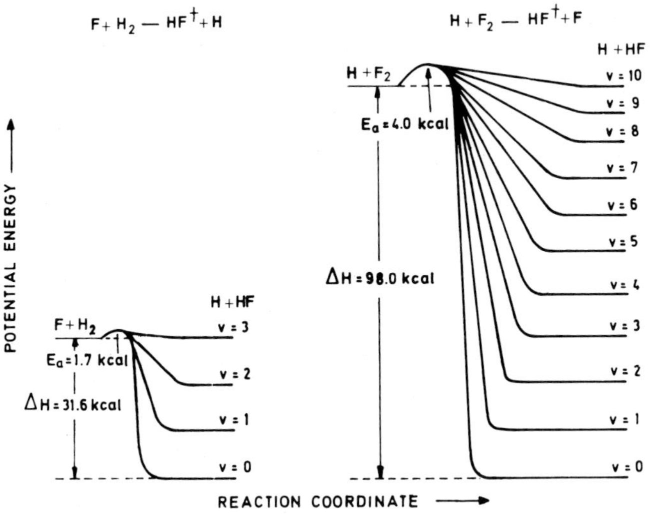

Fig. 4.14. Reaction coordinate diagram for reactions (4.3.8,9) showing the energetically accessible HF vibrational levels (after [4.34]). While the "cold" reaction $F + H_2$ populates the levels $v \leq 3$, i.e., up to the energy threshold (see Fig.4.4), the "hot" ($H+F_2$) reaction significantly populates only the $v \leq 7$ levels

a) that of suitable reaction initiation (generation of the F atoms by photolysis, electron impact, thermal shock, or combustion),
b) the control or the suppression of excessive collisional relaxation and correspondingly the maximization of the energy output, and
c) the optical problems associated with the directional, temporal, and spectral qualities of the output beam.

Having identified this whole array of problems we now proceed to a systematic but brief discussion of the various points, leaving details to the original literature [4.6].

1) *Pulsed* $F + H_2$ *Lasers*

Flash photolysis initiated hydrogen halide lasers contain the same elements as the iodine photodissociation laser described in the preceding section (see Fig.4.13). A flash lamp is positioned alongside the reaction tube which contains a premixed gas filling which has as one of its components a photolyzable fluorine compound (F_2, UF_6, XeF_4, etc.). What rate of F atom production can be expected under these conditions? In a typical flash lamp about 1 kJ of electrical energy may be discharged in a time of about 5 μs. Roughly 60% of this is converted into light output in the spectral range of 2000-10000 Å. Assuming that the fluorine compound used has an absorption band 2000 Å wide, that the quantum yield of dissociation $\Phi_F = 1$, and allowing for 50% losses due to incomplete coupling of the light into the laser tube and to incomplete absorption in the gas, 75 J of useful photolysis light would be available in a volume of typically 100 cm^3. An energy deposition of 0.75 $J/cm^3 \simeq 2 \times 10^{18}$ photons/cm^3 at a photon energy of 4 eV, which corresponds to an initiation rate of 5×10^{23} F atoms/cm^3 s, is thus possible. The rate of F atom production and thus of HF production via the pumping reaction $F + H_2$ can be controlled by arranging the flash lamp characteristics, the initial pressure of the fluorine-containing compound, and the initial H_2 pressure. For typical laser configurations the photon lifetime $\tau_c \sim 10^{-7}$ s. The cross sections of stimulated emission for HF vibrotational transitions are $\sigma \sim 10^{-16}$ cm^2 (cf. Fig.4.7). Hence the threshold value of the population inversion is typically $\Delta N_t = 1/c\sigma\tau_c \sim 10^{13}$ molec/cm^3 [cf. (3.2.13) or (4.2.1)]. With the initiation-pumping rates mentioned above this implies that the time to threshold on the highest gain transition may be in the ns range. For low-pressure lasers, of the kind used in kinetic studies (Sect.4.4), the times to threshold are usually a few μs (see, for example, Fig.4.3).

Electrical discharge initiation, as in the TEA (transverse excited atmospheric pressure) lasers, is usually more efficient, and by varying circuit components the discharge can be made as short as few ns or as long as few ms. Consequently, the time to threshold can be varied over a wider range than in flash photolysis initiation. Thus, as opposed to the case of flash initiation where the duration of the initiating pulse is usually of the order of the laser pulse, in an electrical discharge the initiating pulse can be made considerably shorter than the laser pulse. This allows a better control of the laser efficiency. On the other hand in flash photolysis arrangements gas addition can be easily varied while electric discharges are sensitive to the nature and the pressure of admixtures. Added gases are desirable from the point of view of reducing a possible temperature rise, eliminating possible hot atom effects which can arise in UV photolysis of fluorine compounds and simplifying the kinetic scheme by assuring rapid rotational relaxation (see Sect.4.2 and 4.4). By the same token, however, valuable kinetic information may be given away. A representative electric discharge HF laser is shown in Fig.4.15. For this particular laser also the energy input/output relation has been studied. Results are reproduced in Fig.4.16.

What are the maximum principal efficiencies to be expected in a HF laser exploiting only the "cold" reaction? The discussion on the maximum output in

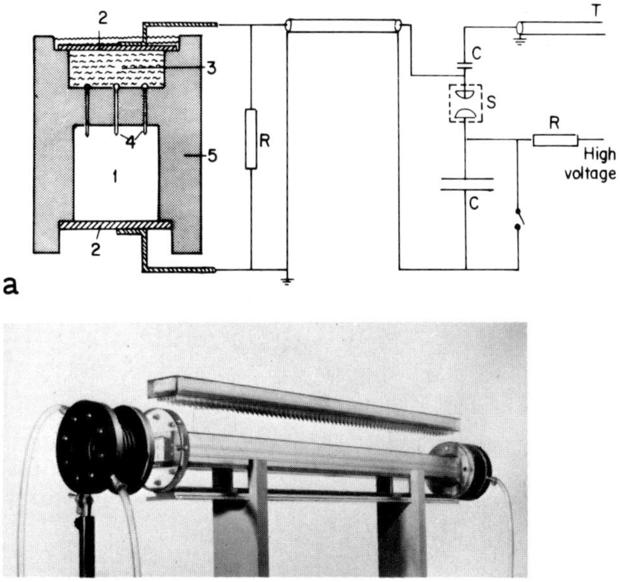

Fig. 4.15. (a) Pulsed discharge-initiated (TEA) HF laser based on reaction (4.3.8). The experimental setup consists of discharge chamber (1) (6×6×100 cm); copper electrodes (2), the upper one being in contact with an electrolyte pre-resistor (3) and having a large number of brass pins (4). The chamber has lucite walls (5). With this design a maximum output energy (on all emission lines) of 5 J can be obtained. As the chambers are of modular design they can be stacked to increase the output if desired (adapted from [4.35]).
(b) Photograph of the laser tube used in the design shown in (a). Top and bottom lifted to show the multiple-pin electrodes

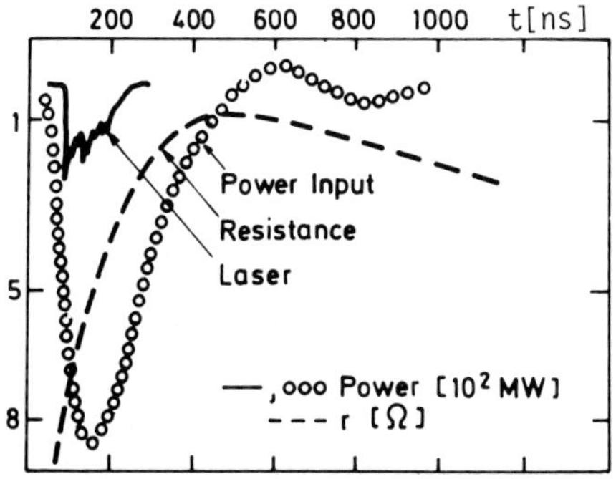

Fig. 4.16. Input/output measurements for a laser as shown in Fig.4.15a,b. The voltage pulse incident on the discharge chamber was 120 kV. Laser gas: 144 Torr SF_6 + 6 Torr H_2. The plot shows the laser output and the electrical input on the same power scale (adapted from [4.35])

Sects.4.2,4.4 shows (at least qualitatively) that on the average 0.5-1 eV (1-2 vibrational quanta) per HF formed can be expected in the form of stimulated radiation. Consequently, if the F atom to start the reaction could be furnished at the expense of a similar amount of energy per atom the overall process efficiency might approach unity. This goal is realistic (the dissociation energy of, say, F_2, is $\Delta H=1.6$ eV), provided that F_2 can be efficiently dissociated. Reality, however, is still far behind this expectation and $\eta \simeq 4$-5% as given in Fig.4.16 seems to be typical for both photolysis and electron impact initiation. The highest output of a nonchain HF laser published was 2500 J in 70 ns, obtained by electron beam bombardment of an SF_6/C_3H_8 mixture [4.36]. As in other types of lasers e-beam initiation or e-beam control of a discharge can be used to overcome some of the stability problems of direct discharges, however at the price of a substantially higher technical effort (see below).

Finally it should be noted that every HF laser can also be operated as a DF laser with very little modification. HCl laser operation is also possible with similar experimental parameters, the most used pumping reaction being (cf. Fig.1.4)

$$Cl + HI \rightarrow HCl(v) + I \quad , \quad \Delta H = -32 \text{ kcal/mole} \quad . \tag{4.3.10}$$

2) *Pulsed $H_2(D_2)/F_2$ Lasers*

There have been several attempts to harness the "hot" reaction (4.3.9) in addition to the "cold" reaction (4.3.8) described above. This is motivated by the kinetic information obtainable in this way but also by the desire to generate new output wavelengths and most of all to increase the laser efficiency. It becomes useful at this point to introduce the distinction between chemical and electrical efficiency, referring to the efficient use of the chemically stored energy in the first case, which can obviously never exceed $\eta = 100\%$, and to the electrical input conversion efficiency in the latter case, which can well be $\eta_e > 100\%$ in the case of a chain reaction. For this to be actually achieved the condition

$$\eta_e = \frac{E}{Q_e} = \frac{\eta Q_c(N/N_0)}{Q_e} > 1 \tag{4.3.11}$$

must be met. Here E is the laser output energy, Q_e is the electrical energy consumed, η is the chemical efficiency, and $Q_c(N/N_0)$ is the total chemical energy released by the chain of pumping processes. N/N_0 is the average chain length, defined as the ratio between the total number of lasing molecules N and the number of primary molecules N_0, i.e., those formed by atoms generated directly by the initiating process. (Thus Q_c is the average chemical energy released per step of the pumping chain.) The averaged chain length in the H_2/F_2 laser has been estimated as 5 [4.37].

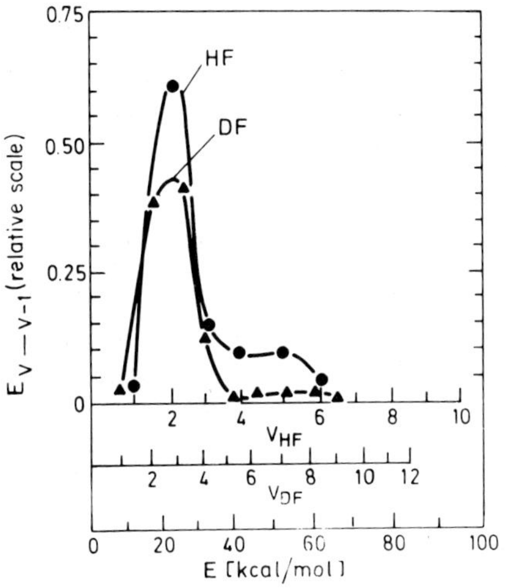

Fig. 4.17. Relative output energies from different vibrational bands in the H_2-F_2 and the D_2-F_2 chain reaction lasers. In both cases most of the emission stems from the cold reaction (4.3.8) (adapted from [4.38])

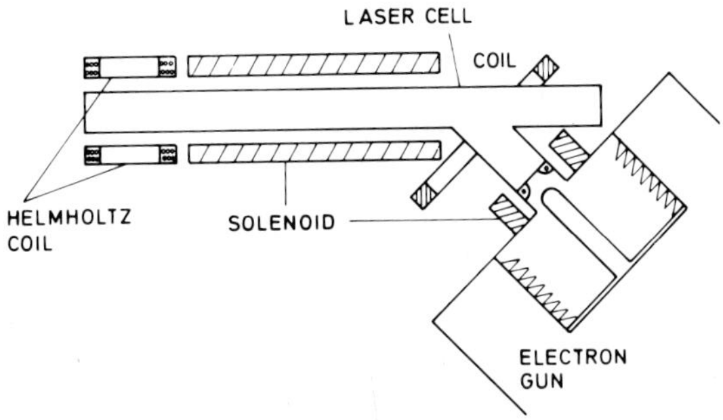

Fig. 4.18. High-power HF laser operating on a chain reaction which is initiated by a high-energy electron beam. In this design, a fluorine-containing compound is dissociated by the high-energy (~2 MeV) electrons. Every electron initiates a large number of dissociation events, gradually loosing its energy. This method provides a rapid ($>10^{26}$ dissociations $s^{-1}cm^{-3}$) uniform pumping mechanism which applies to a wide range of gas pressures (up to a few atmospheres). In this respect it is more efficient and versatile than discharge pumping (adapted from [4.36])

In principle the laser efficiency may be further enhanced by branched chain reactions which greatly accelerate the pumping process. However, the identity and importance of branching reactions in the H_2/F_2 system are not yet definitely clear. When emission from v = 4 was deemed to be absent in the output spectra of $H_2 + F_2$ lasers this was taken as an indication that the branching reaction

$$HF(v{=}4) + F_2 \rightarrow HF(v{=}0) + 2F \qquad (4.3.12)$$

had occurred. Later on this had to be corrected when it was measured that branching reactions in this system have rates which are four orders of magnitude slower than the primary pumping steps. Also, the contribution by the "hot" reaction to the laser output was shown to be disappointingly low; Fig. 4.17 gives some information on the range of vibrational transitions observed together with their relative intensities. Yet, the principle of exploiting a reaction chain – even unbranched – for increasing the electrical pumping efficiency turned out to be technically quite successful as the highest HF output energy reported so far (5000 J in 20 ns) was obtained in this way in an e-beam arrangement with an efficiency $\eta_e = 575\%$ referred to the e-beam energy deposited in the laser volume and a chemical efficiency of $\eta = 10\%$. Figure 4.18 shows the basic scheme of such an e-beam-initiated laser.

3) *Continuous wave HX lasers*

HF laser emission is strongly quenched by the products of the pumping reactions, especially by HF and F [4.39]. Thus, for efficient cw operation the used laser material should be instantly removed from the active volume. This was realized and demonstrated in devices utilizing fast gas flows as early as 1969. Since then, the technical development of cw chemical laser devices has been brought to a high degree of perfection [4.40] (reported laser outputs up to 15 kW). Some remarks on the interaction of gas flow and chemical kinetics will be made in Sect.4.4.4. We shall restrict ourselves here to a short illustration of both high-power supersonic and low-power subsonic devices of this kind. Figure 4.19 shows some principles of a, by now classical, cw chemical laser design where H_2 or D_2 is mixed into an expanding flow of F atoms in some carrier gas. This allows for a very fast gas transport through the active region. The laser axis is transverse to the flow axis. While laser designs of this kind are of more technological interest, the subsonic laser described in Fig.4.20 may be considered as a rather versatile tool for laboratory experiments.

4) *Chemical transfer lasers*

The pumping reactions in HX lasers produce high population inversions. On the other hand, the HX molecules are easily deactivated by collisions. One way around the collisional quenching problems is offered by chemical transfer lasers (CTL) where the excited HX molecules rapidly transfer their energy to another molecule which is less affected by collisional deactivation [4.43]. Chemically pumped HCl(v) or DF(v) can rapidly transfer their vibrational ener-

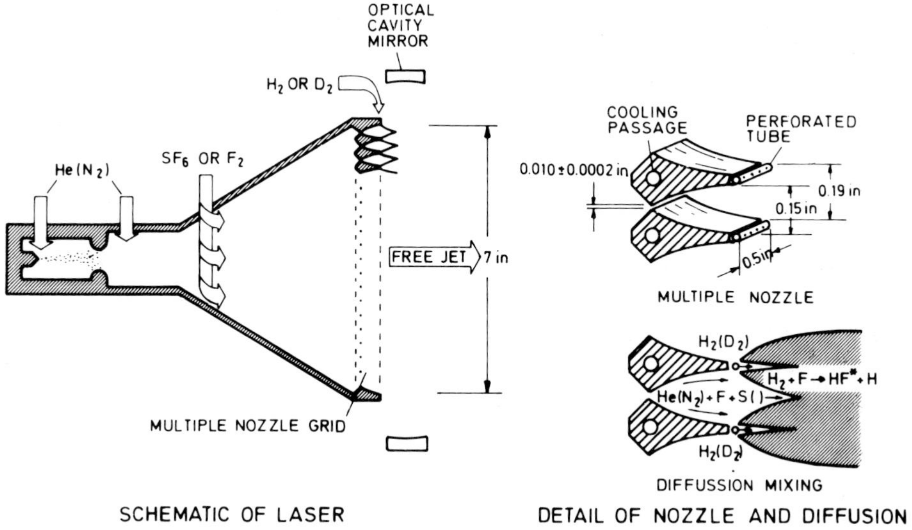

Fig. 4.19. Schematics and details of a 36-slit nozzle for a high-power supersonic hydrogen halide laser (adapted from [4.40,41]). Details of the – rather sophisticated – nozzle design are shown on the right. The purpose of this is to enable very efficient and fast mixing of the reagents so that the laser resonator can be positioned very close behind the nozzles exits, in this way avoiding excessive collisional deactivation prior to laser operation

gy to the antisymmetric stretching mode of CO_2 by the near resonant V-V,(R) processes (cf. Sects.2.4.2, 2.5.4 and 3.3.5),

$$HX(v) + CO_2(00^00) \rightleftarrows HX(v-1) + CO_2(00^01) \quad . \qquad (4.3.13)$$

The vibrational energy gaps involved in $HCl + CO_2$ and $DF + CO_2$ collisions are quite significant, $\Delta E_v \sim 500\ cm^{-1}$. However, most of this energy discrepancy can be compensated by rotational excitation of the HX molecule. Most use has been made for both pulsed and cw operation by the DF/CO_2 transfer scheme which has a rate constant $k \simeq 2\times 10^5\ s^{-1}\ Torr^{-1}$ corresponding to about 40 collisions. The V-R,T deactivation process

$$CO_2(00^01) + DF(v{=}0) \rightarrow CO_2(nm^{\ell}0) + DF(v{=}0) \qquad (4.3.14)$$

is characterized by $k \simeq 2\times 10^4\ s^{-1}\ Torr^{-1}$ corresponding to roughly 400 collisions. This is not much influenced by either DF vibrational excitation or temperature rise. Thus good conditions are provided for a chemically pumped CO_2

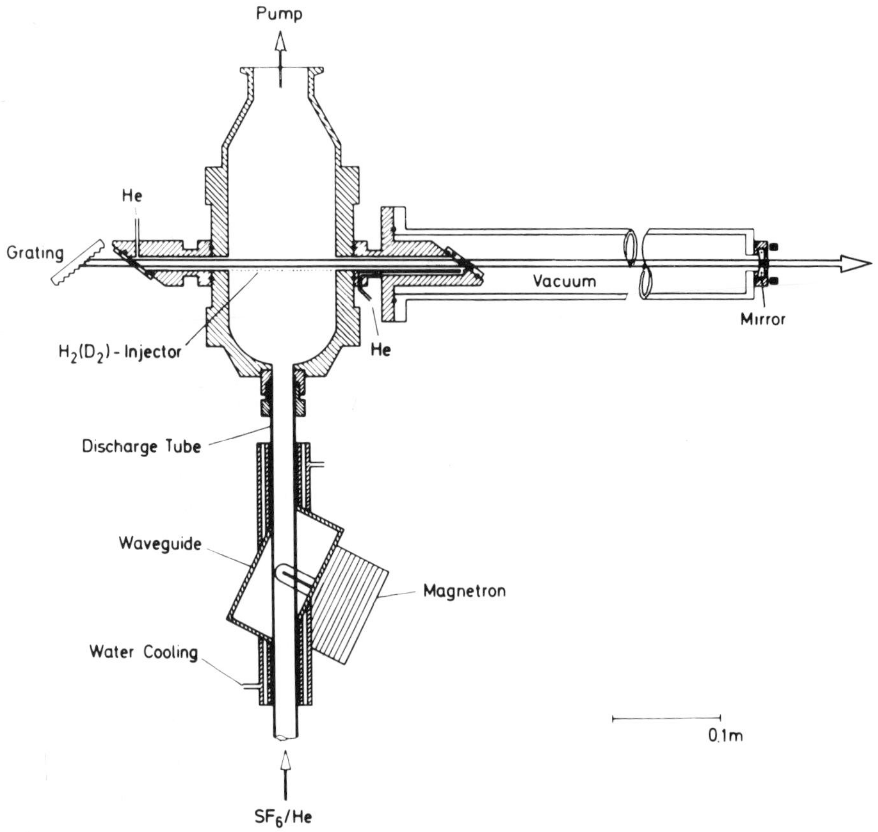
Pump
He
Grating
Vacuum
Mirror
$H_2(D_2)$ - Injector
He
Discharge Tube
Waveguide
Magnetron
Water Cooling
0.1m
SF_6/He

a)

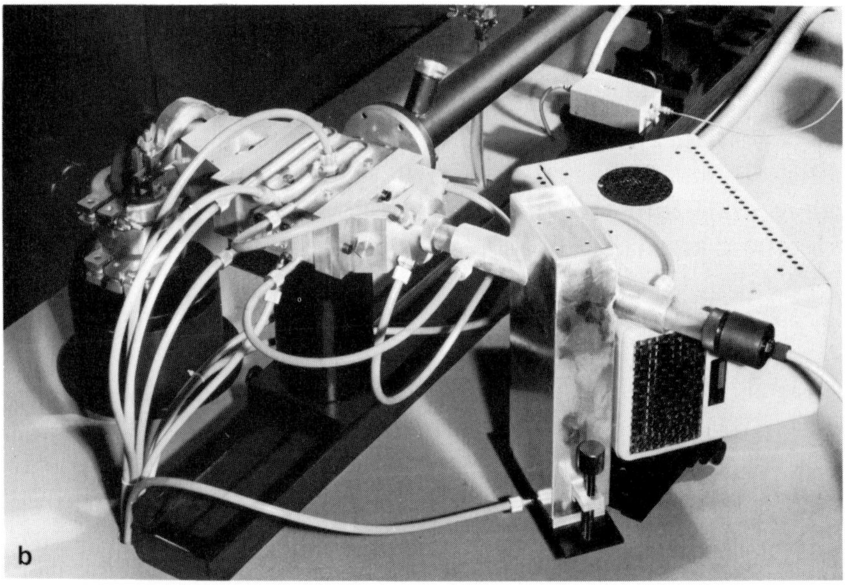
b

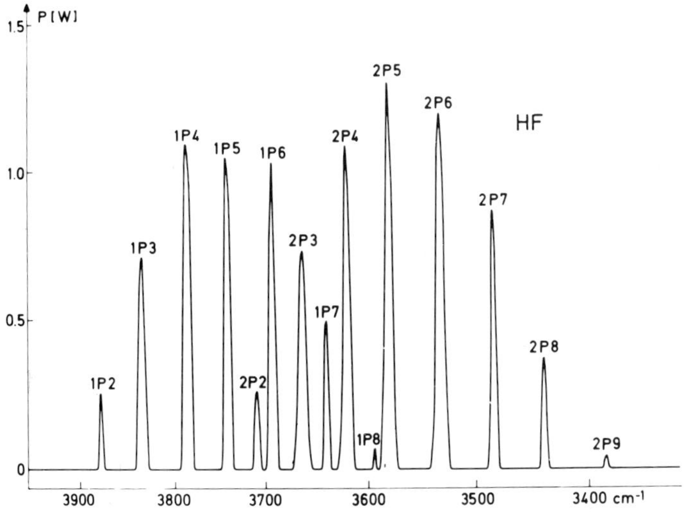

c)

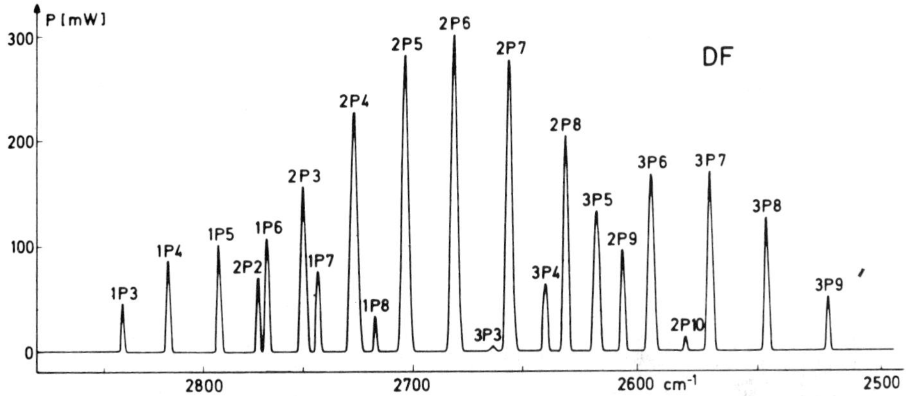

Fig. 4.20a-c. Subsonic cw HF/DF/HCl chemical laser with up to 10 W output power.
(a) Schematic diagram showing the main design parameters. The F/Cl atoms are generated in a microwave discharge and flow into the reaction chamber where they are diffusively mixed with H_2 or D_2 close to the optical axis. As in Fig.4.19 the $H_2(D_2)$ injection and reaction zones are very close to the optical resonator (after [4.42]).
(b) Photograph of the apparatus.
(c) Typical output spectra, here shown for operation as an HF and DF laser. The notation vPJ is for the P branch line, $v,J-1 \rightarrow v-1,J$

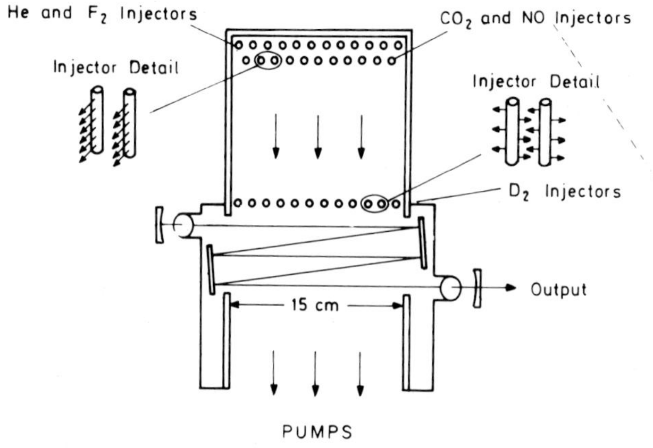

Fig. 4.21. Schematic diagram of a transverse flow subsonic DF/CO_2 laser (adapted from [4.44,45]). The operating characteristics are summarized in Table 4.4

Table 4.4. Operating characteristics of the DF-CO_2 laser system in Fig.4.21

Power output [on P(20) transition]	$\leq$162 W
Chemical efficiency (based on $D_2+F_2 \rightarrow 2DF$)	4.6%
Maximum population of $CO_2(00^01)$ level	1.3×10^{16} cm^{-3}
Static pressure	15.4 Torr
Rotational temperature in reaction zone	400 K
Optimum partial flow rates mmoles/s	He:112;CO_2:57;F_2:7.3;D_2:6.5;NO:1.2

laser. A particular attraction here was the all-chemical laser, generating the necessary F atoms by the reaction [4.43,44]

$$NO + F_2 \rightarrow NOF + F \tag{4.3.15}$$

which are then used to produce DF molecules in the $F + D_2 \rightarrow DF + D$ reaction. An experimental arrangement of a subsonic DF-CO_2 laser based on (4.3.15) is shown in Fig.4.21.

4.3.3 The Chemical CO Laser

The CO laser is an important source of coherent radiation in the 4.7-5.8 μm range [4.45]. It operates on vibrational lines of the v=1→0 to v=16→15 bands. Pumping is usually achieved by the highly exothermic exchange reaction

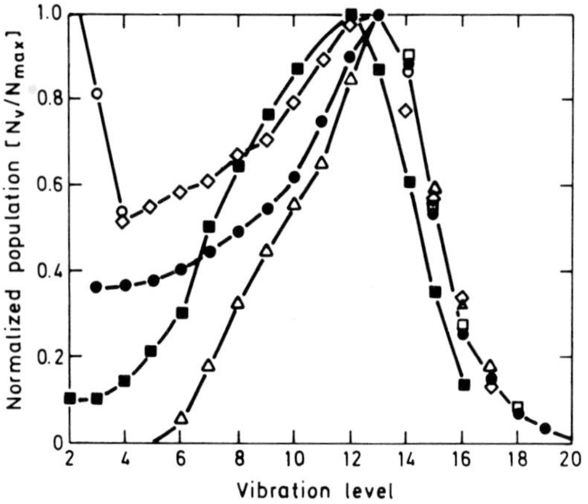

Fig. 4.22. Population distribution of CO produced in reaction (4.3.17) normalized to $N_{max} = 1.0$ versus vibrational level. The different curves represent different measurements (adapted from [4.45] where the original studies are cited)

between an oxygen atom and the CS radical as part of the following reaction sequence:

$$O + CS_2 \rightarrow CS + SO \qquad \Delta H = -22.0 \text{ kcal/mole} \tag{4.3.16}$$

$$O + CS \rightarrow CO(v) + S \qquad \Delta H = -85.0 \text{ kcal/mole} \tag{4.3.17}$$

$$S + O_2 \rightarrow SO + O \qquad \Delta H = -5.6 \text{ kcal/mole} \ . \tag{4.3.18}$$

From an experimental point of view this laser has the disadvantage that it involves a reaction between two unstable species where in particular the oxygen atom generation requires a larger energy expenditure. An advantage from the user's point of view is that it lases on the lower vibrational bands (such as $v=1 \rightarrow 0$) without cooling the laser tube to liquid nitrogen temperature (a necessary measure in electrically pumped CO lasers). The lower transitions are of importance in molecular excitation and probing experiments as described in Chap.5. The relative vibrational state distribution and CO population inversion produced in reaction (4.3.17) were mentioned in Sect.2.5.4. The nascent CO(v) populations are shown in more detail in Fig.4.22.

Unlike the case of hydrogen halides, the V-T relaxation rates of CO are rather small. For example, the process

$$CO(v=1) + CO(v=0) \rightarrow 2CO(v=0) \tag{4.3.19}$$

has a rate constant of 0.01 s^{-1} $Torr^{-1}$. On the other hand V-V exchange processes between CO molecules are very rapid. Rate constants for

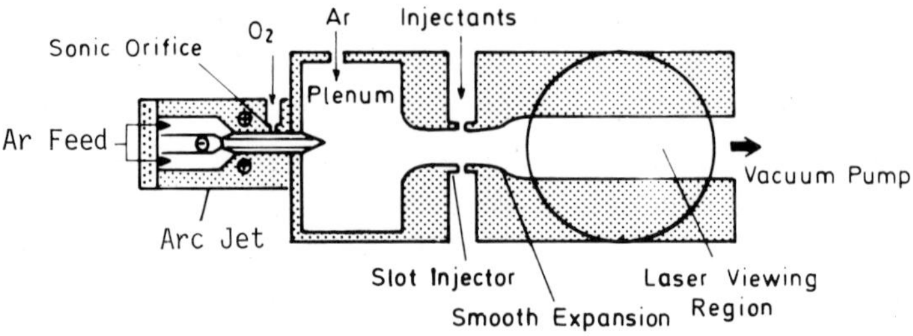

Fig. 4.23. Cross section of the flow arrangement of an arc-excited cw CO chemical laser (adapted from [4.46])

$$CO(v) + CO(v'=0) \rightarrow CO(v-1) + CO(v'=1) \qquad (4.3.20)$$

range from 6.2×10^4 s^{-1} $Torr^{-1}$ (v=2) to 1.5×10^3 s^{-1} $Torr^{-1}$ (v=11) (see Fig.2.53). Processes of this type, which relate to the possibility of V-V up-pumping (Treanor pumping, Sect.2.5.4) constitute the major pumping mechanism of the electrical CO laser and can also be expected to influence the vibrational energy distribution here if the excited CO density is high enough for the excitation transfer to occur on the time scale of the emission.

Although pulsed operation of this laser has been described (in experimental setups similar to Fig.4.15) cw designs have reached a higher state of sophistication. Typical design elements are shown in Fig.4.23. Molecular oxygen is introduced into an arc heater and dissociates at T = 3000 K. By expansion through a nozzle the O atom recombination is frozen and a high O atom concentration preserved. CS_2 vapor in an inert gas stream is then mixed into the flow. One can see the design similarity to the cw HF lasers described earlier in this section (Fig.4.21). 35 W of light output are obtained on roughly 50 emission lines between 5 and 5.6 μm. The chemical conversion efficiency was established to be ~6%. To complete this rough overview of chemical CO laser operation it should be mentioned that cw laser emission can also be obtained in a free-burning O_2/CS_2 flame at a pressure of typically 300 Torr (CS_2:N_2O:O_2=1:216:18) again with power outputs ~10 W [4.46].

4.4 Chemical Laser Kinetics

From a theoretical point of view, understanding the kinetic scheme of chemical lasers provides insights into the nature of chemical systems far from equilibrium. On the practical side, kinetic analyses can be very helpful for the design of efficient laser systems.

The principal tool for a comprehensive kinetic analysis of chemical lasers are the rate (master, balance) equations describing the temporal evolution of molecular level populations and photon densities in a laser cavity. In principle the rate equations for a chemical laser system are an extension of the simplified models described in Sect.3.2.4. Yet, in chemical lasers, owing to the very large number of active levels and allowed radiative and nonradiative transitions between these levels, the solution of the rate equations presents various kinds of difficulties. Some of these difficulties can be overcome by separating between the fast and slow processes and assuming, according to the case, that the fast ones are already equilibrated or the slow ones are still frozen.

After reviewing the structure of the rate equations for a laser operating on vibrotational transitions, in Sect.4.4.1, we shall describe the common and simple case of chemical lasers in rotational equilibrium (Sect.4.4.2). The effects of rotational nonequilibrium will be discussed in Sect.4.4.3. Section 4.4.4 is devoted to a brief kinetic description of cw chemical lasers.

4.4.1 The Rate Equations

The various types of rate processes that deplete and feed the level and photon populations were illustrated schematically in Fig.4.2. If the magnitude of the state-to-state rate constants corresponding to these processes and the optical-mechanical parameters of the laser are known all the information about level populations and output characteristics can be extracted from the laser rate equations. This approach to simulating the behavior of chemical lasers is usually called "modelling" or "computer experiments" [4.47]. Its major present limitation is that for most systems of interest not all the relevant rate constants are accurately known. (For this reason procedures for testing the sensitivity of the solutions of the rate equations to uncertainties in various rate constants have been suggested [4.48].) Yet, in view of the rapid accumulation of experimental and theoretical kinetic data, modelling studies are gradually becoming a more reliable and routine tool. The main practical merit of computer experiments is in being a rather fast and inexpensive diagnostic procedure.

As guiding examples in this discussion we consider chemical lasers pumped by exoergic atom-diatom reactions such as $F + H_2 \rightarrow HF + H$, $H + F_2 \rightarrow HF + H$, $Cl + HBr \rightarrow HCl + Br$, or $O + CS \rightarrow CO + S$. The rate processes constituting the kinetic

Table 4.5. Kinetic scheme of $A + BC \rightarrow AB + C$ lasers[a]

Initiation: e.g., e^- = electrical discharge, $h\nu$ = flash

$$RA \xrightarrow{(e^-, h\nu)} R + A \qquad (R.1)$$

Pumping: $P_{v,J}$

$$A + BC \rightleftarrows AB(v,J) + C \qquad (R.2)$$

Relaxation: $L_{v,J}$

$$\text{R-R,T} \quad AB(v,J) + M_i \rightleftarrows AB(v,J') + M_i \qquad (R.3)$$

$$\text{V-R,T} \quad AB(v,J) + M_i \rightleftarrows AB(v',J') + M_i \qquad (R.4)$$

$$\text{V-V(R,T)} \quad AB(v,J) + AB(m,L) \rightleftarrows AB(v',J') + AB(m',L') \quad , \quad v+m = v'+m' \qquad (R.5)$$

Stimulated ($h\nu$) *and spontaneous radiation:* $R_{v,J}$

P branch, $\chi_{v,J}^{v-1,J+1}$

$$AB(v,J) \xrightleftharpoons{(h\nu)} AB(v-1,J+1) + h\nu_{v,J}^{v-1,J+1} \qquad (R.6)$$

R branch, $\chi_{v,J}^{v-1,J-1}$

$$AB(v,J) \xrightleftharpoons{(h\nu)} AB(v-1,J-1) + h\nu_{v,J}^{v-1,J-1} \qquad (R.7)$$

[a] Molecular species: M_i = RA,R,A,BC,AB,C, and other gases, e.g., Ar. The conventional notation for P and R branch transitions is $P_{v \rightarrow v-1}(J)$ for $v,J-1 \rightarrow v-1,J$ and $R_{v \rightarrow v-1}(J)$ for $v,J+1 \rightarrow v-1,J$.

scheme of these systems are listed in Table 4.5. Assuming that the molecular populations and radiation densities within the cavity are spatially uniform (cf. Sect.3.2.4) the rate equations for the vibrotational populations $N_{v,J}$ (molecules/cm^3) and photon densities $\phi_{v,J}^{v',J'}$ (photons/cm^3) are

$$\frac{dN_{v,J}}{dt} = P_{v,J} - R_{v,J} - L_{v,J} - \frac{\phi_{v,J}^{v',J'}}{\tau_c} \qquad (4.4.1)$$

$$\frac{d\phi_{v,J}^{v',J'}}{dt} = \chi_{v,J}^{v',J'} - \frac{\phi_{v,J}^{v',J'}}{\tau_c} \qquad (v,J \rightarrow v',J'=v-1,J\pm1) \quad , \qquad (4.4.2)$$

where the various contributions to the rate of change of the molecular and photon densities are detailed in Table 4.5 (see also Fig.4.2). The last term in (4.4.2) represents the output coupling; τ_c is the photon lifetime in the cavity.

Equations (4.4.1) and (4.4.2) have to be solved simultaneously with the rate equations for the nonlasing species and the translational temperature;

see below. The number of rate equations may sum up to several dozens, even for relatively simple systems like the $F+H_2 \rightarrow HF+H$ laser. Both the number and the complexity of the rate equations can be substantially reduced with the aid of the rotational equilibrium hypothesis. To assess the significance and the validity of this hypothesis let us first consider the explicit form of the radiation, pumping, and relaxation terms in the rate equations.

Radiation. Every vibrotational level v,J (except for boundary levels where v,J or both are zero) may participate in two R branch and two P branch transitions. Using $\chi_{v,J}^{v',J'}$ to denote the excess of emission over absorption through the $v,J \rightarrow v',J'$ transition, the general expression for the net depletion rate of v,J due to all the allowed transitions is given by

$$R_{v,J} = \chi_{v,J}^{v-1,J+1} - \chi_{v+1,J-1}^{v,J} + \chi_{v,J}^{v-1,J-1} - \chi_{v+1,J+1}^{v,J} , \qquad (4.4.3)$$

where

$$\chi_{v,J}^{v',J'} = c\sigma_{v,J}^{v',J'} \Delta N_{v,J}^{v',J'} \phi_{v,J}^{v',J'} + \alpha A_{v,J}^{v',J'} N_{v,J} . \qquad (4.4.4)$$

The first term here represents the net rate of stimulated radiation in the $v,J \rightarrow v-1,J+1$ transition. The second term accounts for the effects of spontaneous emission. In principle, α should be given different values depending whether χ appears in the photon equation (4.4.2) or the population equation (4.4.1). In the first case, since only photons travelling along the laser axis contribute to $d\phi/dt$, $\alpha \ll 1$ is the effective fraction of photons emitted in the axial direction or, equivalently, the fraction of stable (low loss) cavity modes. In the case of single-mode operation the last term in (4.4.4) is $c\sigma_{v,J}^{v',J'} N_{v,J}$. [See (3.2.24). An estimate for α is provided by $\alpha = \Delta\omega/4\pi$, where $\Delta\omega \sim A/L^2 \sim A/(c^2\tau_c^2)$ is the solid angle subtended by the mirror area A over the average distance $L \sim c\tau_c$, traversed by a photon before leaving the cavity.] On the other hand, $N_{v,J}$ is depleted by spontaneous emission in all directions and the appropriate value for the population equation is $\alpha = 1$. In practice, since vibrotational transitions have relatively long lifetimes ($\sim 10^{-2}$ s) the effect of spontaneous emission in the photon equation represents the source of noise photons necessary to initiate the amplification process. Above the lasing threshold the effect of the second term in (4.4.4) on (4.4.2) is negligible (cf. Sect.3.2.4).

In practically all the modelling studies of chemical lasers it is assumed that the line profile of a lasing transition is saturated homogeneously, even when the major broadening mechanism is an inhomogeneous one, namely, the Doppler effect. The main justification of this assumption is that rapid translational (momentum transfer) relaxation prevents hole burning effects in the line shape of the emitting transition [4.47] (Sect.3.2.3).

An important simplification of the rate equations arises from the fact that during lasing a given v,J level can be the origin of, at most, one transition. The reason is that once lasing starts on one line (the one with higher gain) the intense stimulated emission reduces its gain to its threshold value (cf. Sect.3.2.4). Consequently, the gains of the concurrent lines fall below their threshold values. Thus, on the average, only half of the terms on the rhs of (4.4.4) are relevant. In particular, when rotational equilibrium prevails (Sects.4.2 and 4.4.2), the gain of a P branch transition is always higher than that of the R branch transition originating at the same level so that only the P branch has to be considered.

Pumping and nonlasing species. In the framework of the laser model of Table 4.5 the pumping term in (4.4.1) is given by

$$P_{v,J} = k(\rightarrow v,J;T)[A][BC] - k(v,J\rightarrow;T)[C]N_{v,J} \quad , \tag{4.4.5}$$

where $N_{v,J} = [AB(v,J)]$; $k(\rightarrow v,J;T)$ and $k(v,J\rightarrow;T)$ are the forward and reverse rate constants of the pumping reaction (R.2), respectively. The rate equations for the nonlasing species (which appear in $P_{v,J}$ and $L_{v,J}$) must be solved simultaneously with (4.4.1) and (4.4.2). Their explicit expressions for the kinetic model of Table 4.5 are

$$\frac{d[A]}{dt} = q\psi(t)[RA] - \sum_{v,J} P_{v,J} \simeq q\psi(t)[RA] - \vec{k}(T)[A][BC] \tag{4.4.6}$$

$$\frac{d[RA]}{dt} = -q\psi(t)[RA] \tag{4.4.7}$$

$$-\frac{d[BC]}{dt} = \frac{d[AB]}{dt} = \frac{d[C]}{dt} \simeq \vec{k}(T)[A][BC] \tag{4.4.8}$$

with the initial (t=0) conditions: $[RA]_0 \neq 0$, $[BC]_0 \neq 0$, $[A]_0 = [AB]_0 = [C]_0 = 0$. The function $\psi(t)$ represents the "initiation profile", e.g., the number of light quanta supplied by a flash lamp or the number of electrons supplied by an electrical discharge per unit volume and unit time. q is the efficiency of the initiation process. $\vec{k}(T)$ is the overall forward rate constant of (R.2). [The approximation expressed by the last equalities in (4.4.6) and (4.4.8) corresponds to neglecting the effects of the backward process in (R.2).] Under

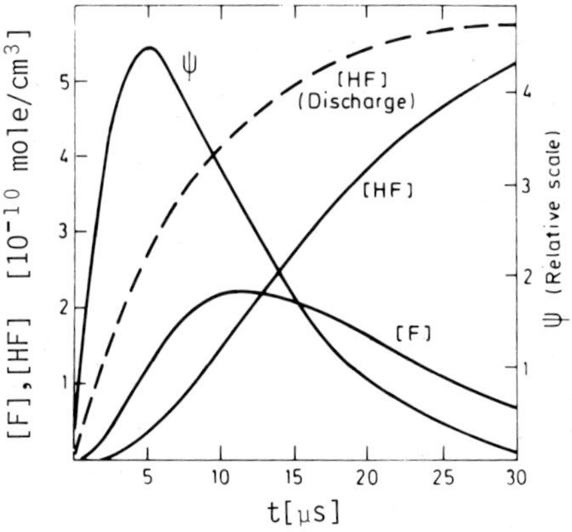

Fig. 4.24. Time dependence of total HF concentration from two initiation methods: I) Fast electric discharge – all F atoms are generated by a very short pulse and then react with H_2 molecules to yield HF molecules with a time profile described by the dashed curve. II) Flash initiation – a flash lamp profile ψ leads to a gradual production of F atoms which react with the H_2 molecules present in the laser cavity to form HF [4.19,20], full curves (adapted from [4.49])

certain conditions, such as fast electrical discharges, the generation of A from RA terminates after a very short time interval, on the time scale of the pumping reaction. In this case one may set $\psi(t)[RA]=0$ and solve the rate equations subject to the modified initial conditions $[A]_0 \neq 0$ and $[RA]_0$ replaced by $[RA]_0 - [A]_0$, with $[A]_0$ being the total concentration of atoms produced in the initiation stage. The time profiles of the total molecular concentrations appropriate to two typical initiation techniques of the $F+H_2$ laser are shown in Fig.4.24. The "discharge" curve corresponds to the modified initial conditions.

Relaxation. The loss term $L_{v,J}$ lumps together contributions from various types of energy transfer processes with different collision partners. Using p to denote the process, e.g., p = R-R,T, or V-V and M_i to denote the collision partner, M_i = A, AB, Ar, etc., we may write

$$L_{v,J} = \sum_p \sum_i L^p_{v,J}(M_i) \quad . \tag{4.4.9}$$

As specific examples consider, say, the rotational (p=R-T) and vibrational (V-R,T) deactivation rates of HF(v,J) molecules in the $F+H_2 \rightarrow HF+H$ laser due to collisions with F atoms (see Sect.2.5),

$$L^{R-T}_{v,J}(F) = [F]\sum_{J'} [k(v,J\rightarrow v,J';T)N_{v,J} - k(v,J'\rightarrow v,J;T)N_{v',J'}] \tag{4.4.10}$$

$$L^{V-R,T}_{v,J}(F) = [F]\sum_{v',J'} [k(v,J\rightarrow v',J';T)N_{v,J} - k(v',J'\rightarrow v,J;T)N_{v',J'}] \; . \tag{4.4.11}$$

Output characteristics. The loss of photons from the cavity due to the useful (coherent radiation) coupling and the dissipative mechanisms (e.g., diffraction, absorption) is represented by the second term of (4.4.2) (cf. Sect. 3.2.2). Using an abbreviated notation where ℓ designates the transition, $\ell = v,J \to v',J'$, and ν_ℓ the frequency of this transition, the outcoupled laser power per unit active laser volume (W/cm^3) corresponding to photons with energy $h\nu_\ell$ is

$$P_\ell = fh\nu_\ell\phi_\ell/\tau_c \quad , \tag{4.4.12}$$

where f represents the fraction of usefully outcoupled photons (cf. Sect. 3.2.4) (f is assumed to be common to all transitions). The total power (per cm^3) is the sum of the individual P_ℓ's. The total pulse energy (joules) is

$$E = V\sum_\ell E_\ell = V\sum_\ell \int P_\ell dt = (fV/\tau_c)\sum_\ell h\nu_\ell \int \phi_\ell dt \quad , \tag{4.4.13}$$

where V is the volume of the active medium.

In most of the kinetic models of chemical lasers the calculation of output intensity is based on the "gain-equal-loss" or "on threshold" assumption (Sect.3.2.4), according to which the gain of photons by stimulated emission is exactly balanced by their losses via (useful and useless) output coupling. The mathematical analogue of this assumption is that $d\phi_\ell/dt = 0$ or, equivalently [cf. (4.4.2)],

$$\chi_\ell = \phi_\ell/\tau_c \quad , \tag{4.4.14}$$

where $\chi_\ell = \chi_{v,J}^{v',J'}$. Neglecting the unimportant (during lasing) spontaneous emission term in (4.4.4), we find

$$\Delta N_\ell = \Delta N_{\ell,t} = 1/c\sigma_\ell\tau_c \quad , \tag{4.4.15}$$

where $\Delta N_{\ell,t}$ is the threshold inversion on the $\ell = v,J \to v',J'$ line. The gain-equal-loss condition is exact for cw lasers. With the exception of systems exhibiting significant relaxation oscillations this condition is very nearly fulfilled also in pulsed chemical lasers.

Temperature. The collisional rate constants $k(n \to n';T)$ as well as the cross sections for stimulated emission σ_ℓ are functions of T. Under conditions which ensure rotational equilibrium T represents the translational-rotational temperature; otherwise T refers only to the translational temperature (assuming, of course, "instantaneous" translational relaxation). The T dependence of reactive and nonreactive rate constants has been discussed in Chap.2. The radiative cross sections depend on T via $\Delta\nu$, the width of the line profile. (For Doppler broadening $\Delta\nu_D \propto T^{\frac{1}{2}}$; Sect.3.1.2. The T dependence of $\Delta\nu_H$, the width due to collision broadening, is more complex. For a gas with constant density $\Delta\nu_H \propto P/T^{\frac{1}{2}} \propto T^{\frac{1}{2}}$, assuming "hard sphere" collisions [4.47].)

In the case of rotational equilibrium the temperature affects the gain not only via $\Delta\nu$, but also through the population difference ΔN. It should be noted, however, that the excess buffer gas, which is usually needed to ensure rotational equilibrium, serves also as a heat bath that moderates temperature variations.

The three major causes for temperature rise in a chemical laser system are:

1) The pumping reaction, which releases part of its exothermicity into the translational (in the case of rotational equilibrium, also rotational) degrees of freedom of the products.
2) The relaxation processes which convert internal to translational (and rotational) energy.
3) The residual initiation energy, e.g., in a Cl + HBr laser where the Cl atoms are generated by flash photolysis of Cl_2, the residual energy is the difference between the energy of the photolyzing photon and the dissociation energy of Cl_2 [4.50], $\varepsilon = h\nu_{flash} - D_0(Cl_2)$.

The temperature equation for the laser model of Table 4.5 is

$$\frac{dT}{dt} = \frac{1}{C_v}\left\{\varepsilon q\psi(t)[RA] - \sum_i E_i \frac{d[M_i]}{dt} - \sum_n E_n \frac{dN_n}{dt} - \sum_\ell P_\ell\right\} , \qquad (4.4.16)$$

where ε is the residual initiation energy, the E_i's are the energies of the nonlasing molecules (except RA), E_n is the energy of a lasing molecule in state n, and P_ℓ is the output power per unit volume, (4.4.12). (Note that E_i and E_n should be measured on the same energy scale.) In the case of rotational equilibirum $n = v$, otherwise $n = v,J$. $C_v = \sum[M_i]c_v^i + \sum N_n c_v^n$ is the heat capacity of the gas; c_v^i and c_v^n are, respectively, the specific heats (per molecule) of species i and of AB(n), $[c_v^{v,J} \simeq (3/2)k$, $c_v^v \simeq (5/2)k]$. The last term in the energy balance equation (4.4.16) is the rate of converting internal to radiation energy.

4.4.2 Rotational Equilibrium

Among the various inelastic energy transfer processes the lasing molecules are exposed to, rotational relaxation is almost invariably the most efficient one. As argued in Sects.1.2 and 4.2 at high buffer gas pressures, rotational relaxation is even faster than stimulated emission and the vibrational populations are given throughout the lasing by (4.2.3). In Sect.2.5.3 we have seen that for times $t > \tau_{R-T}$ the large set of vibrotational master equations reduces to a small and simpler set of vibrational equations. Similarly, the rotational equilibrium hypothesis expressed by (4.2.3) implies a substantial reduction and simplification of the rate equations (4.4.1) and (4.4.2).

Substituting (4.2.3) into (4.4.1), summing over J, and noting $\sum_J p(J|v;T) = 1$, we find

$$\frac{dN_v}{dt} = P_v - R_v - L_v \quad , \tag{4.4.17}$$

where

$$P_v = \sum_J P_{v,J} \quad , \quad L_v = \sum_J L_{v,J} \quad , \quad R_v = \sum_J R_{v,J} \quad . \tag{4.4.18}$$

In addition to reducing the number of population equations the rotational equilibrium assumption leads to simplified expressions for the pumping, radiation, and relaxation terms in (4.4.18). In the first place it is clear that the real (and difficult to measure) R-R,T rates are irrelevant. Formally, L_v contains no contribution from this process since (4.2.3) and detailed balance imply $\sum_J L_{v,J}^{R-R,T} \equiv 0$ (cf. Sect.2.5.3). The vibrational relaxation terms also acquire a simple form; for example, from (4.4.11), (4.2.3), and (4.4.18) we obtain,

$$L_v^{V-R,T}(F) = \sum_J L_{v,J}^{V-R,T}(F) = [F]\sum_{v'} [k(v \to v';T)N_v - k(v' \to v;T)N_{v'}] \quad . \tag{4.4.19}$$

Here T is the translational-rotational temperature and

$$k(v \to v';T) = \sum_{J,J'} p(J|v;T)k(v,J \to v',J';T) \quad . \tag{4.4.20}$$

In complete analogy to the treatment of vibrational deactivation the pumping term (4.4.5) reduces to

$$P_v = \sum_J P_{v,J} = k(\to v;T)[A][BC] - k(v \to;T)[C]N_v \simeq k(\to v;T)[A][BC] \quad . \tag{4.4.21}$$

The rotational equilibrium hypothesis also implies a simple form for the radiation term R_v, as well as a reduction in the relevant number of photon equations [cf. (4.4.2)]. This is a consequence of the special lasing mechanism whose main characteristics — the "single-line operation", the "homogeneous" draining of the Boltzmann rotational profiles, and the "J shifting" of the lasing transition — have been described in Sect.4.2.

Since every vibrational band $v \to v-1$ involves only one lasing (P branch) transition at any instant of time, the radiative term in (4.4.17) [see (4.4.3)] reduces to

$$R_v = R_{v,J(v)} - R_{v+1,J(v)+1} = \chi_{v,J(v)} - \chi_{v+1,J(v)+1} \quad , \tag{4.4.22}$$

where $\chi_{v,J(v)}$ denotes the rate by which the highest gain transition in $v \to v-1$, i.e., $v,J(v) \to v-1,J(v)+1$, drains the population of level v. (Note that for $\bar{v}$, the highest vibrational level participating in the lasing, the second term is identically zero. Similarly for $v = 0$ the first term is zero.) Owing to the single line operation there is only one photon equation [cf. (4.4.2)] per vibrational band. Using the gain-equal-loss ("on threshold") approximation [cf.(4.4.14,15)] the differential photon equations convert to algebraic relations

$$\chi_{v,J(v)} = c\sigma_{v,J(v)}\Delta N_{v,J(v)}\phi_{v,J(v)} = \phi_{v,J(v)}/\tau_c \quad . \tag{4.4.23}$$

Substitution of (4.4.22) into (4.4.17) yields

$$dN_v/dt = P_v - L_v + \chi_{v+1,J(v)+1} - \chi_{v,J(v)} \quad . \tag{4.4.24}$$

Rearranging this set of equations the photon densities are obtained as

$$\phi_{v,J(v)}/\tau_c = \chi_{v,J(v)} = \sum_{v'=v}^{\bar{v}} (P_{v'} - L_{v'} - dN_{v'}/dt) \quad . \tag{4.4.25}$$

Note that the first equality is based on the gain-equal-loss assumption[3] [cf. (4.4.23)]. The numerical algorithm for solving (4.4.24) or (4.4.25) proceeds along the following general lines: One starts to integrate (4.4.24) with all the radiative terms $\chi = 0$. The nonlasing species and temperature equations are integrated simultaneously. At each integration step (4.2.3) is used to cal-

3 The gain-equal-loss assumption replaces the differential photon equations by algebraic equations and thus simplifies the solution of the rate equations. Clearly, it is not necessary to employ this assumption (see below).

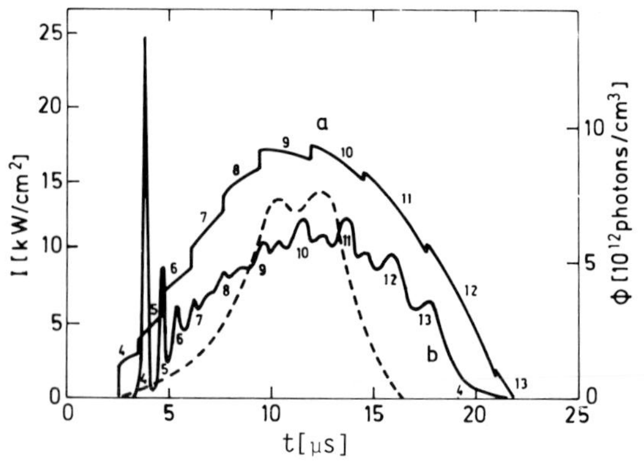

Fig. 4.25. Lasing intensity in the v=1→0 band of the Cl + HBr laser initiated by flash photolysis of Cl_2. The dashed line describes the experimental results; lasing has been observed on the $P_{1\to0}(4\text{-}14)$ lines. (See Fig.2.35 for the energy level diagram of the Cl + HBr → HCl + Br reaction.) The jigsaw curve a represents the results of computer simulation based on the rotational equilibrium assumption and the gain-equal-loss condition. The numbers denote the J value of the highest gain, $P_{1\to0}(J)$, transition. The agreement between theory and experiment with respect to the sequence of lasing transitions, pulse duration, and intensity is fair. It should be noted that although the J shift trend is observed in the experimental output (see [4.50]) lasing occurs on few lines simultaneously, which is a possible indication for incomplete rotational relaxation. The jumps in curve a are due to the numerical treatment of the J shift phenomenon (see text). (Experimental and modelling results adapted from [4.50].) Curve b represents the solutions of the basic laser rate equations (4.4.1,2), i.e., without imposing the gain-equal-loss condition as a constraint (adapted from [4.52])

culate $N_{v,J}$ and the threshold condition (4.4.15) [or (4.4.24)] is tested for all the relevant lines. At the moment that threshold is reached for one of the transitions, say $v,J(v) \to v-1,J(v)+1$, the corresponding population inversion is set equal to its threshold value, $\Delta N_{v,J(v)} = 1/c\sigma_{v,J(v)}\tau_c$ [cf.(4.4.23)]. Using (4.2.3) one then evaluates dN_v/dt and $\phi_{v,J(v)}$ from (4.4.23-25). (For more details see [4.47,50,51].) This procedure is continued until, due to the decrease in dN_v/dt, the highest gain transition in $v \to v-1$ is shifted to the next higher J value (cf. Fig.4.8) and the same procedure is repeated on the new lasing line. Since $p(J|v;T)$ decreases rapidly with J so also is the absolute value of the population inversion $\Delta N_{v,J}$. Eventually, the highest gain transition reaches such a high J value that $\Delta N_{v,J}$ falls below the threshold inversion and lasing terminates.

The application of the procedure described above to determine the time evolution of the photon density, or equivalently the laser power [see (4.4.12) is demonstrated for the Cl/HBr and the H_2/F_2 lasers in Figs.4.25 and 4.26. The

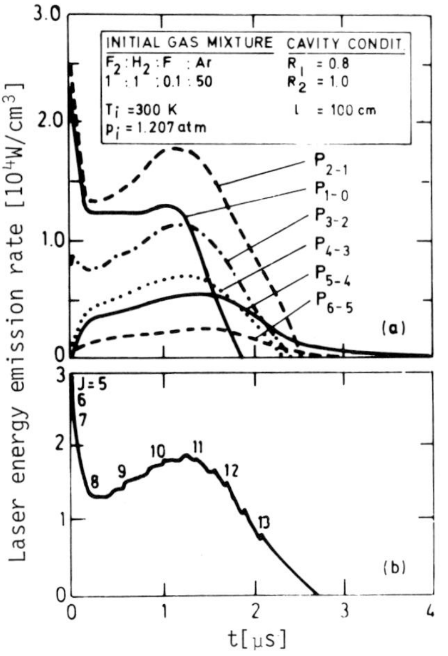

Fig. 4.26a,b. Laser output intensity vs time for the $H_2 + F_2$ chain chemical laser. (a) Laser power in the bands $6 \rightarrow 5$ through $1 \rightarrow 0$. (b) Enlarged output pattern in the $v=2 \rightarrow 1$ band. The numbers indicate the lasing transition. The numerical procedure is similar to that employed in modelling the Cl + HBr laser (Fig.4.25) (adapted from [4.51])

small discontinuities in the laser power whenever a J shift occurs are due to imposing the gain-equal-loss condition and the assumption of single-line operation. In practice there is always some overlap between the lasing transitions, even at high buffer gas pressures (see Fig.4.3). Figure 4.25 shows also (curve b) the output pattern evaluated by direct integration of the basic rate equations (4.4.1,2) rather than using (4.4.23-25). (The central column in Fig.4.3 was also computed in this fashion.) This, more accurate but also considerably more complicated way of solving the rate equations, confirms that the single-line and gain-equal-loss assumptions are adequate approximations for chemical lasers at rotational equilibrium.

4.4.3 Rotational Nonequilibrium

The appearance of pure rotational transitions, R branch lines, and simultaneous lasing on several transitions indicates that the rotational equilibrium assumption is not invariably valid. Rotational nonequilibrium effects are expected, and observed, when the buffer gas pressure does not suffice to ensure that rotational relaxation will be faster than the stimulated emission and pumping processes. Since, as described in Sects.2.4.2 and 2.5.2, the probability of rotational energy transfer is a rapidly decreasing function of J, the nonequilibrium effects will be more pronounced in the region of high J values (see Fig.4.4).

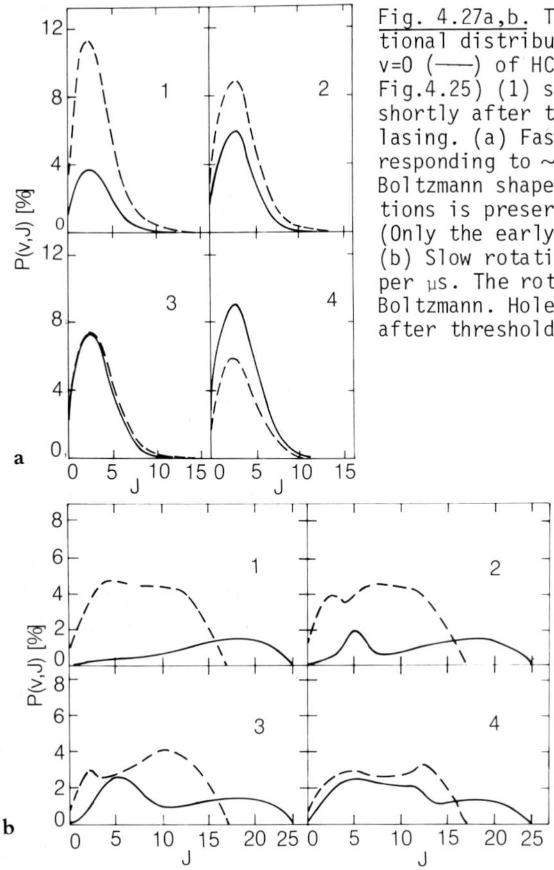

Fig. 4.27a,b. Temporal evolution of the rotational distribution of levels v=1 (---) and v=0 (——) of HCl in the Cl + HBr laser (see Fig.4.25) (1) shortly before threshold, (2) shortly after threshold, (3) and (4) during lasing. (a) Fast rotational relaxation, corresponding to ~2500 collisions per μs. The Boltzmann shape of the rotational distributions is preserved throughout the laser pulse. (Only the early lasing stages are shown.) (b) Slow rotational relaxation, ~50 collisions per μs. The rotational distributions are non-Boltzmann. Hole burning effects are observed after threshold (adapted from [4.52])

Following the detailed description of lasing at complete rotational equilibrium (Sects.4.2 and 4.4.2), we can now easily predict the main features of the lasing mechanism in the limit of slow rotational relaxation. In this limit there is nearly no coupling between rotational levels of the same vibrational manifold. Hence, the various transitions reach threshold, lase, and decay almost independently of each other. Using a pictorial terminology one may classify this behavior as "individual" or "inhomogeneous" as opposed to the "cooperative" or "homogeneous" mechanism which characterizes the lasing at rotational equilibrium. The differences between the two limiting cases are demonstrated in Figs.4.27 and 4.28. These figures display the solutions of the laser rate equations (4.4.1,2) for the Cl + HBr → HCl + Br system (cf. Fig.4.25). As distinguished from the method of solution described in the previous section the results shown in Figs.4.27 and 4.28 have been obtained by solving the full set of detailed rate equations without relying on the rotational equilibrium,

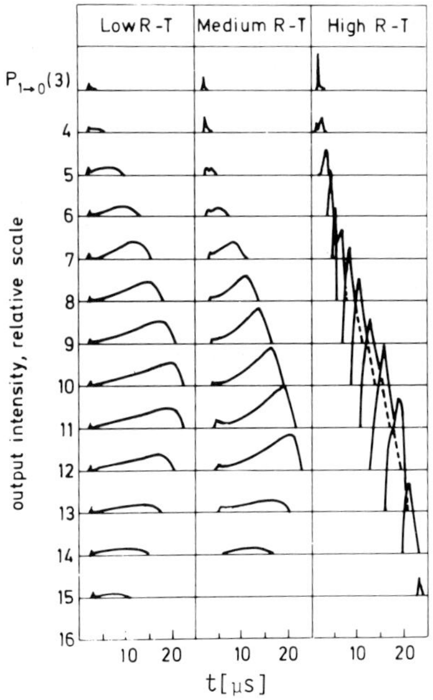

Fig. 4.28. Theoretical pulse shapes for the $P_{2\to1}(J)$ transitions of the Cl + HBr laser. The slow, medium, and high R-T rates correspond to 50, 200, and 2500 collisions per μs, respectively. Individual, multiline operation dominates the lasing at the low R-T limit. Single-line operation and J shifting are clearly seen for the high R-T cases (compare Figs.4.25 and 4.27). The efficiency, as reflected by the area under the pulse profiles increases with R-T rate (adapted from [4.52])

single-line, or gain-equal-loss assumptions[4]. The conditions of fast and slow rotational relaxation were simulated by varying the number of collisions an HCl molecule suffers per unit time (cf. Fig.1.9), which is equivalent to varying the buffer gas pressure. (The low R-T results correspond to initial concentration ratio of Cl_2: HBr=1:3.5 with $[Cl_2]_0+[HBr]_0=2.6\times10^{17}$ molec/cm^3 or, equivalently, about 8 Torr at 300 K which amounts to ~50 collisions in 1 μs. The medium and high R-T results are for 200 and 2500 collisions per μs, corresponding to the addition of ~25 and ~400 Torr Ar.) Figure 4.27 shows that at the low pressure limit the rotational relaxation cannot prevent the hole burning during the lasing. In the high pressure limit the Boltzmann profile is preserved throughout the laser pulse. Figure 4.28 clearly illustrates the extremes of multiline and single-line operation (see also Fig.4.3). The larger areas under the pulse profiles in the high R-T case indicate higher chemical efficiency. A simple qualitative interpretation to this trend can be given as follows: When rotational equilibrium prevails lasing occurs on partially inverted populations and terminates, as argued in Sect.4.2 (see also Sect.2.5.8), when $N_v/N_{v-1}<1$. On the other hand, when rotational coupling

4 A sample of kinetic models which consider the effects of rotational nonequilibrium on the operation of chemical lasers is listed in [4.53].

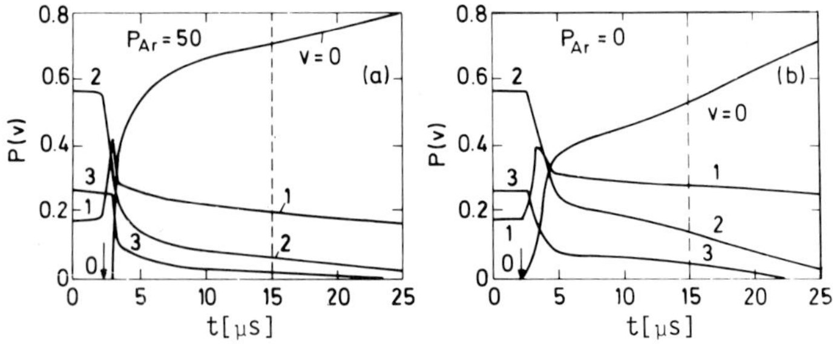

Fig. 4.29a,b. Time development of the vibrational distribution function of the HF molecules in the CF_3I:H_2:Ar laser system. The arrows indicate the threshold and the broken lines the termination of the laser pulse. The figures show the computed results (adapted from [4.20]) for the high (P_{Ar}=50 Torr) and low (P_{Ar}=0 Torr) pressure cases corresponding to the middle and right output patterns of Fig.4.3. In both cases most of the laser energy is extracted from partially inverted populations but the final vibrational population ratios N_{v-1}/N_v and consequently the efficiency are much higher in the case of high buffer gas pressure (fast rotational relaxation)

is weak, every transition ceases to lase when $N_{v,J}/N_{v-1,J+1} = g_J/g_{J+1}$. Since, except for the very low (and poorly pumped) J's, the degeneracy ratio $g_J/g_{J+1} \simeq 1$, it follows that at the end of the pulse $N_{v,J}/N_{v-1,J+1} \simeq N_v/N_{v-1} \simeq 1$. (In practice the limit of no rotational relaxation can never be achieved since even in the absence of a buffer gas some relaxation is caused by the reactive species.) A quantitative display of the time evolution of the vibrational populations in the $F+H_2$ system which contrasts their behavior in the limits of high and low buffer gas pressures is shown in Fig.4.29.

Denoting by $P_i(v)$ and $P_f(v)$ the initial and final vibrational distributions, an upper bound (corresponding to the neglect of vibrational relaxation effects) to the chemical efficiency is given by

$$\bar{\eta} = \sum_v [P_i(v)-P_f(v)]E_v/E \quad , \tag{4.4.26}$$

where E is the exothermicity of the pumping reaction. The larger value of $\bar{\eta}$ at rotational equilibrium is a consequence of the lower $P_f(v)/P_f(v-1)$ ratios.

4.4.4 cw Chemical Lasers

The same chemical and radiative processes which govern the kinetic behavior of a pulsed chemical laser occur, of course, also in the corresponding cw system. In general terms one may say that the main difference between the two laser types is that pulsed lasers involve temporal variations in the gain me-

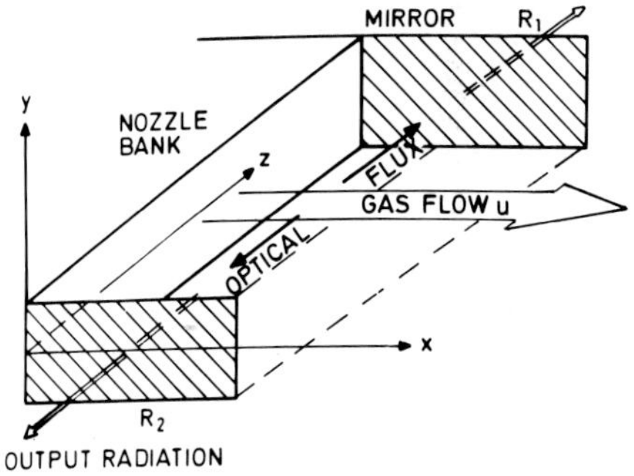

Fig. 4.30. Scheme of a premixed cw laser. The reagents flow along the x direction. z is the optical axis. The reaction starts to take place at $x=0$ (adapted from [4.47]) (see also Fig.4.20)

dium whereas in cw systems the variations are spatial. The simplest cw lasers are those where the reagents are uniformly mixed prior to their entrance into the optical cavity. The rate equations describing such a premixed laser are completely analogous to those of the corresponding pulsed system and can be obtained by a simple variable transformation. For this reason we shall open this section with a brief outline of premixed lasers. Next, we shall describe the extension to a simple mixing model and conclude with a brief discussion of rotational nonequilibrium in cw lasers.

Premixed cw Model

A typical cw chemical laser and a scheme illustrating the premixed model are shown in Fig.4.30. The reactants, F and H_2, and the He diluent flow across the optical cavity with constant velocity u along the x axis. Assume that the F atoms which were produced (e.g., by electrical discharge of SF_6) in the $x<0$ region first meet and react with the H_2 molecules in the $x=0$ plane which we take as the left boundary of the optical cavity. Thus, at $x=0$ $[F]=[F]_0$, $[H_2]=[H_2]_0$ and the concentration of the laser active molecules produced by the pumping reaction $F+H_2 \rightarrow HF+H$ is $[HF]=[HF]_0=0$. The growth of [HF] along the x axis is given by

$$\frac{d[HF]}{dx} = \frac{1}{u}\frac{d[HF]}{dt} = \frac{1}{u}(\vec{k}[F][H_2]-\overleftarrow{k}[H][HF]) \quad , \tag{4.4.27}$$

where $\vec{k}$ and $\overleftarrow{k}$ are, respectively, the forward and reverse rate constants of the pumping reaction. [Recall that the contribution of the reverse reaction to (4.4.27) is usually negligible.] Similarly the equations for $N_{v,J}$, T and the nonlasing species are obtained from the corresponding equations in Sect. 4.4.1 by replacing t with x/u. As we are dealing with cw lasers the level population equations,

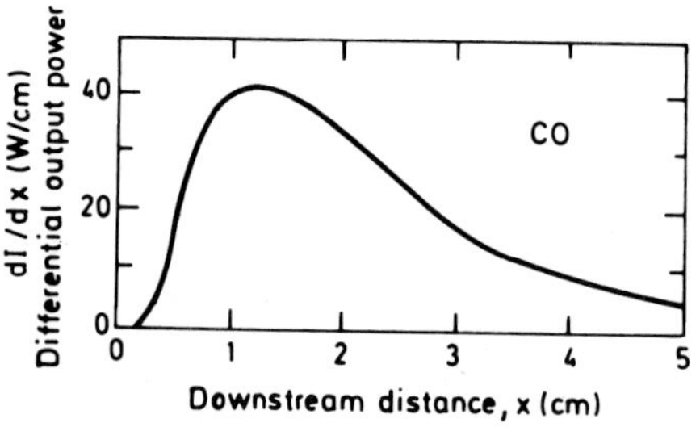

Fig. 4.31. Downstream profile of the output intensity in a CO chain reaction chemical laser as obtained from computer modelling [4.54]. The main reactions in this system are $O + CS_2 \rightarrow CS + O$, $O + CS \rightarrow CO(v) + S$, and $S + O_2 \rightarrow SO + O$ (Sect.4.3.3). The results of the modelling are in good agreement with experimental findings [4.55]. (It should be mentioned that premixing is not assumed in the modelling. Yet the mixing length $x = 1$ cm is relatively small)

$$\frac{dN_{v,J}}{dx} = \frac{1}{u}\frac{dN_{v,J}}{dt} = \frac{1}{u}(P_{v,J} - L_{v,J} - R_{v,J}) \tag{4.4.28}$$

should be solved, subject to gain-equal-loss conditions, for all the lasing transitions. Assuming that only P branch transitions can lase then $R_{v,J} = \chi_{v,J}^{v-1,J+1} - \chi_{v+1,J-1}^{v,J} \equiv \chi_{v,J} - \chi_{v+1,J-1}$ and the gain-equal-loss condition implies that $\chi_{v,J}$ is given by (4.4.23), or more explicitly (4.4.25) with $t = x/u$.

In the case of rotational equilibrium, the multitude of vibrotational equations for $N_{v,J}$ can be replaced by just a few vibrational equations, in the fashion described in Sect.4.4.2. Namely, since $R_{v,J}$ involves only highest gain transitions there is only one threshold constraint $\Delta N_{v,J} = 1/c\sigma_{v,J}\tau_c$ per vibrational band and, in addition, using $N_{v,J}(x) = N_v(x)p(J|v;T)$, (4.4.28) can be summed over all J's to yield vibrational population equations analogous to (4.4.24).

Finally, since the transverse flow coordinate x in the premixed cw model is completely equivalent to the time t in the pulsed model there is a one-to-one correspondence between the t profile of the pulsed laser output and the x profile of the (time-independent) cw output. That is, the laser intensity which starts to grow from some $x_t = ut_t$ reaches a maximum and finally decays to zero at some large downstream distance $x_q = ut_q$. (Note that to "project" a whole pulse with duration of, say, 20 μs in a cw laser with stream velocity of $u = 3 \cdot 10^5$ cm/s the length of the mirrors in the x direction should be at least 6 cm.) To illustrate the correspondence between x and t, Fig.4.31 shows the x profile of the output intensity of a nearly premixed CO chemical laser.

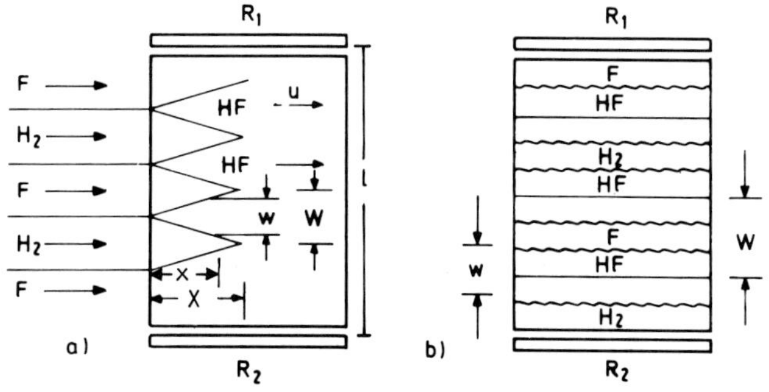

Fig. 4.32. (a) Simple mixing model consisting of few linearly growing mixing zones. (b) Time-dependent, 'unsteady', analogue of the steady flow model shown in (a) (adapted from [4.56])

A Simple Mixing Model

The premixed model is a reasonable approximation when the time scale for mixing is much shorter than that of the kinetic processes. This condition is met in several cases [4.47], e.g., when the parent gases, say SF_6 and H_2, are mixed at $x < 0$ and the reaction is initiated at $x = 0$. The premixed model is not appropriate for many practical systems where, for example, the reactant gases are injected into the cavity through separate streams. In this case the mixing is the rate controlling process. To describe the major effects of the finite mixing rate we shall briefly review here a highly simplified model where, as opposed to the premixed model, the rate of the pumping process is assumed instantaneous relative to the mixing. (This assumption, which emphasizes the role of mixing, is not essential and the model can be easily extended; see below.)

A scheme of the simple mixing model, originally suggested for the $F + H_2$ system, is shown in Fig.4.32 [4.56,57]. The optical cavity is a Fabry-Perot resonator with mirror reflectivities R_1 and R_2. Mixing starts at $x = 0$ in K mixing zones. The width of the zone w is assumed to grow linearly with x, as appropriate for turbulent mixing. (For laminar, diffusive, mixing $w \propto x^{\frac{1}{2}}$ [4.47].) u is the flow velocity and θ is the angle defining the spread of the mixing zone. At $x = X = (\ell/2K)\tan\theta$ all the reactants are consumed and the width becomes constant $w = W = \ell/K$. The model is simplified by assuming that the redistribution of the reactants over the mixing zone and the pumping reaction are instantaneous. Thus, in the special (but common) case of H_2 rich systems, $[F]_0 < [H_2]_0/2$, the species concentrations in the mixing regions are $[F] = 0$, $[HF] = [F]_0/2$ and $[H_2] = ([H_2]_0 - [HF])/2$.

Similar to the premixed model the time-independent (or steady-state) mixing model can be equivalently described by a spatially uniform time-dependent (or 'unsteady') model as illustrated in Fig.4.32b. Again, correspondence between the two schemes is achieved via $u = xt$ so that $w = 2x\ \tan\theta = 2x(v/u) = 2vt$ where $v = u\ \tan\theta$ is the lateral velocity characterizing the expansion of the reaction zone.

The rate equations of the present model are obtained by some straightforward modifications of the rate equations describing the premixed and pulsed models. For the sake of concreteness we assume the existence of rotational equilibrium. (In the original formulation of the model [4.56] it was further assumed, based on an earlier premixed model [4.58], that lasing occurs on a single J line common to all bands. This assumption leads to an approximate closed form solution but is, of course, not essential.) The first modification resulting from the varying width of the mixing zone concerns the gain-equal-loss condition. Since the length of the active medium at the axial location x is Kw, we have

$$\alpha_{v,J} = c\sigma_{v,J}\Delta N_{v,J} = 1/\tau_c = -(c/2Kw)\ln(R_1R_2) \quad . \qquad (4.4.29)$$

This relation follows from using the threshold condition (the equality in) (4.2.1), and the explicit expression $1/\tau_c = (-c/2\ell)\ln(R_1R_2)$ [cf. (3.2.8)] with the length of the active laser medium ℓ replaced by Kw. (This expression for τ_c assumes no dissipative losses, $\chi = 0$ in (3.2.8). τ_c is the effective lifetime of photons in the transverse location x. Using $w = 2x(v/u) = 2vt$, (4.4.29) can be rewritten as an x (or t) dependent constraint on the populations of levels connected by lasing transitions, i.e.,

$$\alpha_{v,J} = gu/2vx = g/4vt \quad , \qquad (4.4.30)$$

where $g = -(c/2K)\ln(R_1R_2)$.

The photon densities in the mixing zone $\phi_{v,J}(x)$ and the corresponding output intensities $ch\nu_{v,J}\phi_{v,J}(x)/\tau_c$ can be calculated from (4.4.25) with modified expressions for P_v and $dN_v/dt = udN_v/dx$: The pumping term P_v should be replaced by $\bar{P}_v$, the net rate of production of HF(v) molecules per unit volume

$$\bar{P}_v = P_i(v)[F]_0\frac{1}{2w}\frac{dw}{dt} - N_v\frac{1}{w}\frac{dw}{dt} = \left\{P_i(v)[HF]-N_v\right\}/t \quad , \qquad (4.4.31)$$

where $P_i(v) = k(\to v;T)/\sum k(\to v;T)$ is the product vibrational distribution in the pumping reaction. The first term in (4.4.31) represents the conversion of re-

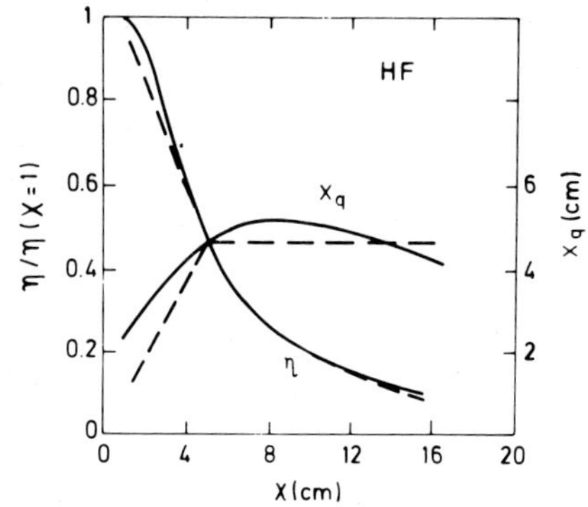

Fig. 4.33. Laser efficiency η and lasing cutoff distance x_q as function of mixing length X. $X \propto \psi$ (see text). The dashed and solid curves represent the results of the simplified [4.56] and a detailed [4.47] mixing model

actants to products due to the expansion of the mixing zone (see Fig.4.32). The form of this term is specific to the assumptions of the model, i.e., the instantaneous redistribution of the reactants in the mixing zone (hence the factor $[F]_0/2w$), and the instantaneous consumption of all the fluorine atoms (for $[F]_0 < [H_2]_0/2$) by the pumping reaction. The second term accounts for the continuous redistribution of N_v over the expanding region. Note the identity of the conservation relations $[HF] = [F]_0/2$ and $\sum \bar{P}_v = 0$. Note also that (4.4.31) applies only to $x \leqq X$ ($w \leqq W$); for $x > X$ (where $w = W$ is constant) $\bar{P}_v \equiv 0$. Hence the cutoff in lasing must occur at $x = x_q \leqq X$.

Finally, let us qualitatively summarize some of the consequences of the finite mixing rate. Approximate analysis of the simple mixing model shows that the important parameter in determining the chemical efficiency η and the cutoff distance x_q is the ratio ψ between the rates of HF-HF deactivation and chemical production of excited molecules [4.56]. These rates are proportional to $k_d[HF] = k_d[F]_0/2$ (k_d is an average vibrational deactivation rate constant) and $u \tan\theta/W = u/2X$, respectively; $\psi \equiv k_d[F]_0 W/u \tan\theta = 2k_d[F]_0 X/u$. The model predicts that when ψ is below some critical value ψ_c, $x_q \simeq X$; otherwise x_q decreases nearly linearly with ψ. The efficiency η of converting the chemical energy stored in the primary reactant streams to laser radiation is predicted to decrease linearly in the region $\psi < \psi_c$ and hyperbolically in the region $\psi > \psi_c$. This behavior of η and x_q is depicted in Fig.4.33 as a function of X; for fixed $[F]_0$ and u, X is proportional to ψ. Also shown are the results of a more accurate mixing model which, among others, allows for J shifting and employs realistic pumping rates. Note that, as expected, the maximum of η occurs in the premixed limit $X \to 0$. The model also predicts that for a constant $[F]_0:[H_2]_0$ ratio and a given X the efficiency decreases with the total pressure, in contrast to the premixed case where η is nearly pressure independent.

The explanation of this trend is that while for both the premixed and mixing controlled cases the vibrational deactivation rate, being a bimolecular process, is proportional to the square of the total pressure, the chemical production rates in the two cases are of second and first order in the total pressure, respectively. Consequently, while in the premixed limit, the total laser intensity $I \propto \eta[F]_0$ increases nearly linearly with the pressure, in the mixing case it first increases and then (above the critical pressure determined by ψ_c) reaches a constant value.

Detailed Mixing Model, Rotational Nonequilibrium

The simplified mixing model described above can be improved and extended by employing a generalized functional form of w and finite (instead of instantaneous) reaction rates and by allowing for rotational nonequilibrium. Referring to the laser scheme of Fig.4.32, the rate equations appropriate to the generalized description are [4.59]

$$\frac{d[Y_s]}{dx} = \frac{[\dot{Y}_s]}{u} + \left([Y_s]^0 - [Y_s]\right)\frac{1}{w}\frac{dw}{dx} \qquad (4.4.32)$$

$$\frac{dN_{v,J}}{dx} = \frac{\dot{N}_{v,J}}{u} - N_{v,J}\frac{1}{w}\frac{dw}{dx} \quad , \qquad (4.4.33)$$

where $[Y_s]$ is the concentration (number density) of the nonlasing species in the mixing zone. $[Y_s]^0$ represents the average concentration of Y_s in the unmixed regions, e.g., for the configuration of Fig.4.32, $[F]^0 = [F]_0/2$ where $[F]_0$ is the actual concentration in the incoming fluorine flow. The first term in both equations reflects the rate of change of the populations due to kinetic processes in the mixing zone.

The two components of the second term in (4.4.32) represent the entrainment of molecules from the primary stream and the continuous expansion of the mixing volume. The effect of this expansion on the lasing species is accounted for by the second term in (4.4.33). Note that the reduction of the present model to the simplified model involves the assumption $d[F]/dx = 0$, $[F] = 0$ so that $\sum \dot{N}_{v,J} = [F]^0(u/w)dw/dx$. The radiative fluxes $I_{v,J}(x)$ are evaluated by solving (4.4.32) and (4.4.33) subject to gain-equal-loss conditions (for all the lasing transitions). As in pulsed lasers, multiline operation characterizes the lasing mechanism at rotational nonequilibrium. Figure 4.34 displays the spectral distribution of P branch transitions in the $v=2 \rightarrow 1$ band of an $F + D_2 \rightarrow DF + D$ laser at selected downstream locations. These are the results of computer modelling based on (4.4.32) and (4.4.33), using "exponential

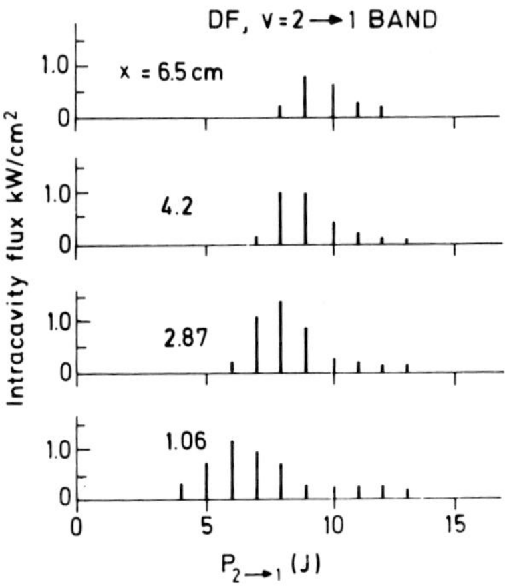

Fig. 4.34. Spectral distribution of P branch emission in the v=2→1 band of the $F+D_2 \rightarrow DF+D$ laser at selected downstream locations. The pressure in the mixed zone is 1 Torr, the mixing length is 12.5 cm, and the flow velocity 2.1×10^5 cm/s (adapted from [4.59])

gap" probabilities for rotational relaxation (cf. Sects.2.4.2 and 2.5.2). The positive J shift in the peak and the decreasing width of the spectral distributions reflect the increase in time $t=x/u$ of the Boltzmann character of the rotational state distributions.

4.5 Some Applications of Chemical Lasers

The concluding section of this chapter is intended to review some typical applications of chemical lasers in reaction kinetics studies and to describe the relevant experimental techniques. The word application is used here in a rather restrictive sense referring primarily to gain probing and output measurements reflecting the kinetic processes taking place within the laser cavity. Another, wider, range of chemical laser applications in chemical kinetics research involves their use for exciting reactants and/or probing products in state-to-state kinetic experiments, e.g., in combination with molecular beam experiments. With the exception of one representative example concerning vibrational energy transfer (Sect.4.5.4) these types of applications will be included in the more general discussion on the impact of lasers on chemical research in Chap.5.

4.5.1 Total Rate Constants, Kinetic Isotope Effects

Chemical laser emission signals can be used to determine the absolute values of total rate constants of pumping reactions. In some of the early studies of this kind [4.60] it was assumed that for certain parts of the laser pulse the emitted signal intensity I(t) is proportional to the pumping rate P(t),

$$I(t) \propto P(t) \quad . \tag{4.5.1}$$

This relation requires quasi-steady-state (gain-equal-loss) conditions of a laser oscillator (Sects.3.2.4 and 4.4.2) and negligible collisional deactivation. As mentioned in Sect.4.4.2, the quasi-steady-state operation is a valid assumption for most pulsed chemical lasers, except possibly during the short interval following the attainment of threshold which may exhibit relaxation oscillations. The second requirement can be (at least approximately) fulfilled by adjusting the initial reactant concentrations such that the pumping rate is much higher than the rate of vibrational relaxation.

Consider chemical lasers operating on vibrotational transitions, and assume rotational equilibrium so that only P branch transitions can lase. When the two requirements listed above are satisfied the intensity $I_{v,J}(t)$ of any lasing transition $v,J \rightarrow v-1,J+1$ is given by [cf. (3.2.34) and note $I \propto P$]

$$I_{v,J}(t) \propto h\nu_{v,J}P_{v,J}(t) = h\nu_{v,J}(d\Delta N_{v,J}/dt) \quad , \tag{4.5.2}$$

where $\nu_{v,J}$ and $\Delta N_{v,J}$ are the frequency and the population inversion of the transition. Note that $P_{v,J}$ denotes here the net pumping rate of the transition (not the level). For chemical lasers of the type $A+BC \rightarrow AB+C$, based on fast generation of A atoms, the pumping rate is (Sect.4.4.1)

$$P_{v,J}(t) = \vec{k}[P(v,J)-(g_J/g_{J+1})P(v-1,J+1)](a-x)(b-x) \quad , \tag{4.5.3}$$

where we have used $\vec{k}P(v,J) = k(\rightarrow v,J)$ (Sect.2.1), $a = [A]_0$, $b = [BC]_0$, $[A] = a-x$, $[BC] = b-x$, and $x = [AB] = [C]$. $\vec{k}$ is the total rate constant of the pumping reaction and $P(v,J) = P(v)p(J|v;T)$ is the effective (rotationally relaxed, vibrationally nonrelaxed) nascent product distribution [cf.(4.2.3)]. [As will soon become clear the term in square brackets in (4.5.3) and thus the rotational equilibrium assumption are not essential, as long as P(v,J) is stationary.] The time dependence of $P_{v,J}(t)$, hence of $I_{v,J}(t)$, is determined by the factor (a-x)(b-x). Choosing the initial concentrations such that $a \ll b$ this factor reduces to (a-x)b and the time dependence of x is governed by the

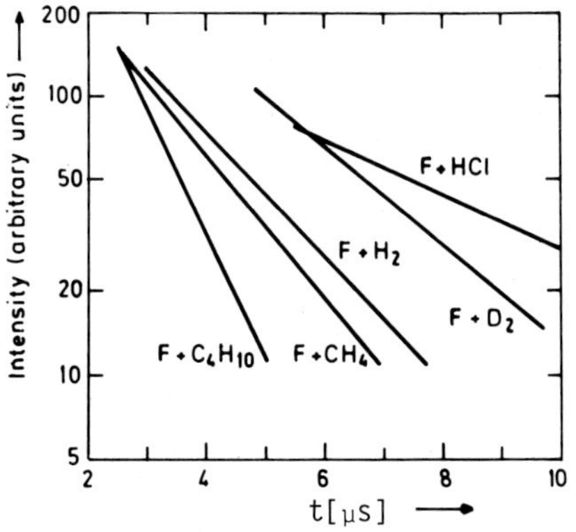

Fig. 4.35. Decay of total emission HF signals in the $F+H_2$, D_2, CH_4, $n\text{-}C_4H_{10}$, HCl lasers (adapted from [4.60])

Table 4.6. Total reaction rate constants from chemical laser experiments

Reaction	$F + H_2$	$F + D_2$	$F + CH_4$	$F + n\text{-}C_4H_{10}$	$F + HCl$
$\vec{k}[10^{13}\ cm^3/mole\cdot s]$	3.8	2.9	4.3	7.6	1.5

simple, pseudo-first-order, kinetic equation $dx/dt = \vec{k}b(a-x)$, so that $(a-x) = \exp(-\vec{k}bt)$. Thus,

$$\begin{aligned} I_{v,J}(t) &\propto h\nu_{v,J}P_{v,J}(t) \\ &= \vec{k}bh\nu_{v,J}[P(v,J)-(g_J/g_{J+1})P(v-1,J+1)]\exp(-\vec{k}bt) \\ &= k'_{v,J}\exp(-\vec{k}bt) \quad . \end{aligned} \qquad (4.5.4)$$

This equation shows that $\vec{k}b$ is simply the slope of $\ln I_{v,J}(t)$ vs t. The total signal intensity $I(t) = \sum I_{v,J}(t)$ has exactly the same time dependence, namely

$$\ln I(t) = \ln I(0) - \vec{k}bt \quad . \qquad (4.5.5)$$

A plot of I(t) vs t for a set of A + BC reactions ($A=F, BC=H_2, D_2, HCl, CH_4, n\text{-}C_4H_{10}$) is shown in Fig.4.35. In all cases $a = [F]_0 \simeq 10^{-2}$ Torr was much lower than $b = [AB]_0 = 0.25$ Torr ($1.3\cdot10^{-8}$ mole/cm^3 at 300 K) and $300\ K \leq T \leq 350\ K$. The rate constants obtained from the slopes of the curves are listed in Table 4.6.

The results of Table 4.6 are based on the assumption that the F atoms production terminates prior to the laser pulse. [This allows one to use $[F] = [F]_0 - x = a - x$ in (4.5.3).] In practice this assumption is not fully justified,

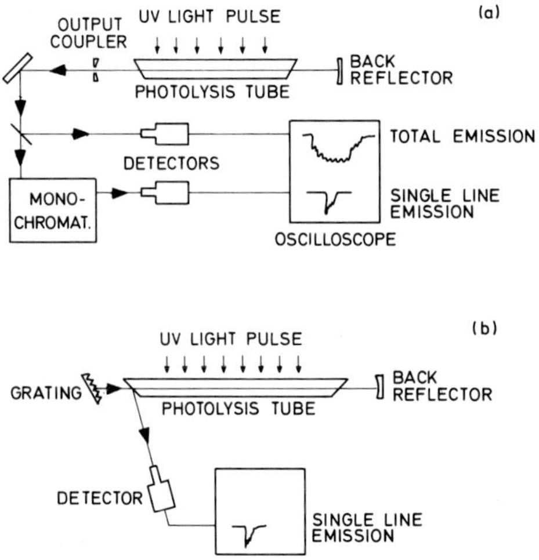

Fig. 4.36. Intracavity mirror (a) and grating tuned (b) chemical laser apparatus. (a) Radiation (→) is coupled out from the cavity through a hole in one of the mirrors. This design which allows the observation of low gain transitions was used to measure the output pattern of the "free running" $F+H_2$ laser shown in Fig.1.6. (b) The grating is used to selectively spoil the gain on all but the desired transition. In this way radiative coupling effects (cascading) are avoided (adapted from [4.19])

leading to an error in the absolute values of the rate constants derived. (Indeed, the data of Table 4.6 deviate up to 50% from those obtained by other methods.) This problem can be circumvented when measuring *relative* rate constants, as described in the next example.

The threshold times of individual vibrotational transitions provide a very sensitive probe of $dN_{v,J}/dt$ and thus of $k(\rightarrow v,J) = \vec{k}P(v,J) = \vec{k}P(v)p(J|v;T)$, when rotational equilibration is completed before threshold. Since threshold is usually reached at the early stages of the pumping process the effects of vibrational deactivation can safely be ignored. These notions have been utilized to determine nascent vibrational populations as well as total reaction rates of chemical laser pumping reactions [4.5,19]. Threshold time measurements of a selected transition involve the use of a grating which prevents the buildup of large photon densities on the other transitions (Fig.4.36). The determination of P(v) will be discussed in the next section. Here we proceed to describe another interesting application of the method, the determination of inter- and intramolecular kinetic isotope effects in the $F+H_2$, HD, D_2 reactions [4.19].

The gain coefficient of, say, the $v,J \to v',J'$ line of HF in the $F + H_2$ laser can be expressed as [cf. (4.2.1)]

$$\alpha_{v,J}^{v',J'} = [HF]c\sigma_{v,J}^{v',J'}[P(v,J)-(g_J/g_{J'})P(v',J')] \equiv [HF]\bar{\alpha}_{v,J}^{v',J'} \quad , \qquad (4.5.6)$$

where we have used $N_{v,J} = NP(v,J)$, $N = [HF]$. $\bar{\alpha}$ is the "gain coefficient per molecule". At buffer gas pressures which ensure rotational equilibrium but do not affect the vibrational populations (at least up to threshold), $P(v,J) = P(v)p(J|v;T)$, where $P(v)$ is the nascent vibrational distribution; the time dependence of $\alpha_{v,J}^{v',J'}$ is governed only by $[HF(t)]$. The experimental conditions (flash intensity and initial concentration of the F-containing compound, e.g., CF_3I) can be adjusted such that $[F(t)] \ll [H_2]_0$ for any t. In this case $[HF(t)]$ is determined by the pseudo-first-order kinetic equation $d[HF]/dt = k_{F+H_2}[H_2]_0[F(t)]$, i.e., $[HF(t)] = k_{F+H_2}[H_2]_0 \int [F(t)]dt$. Similarly, in the $F + HD \to HF + D$ laser, $[HF(t)] = k_{F+HD}^{HF}[HF]_0 \int F(t)dt$.

Threshold on the $v,J \to v',J'$ line is reached at the moment t_t that

$$\alpha_{v,J}^{v',J'}(t_t) = [HF(t_t)]\bar{\alpha}_{v,J}^{v',J'} = 1/\tau_c \quad , \qquad (4.5.7)$$

where it should be remembered that t_t varies from one line to another. Using the same cavity and initiation parameters, τ_c and $F(t)$ are equal for both the $F + H_2 \to HF + H$ and the $F + HD \to HF + D$ lasers. By adjusting $[H_2]_0$ and $[HD]_0$ the threshold time of a given line, common to both systems, can be made the same, so that $\alpha_{v,J}^{v',J'}(t_t)_{F+H_2} = \alpha_{v,J}^{v',J'}(t_t)_{F+HD}$. From (4.5.7) and the explicit expressions for $[HF(t)]$ it now follows

$$k_{F+HD}^{HF}/k_{F+H_2} = ([H_2]_0/[HD]_0)(\bar{\alpha}_{F+H_2}/\bar{\alpha}_{F+HD}^{HF}) \quad , \qquad (4.5.8)$$

where $\bar{\alpha}_{F+H_2}$ is the reduced gain coefficient of the transition observed in the $F + H_2$ system, etc. The $\bar{\alpha}$ values can be evaluated [cf. (4.5.6)] from the known $P(v)$ (next section) and $\sigma_{v,J}^{v',J'}$ for the relevant systems. Using (4.5.8) it was found that $k_{F+HD}^{HF}/k_{F+H_2} \simeq 0.38$ and similarly that $k_{F+HD}^{DF}/k_{F+D_2} \simeq 0.50$ [4.19]. This result and the intermolecular kinetic isotope ratio $k_{F+H_2}/k_{F+D_2} \simeq 1.9$ [4.61-63] lead to an intramolecular isotope ratio of $k_{F+HD}^{HF}/k_{F+HD}^{DF} \simeq 1.4$ at room temperature. This value agrees well with a bulk kinetic study using mass spectrometric analysis of reaction products in a flow system which yielded 1.45 for the same ratio.

A qualitative explanation to the intramolecular isotope effects is provided by a geometric consideration: The center of mass of HD is closer to the D atom. The more rapidly rotating H atom thus shields the heavier D atom and the approaching F atom has a larger probability to react with H than with D (see also Sect.5.2.1).

4.5.2 Vibrational Population Ratios from Threshold Time Measurements

In the previous section it was shown how threshold time measurements can yield, using (4.5.6,7) information on total reaction rate constants. Yet this is just a by-product of this type of experiment. The sensitive dependence of the threshold times upon P(v,J) has been extensively exploited to derive the product vibrational state distributions resulting from various exoergic chemical reactions [4.5] and thus complemented other experimental techniques aiming at the same goal, in particular infrared chemiluminescence measurements.

The preliminary analyses of Figs.4.3 and 4.8 provide the simplest demonstration of the relation between threshold times and nascent vibrational population ratios $N_v/N_{v-1} = P(v)/P(v-1)$. The appearance of $P_{2\rightarrow 1}(3)$ as the earliest and thus the highest gain transition in the free-running $F + H_2$ laser implies that $N_2/N_1 > 1$ (Fig.4.8). [More precisely, $N_2/N_1 > 1.25$ since at $N_2/N_1 = 1.25$ $P_{2\rightarrow 1}(4)$ becomes the highest gain transition.] Clearly, in view of the fact that $N_2/N_1 \simeq 3$ probing only the highest gain transition provides only poor kinetic information. Quantitative kinetic information on vibrational energy partitioning has been derived by various types of chemical laser experiments based on comparing the threshold times, or equivalently, the gain coefficients, of different vibrotational transitions. From the variety of techniques (for a review, see [4.5]), we shall briefly describe the "zero-gain temperature" and the "grating selection" methods.

The basic experimental arrangement employed in the zero-gain temperature method is shown in Fig.4.37. Two laser tubes, a "driver" and a "slave", are mounted within the same optical cavity and a grating is used to select individual lines. The driver which contains a (relatively) high gain reaction mixture drives the slave which contains a (relatively) low gain mixture to oscillate. The presence of the slave shortens the time to threshold of the driver by contributing to the amplification of the photon density. By varying the slave's temperature, thereby its gain coefficients, one can control the time to threshold of the driver (see Fig.4.38). The explicit temperature dependence of the gain coefficient, for rotationally relaxed, vibrationally

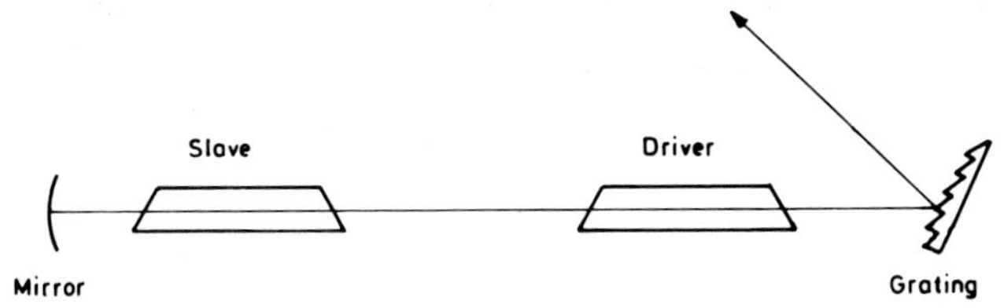

Fig. 4.37. Scheme of the tandem chemical laser setup for a "zero-gain temperature" experiment [4.64]. A driver-slave combination is used together with a grating which ensures oscillations on a single line. The temperature of the slave laser is varied until it becomes transparent to the driver's signal. In this zero-gain point the threshold time of the driver laser is unchanged by the presence of the slave

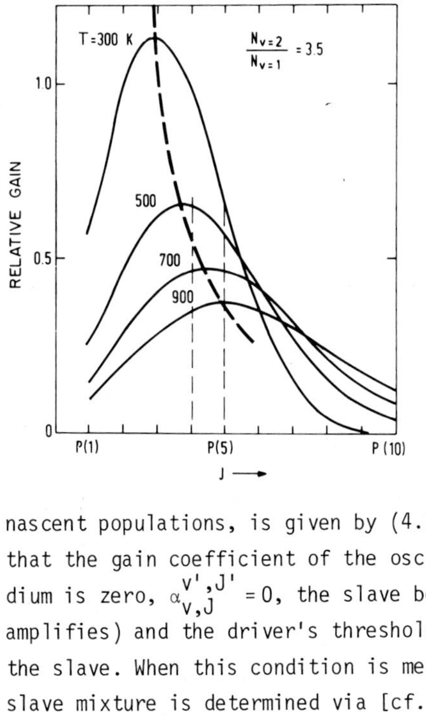

Fig. 4.38. Normalized gain efficients for P branch HF laser transitions as function of the translational-rotational temperature, computed for fixed vibrational inversion N_v/N_{v-1}. (The heavy dashed curve connects the highest gain transitions.) At 700 K, for instance, $P_{2\to1}(4)$ and $P_{2\to1}(5)$ have equal gains. Finding the temperature at which two different lines have equal gain, as reflected by their threshold, is the basis of the equal-gain temperature method, one of the earliest methods used to determine N_v/N_{v-1} ratios [4.64] (adapted from [4.4])

nascent populations, is given by (4.2.10). When the temperature T is such that the gain coefficient of the oscillating transition in the slave's medium is zero, $\alpha_{v,J}^{v',J'} = 0$, the slave becomes transparent (neither absorbs nor amplifies) and the driver's threshold time is unchanged by the presence of the slave. When this condition is met the vibrational population ratio in the slave mixture is determined via [cf. (4.2.10)]

$$
\begin{aligned}
N_v/N_{v-1} &= [p(J'|v-1;T_z) / p(J|v;T_z)](g_J/g_{J'}) \\
&\simeq (B_{v-1}/B_v) \exp\left\{-[B_{v-1}J'(J'+1)-B_vJ(J+1)] / kT_z\right\} \quad , \qquad (4.5.9)
\end{aligned}
$$

where T_z is the zero-gain temperature and $J' = J+1$ and $J-1$ for P and R branch transitions, respectively. The high accuracy of the zero-gain temperature method is reflected by the simplicity of (4.5.9), which involves only spectroscopic constants and the temperature (but does not depend on other parameters such as the line shape of the transition which enters the expression for α). Its accuracy is therefore only limited by the reproducibility and precision of the time to threshold measurement for the driver reaction. Such gain measurements which are all in principle an application of the gain equation (4.2.10) to chemically pumped lasers have been performed for numerous reactions and are comprehensively reviewed in [4.5]. They may be considered as one specific example of intracavity laser spectroscopy, a subject which is treated also in Chap.5.

The grating selection method is based on the fact that the higher the gain coefficient of a given line is the lower is its threshold time [see (4.5.6,7)]. Thus, a hierarchy of the gain coefficients and thus of $P(v,J)$ values [see (4.5.6)] can be constructed by ordering the sequence of relative threshold timings [4.5,19]. Using $P(v,J) = P(v)p(J|v;T)$ this sequence can be transformed into a large set of upper and lower bounds on the vibrational population ratios $P(v)/P(v-1)$. This method has been successfully applied, using the experimental arrangement of Fig.4.36b, to determine the nascent product vibrational populations in the $F + H_2(D_2) \rightarrow HF(DF) + H(D)$ and $F + HD \rightarrow HF(DF) + D(H)$ reactions for all the $v \geq 1$ levels. The v=0 populations were determined by extrapolating the linear surprisal plots obtained for these reactions [4.19]. (For other reactions studied, see [4.5].)

4.5.3 Gain Probing

Optical gain and loss (absorption) can be directly measured with the aid of an external probe beam which passes through the reaction mixture. Thus laser oscillator-amplifier arrangements can be effective in studying details of the pumping and collisional deactivation processes in chemical laser systems. As Fig.4.39 shows this technique uses the signal from one laser, the oscillator, as a diagnostic tool to study the kinetics of another laser having the same active medium, the amplifier. Such measurements can be performed with sufficient time resolution that the processes which populate and depopulate the energy levels involved in the laser operation (and sometimes even of those which are not active in the laser) can be identified and studied.

The ratio between the integrated output and input signals indicated in Fig.4.39 is given by [4.65]

$$I_\ell/I_0 = (h\nu/r\sigma I_0)\ln\left\{1+[\exp(r\sigma I_0/h\nu)-1]\exp(\sigma\ell\Delta N)\right\} \quad , \qquad (4.5.10)$$

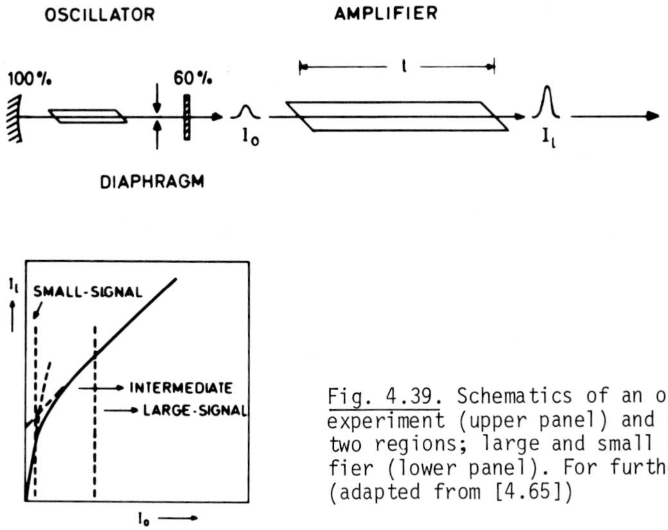

Fig. 4.39. Schematics of an oscillator-amplifier experiment (upper panel) and a graph showing the two regions; large and small signal of an amplifier (lower panel). For further details see text (adapted from [4.65])

where $r = 1 + g_f/g_i$ with g_f and g_i denoting the upper and lower level degeneracies and I is the (time-integrated) pulse intensity. The other symbols have the usual meaning. Figure 4.39 shows two distinct regions of operation: the small-signal region with $r\sigma I_0/h\nu \ll 1$, and the large-signal region with $r\sigma I_0/h\nu \gg 1$. In the first case (4.5.10) reduces to an exponential, Beer-Lambert, type expression (cf. Sect.3.2.1)

$$I_\ell/I_0 = \exp(\sigma\ell\Delta N) \qquad (4.5.11)$$

while in the latter case I_ℓ increases linearly with ℓ,

$$I_\ell/I_0 = 1 + (\ell\Delta N h\nu/rI_0) \quad . \qquad (4.5.12)$$

Thus by combining large and small signal gain measurements the cross section for stimulated emission σ can be measured. Such information is not only of value in calibrating the absolute ΔN values but also of spectroscopic interest.

Time-resolved gain spectroscopy, as this approach may be called, has been applied to the HCl [4.66] as well as to the HF [4.67] chemical lasers. The approach has been successfully used also to study the decay of excited iodine atoms in the photochemical iodine laser [4.65]. Figure 4.40 shows some of the results. After about 3 ms there is no further generation of I^* by the flash

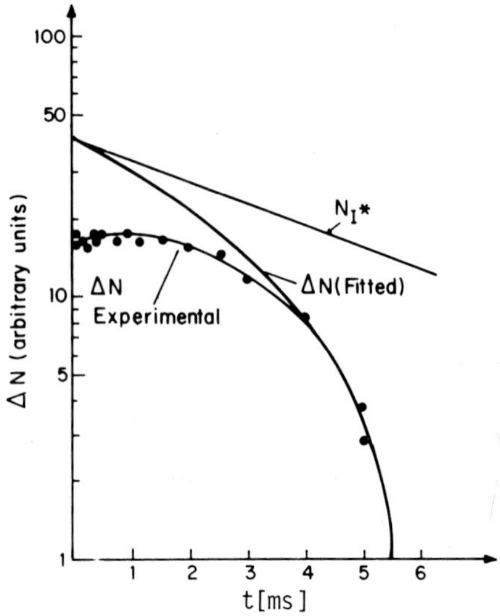

Fig. 4.40. Decay of the population inversion $N = N_{I^*} - (1/2)N_I$ and of the concentration of excited iodine atoms N_{I^*} after flash photolysis of 20 Torr of CF_3I; see Sect.4.3.1 (adapted from [4.65])

lamp and the decay of excited iodine atoms is governed by the various bimolecular deactivation processes (cf. Table 4.2). These can be represented by

$$I^* + M \xrightarrow{k_q} I + M \quad , \tag{4.5.13}$$

where k_q is an effective quenching rate constant which lumps together the contributions from the various collisional deactivation processes.

Assuming that at $t = 0$ all the iodine atoms are excited and that $N_M = [M] \gg N_{I^*}$ we find

$$\Delta N = N_{I^*} - N_I/2 = (N^0_{I^*}/2)[3\ \exp(-k_q N_M t)-1] \quad , \tag{4.5.14}$$

where $N^0_{I^*}$ is the initial I^* concentration.

The time τ at which ΔN has dropped to zero is $\tau = \ln 3/k_q N_M$ (Fig.4.40). This yields a rate coefficient $k_q = 2.55 \cdot 10^{-16}$ cm^3/molec s. The various contributions to k_q may be separated by controlled concentration changes of the various deactivating species.

Another demonstration for the usefulness of direct gain measurements in chemical laser research concerns the reaction $O + CS \rightarrow CO + S$ [4.68], the pumping reaction of the CO laser (Sect.4.3.3). In an arrangement similar to Fig.4.39 mixtures of CS_2, CO, and Ar were photolyzed in the amplifier section. As a probe a cw electrically discharge CO laser cooled in liquid nitrogen was employed. The results obtained in this study correspond to the open triangles in Fig.4.22.

4.5.4 Energy Transfer Measurements: An Example

Some uses of chemical lasers as excitation sources and diagnostic means in state selective kinetic experiments are discussed in Chap.5. The emphasis there is on reactive processes. In view of the fact that chemical lasers have also been extensively applied to derive kinetic information on inelastic non-reactive collisions we shall now describe one example of this kind.

The vibrational spacing of NO is about one-half of the vibrational spacing of HF. Thus an intense HF radiation can directly pump NO(v=0) molecules, through the v=0→2 overtone band, to the v=2 level. A pulsed HF laser emitting on the $P_{2\,v=1\rightarrow 0}(6)$ transition was used to pump the $P_{0\rightarrow 2}(17/2)$ line[5] in ground state ($^2\Pi_{1/2}$) NO [4.69]. Subsequently, a cw, electrically excited, CO laser probed the v=1→2 and v=2→3 transitions of both ground and excited ($^2\Pi_{3/2}$) NO states. Figure 4.41 shows the relevant NO level scheme. To improve the

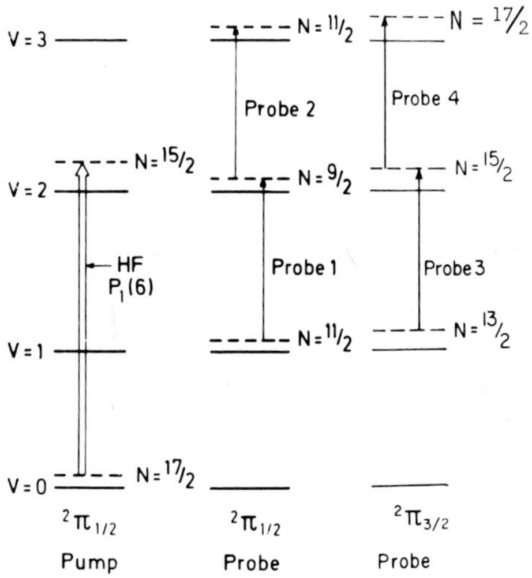

Fig. 4.41. Infrared-infrared double resonance study of spin exchange processes and vibrational relaxation in the NO molecule [4.69]

frequency match argon was added to broaden the NO absorption lines. The study yielded information on spin exchange processes of the type

$$NO(^2\Pi_{1/2}, v=1) + M \rightarrow NO(^2\Pi_{3/2}, v=1) + M \qquad (4.5.15)$$

5 $P_{0\rightarrow 2}(17/2)$ stands for the transition v=0,N=17/2→v=2, N=15/2. N is defined as follows: For diatomic molecules whose ground electronic state is not $^1\Sigma$ the symbol $J = |\underline{J}|$ is the absolute value of the vector sum $\underline{J} = \underline{N} + \underline{S}$. Here $\underline{N}$ is the sum of the rotational and the electronic angular momenta.

and similarly for the NO in the state v=2. These processes were shown to occur on time scales of 50 ns with 500 Torr Ar. In addition, by monitoring the decay of the NO(v=1→2) absorption a rate constant $k = 2.6\cdot10^{3}\ s^{-1}\ Torr^{-1}$ was obtained for the process

$$NO(v=1) + NO(v=0) \rightarrow 2NO(v=0) \quad , \qquad (4.5.16)$$

in good agreement with results of an infrared fluorescence study [4.69].

4.4.5 An Industrial Diagnostic Application

At the end of this section we shall point to one particular application of the HF chemical laser in the area of environmental pollution monitoring. The necessity for effective control of pollutants has become obvious in recent years and the laser has proved to be a powerful tool in this field with the particular advantage of long-distance measurements. The damaging influence of HF from brickworks and other plants on the environment has long been known. The concentrations of HF contained in the exhaust of an aluminum plant have been monitored using a compact cw HF laser of the type shown in Fig.4.20b [4.70]. With the laser emission spectra given in Fig.4.20c absorption cross sections for H_2O vapor, CO_2, and HF have been measured. These range around 10^{-21} cm^2 for 10 Torr H_2O, up to $5\cdot10^{-20}$ cm^2 for atmospheric CO_2 concentrations and up to 10^{-17} cm^2 for HF. The laser was used for field measurements above the electrolytic reduction cells in which alumina (Al_2O_3) dissolved in

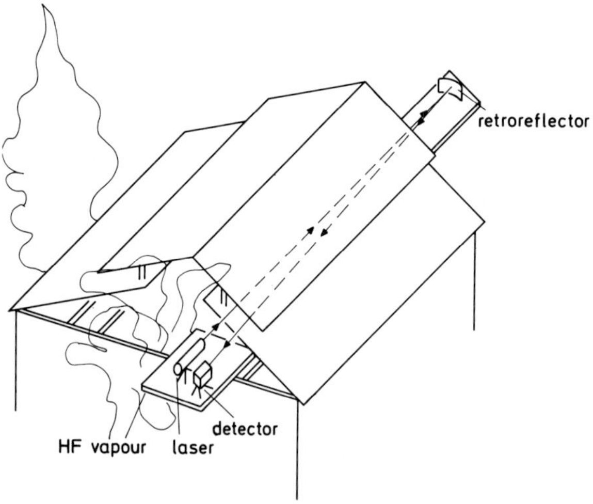

Fig. 4.42. Scheme of the HF-pollutant monitoring experiment (based on [4.70])

cryolite (Na_3AlF_6) is electrolyzed to yield aluminum metal which accumulates at the cathode. At the carbon anode oxidation takes place to form CO and CO_2. The loss of HF from the cells, due to evaporation of Na_3AlF_6 and water, increases with increasing temperature. The laser system was installed under the rooftop, close to the exhaust of a row of electrolysis cells. The location of the optical absorption path with respect to the electrolysis cells is shown in Fig.4.42. HF, CO_2, and H_2O were simultaneously monitored.

The time correlation between HF and CO_2 emission was established while the H_2O concentration was independent. The pollutant concentrations could be traced to certain manipulations at the electrolysis cells. It was demonstrated then that the chemical HF laser can be used to determine the HF pollution in the vicinity of industrial plants. This suggests usage on other pollutants as well, and to do so not only by absorption but also by scattering and laser-induced fluorescence measurements [4.71].

5. Laser Chemistry

This chapter examines the practice of laser chemistry, in the gas phase. It deals mostly with the use of lasers as means for the preparation of a selectively excited initial state and also documents their application as efficient diagnostic tools. Several case studies are discussed in some detail and attention is given to the principles of the design of the experimental arrangements. The novel type of observations made possible by the use of lasers and the improved spectral and temporal resolution are illustrated by specific applications to concrete problems. Section 5.1 presents an overview of the achievements and prospects of laser chemistry. Bimolecular reactions are discussed in Sect.5.2, starting with the most elementary atom-diatom exchange reactions $A + BC \rightarrow AB + C$, in which one of the reactants is excited by a laser. In a bimolecular reaction the relative initial and final velocity distributions of reactants and products are best monitored using the molecular beam method. Experiments combining molecular beams with laser excitation are thus most suitable for analysis and direct comparison with theory. This advantage is often offset by experimental difficulties, necessitating the use of bulk (non-beam) conditions, also discussed in Sect.5.2. Reactions of polyatomic molecules are taken up in Sect.5.3 (single-photon electronic excitation) and in Sect.5.4 (multiple-photon excitation). The emphasis in both sections is on unimolecular reactions, i.e., on processes where the energy deposited in the molecule suffices to cause a chemical change such as dissociation, isomerization, and ionization (cf. Sects.1.3 and 2.3.6). In the absence of collisions, lasers can be used to "heat up" a selected molecule in a mixture. The relatively extended discussion of multiphoton absorption reflects the widespread use of this unique laser-based method of excitation, which might perhaps eventually lead to practical mode-selective chemistry.

Section 5.5 is a summary of laser-related techniques used both in excitation and diagnostics. The latter is of the utmost importance in modern chemistry, as laser-based diagnostics have extended the range of detection almost down to the single atom (molecule) level. The chapter is concluded on a practical note suggesting some large-scale applications of lasers in chemical research and development. Throughout, the discussion is illustrative rather than exhaustive and much of the excellent work in the field is not mentioned explicitly. It is hoped, however, that the examples and references will stimulate the reader to consult the original literature for the many additional exciting and innovative uses of lasers in chemistry.

5.1 The Laser Evolution

To provide a systematic account of the past achievements, future potential, and present-day limitations of laser chemistry, we compare the possibilities offered by the laser with the requests made by the kineticists and note where a reasonable overlap exists. In laser chemistry the following four laser properties can be utilized (cf. Sect.3.2):

1) High energy content: At many frequencies, lasers are the only sources that provide sufficient energy flux to carry out chemical conversions. As a corollary, lasers open up new possibilities for the excitation of forbidden transitions (including multiphoton transitions) and also provide the large number of photons necessary for conversion of reagents to products on a preparative scale.
2) Excellent monochromaticity: This property allows selective absorption and controlled energy deposition.
3) Availability of short pulses: Very fast processes in the ns, ps, and even sub-ps range are now accessible for direct observation.
4) Good beam quality: This provides good directionality, easy maintenance of a collimated beam over a large distances, and tight focussing possibilities.

Figures 5.1 and 5.2 show schematically the excitation and probing problems associated with state-selected bimolecular reactions under single-collision conditions (Fig.5.1) and with unimolecular processes (Fig.5.2). In Fig.5.1 we are preferentially dealing with small molecules, where a limited number of states and degrees of freedom are involved in the reactions. In Fig.5.2 interest is focussed more on polyatomics with their associated features of internal energy coupling. By comparing now the list of laser features with the different types of laser molecule interaction types shown

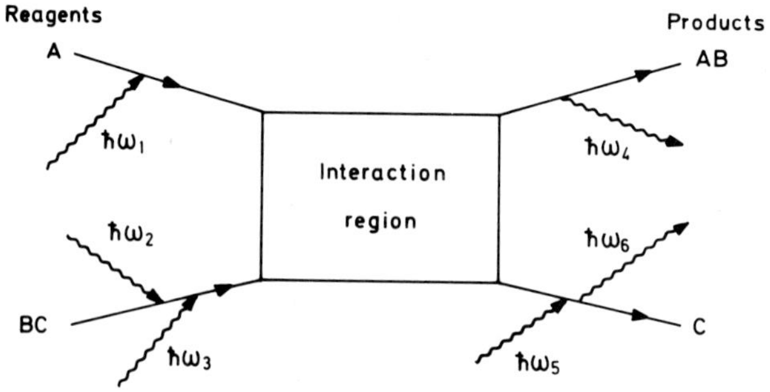

Fig.5.1. Types of laser interactions with bimolecular reaction systems. Cases shown refer to pumping by a single photon $\hbar\omega_1$ or by a combination of photons $\hbar\omega_2 + \hbar\omega_3$. Product state analysis can for instance be performed by observing $\hbar\omega_4$ in spontaneous or stimulated emission (chemical lasers) or by laser-induced fluorescence $\hbar\omega_6$ after pumping by $\hbar\omega_5$. Multiphoton excitation would correspond to $(\hbar\omega_2)_n$. Not all types of laser probing are indicated here (see (Sect.5.5)

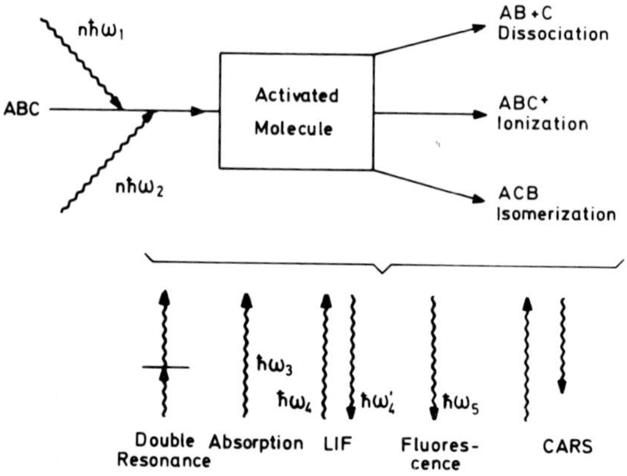

Fig.5.2. Laser interaction with a polyatomic molecule leading to unimolecular processes like dissociation, ionization, or isomerization. Excitation is achieved by single - or multiple-photon absorption using one laser, $\hbar\omega_1$, or a combination of lasers $\hbar\omega_1$, $\hbar\omega_2$. The excited molecule and the possible products can be probed by observing spontaneous or laser-induced fluorescence (LIF). Likewise absorption (internal or external to the probe laser cavity) or coherent anti-Stokes Raman scattering (among other techniques not mentioned) may be used

in Figs.5.1 and 5.2 one can conclude the following. The laser properties mentioned under 1) and 2) are generally desired for both reagent excitation and product probing. Of course energy-selective reagent excitation was done prior to the advent of lasers, for instance vibrational excitation by chemical activation (Sect.2.3.5) or by the use of conventional optical excitation to reach electronic states as well as for vibrational states (by such mechanisms as Franck-Condon pumping or internal conversion, Sect.2.4.9). However, none of these methods provides the same flexibility and freedom of choice of the desired molecular state as does the laser. Similar to reagent excitation, product internal energy probing is also possible without the aid of lasers, e.g., by observing spontaneous infrared or visible fluorescence and to some degree by absorption measurements (e.g., flash spectroscopy or rapid scan infrared spectrometry). However, since they have increased the ranges of sensitivity, precision, and wavelengths considerably, lasers are widely used for this application.

The properties mentioned under 3) and 4) have many uses (sometimes indirectly): High intensity and short pulses have kinetic consequences as competing processes (relaxation, heat conduction, diffusion to the wall) can be suppressed. High intensity also allows stepwise or direct multiphoton processes to come into play. Often high intensity is a side effect associated with short pulses and good spatial coherence (focussability). These features also provide for potential interesting applications beyond the range of conventional gas phase kinetics in high pressure reaction systems. Some discussion of these more practical aspects is given in Sect.5.6.

5.2 Bimolecular Reactions

This section deals with the experimental determination of the rates of the most elementary reactive collisions,

$$A + BC \xrightarrow{h\nu(\text{laser})} AB + C \qquad (5.2.1)$$

$$AB + CD \xrightarrow{h\nu(\text{laser})} AC + BD \qquad (5.2.2)$$

where AB and CD are diatomic or triatomic molecules.

Lasers can be used to fully select the internal state of the reactants and to probe the internal state of the products. The more ambitious current experiments (Sect.5.5.2) use lasers to determine also the velocity of the products. When the translational and internal energies are well specified, one can obtain the detailed rate constant $k(n,m \rightarrow n',m';E)$, corresponding to the process

$$A(n) + BC(m) \rightarrow AB(n') + C(m') \quad , \qquad (5.2.3)$$

where the internal states of all species and the total energy are well defined. Thus far, this degree of detail has not been realized. Experimental feasibility is largely dependent on the ratio of observed signal to noise that scales roughly as $N^{\frac{1}{2}}$, N being the number of counts of the detector. A simple way to increase N is to increase the pressure. This, however, may bring about other side effects such as A+A or BC+BC collisions. The alternative is to use powerful detection methods such as laser-induced fluorescence (LIF, Sect.5.5.2). Here the number of counts is increased by using the more efficient visible fluorescence. Another option is to reduce the resolution of the final states so that more events are counted together in one bin. In particular, one often gives up the resolution of the final velocity. It should also be noted that reagent state selection is sometimes achieved by rejection of the unwanted components and this is particularly true for the selection of the initial velocity (which is possible, using molecular beams). By not attempting to select by elimination, one is again increasing the number of collisions and hence the number of final counts. The net result is that while what we want to measure is $k(m,n\to m',n';E)$, what we typically can measure is $\langle k(m,n\to m',n';E)\rangle$ where (cf. Sect.2.4.1) the brackets denote an average over those states of the reagents and a sum over those states of the products which are not resolved in the experiment. In particular, using thermal reagents without velocity selection, one can, at best, determine the average of $k(m,n\to m',n';E)$ over a thermal translational distribution

$$k(m,n\to m',n';T) = \langle k(m,n\to m',n';E)\rangle \quad . \tag{5.2.4}$$

For further discussion and, in particular, for the effect of translational energy on reaction rates, see Sect.2.2.8.

We now examine the current progress in measuring detailed rate constants

$$k(\Gamma\to\Gamma';T) = \sum_{\{n,m\}\varepsilon\Gamma} p(n,m|\Gamma) \sum_{\{n'm'\}\varepsilon\Gamma} k(n,m\to n',m';T) \quad , \tag{5.2.5}$$

where $p(n,m|\Gamma)$ is the fraction of reagents in the group Γ which are in the internal state n,m [see, e.g., (2.1.14)]. We list below, in increasing order of detail, some of the rate constants of the type (5.2.5) that have been measured:

a) The easiest to obtain is the *total* rate constant (reactive and nonreactive processes) *out* of the prepared state. An example is $k(v,J\to\ ;T)$, the rate constant for depletion of a given v,J state of a diatomic molecule (Sect.2.1). In practice, even that modest goal is not trivial to attain as rotational relaxation is very rapid, so that often only k(v→ ;T) can be obtained.

b) On the next level, we select the reaction of interest, say A + BC(v,J) ⟶AB + C, and measure its rate by following the time evolution of one of the products. As shown below (Fig.5.6), one often measures the total decay rate of the initially prepared state simultaneously with the formation rate of the product, allowing direct assessment of the contribution of the particular reaction channel to the overall rate. Ever-present additional processes are vibrational and rotational relaxation, while often other reaction channels may also be operative.

c) The most detailed experiment determines the state-to-state rate, e.g., monitoring the rate of formation of different vibrational states of the products as a function of vibrational excitation of one of the reactants. This can be done either by resolving the spontaneous emission from the excited products, or by using a second light source capable of selectively monitoring a given state in one of the products (Sect.5.5).

We start by describing some experiments using the crossed beam configuration (Sect.5.2.1) and with the next best choice: a laser-excited atomic or molecular beam interacting with low-pressure stationary gas samples (beam-bulk configuration). In Sect.5.2.2 we proceed to discuss some bulk reactions, less amenable to straightforward theoretical interpretation, but of considerable practical importance. Commercial use of lasers requires handling of large quantities of material, often leading to a complicated reaction pattern necessitating a detailed knowledge of many rate constants. This is illustrated for the case of Br isotope separation using the often cited (Chaps.1 and 2) HCl(v=2) + Br →HBr + Cl1 reaction.

5.2.1 Molecular Beam Studies

A prominent example of laser-enhanced reaction in crossed molecular beams is

$$K + HCl(v) \rightarrow KCl + H \quad \Delta E = 1\ \text{kcal/mole} \quad . \tag{5.2.6}$$

An HCl chemical laser was used to prepare HCl(v=1), without J resolution at this stage [5.1]. KCl, formed by reaction with the potassium atom beam, was observed as a function of scattering angle θ. Excitation to v=1 did not change the angular distribution of reactively scattered KCl, but increased the yield by approximately two orders of magnitude, as expected for endothermic reactions. Reaction (5.2.6) is the only one for which the rotationally resolved k(v,J→ ;T) rate constant has been measured by laser excitation [5.2]. Using

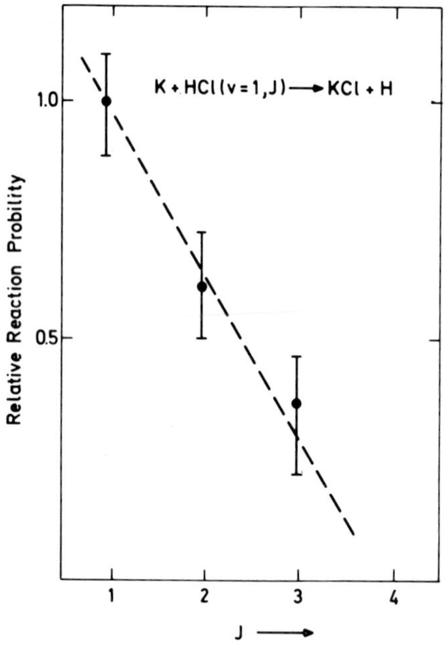

Fig.5.3. The relative cross section for the reaction K+HCl(v=1,J)→ KCl+H as a function of J. Single rotational states were pumped using a grating tuned HCl chemical laser. The data were obtained by integrating over the entire angular distribution (adapted from [5.2])

a grating tuned HCl laser, low J states (J=1-4) were selectively populated. Relative rate constants could be obtained by knowledge of laser power, population in the appropriate rotational state of v=0, and the absorption cross section. As Fig.5.3 shows, reactivity decreases with J, a trend also observed in similar nonlaser initiated reactions [5.3,4]. Simple geometrical considerations can qualitatively account for this effect. In order to react, the approaching K atom must see the chlorine atom. In a rotating HCl molecule, the Cl atom is essentially stationary, while the light H atom whirls around it. Low J's mean slow H atom motion, and less effective shielding of the Cl atom, making reaction more probable. (See Sect.4.5.1 for a similar effect in the F + HD reaction). At some point, further increase in J has littl effect on the shielding efficiency [5.2,3]. It should be stressed that in the nonlaser experiments, rate constants are often obtained from the overall depletion of the particular v,J state. It is *assumed* that this is caused mainly by reaction rather then by, say, rotational relaxation. Only in a laser experiment can a single v,J state excitation be combined with direct monitoring of the products, allowing a straightforward interpretation.

Turning now to the determination of state-to-state (v→v' in this case) rate constants, we find that rather than crossed molecular beams a beam-bulk configuration was employed, using the higher yield of the latter method to improve (S/N) — the signal-to-noise ratio.

In a series of experiments, various excitation modes of the reactions

$$M + HF(v) \rightarrow MF(v') + H \quad M = Ca,\ Sr,\ Ba \qquad (5.2.7)$$

were investigated [5.5,6]. The reaction with Ba is exothermic, while with Ca and Sr, endothermic. Correspondingly, excitation of HF to v=1 has a large effect (the rate increases by four orders of magnitude) for Sr, and a much smaller effect on Ba. Product vibrational distributions were followed by laser-induced fluorescence. A feeling for the experimental difficulties involved can be obtained by considering Fig.5.4. The total fluorescence intensity was monitored as a function of excitation wavelength. Relative populations of BaF(v) states were obtained from the excitation spectrum by correcting for the absorption coefficients and fluorescence quantum yields. The fact that the fluorescence was not spectrally resolved (that would mean a prohibitively large attenuation of the signal) created a major problem as scattered light from the excitation laser reached the detector. In this case the problem was solved by constructing an efficient light collecting optical system, which allowed the observation of about 20 LIF photons per dye laser pulse, without interference from the $\sim 10^{14}$ photons which this pulse contained.

In the case of Ba + HF, the HF(v=0) reaction is substantial and tends to obscure the effect of vibrational excitation. Close inspection of Fig.5.4a reveals the appearance of new bands at longer wavelengths. These become much more prominent on recording the difference spectrum: LIF is observed with the HF laser "on" and "off" alternatively, at a fixed delay after the dye laser pulse (see Fig.5.4 for details). In this way the new vibrational bands are clearly displayed (Fig.5.4b). This case demonstrates a typical experimental difficulty: Even with laser excitation, the absolute number of vibrationally excited molecules is often small compared to their total number density. In part this is due to the fact that the laser interrogates only molecules that happen to be in the one particular J state of the v=0 manifold, which absorbs the laser line. Since typically the laser fires once per second for about one μs, its effect can be overwhelmed by the continuously ongoing thermal reaction of v=0 molecules. The standard way to compensate for this effect is to gate the detection electronics in such a way that fluorescence is recorded only during a short interval of time right after the laser pulse. This technique was employed in obtaining Fig.5.4b.

In the case of Ca or Sr, the v=0 reactions are endothermic, and practically do not obscure the exothermic v=1 reaction. The reaction rate for HF(v=1) is four orders of magnitude larger than for HF(v=0). With DF instead

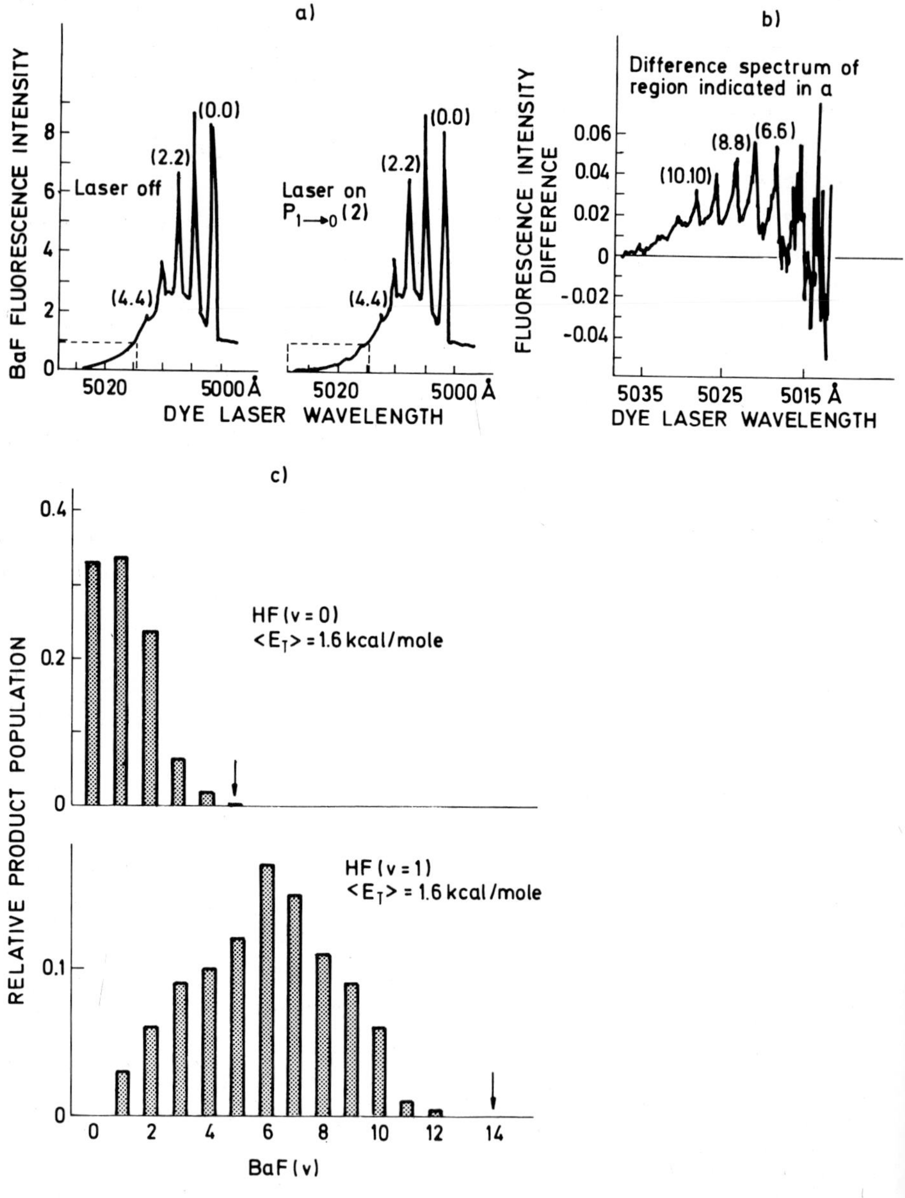

Fig.5.4a-c. Effect of vibrational excitation on the Ba + HF reaction. The laser-induced fluorescence excitation spectrum of BaF is shown. Since HF(v=0) reacts quite readily with Ba atoms, a strong signal is obtained even without laser excitation

of HF (using DF laser for excitation) the reaction with Sr becomes slightly exothermic but no reaction was observed.

The average fraction of the available energy appearing as product vibrational excitation in the $M + HF(v=1) \rightarrow MF + H$ reaction is $\langle f_v \rangle = 0.29$, 0.33 and 0.40, for M = Sr, Ba and Ca, respectively [5.5,6]. In the $Ba + HF(v=0) \rightarrow BaF + H$ reaction $\langle f_v \rangle = 0.12$. The increase in product vibrational excitation upon reagent vibrational excitation is in line with the qualitative considerations of Sects.1.3.1 (cf. Fig.1.13) and 2.3.4. It has also been suggested [5.5] that reagent vibrational excitation may lead to "mixed energy release", [5.7]. This term is used to describe reactions where the typical trajectories "cut the corner" near the saddle point of the potential energy surface. That is, the trajectory enters the products valley from the side rather than along the minimum energy path, thereby enhancing product vibrational excitation, (cf. Figs.2.25,27).

A different mode of energizing of reaction (5.2.7) is by electronic excitation of the metal atom. This was done for Sr by populating the $^3P_1$ state [5.8]

$$Sr(^3P_1) + HF \rightarrow SrF(v') + H \tag{5.2.8a}$$

$$Sr(^3P_1) + HCl \rightarrow SrCl(v') + H \quad . \tag{5.2.8b}$$

Fig.5.4. (a) An overall excitation spectrum with and without HF laser excitation: the effect of v=1 reaction can be observed mostly in the long wavelength tail of the spectrum, but is barely discernible. The peaks in the spectrum correspond to Q branch heads formed at low J values of the $\Delta v=0$ sequence of the BaF $C^2\Pi_{1/2} - X^2\Sigma^+$ transition.
(b) Using gated electronics to limit observation only to a short period of time immediately after the laser fires, a difference spectrum is generated by subtracting the spectrum obtained with the laser blocked from that obtained when the laser irradiates the sample. The increase in noise to the low wavelength side is due to higher yields of nonlaser reaction, leading to larger fluctuations in the difference spectrum (signal increment due to laser excitation is of the same order of magnitude as fluctuations in the thermal reaction).
(c) A bar graph showing state to state relative reaction probabilities. Population ratios for different Ba(v) states were obtained from Fig.5.4b by knowledge of the appropriate Franck-Condon factors. It is seen that the probability of forming higher vibrational states of the product increases on vibrationally exciting the reactant. The graph was calculated assuming that the fraction of detected BaF product produced by HF(v=1) is 1%. The rest is formed by HF(v=0). The arrows mark the highest vibrational level of BaF that can be populated by the reaction (adapted from [5.5])

Upon electronic excitation the reaction becomes highly exothermic, and is accompanied by substantial vibrational excitation. For SrCl, (5.2.8b), a very narrow vibrational distribution of SrCl(v) is obtained—only v states in the range 27 to 31 are observed. This is a further example of the selectivity that can be achieved by preparing the proper initial state.

The effect of orientation on the result of reactive scattering was mentioned in Sect.1.3.1 for HF vibrational excitation. It has also been studied for electronic excitation in some cases [5.10]

$$I_2(B^3\Pi) + In \rightarrow InI^* + I \qquad (5.2.9a)$$

$$I_2(B^3\Pi) + Tl \rightarrow TlI^* + I \quad . \qquad (5.2.9b)$$

Reactions (5.2.9a) and (5.2.9b) yield electronically excited products that are conveniently monitored by their luminescence. Figure 5.5 shows that polarization has a considerable effect on the reaction rate, indicating a preferred orientation in the transition state, in agreement with theoretical predictions [5.9].

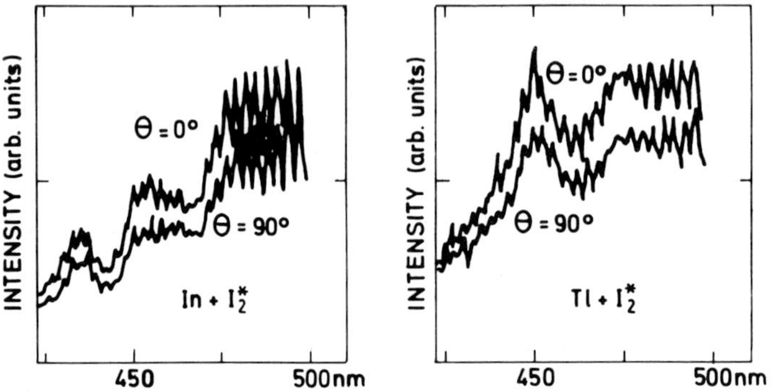

Fig.5.5. Luminescence intensity of electronically excited MI produced in the reaction $M + I_2^* \rightarrow MI^* + I$, M=In, Tl. Iodine was excited into the $B^3\Pi$ state using an Ar laser at 514.5 nm. Varying the polarization of the laser light with respect to the direction of the metal beam shows that reaction in the $\theta=0°$ direction (head-on collision) is faster than in the $\theta=90°$ direction (broadside approach) (cf. Fig.1.14) (adapted from [5.10])

5.2.2 Reactions in the Bulk

The simplest reaction of the type

$$A + BC \rightarrow AB + C \tag{5.2.10}$$

is the hydrogen reaction $H' + H_2 \rightarrow H'H + H$. The potential energy surface for this reaction is well known [5.11], making it the best candidate for comparison of theory and experiment. Of particular interest in the context of this section is the effect of vibrational excitation. Isotopic substitution allows relatively easy monitoring. $H_2(v=1)$ was populated by energy transfer from HF(v=1) which was directly pumped by an HF laser [5.12]. The reaction

$$D + H_2(v=1) \rightarrow HD(v=1,0) + H \tag{5.2.11}$$

was followed in real time by monitoring H atoms using Lyman α resonance absorption. A huge increase in rate was recorded—a factor of 4×10^4 compared to ground-state molecules. This result is a challenge to theories, as quantum mechanical estimates and classical trajectory calculations predicted a substantially smaller rate enhancement.

Many other reactions of the $A + BC(v)$ type were studied using resonance atom excitation and vibrational fluorescence of the diatomic molecule. A representative experimental setup for a kinetic study of this type is shown in Fig.5.6.

Gas flow systems are of advantage in these experiments, helping to maintain a stationary concentration of the (highly labile) reactants at the observation points. BC molecules are excited upstream and then mixed with the atoms that are often prepared in a microwave discharge. Infrared luminescence can be used to follow the overall decay rate (reactive and nonreactive) of BC(v=1) molecules. Products can be monitored by resonance fluorescence from the atoms, as shown in the figure, and by other methods, e.g., mass spectrometric analysis, LIF of the molecule AB, or by infrared fluorescence (when AB is vibrationally excited) (cf. Fig.5.1).

The setup of Fig.5.6 can be used to discriminate between the reaction of interest and competing processes. In the case

$$O(^3P) + HCl \rightarrow OH + Cl \quad \Delta E \sim 1\ \text{kcal/mole} \tag{5.2.12}$$

the activation barrier is found to be smaller than the energy of a single vibrational quantum of HCl. Cl production rate is determined by k_{13a}, while total disappearance rate of HCl(v=1) due to collisions with O atoms is determined by $k_{13a} + k_{13b}$

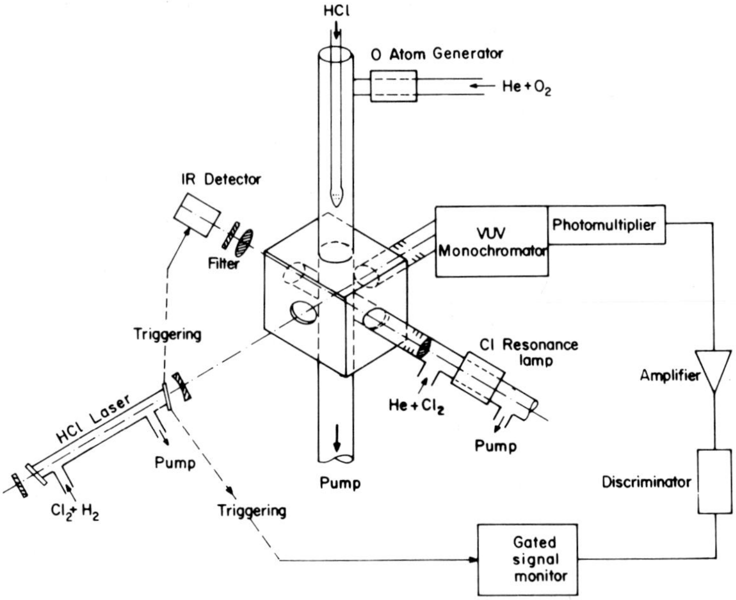

Fig.5.6. Experimental setup for measuring the rate of the reaction $O + HCl(v=0,1) \rightarrow OH + Cl$, illustrating the general methods of following $A + BC(v)$ type reactions. HCl and O atoms are premixed and flown across the beam of a pulsed chemical HCl laser. The depletion of $HCl(v > 0)$ is monitored by the infrared emission of these molecules, using an InSb infrared detector. A resonance chlorine atom lamp, operated in the vacuum ultraviolet, selectively excites the Cl atoms produced and the resulting fluorescence reflects the rate of Cl atom production. The fluorescence is dispersed and detected with a solar blind photomultiplier. The infrared and VUV signals are detected synchronously, using a gate to discriminate against v=0 reaction: The gate was activated by a trigger from the laser, so that Cl resonance fluorescence was recorded either immediately (5 μs) after the laser pulse, or 100 ms later. The signals obtained are due to reaction of HCl(v=0,1) in the first case, and HCl(v=0) only in the second (adapted from [5.13])

$$HCl(v=1) + O \rightarrow OH + Cl \qquad (5.2.13a)$$

$$HCl(v=1) + O \rightarrow HCl(v=0) + O \quad . \qquad (5.2.13b)$$

The apparatus in Fig.5.6 allows separate measurements of $k_{13a} + k_{13b}$ and k_{13a}. It was found that vibrational excitation of HCl to v=1 increased the reaction rate by two orders of magnitude. Notwithstanding, the major decay route in this case is the nonreactive pathway.

Another example demonstrating the use of lasers to extract detailed kinetic information on a complex chemical system is the study [5.14] of the various reactive and inelastic pathways of $Cl + HCl(v=0,1)$ collisions. The atom-diatom exchange reaction with ground state HCl,

$$^{37}Cl + H^{35}Cl(v=0) \rightarrow H^{37}Cl(v=0) + {}^{35}Cl \qquad (5.2.14)$$

is quite slow: $k \simeq 1.7 \times 10^{-15}$ cm^3 molec^{-1} s^{-1} (T = 300 K), $E_a \sim 5$ kcal/mole, indicating the existence of a rather high potential energy barrier. Hence, since one vibrational quantum of HCl is $E_1 - E_0 \simeq 8$ kcal/mole, the nearly thermoneutral reaction (ΔE_0=2 cm^{-1}),

$$^{37}Cl + H^{35}Cl(v=1) \rightarrow H^{37}Cl(v=1) + {}^{35}Cl \qquad (5.2.15)$$

is expected to proceed at a much higher rate. In order to measure k_{15} a mixture of naturally abundant (^{35}Cl:^{37}Cl=3:1) Cl atoms (generated by a microwave discharge of Cl_2) and HCl molecules was irradiated by an HCl chemical laser in a set-up similar to that of Fig.5.6 [5.14]. The HCl laser was operated with a natural Cl_2/H_2 mixture. Due to the higher gain of the more abundant isotope and the fast V-V energy transfer

$$H^{37}Cl(v=0) + H^{35}Cl(v=1) \rightarrow H^{37}Cl(v=1) + H^{35}Cl(v=0) \quad , \qquad (5.2.16)$$

the laser emission occurs mostly on the $H^{35}Cl$ lines. (When energy transfer between two types of inverted populations is fast, lasing occurs on the one with the higher gain, cf. Sects.4.2,4). Consequently, the laser excites predominantly $H^{35}Cl$ molecules. Besides (5.2.15) these molecules participate in several other processes; the more important ones are

$$^{37}Cl + H^{35}Cl(v=1) \rightarrow H^{37}Cl(v=0) + {}^{35}Cl \qquad (5.2.17)$$

$$^{37}Cl + H^{35}Cl(v=1) \rightarrow {}^{37}Cl + H^{35}Cl(v=0) \qquad (5.2.18)$$

$$^{37}Cl(^2P_{3/2}) + H^{35}Cl(v=1) \rightarrow \begin{cases} ^{37}Cl(^2P_{1/2}) + H^{35}Cl(v=0) & (5.2.19) \\ ^{35}Cl(^2P_{1/2}) + H^{37}Cl(v=0) & (5.2.20) \end{cases}$$

and the analogous collisions with ^{35}Cl atoms. In view of the minor mass differences the rate constants for processes with different atomic isotopes may be assumed identical, thereby simplifying the kinetic analysis.

Reactions (5.2.15,17,20) lead to enrichment of the $H^{37}Cl$ content. In the experiment the increase in $H^{37}Cl$ concentration was measured by leaking a small fraction of the reaction mixture into a mass spectrometer and monitoring the mass composition shortly after the laser pulse. The change in $H^{35}Cl(v=1)$ concentration due to (5.2.15-20) was measured by monitoring the infrared fluorescence of these molecules. Independent measurements yielded the rate constants of (5.2.16,19,20). Combining all these data one can evaluate $k_{15}+k_{17}$ and the total (inelastic and reactive) V-T rate constant $k_{17}+k_{18}$.

It is found that

$$k_{17} + k_{18} = (6 \pm 1) \times 10^{-12}\ \text{cm}^3\ \text{molec}^{-1}\ \text{s}^{-1}$$

while

$$k_{15} + k_{17} = (1.1 \pm 0.7) \times 10^{-11}\ \text{cm}^3\ \text{molec}^{-1}\ \text{s}^{-1}\quad .$$

While a complete determination of all the individual rate constants was not achieved, vibrational excitation clearly leads to considerable ($\sim 10^3$) rate enhancement, as expected.

The propensity rule for enhancement of an endothermic reaction (or one characterized by an energy barrier) may be used in practical applications, such as isotope separation or selective photochemical synthesis. Considering for instance the reaction (Sects.1.3.1 and 2.4.1)

$$\text{Br} + \text{H}^{35}\text{Cl}(v{=}2) \rightarrow \text{HBr}(v{=}0) + {}^{35}\text{Cl} \qquad \Delta E = 245\ \text{cm}^{-1}\quad , \tag{5.2.21}$$

it is expected that only HCl molecules that have been promoted to $v=2$ will react: the endothermic reactions $\text{Br}+\text{HCl}(v=0,1)$ are orders of magnitude slower (cf. Fig.1.12).

Population of $\text{HCl}(v=2)$ is possible either by sequential absorption $0 \rightarrow 1 \rightarrow 2$ using a chemical HCl laser, or by direct population (overtone absorption) using an optical parametric oscillator, or a similar tunable laser. Isotopic selectivity of the initial state is ensured by the laser narrow bandwidth. However, prospects for macroscopic chlorine isotope separation must take into account the considerable complexity of the overall reacting system. Evidently, the isotopically selected chlorine atom ensuing from (5.2.21) is a very reactive species and must be properly scavenged, in order to preserve selectivity. Otherwise, it might rapidly exchange according to (5.2.15) or (5.2.20). This and other complications are summarized in Fig.5.7.

Inspection of the figure shows that a major problem is V-V exchange (5.2.16) which occurs on the average once every ten collisions. By comparison $k_{21}=2.1 \times 10^{-11}\ \text{cm}^3\ \text{molecule}^{-1}\ \text{s}^{-1}$ [5.15] corresponds to one reaction on every 35 collisions. It is obvious that in order for the wanted reaction to dominate, the concentration of Br atoms must be very high.

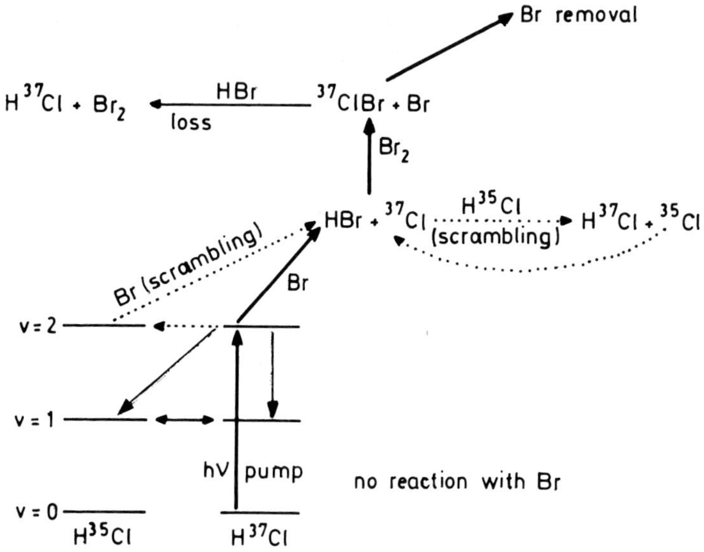

Fig.5.7. Flow diagram showing the processes involved in laser isotope separation of chlorine using reaction (5.2.21). The heavy arrows indicate the desired reaction path and dotted lines show the scrambling processes. The final stable compound is ClBr (adapted from [5.15])

After an initial unsuccessful attempt [5.15] limited isotopic enrichment was obtained [5.16], but in spite of the huge acceleration of the first isotopic selective reaction rate, practical yields are low, and overall enrichment factors are modest (factor of two).

Vibrational excitation of infrared inactive molecules is possible by using the stimulated Raman effect. When a high-power laser beam is transmitted through a sample of polarizable molecules, the outcoming beam contains new frequencies that are shifted with respect to the original laser frequency ν_ℓ

$$\nu_s = \nu_\ell - \nu_i \quad ,$$

where ν_i is the frequency of a Raman active vibration. Simultaneously, measurable populations in the v=1 state of that mode are prepared. A particular merit of the method is that it also applies to homonuclear molecules, such as H_2.

The gas phase metathesis reaction

$$H_2 + D_2 \rightarrow 2HD \qquad (5.2.22)$$

was studied in high-pressure mixtures of $H_2 + D_2$ (~20 atm) [5.17]. Significant production of HD was observed, a result explainable as due to collisional vibrational up-pumping (Sect.2.5.4) followed by reaction

$$H_2(v=3) + D_2(v=0) \rightarrow 2HD \tag{5.2.22a}$$

$$D_2(v=4) + H_2(v=0) \rightarrow 2HD \quad . \tag{5.2.22b}$$

However, reactions of the type

$$H_2(v=1) + D_2(v=3) \rightarrow 2HD \tag{5.2.22c}$$

and even atom-molecule reactions could not be excluded.

This laser-induced reaction must be considered in relation to complementary shock tube results [5.18] as well as extensive quantum mechanical calculations [5.19]. Considerable disagreement exists in this respect between experiment and theory concerning activation energies for this process (40 kcal/mole experimentally vs. 109 kcal/mole theoretically). It has been suggested [5.20] that this discrepancy may be resolved if the transition state has a T- or a Y-shaped structure. Another possible mechanism involves a thermolecular, six-center reaction path [5.21], for which the calculated potential energy surface lies below that of $2H_2 + 2H$.

The most extensively studied case of a bimolecular reaction involving vibrational excitation is that of nitric oxide with ozone with excitation in either one of the reagents. The enhancement of the reaction rate following excitation of $O_3(001)$ was first reported in [5.22] and then studied in greater detail in [5.23]. Conventional kinetic experiments [5.24] established that at least two potential energy surfaces are involved, one (5.2.23) leading to the electronic ground state of NO_2, and another producing a mixture of electronically excited states (5.2.24)

$$O_3 + NO \rightarrow NO_2(^2A_1) + O_2 \tag{5.2.23}$$

$$O_3 + NO \rightarrow NO_2^*(^2B_{1,2}) + O_2 \quad . \tag{5.2.24}$$

($^2B_{1,2}$ is a shorthand notation for a combination of $^2B_1$ and $^2B_2$ states). The temperature-dependent rate constants were established to be (in cm^3 $molec^{-1}$ s^{-1}, activation energies in cal $mole^{-1}$),

$$k_{23} = (7.14 \pm 0.17) \times 10^{-13} \exp[(-2350 \pm 150)/RT]$$

$$k_{24} = (1.26 \pm 0.25) \times 10^{-12} \exp[(-4200 \pm 300)/RT] \quad .$$

Molecular beam studies [5.25] provided enough energy resolution to allow a detailed study of translational energy effects. The results were interpreted as indicating specific fine structure reactivity of NO: The ground $^2\Pi_{1/2}$ state leads mostly to ground state NO_2, while the $^2\Pi_{3/2}$ component, 120 cm^{-1} higher in energy, leads to excited state NO_2. Of interest in this discussion are therefore the corresponding reactions with vibrationally excited ozone or NO. Reactions (5.2.25) and (5.2.26) require specification of the energy localization among the vibrational modes ν_1, ν_2, and ν_3. Vibrational energy transfer studies have shown [5.26] that collisional energy transfer between the stretching modes ν_1 and ν_3 is very fast, because $\nu_1 \cong \nu_3$, Table 5.1. Therefore these modes are kinetically indistinguishable and are characterized by a Boltzmann-like energy distribution (see Sect.2.5.5). The transfer of vibrational energy to the bending mode is, however, slow enough that two separate excitation conditions at shorter and longer times after the CO_2 laser pulse must be considered. The V-T relaxation proceeds via the bending mode according to (5.2.27) and (5.2.28).

$$\begin{bmatrix} O_3(001) \\ O_3(100) \end{bmatrix} + NO \rightarrow NO_2^*(^2B_{1,2}) + O_2 \tag{5.2.25}$$

$$\begin{bmatrix} O_3(001) \\ O_3(100) \end{bmatrix} + NO \rightarrow NO_2(^2A_1) + O_2 \tag{5.2.26}$$

$$\begin{bmatrix} O_3(001) \\ O_3(100) \end{bmatrix} + NO \rightarrow O_3(010) + NO \tag{5.2.27}$$

$$O_3(010) + NO \rightarrow O_3(000) + NO \quad . \tag{5.2.28}$$

Table 5.1. Vibrational frequencies of NO and ozone [cm^{-1}]

NO	Ozone		
	ν_1(symmetric stretch)	ν_2(bending)	ν_3(antisymmetric stretch)
1904	1071	715	1075

Decay processes for electronically and vibrationally excited NO_2 must be included in the kinetic analysis. The total decay rate k_{tot} of the laser-induced chemiluminescence signal in this system shows a strong non-Arrhenius temperature dependence. This rate is the sum of contributions from the

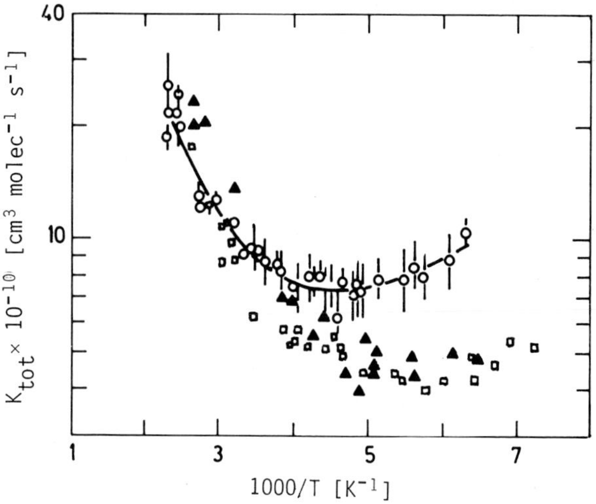

▲ Fig.5.8. Total decay rate coefficient in the reaction system (5.2.25-27) according to several sources; (○) [5.27], (□) [5.28], (▲) [5.23b]. The solid curve is a least-squares fit according to [5.27]. k_{tot} is the total rate constant for disappearance of $O_3(001)$ or (100) as obtained from the decay rate of $NO_2(^2B_{1,2})$

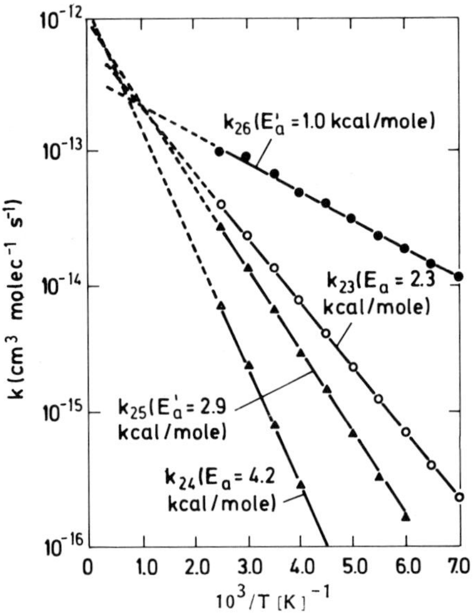

◀ Fig.5.9. Arrhenius plots for reactions (5.2.23-26). The plots show that the activation energies for the laser-induced reactions are approximately 1.3 kcal/mole lower than for the thermal reactions. This result indicates that vibrational energy is only partially (about 50%) effective in overcoming the reaction barrier (adapted from [5.28])

laser-enhanced luminescent reaction rate (5.2.25), the laser-enhanced non-luminescent reaction (5.2.26), and the vibrational relaxation of ozone by nitric oxide (5.2.27,28).

$$k_{tot} = k_{25} + k_{26} + k_{27} \quad . \tag{5.2.29}$$

The temperature dependence of k_{tot} is shown in Fig.5.8. A separation of the individual rates in a rather comprehensive kinetic analysis has been attempted and results are shown in Fig.5.9. The rate constants for (5.2.23-26) were found to closely obey an Arrhenius-like behavior (Fig.5.9). The activation energy of vibrationally excited molecules is 1.3 kcal/mole smaller than for ground state ozone. Thus only part of the initial vibrational energy deposited (2.7 kcal/mole) was utilized in the reaction. Deactivation (5.2.27) shows a non-Arrhenius temperature dependence, with a pronounced minimum at about 230 K (RT ~460 cal/mole), suggesting that attractive forces (complex formation) are important in the nonreactive channel. The minimum is reflected in the temperature dependence of k_{tot} (Fig.5.8).

The above results can be compared to rate data obtained [5.29] for the same reaction, this time, however, putting laser energy into the NO rather than the O_3 reaction partner. In this fluorescence study only the chemiluminescent reaction (5.2.30) could be observed directly. The reactions of interest are the following:

$$NO(v=1) + O_3 \rightarrow NO_2^*(^2B_{1,2}) + O_2 \quad (5.2.30)$$

$$NO(v=1) + O_3 \rightarrow NO_2(^2A_1) + O_2 \quad (5.2.31)$$

$$NO(v=1) + O_3 \rightarrow NO(v=0) + O_3 \quad . \quad (5.2.32)$$

The rate enhancement k_{30}/k_{24} was found to be 4.7. This is nearly the same value as reported above for the vibrational excitation of ozone where k_{25}/k_{24} = 4.1 was found. For the enhancement of reaction (5.2.31) an indirect figure is obtained, giving $(k_{31} + k_{32})/k_{23}=18$ while for excitation of ozone $(k_{26} + k_{27})/k_{23}=17$ was reported. Thus for reaction into this channel pumping the NO bond does not result in a greater enhancement than pumping O_3. In view of the fact that NO vibrational excitation provides nearly twice the energy of O_3 excitation, this suggests that energy transfer from NO to O_3 precedes the reaction. This conclusion is consistent with the interpretation of the minimum in k_{tot} as shown in Fig.5.8.

The lack of sensitivity of the rate constant towards the identity of the initially excited state, and the fact that the decrease in Arrhenius activation energy is about the same as a vibrational quantum of the bending mode of ozone, are consistent with the following sequence: Vibrational excitation is followed by rapid V-V transfer, resulting in population of the lowest vibrational mode in the system (O_3 bending). This mode is the one that is directly effective in overcoming the energy barrier for reaction.

5.3 Electronic Excitation of Polyatomic Molecules

Electronic transitions of molecules occur mostly in the visible or ultraviolet. Consequently, activation by one-photon absorption often suffices to overcome the barrier for a chemical reaction. Photochemical transformations following electronic transitions are indeed very common and are extensively discussed in standard textbooks and monographs [5.30,31]. Lasers have been widely used in many reactions schemes, making complete coverage of the field a hopeless task, even within our self-imposed limitations (discussion of low-pressure gas phase reactions only). Rather than dealing with general aspects, we chose to present three case studies in some detail. These include direct *photodissociation* (i.e., excitation to a repulsive level), *photopredissociation* (excitation to a bound level coupled to a repulsive one), and excitation into a bound, stable state. As we shall see, laser methods played a key role in the elucidation of the photophysical and photochemical processes involved. The basic advantage of the laser in these applications is that it provides a large concentration of excited molecules with a well-defined energy content. Often, $\Delta E/E < 10^{-6}$, where E is the total energy and ΔE the energy uncertainty (bandwidth) of the laser. The actual energy uncertainty is thus usually determined only by the sample initial temperature. Lasers are the only light sources that combine such "energetic purity" with large intensities and good time resolution. Of special interest to chemists are dye lasers, which may be tuned to exactly match a molecular transition. Thus lasers opened up the domain of highly selective, very low-pressure, gas phase photochemistry, allowing for the first time and in a large number of cases the observation of truly unimolecular processes.

5.3.1 Direct Photodissociation: The A State of ICN

Energy Considerations

The energy level diagram and UV absorption spectrum of ICN are shown in Fig.5.10. The transitions marked as A and α are diffuse even under high resolution conditions, indicating that the upper potential surfaces are repulsive (Sect.1.3.4), while the higher lying B and C lead to bound states. We shall restrict our discussion to the A state only.[1]

---

1 Recently laser emission was obtained from vibrationally excited CN radicals formed by photodissociation of ICN at wave lengths between 1550-2000 Å [5.32 The reaction abviously involved higher electronic excited states than the A state.

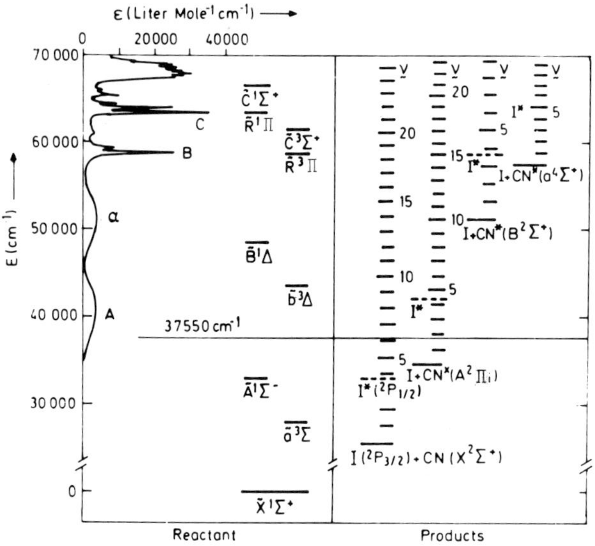

Fig.5.10. Energy level diagram of ICN and of primary photodissociation products. The absorption spectrum of ICN is shown on the left. Excitation at 37550 cm^{-1} (the frequency of a quadrupled Nd:YAG laser) populates the dissociative $A^1\Sigma$ state of ICN. It can be seen from the figure that at this energy, low-lying electronic states of I or of CN may be populated, as well as several vibrational states of CN [5.32,33]

As Fig.3.1 shows, excitation into the A state continuum provides a large energy excess over that required for I-CN bond rupture (25500 cm^{-1}). The energy available for distribution among the degrees of freedom of the fragments and the relative translational motion is given by (cf. Figs.1.2,23)

$$E = E_{int} + E_T = E'_{int} + h\nu - D \qquad (5.3.1)$$

where E_{int} and E_T are the internal and translational energies of the products, E'_{int} is the internal energy of ICN before light absorption, $h\nu$ is the energy of the laser photon, and D is the dissociation energy (the C-I bond energy).

Assuming random distribution of energy into all available channels, one predicts (Sect.2.2.6) considerable vibrational excitation of CN and some electronic excitations of the iodine atom (Fig.5.10). Prelaser studies failed to detect any appreciable excitation in either mode [5.34,35], a result qualitatively understood as indicating the stiffness of the CN bond. The process may be envisioned as a "half collision" event [5.36], in which the absorbed photon "turns on" a strong repulsive force along the I-CN bond. The near absence of vibrational excitation of CN necessarily indicates considerable rotational and/or translational excitation. Both were subsequently studied using laser techniques, rotational distribution by laser-induced fluorescence, and translational energies by photofragement spectroscopy (Sect.5.5).

Photofragmentation Studies

Figure 5.11 shows the flux of photofragments as a function of E_T in (5.3.1). Two peaks are observed, indicating two different values in the initial translational energy, which are better resolved than the thermal spread caused by E'_{int}. By (5.3.1) they correspond to species with low and high internal energy, designated as $\downarrow E_{int}$ and $\uparrow E_{int}$, respectively. The angular distribution of the CN fragment is shown in Fig.5.12. It is seen that both the $\downarrow E_{int}$ and $\uparrow E_{int}$ angular distributions peak near $\theta = 0°$, indicating the predominance of a parallel transition and the lack of rotational motion prior to dissociation.

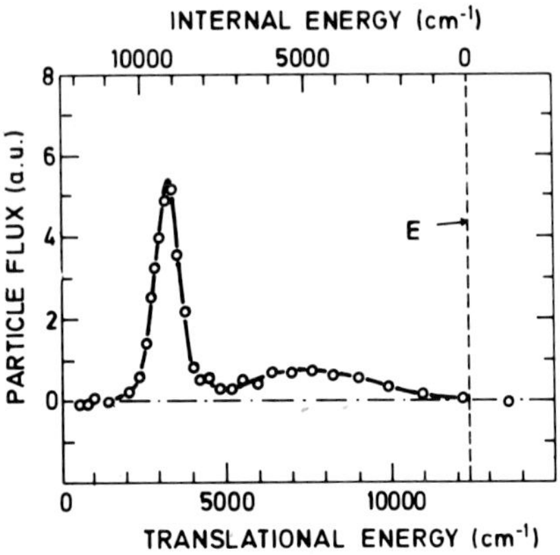

Fig.5.11. The photofragment spectrum of ICN, taken at 37550 cm^{-1}. E is the energy available for partitioning into translational and internal degrees of freedom (see text). The appearance of two peaks indicates two distinct populations with different internal energies (adapted from [5.37])

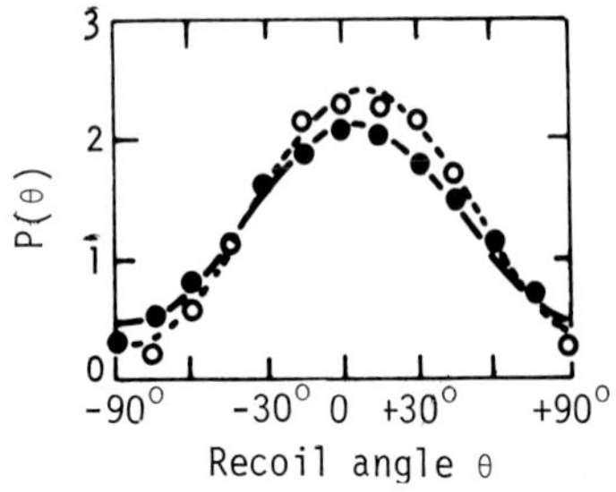

Fig.5.12. The angular distribution P(θ) of recoiling CN fragments in the laboratory coordinate system, on photolysis of ICN at 37550 cm^{-1}. Open and solid circles are for fragments with high and low internal energy, respectively. These data are consistent with the assignment of the transition as parallel to the molecular axis and a short-lived upper state, with a lifetime of $\sim 10^{-13}$ s. The dissociation occurs long before a rotation period can be completed and is an example of a direct dissociation, not involving a long-lived intermediate (adapted from [5.37])

In Fig.5.12, θ is the angle between the laboratory recoil direction and the direction of the electric vector of the laser light. Thus, the angular distributions of the $\downarrow E_{int}$ and $\uparrow E_{int}$ peaks maximize near $\theta=0^{\circ}$. The major conclusions from this study were as follows [5.37]:

1) Both $\downarrow E_{int}$ and $\uparrow E_{int}$ peaks arise from transition(s) that are predominantly parallel to the molecular axis. This is in contrast with previous, non-laser assisted, analyses of the absorption spectrum [5.33,38], which led to assignment of the transition as a perpendicular one.

2) The lifetime of the excited state(s) involved is very short, $\sim 10^{-13}$ s for both $\downarrow E_{int}$ and $\uparrow E_{int}$, though only an upper limit could be determined for the latter. These values (of the order of one vibrational period, much shorter than a rotational period) are in line with the diffuse character of the absorption spectrum.

3) The fragments are formed with considerable translational energies corresponding to $\uparrow E_{int} = 9000\ cm^{-1}$ and $\downarrow E_{int} = 5000\ cm^{-1}$, respectively. These values are much larger than the initial translational energy of ICN ($\sim 200\ cm^{-1}$). Large as it is, translational energy alone cannot account for the difference between available energy ($\sim 37550\ cm^{-1}$) and dissociation energy ($25500\ cm^{-1}$; see Fig.5.10). The photofragmentation data do not allow the determination of internal energy distributions between the electronic, vibrational, and rotational degrees of freedom but partial information can be derived from energy considerations. Namely, electronic excitation is excluded for the $\downarrow E_{int}$ fragments, but is feasible [both as $I^*(5^2P_{1/2})$ or as $CN^*(\tilde{A}^2\Pi_i)$] for the $\uparrow E_{int}$ peak (see Fig.5.11).

Laser-Induced Fluorescence Studies

Part of the missing information on the branching ratios can be gained by laser-induced fluorescence studies (Sect.5.5.2). ICN was dissociated by a flash lamp [5.39] or by a laser [5.40] (at the same frequency used in the photofragmentation study). The CN vibrational and rotational distributions were monitored by fluorescence excitation using a narrow band dye laser. In both studies no vibrational excitation was observed, nor could $CN^*(\tilde{A}^2\Pi_i)$ be detected. In contrast, strong rotational excitation was clearly observed (Fig.5.13). The distribution was probed by obtaining an excitation spectrum, i.e., populating an electronic excited state of CN and observing the total fluorescence as a function of excitation wavelength. In the presence of a large excess of helium, a rotationally relaxed population is obtained (Fig.5.13a). With little added helium a large degree of rotational excitation

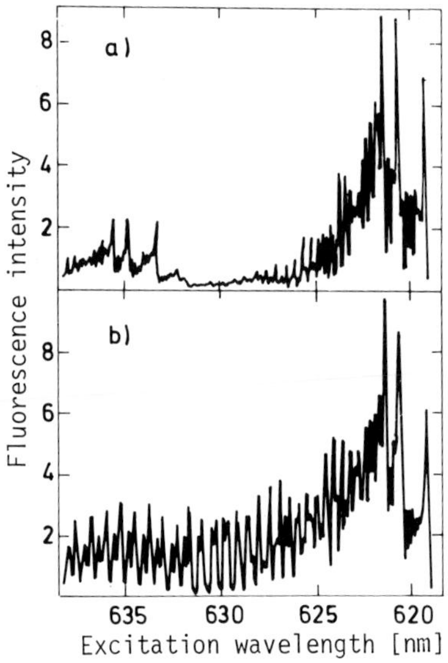

Fig.5.13a,b. Rotational fine structure in the laser-induced fluorescence excitation spectrum of CN, formed by photodissociation of ICN. (a) In the presence of a large excess of helium, where rotational relaxation is fast. (b) With very little helium added; extensive rotational excitation is revealed in (b) (adapted from [5.40])

is revealed by the strong features between 625 and 635 nm, due to transitions from high J states (Fig.5.13b). In fact, only a lower limit for rotational excitation could be determined as the reaction was not carried out under truly collision-free conditions, and some rotational relaxation may have taken place.

The predicted extensive rotational and translational excitations are thus confirmed. The absence of vibrational excitation is disturbing, as it leaves the appearance of the peaks in the photofragmentation spectrum unaccounted for. A redetermination of I^*/I ratio, using laser source excitation, indicates that it is as high as 2:3 on excitation at 266 nm [5.41]; the failure to detect I^* by the flash photolytic method [5.35], is probably due to fast quenching of I^* by ICN.

The large extent of rotational excitation indicates a bent configuration in the excited state. This fact and the absence of vibrational excitation should be considered in future models of the A state photodissociation.

In summary, laser studies allowed a much more complete investigation of ICN photodissociation. They established it as a "direct" event, involving no intermediate states or species. The A transition was found to be predominantly parallel, contrary to previous tentative assignments based only on absorption measurements. The distribution of translational energies was studied for the first time, and the fragments internal energies restudied with much better time resolution than previously available.

5.3.2 Photopredissociation: Formaldehyde

The $^1A_g(S_0) \rightarrow {}^1A_u(S_1)$ $(n \rightarrow \pi^*)$ transition of formaldehyde was discussed in Sect.3.3.7. The low resolution spectrum (Fig.3.36) does not provide any dynamical information. Higher resolution reveals an extensive rotational structure which becomes diffuse at wavelengths below 2750 Å. This is an example of a predissociative transition (Sect.3.3.3). As the energy level diagram of formaldehyde (Fig.3.37) shows, the lowest energetically allowed dissociation processes are

$$H_2CO^* \rightarrow H_2 + CO \quad \text{molecular dissociation} \tag{5.3.2}$$

$$H_2CO^* \rightarrow H + HCO \quad \text{radical dissociation} \quad . \tag{5.3.3}$$

Extensive radical scavenging experiments [5.42] (using nonlaser sources) showed that both dissociation channels are operative, the molecular one predominating at lower excitation energies, the radical one at higher energies. In all experiments only the n-π^* transition was excited. (By higher energiés we mean excess of vibrational energy above the origin of the transition). Further evidence for a gradual change in the photodynamics as a function of excess energy was obtained from fluorescence lifetime studies: The measured lifetime decreases monotonically as excess energy is increased, with concurrent decrease in fluorescence quantum yield [5.43]. A large isotope effect was observed in the fluorescence lifetime and efficiency of a given vibronic level: Protonated formaldehyde is a much poorer emitter than its deuterated analogue. This is an example of an isotope effect on radiationless transitions (Sect.2.4.9) in which high-frequency vibrations are more efficient accepting modes for electronic excitation than low-frequency ones (see the discussion of the energy gap law in Sect.2.4).

The predissociation mechanism appeared to be quite straightforward. Initial excitation to S_1 is followed by vibrationally assisted internal conversion to S_0 which then immediately dissociates (see Fig.3.38). As excitation energy increases, rates become larger due to higher density of states in the accepting mode. Also, the radical dissociation channel opens at some (not yet well-defined) excitation energy, and eventually predominates.

Laser chemistry of formaldehyde [5.44] was started in 1972 with an isotope separation experiment: The very large isotopic shift between H_2CO and D_2CO coupled to an instantaneous unimolecular separation mechanism suggested easy photochemical separation. A laser was needed primarily because of the small extinction coefficient ($\varepsilon \sim 1$-$10\ M^{-1}\ cm^{-1}$), making the use of conventional light sources impractical. Eventually, lasers were used to achieve the much harder job of separating carbon [5.45] and oxygen [5.46] isotopes. Optimization of the separation process required detailed knowledge of formaldehyde photophysics. In particular, factors affecting the branching ratio of reactions (5.3.2) and (5.3.3) were important because molecular dissociation does not involve isotopic scrambling (leading, as it does, to stable primary products), and should be enhanced, if possible.

The high radiation intensities provided by lasers allowed reduction of the pressure to the 10^{-3} - 10^{-4} Torr range, and eventually the use of molecular beams. In addition, they afforded for the first time real time monitoring of the products—leading to a completely new outlook on formaldehyde photochemistry, as detailed below.

An immediate result of the low-pressure studies was the discovery of non-exponential decay curves. Thus the "lifetimes" quoted in Fig.5.14, obtained at a low spectral resolution and relatively high (0.2-1 Torr) pressures, are some ill-defined averages of many decay processes. (An important experimental argument involved is that when the ratio of the amplitudes of two components in a decay curve exceeds ~10, the smaller one is difficult to detect). In this application, the necessary signal-to-noise ratio requires a nanosecond light source that delivers $\sim 10^{12}$ photons in a spectral range of the order of the absorption linewidth (0.06 cm^{-1}). The only such source is a laser.

Figure 5.15 displays an example of the observed biexponential decay curves [5.47]. At the low presssures employed (5×10^{-4} Torr) one suspects that they indicate population of at least two different states. Another possibility is that energy transfer in the system is extremely efficient (2-3 orders of magnitude larger than gas kinetic).

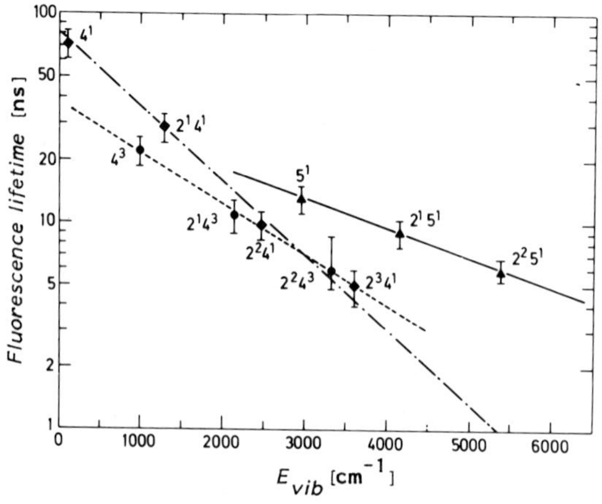

Fig.5.14. The measured lifetimes of some vibrational levels of S_1 of formaldehyde. The notation used, $n^a m^b$, indicates excitation of the nth and m^{th} vibrational modes to the a'th and b'th overtone, respectively (adapted from [5.43])

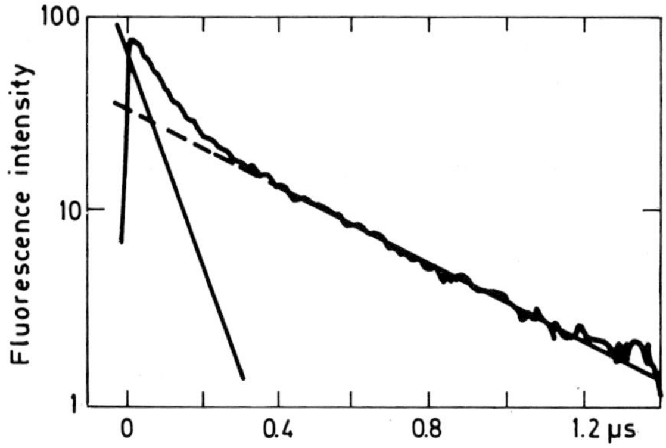

Fig.5.15. Biexponential decay of formaldehyde S_1. Excitation is to the 4^1 vibrational level, using a narrow bandwidth (0.8 cm^{-1}) laser source. Pressure is 5×10^{-4} Torr (after [5.47])

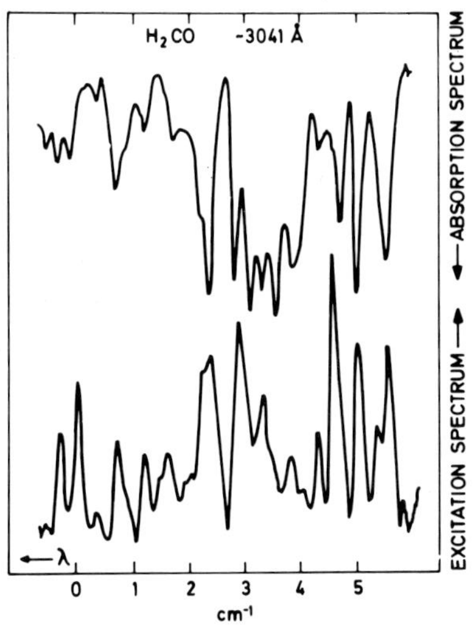

Fig.5.16. Comparison of the excitation and absorption spectra of H_2CO around 3041 Å. The same dye laser was used to record both spectra. Different relative intensities of absorption and excitation spectra indicate a rotational dependence of the fluorescence quantum efficiency (after [5.50])

In either case, one expects a large *rotational* effect on the decay curves, as the separation between two vibrational levels is usually too large to allow simultaneous excitation by a narrowband laser, or to lead to efficient energy transfer (cf. Sect.2.4).

The expected rotational dependence was indeed subsequently confirmed by experiment. Figure 5.16 shows a simultaneous recording of the high-resolution absorption and excitation spectra of H_2CO around 3041 Å. The spectrum is obtained by monitoring the total fluorescence of the sample as the excitation wavelength is scanned.

At the small optical densities involved one expects the excitation spectrum to reproduce the absorption spectrum provided that the quantum efficiency is constant.[2] The deviations observed in the figure must thus be ascribed to different lifetimes (or quantum yields) of different rotational states within a single vibronic level. This was later confirmed by extensive direct lifetime measurements: Zero pressure lifetimes (as measured around 10^{-4} Torr) vary, within a single vibronic state, by as much as a factor of 200 [5.48]. A preliminary limited molecular beam study confirmed these results [5.49].

It should be remarked that biexponential decay has been observed in a large number of laser excitation studies on molecules such as SO_2, NO_2, pyrazine, and biacetyl, to name a few. This topic will be further discussed in Sect.5.3.3, dealing with radiationless transitions between bound states. Formaldehyde offers a unique opportunity to study rotational dependence of radiationless transition probabilities, in view of the availability of a rather complete assignment of the high-resolution absorption spectrum. The variation of fluorescence quantum yield between different rotational states must be ascribed to some nonradiative processes, as these are known to dominate the fate of excited states of formaldehyde. Rotational effects have been largely ignored in radiationless transition theories, but in view of these and other laser excitation results must evidently be incorporated into future models. In any case, this may be looked upon as a demonstration of constraints introduced by the preparation of the initial state and their effect on the eventual dynamics (Sect.2.3.7).

According to the simple model of formaldehyde predissociation, the decay of S_1 states is the rate determining step. Once a highly excited S_0 is produced, it dissociates within the energy redistribution lifetime ($\sim 10^{-11}$ - 10^{-12} s; Sect.2.4.9). Laser techniques, as mentioned before, allowed for the first time a real time monitoring of the appearance rate of the products. This was done by monitoring CO produced by the UV laser either by infrared luminescence (vibrationally excited CO) or by absorption of a cw CO laser radiation. The results clearly demonstrate the harzards of deriving a mechanism on the basis of incomplete information: The appearance rate of CO was found to be much slower than the decay of S_1, for a given excitation frequency [5.51]. In the case of the nitrogen laser frequency ($\sim$ 29665 cm^{-1}) the

2 The fluorescence intensity I_f is related to the absorption coefficient ε via $I_f = I_0\,[1-\exp(-\varepsilon C\ell)]\phi$ where ϕ is the quantum efficiency. At low optical densities $\varepsilon C\ell \ll 1$, $I_f = I_0 \varepsilon C\ell\phi$.

average S_1 lifetime is 46 ns. The appearance time of CO production was found to increase on decreasing the pressure, indicating a collision-induced process. At the lowest pressure used (0.1 Torr) it exceeded 4 μs, i.e., almost a factor of 100 slower than the unimolecular decay time of S_1.

This result required a complete reassessment of the mechanism of formaldehyde photochemistry. It implies the existence of a long-lived intermediate species, formed from S_1 and serving as the precursor of CO. Furthermore, at the quoted and lower frequencies, it is conceivable that CO may only be produced by collisional processes.

The nature of the intermediate is still a matter of debate. It might be another electronic state of formaldehyde. The only known, energetically accessible states are S_0 and T_1 (see Fig.3.37). The triplet is practically ruled out by the observed fast S_1 decay: its density of states is too low to account for such rates. Model calculations [5.52] show that highly vibrationally excited S_0 is a reasonable possibility, in line with the early model (see above). However, these calculations set the barrier for dissociation at 4000 cm^{-1} above the origin of S_1. This, in turn, is not compatible with the observed shortening of fluorescence lifetime at much lower excitation energies.

Another possible approach considers the existence of a metastable chemical species rather than a metastable excited state of formaldehyde. An interesting suggestion is the biradical HCOH, formed by hydrogen migration from the carbon to the oxygen atom. A possible mechanism would thus be

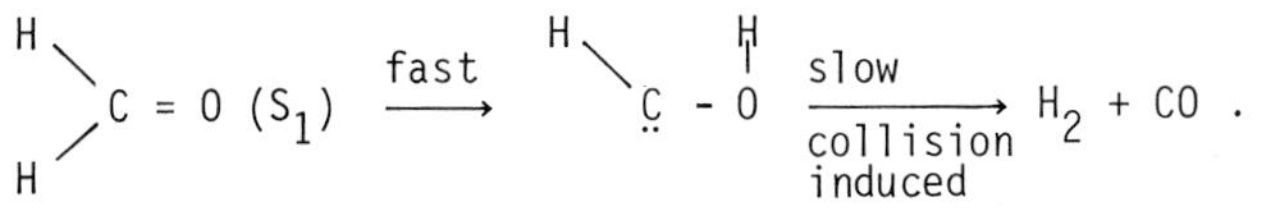

$$\begin{array}{c} H \\ \diagdown \\ \diagup C = O \ (S_1) \\ H \end{array} \xrightarrow{\text{fast}} \begin{array}{c} H \\ \diagdown \quad\; \overset{\displaystyle H}{\underset{..}{|}} \\ \; C - O \end{array} \xrightarrow[\substack{\text{collision}\\ \text{induced}}]{\text{slow}} H_2 + CO \ . \qquad (5.3.4)$$

This possibility is supported by a low-temperature matrix isolation study [5.53]. Formaldehyde was embedded in an Ar matrix and photolyzed using a broadband source (Hg high-pressure lamp). In a matrix, labile transient species are trapped, and can be observed for example by their infrared spectrum. One of the species revealed in this way was glycolaldehyde, which is presumed to be formed by the following reaction:

$$\begin{array}{c} H \\ \diagdown \\ \diagup C: \\ HO \end{array} + \begin{array}{c} \overset{\displaystyle H}{\underset{\displaystyle}{|}} \\ C = O \\ \diagup \\ H \end{array} \rightarrow \begin{array}{c} \overset{\displaystyle H}{\underset{|}{|}} \quad \overset{\displaystyle H}{\underset{|}{|}} \\ HO - C - C = O \\ \overset{|}{H} \end{array} \ . \qquad (5.3.5)$$

hydroxy   formalde-   glycolaldehyde
methylene hyde
radical

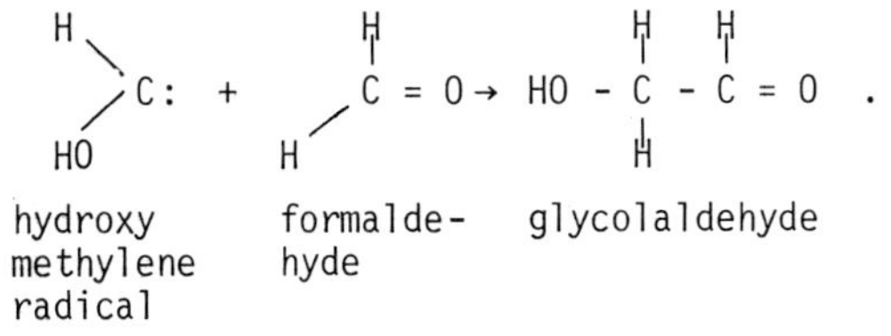

A theoretical study of the reaction dynamics using available energy surfaces [5.54] considered the possibility of tunneling of an H atom as a primary step.

A further important result obtained in the real time monitoring of CO production [5.51] is worth mentioning. The experimental technique involved detection of carbon monoxide by absorption, using a cw CO laser as a light source. A unique feature of this technique, apart from its high selectivity and sensitivity, is the fact that vibrationally excited CO can be easily monitored, by tuning the laser to the appropriate transition. Recalling the large amount of available energy ($\sim 28000\ cm^{-1}$) one expects on prior grounds (Chap.2) considerable vibrational excitation of either CO or H_2, or both. (In the classical RRHO approximation the prior vibrational distribution of each of the diatomic products is $P^0(v) \propto (E - E_v)^{7/2}$, where E is the available energy, cf. App.2A). The vibrational frequencies of CO and H_2 are 2170 and 4395 cm^{-1}, respectively. Prior predictions were not sustained: little vibrational excitation of CO was found, with v=1 population being only ~12% of v=0 population. This result is reminiscent of the ICN case, discussed above, and likewise suggests considerable translational and/or rotational excitation. The absence of vibrational excitation also calls for reevaluation of the biradical intermediate theory: The CO bond is expected to undergo a considerable contraction in the reaction. The bond order is 1 in the radical and 3 in CO. Such changes often lead to vibrational excitation, as discussed in Sect.2.3.3.

In summary, laser studies allowed for the first time the practical determination of rotational effects on the lifetime and quantum efficiency of S_1 fluorescence in formaldehyde. The results indicate a strong rotational dependence of the radiationless process underway. Another feature revealed by laser studies is the existence of an intermediate species, forming the molecular products. In fact, it is not at all clear at the moment whether these are produced by a unimolecular mechanism, or are formed only by a collision-induced process. A partial mapping of the energy distribution in the products has been obtained, but it accounts only for a small fraction of the considerable energy that must be dissipated ($\sim 28000\ cm^{-1}$ for $H_2 + CO$ formation).

Laser experiments are expected to help in elucidating some of the remaining unresolved problems concerning formaldehyde photodissociation: the nature of the fast process depleting S_1, particularly near the origin; the nature of the intermediate; the fate of the released energy. An obvious experiment involves the determination of translational energy distribution by photofragment spectroscopy. Another one is real time monitoring of HCO or H, the two remaining primary products. In fact, the former has been recently monitored by intracavity absorption (Sect.5.5). Time resolution was not sufficient in these experiments to determine whether HCO production is direct or proceeds via a long-lived intermediate. Advances in laser technology are expected to overcome this shortcoming in the near future.

5.3.3 Excitation of Bound Electronic States: Biacetyl and Glyoxal

The biacetyl (H_3C-CO-CO-CH_3) absorption spectrum in the visible is due to n-π^* transitions (Sect.3.3.7). An energy level diagram is shown in Fig.5.17. Two n-π^* states are expected for this bicarbonyl compound, designated as $^{1,3}A_u$ and $^{1,3}B_g$. The former pair corresponds to the lowest singlet and triplet states of formaldehyde discussed in Sect.3.3.7.

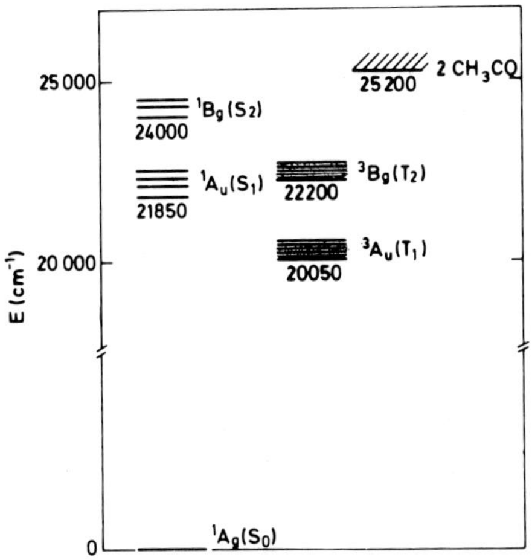

Fig.5.17. Energy level diagram of biacetyl. The exact locations of the $^{1,3}B_g$ states are not known

Biacetyl is known to be photolabile: On irradiation in the n-π^* transition band, radical reactions (initiated by CH_3CO radicals) are observed. This cannot be due to unimolecular decomposition as the lowest dissociation energy (to form 2 CH_3CO) is 25200 cm^{-1}, well above the band origin (Fig. 5.17). It turns out that radical formation is due to interaction between two excited molecules in the triplet state. The triplet is a long-lived state, reflecting the fact that the $T_1 \rightarrow S_0$ transition is spin forbidden. The measured lifetime of triplet biacetyl is 1.8 ms, leaving ample time for collisions between two such molecules. This process is called triplet-triplet annihilation [see (5.3.13)] and leads to formation of an energy-rich molecule, capable of dissociation. Long-lived triplet states play a dominant role in the photochemistry of organic molecules in general, and carbonyl compounds in particular [5.30,31,55].

It is clear that collision-free studies concerning biacetyl photophysics must be conducted at very low pressures in order to avoid triplet-triplet collision-induced processes. Prelaser studies led to the following kinetic scheme:

$$S_0 \underset{k_f}{\overset{k_a}{\rightleftharpoons}} S_1(v) \qquad \text{(absorption/fluorescence)} \qquad (5.3.6)$$

$$S_1(v) \xrightarrow{k_d} \text{dissociation products} \qquad (5.3.7)$$

$$S_1(v) \underset{k_{TS}}{\overset{k_{ST}}{\rightleftharpoons}} T_1(v') \qquad \text{(intersystem crossing ISC)} \qquad (5.3.8)$$

$$S_1(v) \xrightarrow{k_{SS}} S_0(v') \qquad \text{(internal conversion)} \qquad (5.3.9)$$

$$S_1(v) + M \xrightarrow{k_c^S} S_1(v') + M \qquad \text{(collisional deactivation)} \qquad (5.3.10)$$

$$T_1(v) + M \xrightarrow{k_c^T} T_1(v') + M \qquad \text{(collisional deactivation)} \qquad (5.3.11)$$

$$T_1(v) \xrightarrow{k_p} S_0(v') + h\nu \qquad \text{(phosphorescence)} \qquad (5.3.12)$$

$$T_1(v) + T_1(v') \xrightarrow{k_{TT}} S_0(u) + S_1(u') \quad (\text{or } S_2(u')) \qquad \text{(triplet-triplet annihilation)} \qquad (5.3.13)$$

$$S_2(v) \xrightarrow{k'_d} \text{dissociation products} \quad . \qquad (5.3.14)$$

Here states belonging to the lowest excited singlet are designated as S_1, and to the lowest lying triplet as T_1; (v) or (u) and (0) represent vibrationally excited and vibrationally relaxed states, respectively. S_2 is a higher lying singlet, and M a collision partner (including, possibly ground state biacetyl). In this kinetic scheme rate constants are also assigned to radiationless transitions. In prelaser studies, no distinction was made between $v > 0$ and $v=0$ in process (5.3.8), i.e., the ISC rate was thought (sometimes implicitly) to be independent of vibrational excitation (see Sect.2.4.9 where this assumption is challenged).

Biacetyl is a favorable case for the experimentalist, as it exhibits strong fluorescence and phosphorescence in gas phase, liquid solutions, and as a solid. It was established that both fluorescence and phosphorescence are not dominated by the radiative rate constants (k_f and k_p, respectively) but rather by the radiationless processes. This is reflected by the fact that the fluorescence quantum yield is 2.5×10^{-3} [5.56] and the phosphorescence quantum yield is 0.15 [5.57]. Decay rate constants 10^8 s^{-1} for $S_1(0)$ and 5.5×10^2 s^{-1} for $T_1(0)$ were thus largely ascribed to k_{ST} and k_{TS}, respectively.

Tunable dye lasers are ideal to the study of biacetyl photophysics. They allow fast and selective excitation of a given vibrational state of S_1 at much lower pressures than nonlaser light sources. Thus they were used in the first detailed studies of processes such as (5.3.8), for which the rate constants are expected to be energy dependent. Conservation of energy requires that formation of vibrationless triplet states is due to collision-induced processes (see Fig.5.17—at least 1800 cm^{-1} have to be dissipated as translational energy). One of the early successes of laser studies was the measurement of the relevant rate constants requiring operation below 10^{-2} Torr, even with nanosecond sources.

The detailed kinetic scheme revealed by these laser studies is much more complex than suggested by (5.3.6-14). The results can be analysed using the theory of radiationless transitions introduced in Sect.2.4.9 and are believed to be characteristic of the photophysics of medium-sized molecules. It is thus instructive to consider the case in some more detail. At a low enough pressure, several decay patterns of the initially excited state were recognized, depending on excitation energy. They were divided into four major categories: fast and slow fluorescence and "hot" and thermalized phosphorescence. The fast fluorescence is the familiar fluorescence mentioned before, with a lifetime of ~10 ns (decay rate constant $\sim 10^8$ s^{-1}) governed primarily by k_{ST} and k_{SS}. The slow fluorescence was overlooked in prelaser studies. Its decay rate is strongly dependent on both pressure and excitation wavelength. Very low pressure ("zero pressure") rates vary from 10^4 to 10^7 s^{-1}, see Fig.5.18. The self-quenching rate constant, which determines the decay rate at higher pressures, also varies with initial excitation energy (Fig.5.18b). Note particularly the sharp drop of this rate constant at about 22000 cm^{-1}. The "blue" (short wavelength) part of the spectrum of the slow emission was found to be identical with that of the fast fluorescence. The relative intensity on the "red" (long wavelength) part was always larger than that of the fast fluorescence. This extra red

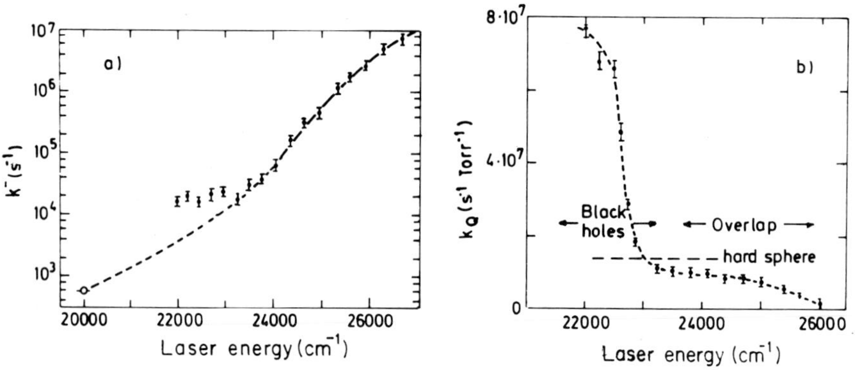

Fig.5.18. (a) (●) Observed zero pressure slow fluorescence rate as a function of excitation energy. (○) Rate of thermalized phosphorescence. k^- is
given by $k^- = k_T + k_S k_{ST}/N(k_S + k_{ST})$, where $N = k_{ST}/k_{TS}$ is the ratio of triplet
to singlet densities of states. $k_T(k_S)$ is the total triplet (singlet) decay
rate.
(b) Experimental fluorescence self-quenching constants k_q as a function of
excitation energy. See Fig.5.19 for definition of "black holes" and "overlap"
(adapted from [5.58])

emission was attributed to phosphorescence from $T_1(v)$-vibrationally excited
triplet states, being the "hot" phosphorescence mentioned above. The fourth
type had the same emission spectrum as the "hot" phosphorescence, but was
characterized by a long delay time (milliseconds). This emission, previously
refered to as biacetyl phosphorescence, will henceforth be termed-thermal
phosphorescence. The results can be rationalized using the following model
(see Sect.2.3.7). The true eigenstate of the molecule at any given energy is
given by the linear combination of singlet (S) and triplet (T) states. Only
singlet states may be reached by light absorption. Singlet states will be
populated if their energy falls within the bandwidth of the exciting source
$\Delta\nu = \tau_c^{-1}$, τ_c being the duration of the exciting laser pulse. Onc can distinguish
two extreme cases:

a) Isolated singlets. Here the density of singlet states is small, and
their separation is larger than their zero-order width. This is the case
near the origin. These states couple with resonant triplet levels that may
subsequently radiate. The situation may be described by the following scheme
[cf. (5.3.6-14)]:

$$
\begin{array}{ccc}
\overset{k_{SS_0}}{\underset{k_c^e\ \ k_f}{\nwarrow\swarrow}}\ S & \underset{k_{TS}}{\overset{k_{ST}}{\rightleftarrows}} & T\ \overset{k_{TS_0}}{\underset{k_p}{\nearrow k_c^e \searrow}}
\end{array}
\qquad (5.3.15)
$$

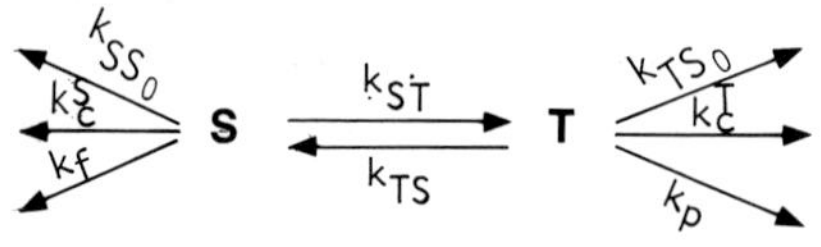

The initially excited S state is depleted by radiationless transitions to the triplet (described by the rate constant k_{ST}) and to S_0 (rate constant k_{SS_0}), by fluorescence radiation (k_f) and by collisions K_c^S. The resonant T states are populated by k_{ST} and depopulated by the reverse radiationless transition (k_{TS}), by transition to S_0 (k_{TS_0}), by phosphorescence radiation (k_p), and by collisions (k_c^T). Most triplet states are out of the range of optical excitation (due to the small width of S) and may be populated only by collisions. Once populated, return to S is impossible (except by collisions). These regions, outside the interaction width of zero order S states, were called "black holes".

b) Regions where singlet density ρ_S (Sect.2.2.2) is large, i.e., high vibrational excitation of S_1. In that case coupling between S and T states is practically ensured for any energy of excitation, in both ways. This is called the "overlap" zone. A graphic description of the black holes and overlap zones is given in Fig.5.19.

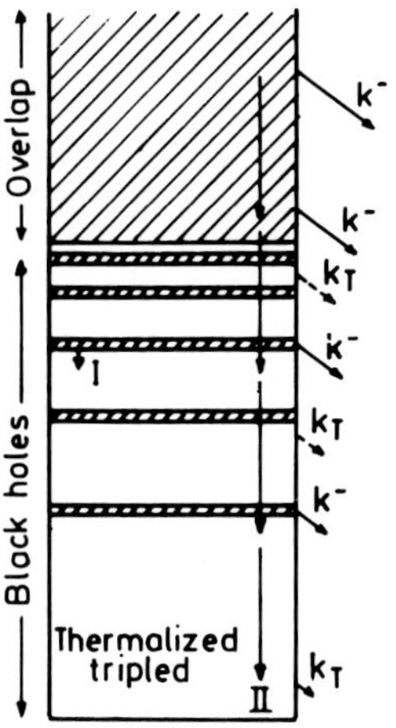

Fig.5.19. A schematic representation of the excited states of biacetyl. The fluorescent states are contained in the hatched regions, nonfluorescent states in the white regions. Vertical arrows represent collisional deactivation, k_T is the total decay rate of T_1. k^- is defined in Fig.5.18. Both increase as excitation energy is increased (shown schematically by different arrow lengths). I denotes a long-range, "soft", collision leading to efficient fluorescence quenching in the black hole region; II shows a series of "hard" collisions, transforming a molecule from the overlap region to the thermalized triplet (adapted from [5.58])

The scheme described above can account for the observed emission behavior. The singlet state decay, monitored by fluorescence, is found to consist of two exponential components, a fast one due primarily to k_{ST} and a slow one due to irreversible energy loss by collisions. The triplet is rapidly populated by intersystem crossing, and then slowly decays by radiative and nonradiative processes. These kinetic features are reflected by a fast rise and a slow decay of the phosphorescence of vibrationally excited triplet molecules. The decay rate is equal to that of the slow fluorescence. Thus the only way to distinguish between the two is by their spectral response.

Collisions are treated according to whether they affect the black hole zone or the overlap zone. In the black hole zone, a very slight perturbation, of the order of the width of S_1 ($\sim 10^{-3}$ cm^{-1}), is capable of moving the molecule to a nonfluorescent state. Thus, one expects an extremely large pressure effect in this range, (process I in Fig.5.19). On the other hand, a substantial amount of energy must be removed from a molecule in the overlap zone in order to move it to a nonfluorescent range. Such strong collisions are possible only in essentially head-on encounters, and a much smaller quenching cross section is expected. Finally, all collisions will eventually yield the thermalized triplet. The results depicted in Fig.5.18 are clearly in qualitative agreement with this scheme. Furthermore, the rate constants derived from the experimental results are consistent with calculated densities of triplet and singlet states.

We conclude this section on laser studies of radiationless transitions by a discussion of glyoxal, a 6-atomic molecule that is structurally similar to biacetyl. (The methyl groups are replaced by hydrogen atoms.) The energy level diagram is schematically the same as that shown in Fig. 5.17. A major difference is revealed when considering the density of states. For 3000 cm^{-1} excess energy in S_1, the triplet density of states is calculated to be 1.5 cm and 700 cm for glyoxal and biacetyl, respectively [5.59]. At higher energies the density of states in glyoxal is also very large. Thus, this molecule provides a possible link between a small molecule behavior and a large molecule behavior, in the context of radiationless transition theory (Sect.2.4.9).

Figure 5.20 shows that the zero pressure fluorescence lifetime strongly depends on excitation energy. These results may be interpreted on the basis of a similar model to that of biacetyl. Excitation at relatively low energies, near the singlet origin, populates a state that is weakly coupled to the zero-order triplet states, because of the low density of states.

Thus glyoxal is definitely in the "small molecule" limit of radiationless transition theory, for low excitation energies. At somewhat higher excitation one obtains a double exponential, as in biacetyl, as singlet-triplet interaction becomes stronger. Finally, at still higher energies, the fast decay dominates, due to fast relaxation of the triplet by either a photochemical process (dissociation threshold is $\sim$25000 cm^{-1}) or by decay to S_0 (compare with formaldehyde, Sect.5.3.2).

Recently, laser excitation of glyoxal was studied in molecular beams [5.60]. The laser was tuned to one of the low lying vibrational states of S_1. A second laser was fired after a short delay in order to probe the

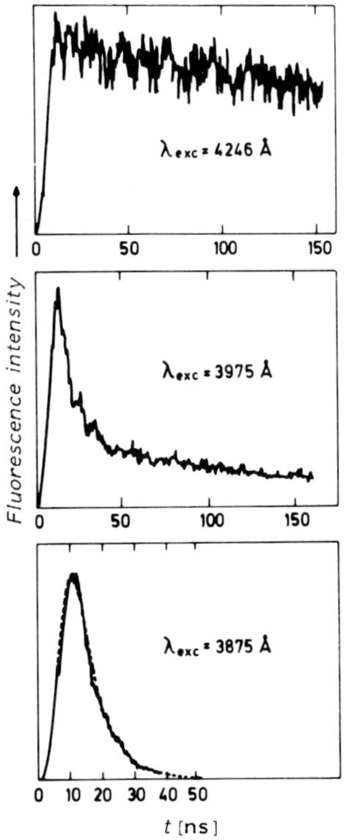

Fig.5.20. Fluorescence decay of glyoxal as a function of excitation wavelength. At energies near the origin of S_1 (λ_{exc}=4246 Å) the decay is slow, governed by the radiative lifetime. With large excess energy (λ_{exc}=3875 Å) the lifetime is very short due to efficient intramolecular energy transfer. Excitation at 3975 Å leads to an intermediate case involving a biexponential decay. The dashed curve in the lowest section is a fit to a decay lifetime of 6 ns (adapted from [5.59])

prepared state. In the case of the 4^1 state, the decay lifetime is shorter than the radiative lifetime [5.61], indicating strong $S_1 - S_0$ coupling (the energy is too low for unimolecular dissociation). For this state, it is believed that initial population of S_1 leads rapidly to dissipation of energy into high vibrational levels of S_0 ($\rho \sim 10^8$ cm). If no constraints are introduced by the initial preparation, each of the 10^8 states/cm^{-1} has an equal population probability. In that case one does not expect the probe laser to reveal a vibrational structure, but rather a broad, diffuse spectrum. In practice some surprisingly narrow features in the fluorescence excitation spectrum of the probe laser are revealed near the frequency of the pump laser (Fig.5.21). These features were interpreted as indicating that constraints are strongly imposed by the pump laser, and that in fact energy is not randomly distributed among all energetically possible vibrational levels of S_0. This conclusion must await independent verification, as monitoring only the excitation spectrum may not be sufficient to characterize the states involved in the transition. Still, the conclusion can have a major impact on the interpretation of experiments designed to populate highly

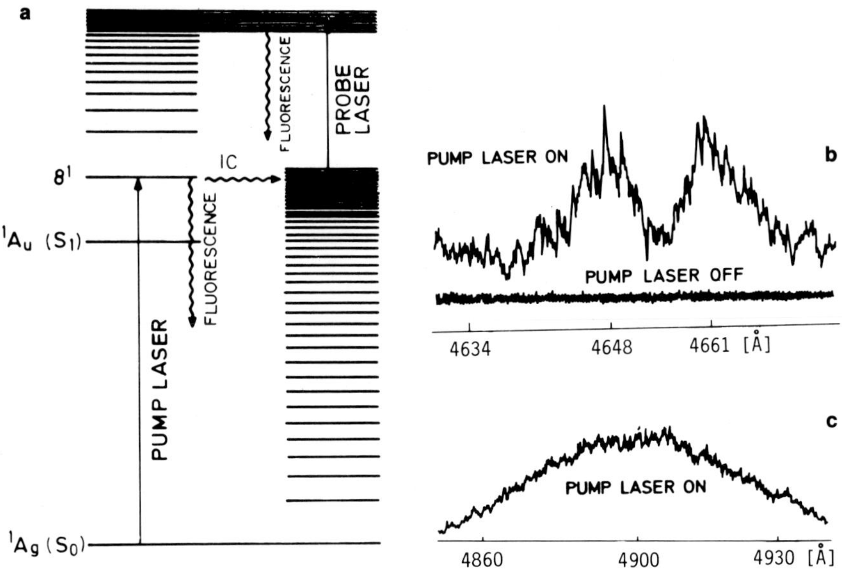

Fig.5.21. (a) The principle of the pump probe experiment: A vibrational level of S_1 (^1A_u) of glyoxal is populated by the pump laser. Rapid internal conversion spreads the energy among high lying vibrational states of S_0 (^1A_g). The probe laser, properly delayed, induces transitions from these states to high lying states of S_1. Throughout the experiment, total fluorescence is monitored. (b) Excitation spectrum obtained by scanning the probe laser using a 4403 Å pump laser [see (a)]. The lower trace is obtained when the pump laser is blocked. Note the sharp features, indicating well-separated transitions. (c) As in (b), with the probe laser operated at a longer wavelength. Only a broad spectrum is obtained (adapted from [5.60])

vibrationally excited states of S_0 in molecules by a different laser method, multiphoton absorption of infrared radiation. This topic is discussed next. A similar interpretation has been advanced to account for sharp vibrational features appearing in the two-photon excited fluorescence of 1,2,5,6 dibenzanthracene [5.62].

5.4 Multiphoton Activation and Fragmentation

Multiphoton activation leading to dissociation or ionization has been mentioned in earlier chapters (cf. Sects.1.3.2 and 2.3.5). Lasers are so far the only light sources that can provide the high photon flux required for activation of an isolated molecule by absorption of a large number of photons.

This section discusses the available experimental data and explores two main aspects of the phenomenon. First, what is the mechanism by which a molecule can absorb many photons of essentially the same frequency, even though anharmonicity tends to spoil the necessary resonance condition (Sect.1.3.2), and the extent of specificity and selectivity that can be achieved in this activation process. Section 2.3 provides the necessary background, and should be consulted for details. Briefly, one distinguishes between species selectivity, i.e., preferential activation of one species in a mixture, and mode selectivity, i.e., the possibility that not all isoenergetic states of a given molecule are equally accessible by multiphoton absorption. Thermal activation will equally populate all these states and the possibility of introducing constraints by multiphoton pumping is of considerable interest. If such constraints can be maintained up to and beyond the reaction threshold, the simple RRK (or RRKM) treatment is not applicable and mode-specific chemistry is possible.

Single-photon absorption can be understood in terms of the formalism of Chap.3 (Sects.3.1 and 3.3, mostly). Sections 5.2 and 5.3 discussed the use of lasers to promote reactions following a single (or very few) photon absorption. The advantage of lasers in this application stemmed from the fact that they can create a large population in an energetically well-defined excited state. Selection rules developed for conventional sources are applicable for laser sources. However, the perturbation approach of Sect.3.1 can be extended to higher orders to account for the simultaneous absorption of several photons by a molecule. In the simplest case of two-photon absorption, the equivalent of Beer's law is

$$dI/dx = \delta C I^2 \quad , \qquad (5.4.1)$$

where C is the concentration (density) of the absorber and δ (units: cm^6 s $photon^{-1}$ $molecule^{-1}$) is the two-photon absorption coefficient. Equation (4.1) should be compared with the usual Beer's law,

$$dI/dx = \varepsilon C I \quad . \qquad (5.4.2)$$

Similarly, the probability of the simultaneous absorption of n photons is proportional to the nth power of the intensity. By analogy with the discussion of Sect.3.1 the process can be thought of as a collision between $n+1$ particles: as n increases, the probability of many particle encounters decreases and absorption cross sections are correspondingly small.

Typical values of δ are about 10^{-50} (for strictly direct two-photon transitions without any resonant enhancement), explaining the use of lasers to bring about an observable effect. Consider, for example, a gas sample at a pressure of a few Torrs ($\sim 10^{17}$ molecules/cm^3). In order to observe a dI/I ratio as small as 10^{-6} over 1 cm path length, the light intensity must be 10^{27} photons s^{-1} cm^{-1} — an intensity which is available only from laser sources. The cross section for a transition between initial state i and final state f may be written as [5.63]

$$\delta_{if} = 128\pi^3\alpha^2\nu^2|S_{if}|^2 g(2\nu) \quad , \tag{5.4.3}$$

where α is the hyperfine constant (1/137), ν the frequency employed, $g(2\nu)$ the line shape function (Sect.3.1.2), and S_{if} is the composite matrix element of a two-photon transition. In the case of two photons with the same frequency, we have (for electric dipole transitions)

$$S_{if} = \sum_k \frac{(\underline{e}\cdot\underline{\mu}_{ik})\,(\underline{\mu}_{kf}\cdot\underline{e})}{\nu_{ki} - \nu} \quad , \tag{5.4.4}$$

where $\underline{e}$ is the polarization vector of the photon and $\underline{\mu}_{ij}$ the familiar dipole operator (Sect.3.1). Summation is carried over all available intermediate states k.

Equation (5.4.4) suggests that if an intermediate state k exists, such that $\nu_{ki} = \nu$, the transition probability S_{if} will be greatly enhanced. In the case of exact resonance, we have the simple case of two successive one-photon absorption events. However, near resonant enhancement of the coherent two-photon absorption probability occurs even for a mismatch of a few cm^{-1} [5.64]. The two-photon absorption cross section of the $3S(F=2) \rightarrow 4D_{5/2}$ or $4D_{3/2}$ transition in atomic sodium was found to increase by up to six orders of magnitude when the laser frequency was tuned close to the intermediate $3P_{3/2}$ states.

The absorption cross section becomes successively smaller as n, (the number of absorbed photons) increases in the absence of intermediate resonances, making even laser sources impractical for simultaneous n photon absorption when $n \gtrsim 10$. However, many-photon absorption in an isolated molecule is not necessarily a simultaneous event. As Fig.5.22 suggests, under suitable conditions, molecules can sequentially absorb a large number of photons. Symbolically, one might distinguish between coherent and sequential absorption by writing them as

$$A(v=0) \xrightarrow{nh\nu} A(v=n) \qquad \text{coherent}$$

$$A(v=0) \xrightarrow{h\nu} A(v=1) \xrightarrow{h\nu} \dots \xrightarrow{h\nu} A(v=n) \qquad \text{sequential} \quad .$$

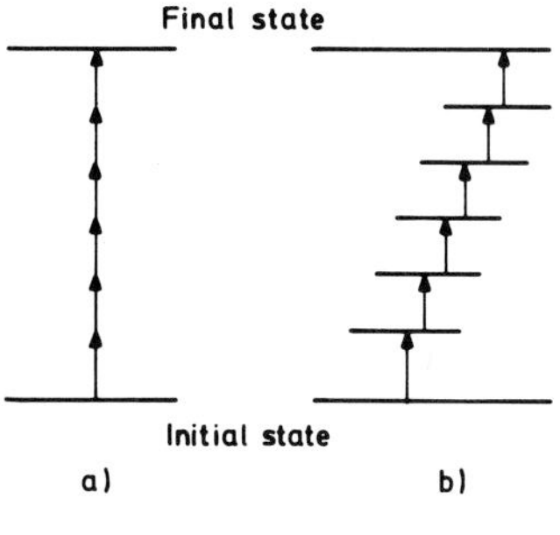

Fig.5.22. Simultaneous, or coherent absorption, of many photons (a) and sequential absorption of the same number of photons using intermediate states (b). Process (a) is similar to the two-photon absorption discussed in the text [cf. (5.4.1)], while the mechanism of process (b) is the subject matter of the rest of this section

Coherent two-photon excitation found many spectroscopic applications, as the selection rules governing the transition probability are different from those pertaining to one-photon transitions [5.63]. Most of its applications involved electronic transitions using visible or UV laser sources. It is often encountered as an intermediate step in multiphoton ionization (Sect.5.4.4). On the other hand, sequential absorption requires real intermediate states, is governed by the usual one-photon selection rules, and has been observed largely in connection with vibrational transitions. Indeed, it is found to be a widespread phenomenon and absorption of at least 30 or 40 photons leading to dissociation is a rather common occurrence with high-power laser sources. As mentioned briefly in Sect.1.3, reactions following multiphoton absorption can be made species selective, by proper choice of the laser frequency. An often quoted practical application is the separation of isotopes which was demonstrated for a variety of molecules (Sect.5.6). The power of the method may be appreciated from the single-step 1400-fold enrichment factor reported for deuterium in the multiphoton dissociation of CF_3CHCl_2 [5.65]. Extensive discussion of experimental results and applications is given in several monographs and reviews [5.66-69], which should be consulted for more complete coverage.

5.4.1 The Nature of Multiphoton Excitation

Infrared multiphoton absorption was so far observed only in polyatomic molecules. We shall thus begin our discussion by reference to a schematic energy level diagram of a polyatomic molecule, shown in Fig.5.23. The normal mode description was seen in Sect.3.3 to adequately account for the observed infrared absorption spectrum of polyatomic molecules when low-power light sources are employed. The energy levels of one such mode are shown in the figure as manifold (a). Due to anharmonicity, the separation between adjacent levels becomes smaller at higher energies, with eventual transition to a continuum at the dissociation limit. In an N atomic molecule, 3N-6 (3N-5 for linear molecules), such manifolds must be considered. The density of states becomes quite large at energies appreciably larger than the lowest characteristic frequencies. At these energies it is convenient to use the concept of energy level density ρ (cf. Sect.2.2.2). When ρ, which equals D^{-1} (D is the average separation between adjacent levels), becomes "large enough" one refers to a quasicontinuum of states (Sect.1.3.2). The actual measure is determined by the resolution of our probe. A convenient yardstick is given, for example, by the linewidth $\Delta\nu_\ell$ of the light source used to excite the molecule. The quasicontinuum onset may be taken as the energy E at which

$$D(E) = [\rho(E)]^{-1} < h\Delta\nu_\ell \quad . \tag{5.4.5}$$

In order to preserve selectivity one must keep $\Delta\nu_\ell$ smaller than the spectral shift $\Delta\nu_s$ between the absorption band of the substrate under study and other species. Otherwise the process will lead to excitation of many species in the sample resulting in higher reactivity of them all. In general, therefore, the laser width is subject to the condition

$$D(E)/h < \Delta\nu_\ell < \Delta\nu_s \quad . \tag{5.4.6}$$

for efficient, selective multiphoton excitation. The quasicontinuum is represented in Fig.5.23 by manifold (b). In typical experiments, the laser power is $10^6 - 10^8$ W/cm^2, leading to a power-broadened linewidth $\Delta\nu (=\mu E/\hbar$, Sect.3.1.3) of 1 - 10 cm^{-1} for a dipole moment μ of $\sim$1 Debye (see Figs.1.20 and 3.5). This linewidth determines the onset of the quasicontinuum, which in the case of diatomic molecules would be very near the dissociation limit, whereas in the case of medium-sized molecules, such as SF_6, it is reached at 2-3 $\times 10^3$ cm^{-1} above zero-point energy. In the case

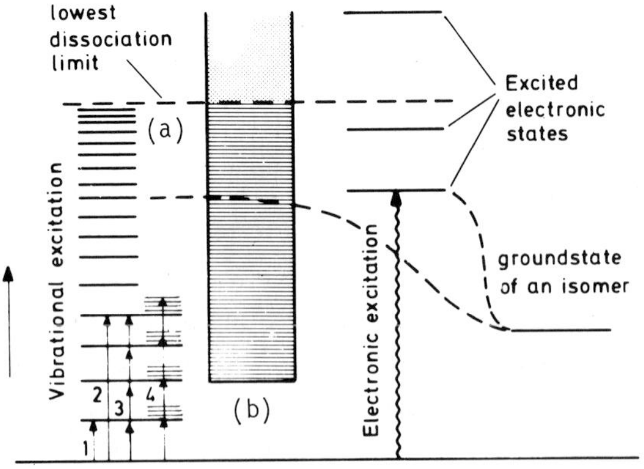

Fig.5.23. Energy level diagram of a polyatomic molecule and possible light absorption processes. (For details, see text)

of large molecules ($n \gtrsim 20$ atoms) it can be found at energies as low as 500-1000 cm^{-1} (see Fig.1.22). Note that for large molecules the zero-point energy ($\sum \frac{1}{2}\hbar\omega_i$, where ω_i are the normal mode frequencies) can be quite high and thermal excitation at moderate temperatures can lead to partial population of quasicontinuum states.

Consider now the absorption spectrum of the molecule at different incident power densities. At very low power levels, $\mu E/\hbar$ is much smaller than Doppler or pressure broadening. Typically, the absorption is described by processes 1 or 2 in Fig.5.23. Process 1 leads to the conventional infrared absorption spectrum. Process 2 is due to a $\Delta v=4$ transition, and as an overtone transition, is forbidden in the harmonic oscillator approximation (Sect.3.3). As discussed in Sect.1.3.2, such transitions can be observed with laser sources (cf. Figs.1.18 and 2.32). In many cases in polyatomics the observed linewidth is very large (~ 100 cm^{-1})—an indication of intramolecular coupling between manifolds (a) and (b) (Sect.2.3.7).

Process 3 is a simultaneous four-photon absorption of the type referred to in the previous section. Note that the resonance condition is $\Delta E=E(v=4) - E(v=0)=4h\nu$. The transition probability is proportional to I^4 and is controlled by four-photon selection rules [5.63]. As stated above [cf. (5.4.3)], the cross section for this process is usually small, but may be considerably increased by intermediate levels which are in near resonance with the laser frequency. In polyatomic molecules, the occurrence of such near resonances is quite common due to the high density of vibrational energy levels.

A fourth possibility, shown in the figure as process 4, is that of sequential absorption. Its nature requires a somewhat more detailed discussion. It was stated that due to anharmonicity, a photon in resonance with a $v \rightarrow v+1$ transition cannot resonantly interact with the $v+1 \rightarrow v+2$ transition. A mechanism compensating for the anharmonic effect must, therefore, be operative to allow process 4 to occur. One such mechanism is rotational compensation [5.71], shown in Fig.5.24.

It was shown in Sect.3.3.6 that for a given vibrational transition, the P, Q, and R branches appear at increasingly higher frequencies. As Fig.5.24a shows, it is possible to find a rotational quantum number J, for which the frequencies of the transitions $P_J(0\rightarrow1)$, $Q_{J'}(1\rightarrow2)$, and $R_{J''}(2\rightarrow3)$ are equal within the bandwidth of the laser. Once again, note that power broadening is instrumental in fine tuning the match between laser frequency and $\Delta E/h$ (Fig.5.24c). Figure 5.24a applies to a diatomic molecule. In polyatomic molecules, J is not the only quantum number that should be considered. In

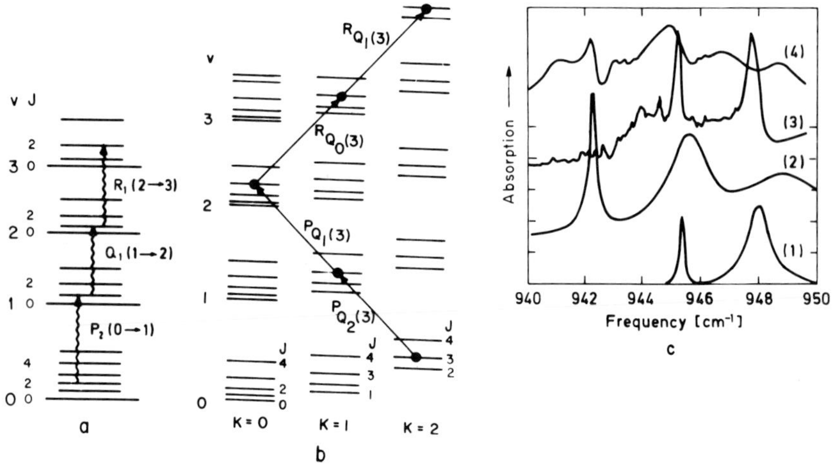

Fig.5.24a-c. Rotational compensation and power broadening mechanisms in multiphoton excitation. (a) Rotational compensation principle in a diatomic molecule. (b) Rotational compensation in a symmetric top; note that J remains unchanged throughout the excitation. (c) Model calculations on SF_6: The average number of absorbed photons <n> is plotted vs excitation frequency (adapted from [5.70]).
(1) Low laser power, no rotational compensation;
(2) high laser power, no rotational compensation;
(3) low laser power, with rotational compensation;
(4) high laser power, including rotational compensation.
The last spectrum may be compared with experimental results (Fig.5.30)

the case of the symmetric top (Sect.3.3.6) extension of the argument is straightforward. Figure 5.24b shows a sequence of possible vibration-rotation transitions on $^{P,R}Q_k$ branches of a perpendicular transition of a symmetric top (compare Fig.3.32). Throughout the sequence, the selection rules (3.3.33) $\Delta J=0$, $\Delta K=\pm1$ are obeyed. Detailed consideration of molecules with a well-known energy level diagram, such as SF_6 (Sect.3.3.6), shows that anharmonic splitting and Coriolis interaction can provide a ladder for single-photon resonances up to v=5-10.

Consider an isolated two-level system, where absorption is proportional to $\Delta N=N_i-(g_f/g_i)N_f$ (cf. Sect.3.1.1). As incident power increases ΔN approaches zero, resulting in decreased power absorption by the molecule— the transition becomes saturated. In the context of sequential multiphoton excitation, this phenomenon is sometimes referred to as the "bottleneck effect". Starting with the situation where most molecules are in the ground state, absorption probability is independent of laser power at very low power densities, decreases as saturation sets in, and increases again when the isolated two-level approximation breaks down due to coupling to other energy levels in the molecule. See below (Figs.5.26 and 5.30) for discussion of models of this coupling.

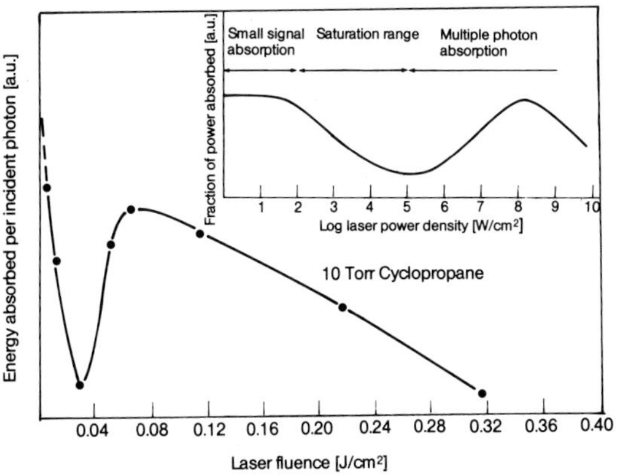

Fig.5.25. Absorption characteristics of polyatomic molecules at a fixed frequency as a function of laser power. Experimental results for cyclopropane at 3.22 μm are shown (adapted from [5.95]). The inset shows a schematic representation according to the discussion in the text, covering a much larger intensity range (note the logarithmic scale)

A schematic representation of this sequence is given in Fig.5.25, where the different absorption characteristics are displayed as a function of laser power. The transition from the saturation zone [giving rise to passive Q switching (Sect.3.2.4)] to the multiple photon absorption is characterized by a rise in the fraction of photons absorbed. Further increase of incident power leads to decreasing fractional absorption, as practically all available molecules are utilized. Increasing the power beyond a certain limit (10^{10} - 10^{11} W/cm^2) usually leads to nondiscriminatory ionization and fragmentation of molecules, followed by a complicated sequence of chain reactions. The process is often accompanied by a visible spark and is known as dielectric breakdown.

The sequence leading to multiple photon absorption through the participation of nonresonant modes is sometimes referred to as the "pumping dumping" mechanism: Energy is initially pumped into a well-defined mode, whose frequency is close to that of the laser, and is subsequently dumped to the large energy reservoir (the "bath") formed by the other vibrational modes (Fig.5.26).

In Sect.2.3.7 we found that whenever a discrete level is degenerate with a dense manifold of levels, initial excitation of the discrete level is eventually spread over the whole manifold. The exact dynamics of the process are essentially determined by the initial conditions, namely the excitation mechanism. A model calculation for the decay of excitation in a discrete level, coupled to a dense manifold, is shown in Fig.5.27.

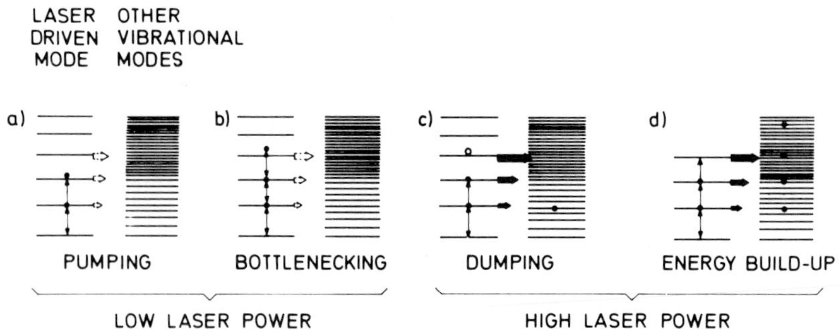

Fig.5.26a-d. Graphical illustration of the pumping dumping mechanism. The entire molecular vibrational manifold is divided to one laser-driven mode, the other modes being grouped together to constitute a heat bath.

(a) Pumping starts by absorption of laser light in the laser-driven mode, utilizing rotational compensation to reach $v > 1$ levels. (b) At low fluence levels, up-pumping is effectively blocked after a few (here, three) quanta are absorbed. Intramolecular coupling, shown by horizontal arrows, increases with energy but is too weak to overcome the bottleneck effect. (c) On increasing laser power, intramolecular coupling is assisted by power broadening, energy being dissipated into the unpumped manifold. This allows further photon absorption in the pumped mode. (d) Repeating the sequence, energy is accumulated in the molecule

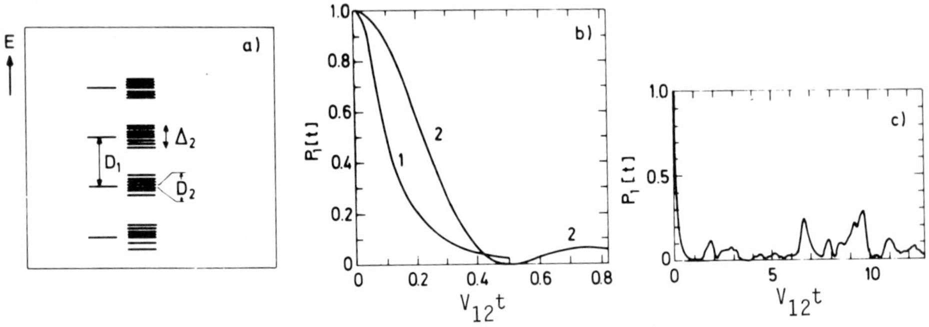

Fig.5.27a-c. Time evolution of a level belonging to the discrete region (region I) due to coupling to a group of states in the quasicontinuum region (region II). Level spacing is D_1 and D_2 in regions I and II, respectively. All region II states in a group of width Δ_2 are assumed to be coupled to a region I state with the same coupling constant V_{12}. The rate constant for depletion of the population of the discrete level k_{12} is given by (2.2.15), $k=h^{-1}V_{12}^2\rho_2$, where $\rho_2=\Delta_2^{-1}$ is the density of states in region II. Assuming that $D_1 \gg V_{12} \gg D_2$ and $\Delta_2 > [V_{12}]^2\rho_2$, the population P_1 of the discrete level is calculated as a function of time.

(a) Energy level diagram. (b) $P_1(t)$ vs time for two cases: (1) — coupling to 61 states in the quasicontinuum, (2) — coupling to only 13 states. Curve (1) is indistinguishable from an exponential decay. In curve (2) the decay is nonexponential and slower. (c) $P_1(t)$ on longer time scales for coupling to 61 states. Appreciable population is seen to recur in the discrete level at $V_{12}t \sim 7$. See Sect.2.3.7 for discussion of the recurrence time, given here by $t \sim 2\pi D_2^{-1}$ (adapted from [5.72])

Once excitation is spread over the molecular manifold, it becomes easy to meet the resonance condition. In manifold (b) (Fig.5.23) it is always possible to find a pair of levels i and f, such that $h\nu_\ell = E_f - E_i$. Note, however, that absorption probability depends on molecular symmetry and selection rules, and even in the quasicontinuum may be strongly dependent on ν_ℓ.

The physical picture of multiphoton absorption may be summarized using the often quoted "three-region" model (Fig.5.28). Initial interaction between the laser field and the molecule can be described in terms of the usual spectroscopic methods. Absorption in this region (region I) is subject to strict one- or few-photon resonance conditions, as energy level spacing is large compared to $\Delta\nu_\ell$ [cf. (5.4.6)]. The restrictions are somewhat relaxed by rotational compensation and power broadening (Fig.5.24). The states populated by laser radiation at this initial stage of the process are sometimes termed "doorway states". As soon as coupling of the doorway state to the quasicontinuum (region II) takes place, the resonant condition is relaxed, and further excitation can be achieved by sequential absorption along manifold (b). It follows that a molecule, that is transparent at some particular frequency when in its ground state, absorbs light at that frequency when region II is populated. A convincing demonstration of this property is provided by two-laser experiments (Fig.5.29): UF_6 absorbs at about 16 μm, and is virtually transparent (at room temperature) around 10 μm. A CF_4 laser at 16 μm was used to dissociate UF_6, but yields were rather low. A CO_2 laser operating at 9.6 μm is much more powerful than the CF_4 laser. [In the experiment the CF_4 laser was optically pumped by another CO_2 laser (Fig.5.28) and was an order of magnitude weaker]. Irradiation of the sample with the CO_2 laser only did not cause any dissociation, but use of both lasers simultaneously increased the yield dramatically.

In most experiments to date, a single laser frequency was used. In that case the overall absorption is determined by a weighted average of individual cross sections of the consecutive steps. As Fig.5.23 and the discussion following it show, maximum absorption is expected to shift to lower frequencies compared to single-photon absorption. This can be traced back to the fact that anharmonicity tends to decrease the separation between consecutive vibrational levels as the vibrational quantum number increases (Sect.3.3). Changes of overall absorption cross sections as a function of laser fluence level are shown in Fig.5.30. At higher fluence levels, states with a larger energy content are more heavily weighted, leading to the observed gradual red shift.

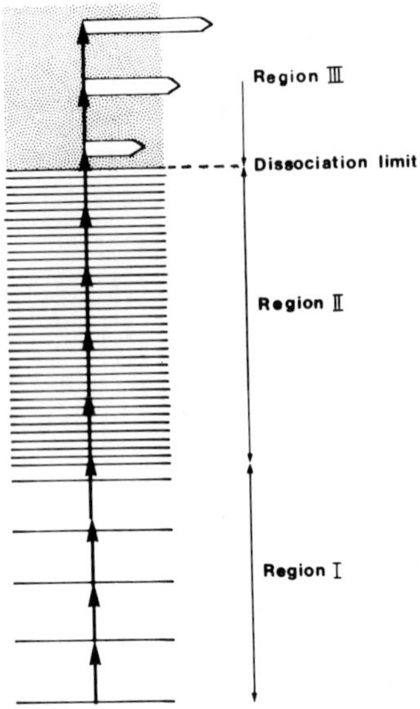

Fig.5.28. Energy level diagram displaying the three-region model of multiphoton dissociation. Initial excitation is in region I, in which energy level spacing $\Delta E(I)$ is large compared to the laser bandwidth $\Delta\nu_\ell$. Region II is the quasicontinuum region where $\Delta E(II) < \Delta\nu_\ell$. It is populated by intramolecular coupling from states in region I. Further energization of the molecule by incoherent absorption in region II leads eventually to region III, beyond the dissociation limit (following [5.75])

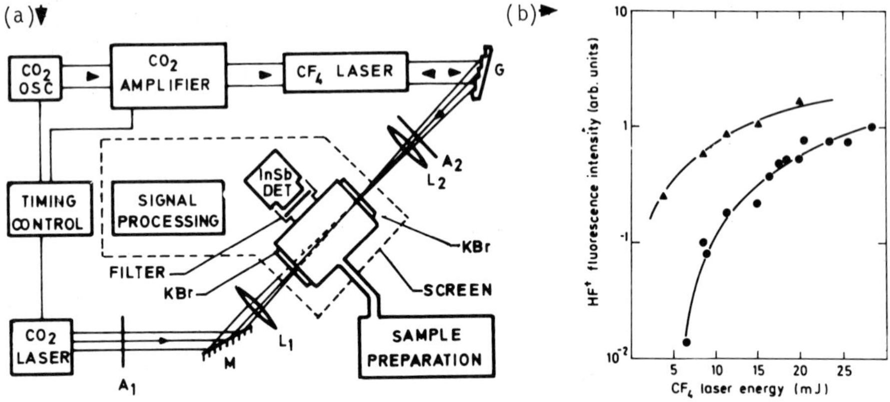

Fig.5.29a,b. Multiphoton dissociation of UF_6, illustrating the yield enhancement using two frequency excitation. A relatively weak laser excites molecules to region II, while a more powerful one, ineffective in region I, efficiently promotes the partially excited molecules to beyond the dissociation limit.

(a) The experimental set-up. (b) Dissociation yield as monitored by HF vibrational fluorescence. HF was produced by the reaction $F + H_2 \rightarrow HF(v) + H$. lower-curve – signal due to CF_4 laser irradiation; upper curve – signal when both lasers are used. CO_2 laser energy was 0.7 Joule per pulse at 1077 cm^{-1} (adapted from [5.73])

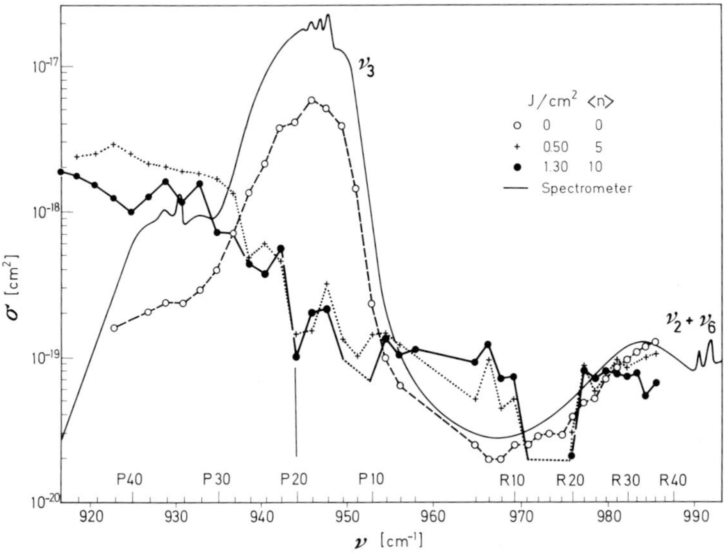

Fig.5.30. SF_6 absorption spectra taken before and immediately after excitation by a high-power CO_2 laser. The spectral range scanned is given by the available CO_2 laser probe lines. Pumping is always by the P(20) line of the 001-100 laser transition (944.2 cm^{-1}) whose spectral position is indicated in the figure. The smooth curve is a spectrometer recording. (-∘-∘-) corresponds to the spectrum taken with the probe laser without pump laser. The two other curves denote absorption with increasing average excitation $\langle q\rangle = 4$, 12 referring to the average number of absorbed quanta. The probe laser fluence is ~1 mJ/cm^2 (adapted from [5.74])

Eventually excitation carries the molecule to energies beyond the dissociation limit. In this energy range, region III, dissociation will occur, unless collisional deactivation takes place.

As discussed in Chap.2 (Sects.2.2.6, 2.3.5, and 2.3.6 in particular), the preparation of the initial state determines the constraints governing the time evolution of the molecule. Here we are interested in highly vibrationally excited molecules which can undergo dissociation. The overall dissociation rate constant can be expressed as

$$k_d = \sum_n P(n) k_d(n) \quad , \tag{5.4.7}$$

where $k_d(n)$ is the dissociation rate constant for molecules in state n and P(n) is the fraction of these molecules. If $k_d(n)$ depends only on the energy content of the molecule (as is the case in the RRK or Slater formulations) then we can write

$$k_d = \int dE P(E) k_d(E) \quad , \tag{5.4.8}$$

where the passage to the integral is legitimate because of the high density of states in the quasicontinuum. P(E) is determined by the preparation of the initial state. For thermal ("Bunsen burner heated") reactants P(E) is given by

$$P(E) = \rho(E)\exp(-E/kT)/Q \quad , \tag{5.4.9}$$

where $\rho(E)$ is the density of states of the molecule and T is the temperature. This form implies that $\langle E \rangle$ is the only constraint on P(E), so that all states in a small energy shell E, E + dE [whose number is $\rho(E)dE$], are equally probable and the relative populations of different energy shells are determined by T which is uniquely related to $\langle E \rangle$ [cf. (2.3.20)]. The thermal distribution (5.4.9) is due to collisions which enhance energy redistribution within each molecule as well as energy transfer from one molecule to another. Thus, when the excited molecules are prepared by multiphoton absorption under collision-free conditions, it is not at all obvious that P(E) will be a canonical distribution like (5.4.9). That is, the preparation of the initial state may be more selective, or in other words P(E) might be governed by more than the single constraint $\langle E \rangle$ (equivalently $\langle E \rangle_{abs}$ the average radiation energy absorbed by the molecules). In view of (5.4.8) selectively may be manifested in two forms. First, not all states $\rho(E)dE$ of a given energy are necessarily equally populated. Second, the partitioning of molecules among different energy levels is not necessarily characterized by the canonical (or "thermal") form (5.4.9). In addition, (5.4.8) itself may be only approximate in that $k_d(n)$ may depend not only on the total energy but also on the initial excitation process (Sect.2.3.7). The subsequent discussion intends to examine the extent, conditions, and consequences of selectivity in multiphoton excitation. As in thermal reactions, constraints can become apparent from the distribution of final product states. In the case of multiphoton excitation, a further constraint is imposed by the frequency-selective pumping method whose rate may conceivably compete with dissociation rate and perhaps also with energy randomization rate.

5.4.2 The Rate Equation Approach

In view of the concluding paragraphs of the previous section, one expects to reveal constraints introduced by the multiphoton excitation (MPE) process by considering the distribution of products, the energy disposal in the products, their rates of formation, etc. It turns out that a large body of experimental data can be accounted for by assuming that the rate constant is only a function of the total energy [cf. (5.4.8)] and moreover that the distribution of the total energy is closely approximated by (5.4.9). In order to develop a simple model for the distribution of excitation energies, let us consider first a limit where intramolecular energy migration is very rapid. On the time scale of up-pumping all isoenergetic states are then taken to be equiprobable. The only variable is the energy content or, equivalently, the number of absorbed photons. Furthermore, absorption in the quasicontinuum accounts for most of the acquired energy. Consequently coherent effects involving the simultaneous absorption of several photons [cf. (5.4.1)] may often be neglected. One can then describe the absorption by the use of rate equations (cf. Sects.2.5, 3.2 and 4.4). We shall use this approximation as our starting point, as it provides a convenient reference, particularly when comparison to other activation procedures is made. Where appropriate, we point out evidence showing its shortcomings. Since these usually entail some sort of specificity, they are of prime interest to laser chemists.

The rate equation describing the time evolution of molecules after absorption of n photons is [5.76]

$$\frac{dN_n}{dt} = \sigma_{n-1}[N_{n-1} - (g_{n-1}/g_n)N_n]I$$
$$-\sigma_n[N_n - (g_n/g_{n+1})N_{n+1}]I - k_nN_n \quad . \tag{5.4.10}$$

Here N_n and k_n are shorthand notations for $N(E_n)$ and $k(E_n)$ where $E_n=nh\nu$. Similarly, $\sigma_n=\sigma_{n\to n+1}$ is the absorption cross section of molecules after having absorbed n photons, and is only a function of E_n. By writing (5.4.10) we imply (5.4.8), namely, that the kinetics of the molecules are determined only by their energy content. Thus, mode selectivity is excluded. The first two terms on the rhs of (5.4.10) denote absorption and induced emission; spontaneous emission is neglected (cf. Sect.4.4). Detailed balance (Sect. 2.1.3) is satisfied by the condition $g_n\sigma_{n\to n-1}=g_{n-1}\sigma_{n-1\to n}$. k_n is a unimolecular rate constant for depletion of level n (at energy E_n) by dissociation [k_d, cf. (5.4.8)], isomerization, or internal conversion to another electronic state.

The usefulness of the rate equation approach depends on the information that one is trying to obtain. Obviously the very many rate constants appearing in (5.4.10) are generally unknown. In the absence of detailed knowledge, one can try to compute the average energy $\langle E\rangle$ deposited in a sample at a given incident fluence. Fluence (energy per unit area) rather than intensity (power per unit area) is the appropriate quantity when calculating overall yields. This follows from the assumption that the molecule is unperturbed by collisions in the interval between consecutive photon absorption

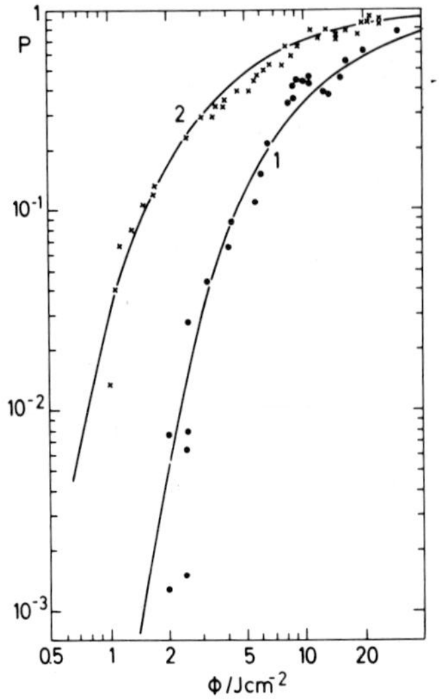

Fig.5.31. Dissociation probabilities P of SF_6 vs incident fluence of CO_2 laser radiation. Curve (1) static gas measurements, curve (2) molecular beam experiment. The dots are experimental data, the curves are a theoretical fit based on a rate equation formalism. Analysis of incremental absorption as a function of laser fluence revealed only a weak dependence, so that σ_n was taken to be constant. The degeneracy is of the form $g_n=(E_n)^{s-1}$, s being the number of modes among which energy randomization is rapid. The fit shown in the figure was obtained for $s\sim 2$ (adapted from [5.77])

events, and that energy is rapidly distributed between all available degrees of freedom. Note, however, that the rate of an individual step is proportional to the light intensity. The linear dependence of the rate of each individual step on I results in an overall dependence of the yield on $\phi=\int I dt$. Integration is carried out over the total duration of the laser pulse. One such computation is compared in Fig.5.31 with an experiment showing that the model can reasonably well account for the experimental results. We note with interest that an extremely small number of "active" degrees of freedom (actually, two out of fifteen) are needed to reproduce the shape of the yield vs fluence curve. This result may seem to indicate severe constraints in the system, as was actually suggested for CCl_2F_2 dissociation [5.78]. An example of such a constraint may be visualized as follows. If, as is often the case, the condition $\sigma_{n+1} < \sigma_n$ holds, a bottleneck will be formed at E(n+1), i.e., the fraction of molecules with energy E(n+1) will be larger than when σ is constant and not dependent on E. In extreme cases, molecules may be channelled not only to particular energy levels, but also to particular modes, especially at low energy levels. It should be cautioned, however, that conclusions as to the "nonthermal" (cf. Sect.2.3.5) character of multiphoton dissociation based on such analyses are premature at best. More detailed information on the intermediate stages is required. Thus, a different form of σ (cf. legend to Fig.5.31) may lead to an entirely different s value.[3]

Most experimental data are concerned with the end result of the excitation. According to Fig.5.23, the following processes are energetically possible:

a) Inverse internal conversion (Sect.5.3.3) to form an electronically excited state.
b) Isomerization, i.e., a rearrangement of chemical bonds with no concurrent permanent separation of fragments.
c) Dissociation, i.e., elimination of one or more atoms from the parent molecule.

Multiphoton-induced dissociation is by far the most extensively studied case and will now be discussed in some detail in this section. Electronic excitation and isomerization will be taken up in Sect.5.4.4.

An important feature concerning MPD is the competition between dissociation and further excitation (see also Sect.1.3.2). Considerable excess energy beyond the dissociation limit can be stored in the molecule provided k_n is smaller than the rate of light absorption by the state n. This property is well accounted for by (5.4.10) as can be illustrated by an example. In the quasicontinuum, σ may be tentatively assumed to be 10^{-18} cm^2 for $h\nu$ = 1000 cm^{-1}. The rate of light absorption for a 10^9 W/cm^2 laser source is about 10^{11} s^{-1}. This rate is of the same order of magnitude as energy redistribution in the molecule (Sect.2.3.6) and possibly larger than dissociation rates.

The suggested mechanisms for the energy acquisition by the molecules and the rate equation approach will now be used to account for some experimental observations.

Isotopic Selectivity

This is determined by the species-selective absorption in region I. In many cases this region consists of several (2-4) vibrational levels in the doorway mode. Anharmonicity results in a gradual shift of the absorption maximum to lower frequencies (cf. Fig.5.23). Thus one expects to obtain maximum dissociation efficiency for a given isotope at a lower frequency than that observed in the usual infrared spectrum. This effect is indeed observed (Fig.5.32).

---

3 The rate equations as written above apply to the three-region model. They also apply to the "pumping-dumping" model, provided that the "dumping" rate is much faster than the pumping rate, for any excitation level. In the more general case, one can introduce energy-dependent cross relaxation terms and obtain an intensity dependence of the dissociation rate. For details, see, e.g., [5.79].

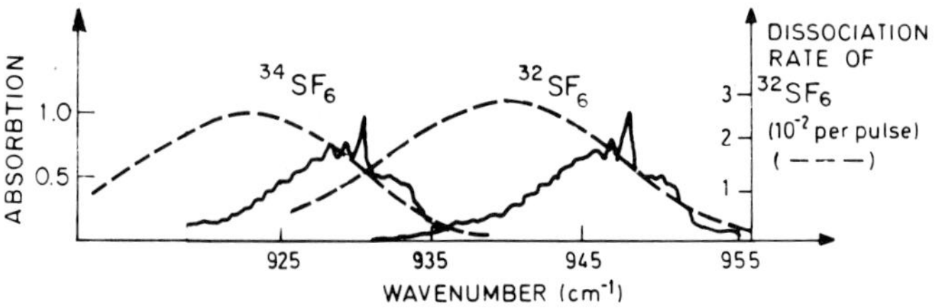

Fig.5.32. Low-power, low-resolution absorption spectra of $^{32}SF_6$ and $^{34}SF_6$ (——) and the dissociation probability per pulse as a function of frequency in high-power MPD experiments (----). The latter shows a clear red shift compared to the former. It is due to consecutive absorption in region I, where anharmonicity tends to decrease the frequency of the most probably absorbed laser photon (adapted from [5.80])

Power and Fluence Dependence

Typical power dependence curves are shown in Fig.5.33. The usual way to vary power is by inserting an attenuator in the laser beam. This technique cannot distinguish between the power and fluence dependencies: In the older SF_6 study [5.81] referred to in Fig.5.33 the dissociation yield was plotted versus laser power for the abscissa, while the more recent CF_2Cl_2 study [5.82], in line with the implications of (5.4.10), used fluence. Figure 5.33a emphasizes the threshold phenomenon: No dissociation is discernible below a certain power level. Exact determination of the threshold (fluence or intensity) depends on the sensitivity of the detection method. This is illustrated in Fig.5.33b, where smooth extrapolation to lower dissociation energies is suggested, rather than an abrupt change. A saturation effect is seen to set in above 10 J/cm^2 for CF_2Cl_2 dissociation. (Note that Fig.5.31 displays essentially the same data as Fig.5.33 and supports the view expressed here in connection with Fig.5.33b). An experimental demonstration of the role of fluence in determining overall yields is given by Fig.5.34.

In this experiment the laser pulse energy was kept constant, while the pulse length was changed by a factor of 200. As seen from the figure a rather small effect on the overall conversion yield was observed.

The absence of intensity dependence in this and similar experiments points to a single photon absorption sequence, starting in region I, reaching all the way up to the dissociation limit, and justifying the use of (5.4.10). In the absence of an appropriately spaced ladder of vibrotational states, laser intensity may affect the dissociation yield, e.g., due to simultaneous two-photon absorption steps. This situation can be expected to arise, for instance, when low to intermediate power levels are employed, causing in-

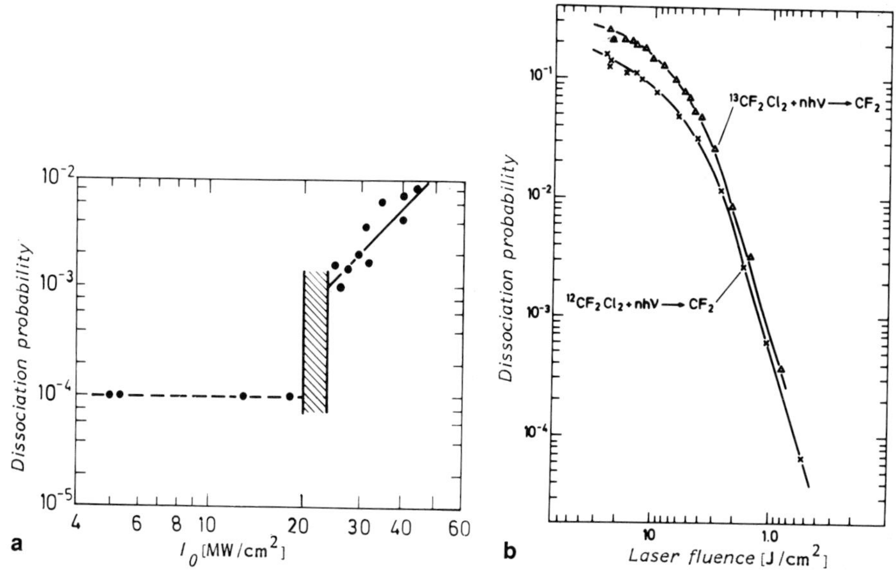

Fig.5.33. (a) Dependence of SF_6 dissociation rate on laser power density. A clear threshold is observed. The experimental method of varying the intensity could not distinguish between intensity and fluence effects (adapted from [5.81]). (b) The dependence of CF_2Cl_2 multiphoton dissociation rate on laser fluence. The threshold is clearly seen to depend on detection sensitivity (adapted from [5.82])

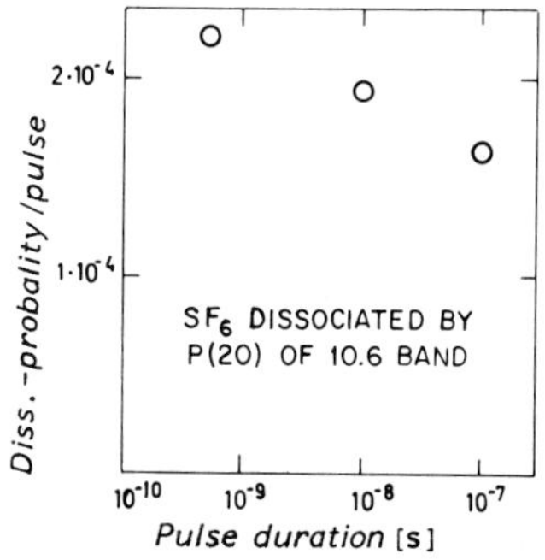

Fig.5.34. The fraction of SF_6 molecules per laser pulse as a function of pulse duration. Pulse energy was kept constant at 0.1 J. A 200-fold increase in peak power leads to only 30% increase in yield. This was taken as proof that fluence is the laser property determining the overall yield (adapted from [5.83])

sufficient power broadening. Using a gas phase "titration" technique, intensity rather than fluence was found to affect the yield of multiphoton dissociation of CF_3I [5.84]. Comparing two 0.4 J pulses, one 100 ns long and the other 1.5 μs long, the short pulse was found to give rise to a yield higher by a factor of seven. While intensity effects in region II cannot be completely ruled out, a possible interpretation is that the higher intensity of the short pulse is instrumental in overcoming the rotational bottleneck effect in region I. This can be brought about by power broadening of single-photon transitions or by near-resonance enhanced coherent n-photon absorption (n=2-4). Any interpretation should account also for the high conversion yields observed—approaching 100% at high intensity (or fluence) levels in some cases. The population of a given v,J,K state of a polyatomic molecule at room temperature is usually less than 10^{-2} of the total population. Appreciable dissociation can take place only if a large fraction of the molecules directly interact with the electromagnetic radiation. Several different modes of initial interaction, necessarily involving a several-photon initial step, are thus indicated, strongly suggesting that in region I one-photon absorption alone (even considering rotational compensation) cannot account for truly collision-free dissociation.

Intensity effects were subsequently also observed in the MPD of other molecules [5.85] by comparing the effects of pulses with equal energy content but different intensity distribution. This was achieved experimentally by spatially splitting the laser beam into two separate parts with equal energy content, slightly delaying one of them in time, and recombining them in the photolysis cell. In this way the peak intensity was reduced by a factor of two while keeping the total energy (fluence) constant. Higher intensity was found to lead to higher dissociation yields for a number of molecules (CF_3I, C_3F_7I, CrO_2Cl_2, trans 1,2 $C_2H_2Cl_2$, and CF_3COCF_3). It is concluded that in general the rate equation (5.4.10) should include terms with I^n ($n \geqq 2$) (see also Fig.5.37 and [5.88]).

Time Dependence of Product Appearance

Consider an isolated large molecule in the presence of a laser field with the appropriate frequency. According to the rate equation approach, the molecule will eventually dissociate even at very low power levels if bottlenecking is avoided and that provided the molecule is kept isolated for the whole duration of the experiment. In practice one has to take into account the spontaneous emission rate, neglected in (5.4.10). This limits the allowed duration of the experiment to a few seconds, at most. One practical way to keep molecules under almost collision-free conditions for relatively long time intervals is to put molecular ions in an ion cyclotron resonance trap [5.86]. It has been reported that with this arrangement dissociation of a molecular dimer (5.4.11) can be achieved using a cw laser with an average power of only 4 W/cm^2 (Fig.5.35).

$$[(C_2H_5)_2O]_2H^+ \rightarrow (C_2H_5)_2OH^+ + (C_2H_5)_2O \quad . \qquad (5.4.11)$$

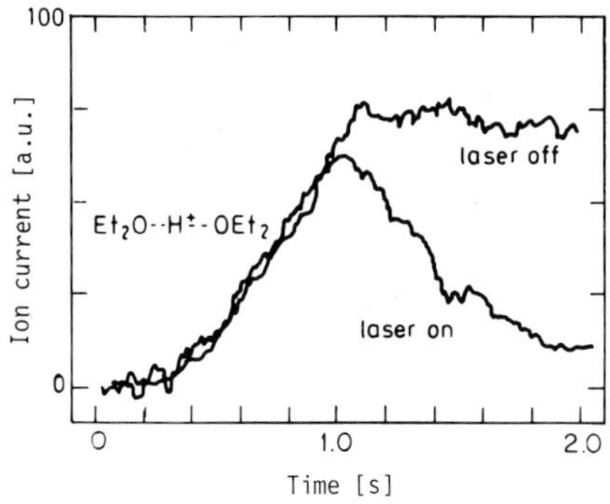

Fig.5.35. Multiphoton dissociation of the $[(C_2H_5)_2O]_2H^+$ dimer by a cw low power CO_2 laser. In the absence of laser radiation, a steady state concentration is reached after 1 s, as reflected by the constant signal intensity. The decrease in dimer concentration when the laser is on is due to dissociation (adapted from [5.86])

The analysis of this experiment requires some caution, as the ion dimer $(E_tO)_2H^+$ is produced by collisions, indicating that on the time scale of the experiment collisions may have a role in the dissociation mechanism as well.

Collisional interference can be controlled using pulsed lasers and varying the pressure. Time-resolved monitoring of the products is required to ensure the absence of collisional processes. One example is the multiphoton dissociation of tetramethyldioxetane, a molecule yielding electronically excited acetone on thermal dissociation,

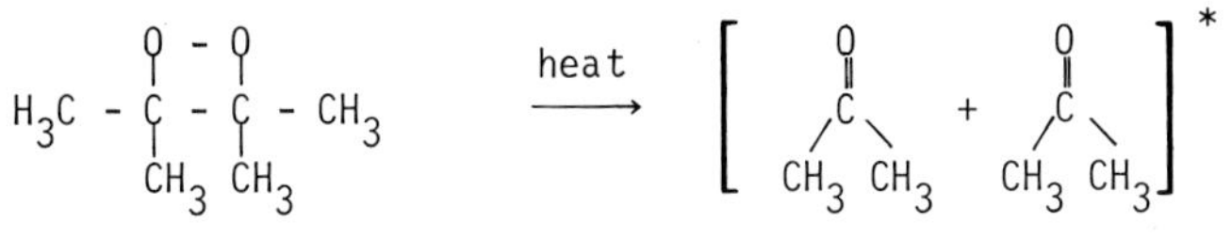

$$
\begin{array}{c}
\text{H}_3\text{C} - \overset{\overset{\text{O} - \text{O}}{|\quad|}}{\underset{\underset{\text{CH}_3 \ \text{CH}_3}{|\quad|}}{\text{C} - \text{C}}} - \text{CH}_3
\end{array}
\quad \xrightarrow{\text{heat}} \quad
\left[
\underset{\text{CH}_3 \ \text{CH}_3}{\overset{\overset{\text{O}}{\|}}{\text{C}}} \ + \ \underset{\text{CH}_3 \ \text{CH}_3}{\overset{\overset{\text{O}}{\|}}{\text{C}}}
\right]^*
\qquad (5.4.12)
$$

Tetramethyldioxetane   Two acetone molecules, one of
                                        which is electronically excited

As Fig.5.36 shows, the onset of chemiluminescence is delayed with respect to the laser pulse, i.e., luminescence starts only after the laser pulse passed its peak power. Increasing the fluence resulted in a decrease in the delay period. One also finds that the energy content of the laser pulse during the delay period is constant, i.e., independent of the intensity during that period. These observations are in qualitative agreement with the predictions of the rate equation model, provided that the overall rate is determined by the sequential absorption processes in region II and not by dissociation processes in region III.

Similar observations were made in other cases. One of them is the elimination of HCl from CF_2CHCl [5.88]. Real time monitoring was provided by laser-induced fluorescence from CF_2 excited by a dye laser. In this study the product yield is found to be a function of the laser *intensity* at a

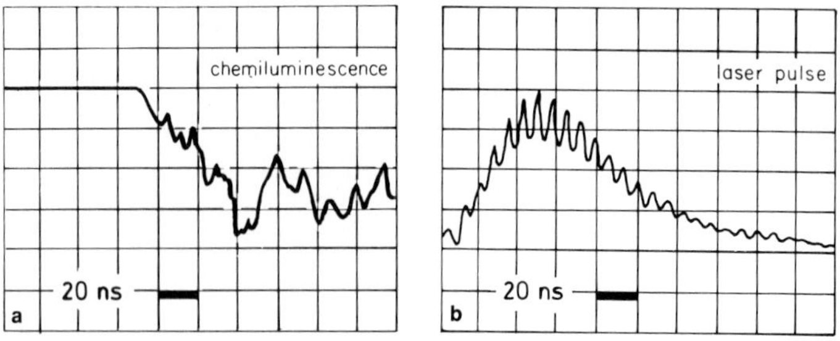

Fig.5.36. A synchronous recording of the CO_2 TEA laser pulse (a) and the chemiluminescence resulting from the multiphoton dissociation of tetramethyldioxetane (b). The luminescence is seen to be delayed with respect to the laser by about 70 ns (adapted from [5.87])

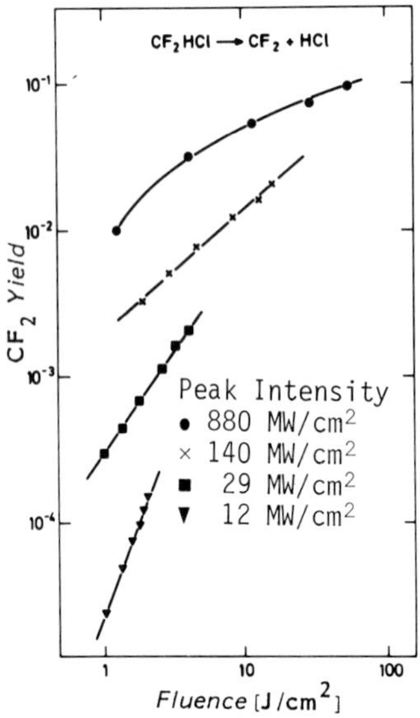

Fig.5.37. Intensity effects on the yield in the multiphoton dissociation of CF_2HCl. The data points for a given laser pulse represent the yield of CF_2 observed at various times τ_D during the pulse, plotted against the fluence $\phi(\tau_D)=\int_0^{\tau_D}\{I(t)dt\}$. For the most intense pulse, there is 400 times more product formed during the first 25 ns than during 155 ns irradiation by the weakest pulse. The absence of data at fluences below 1 J/cm^2 is due to an induction period similar to that observed in Fig.5.36 (adapted from [5.89])

given fluence (Fig.5.37). This is in contrast to the results obtained with SF_6 (Fig.5.34). One possibility is that in small molecules such as CF_2HCl bottlenecking in region I determines the overall rate, and power broadening due to increased intensity is the decisive factor. Indeed, the addition of a large excess of argon eliminates intensity effects on the yield [5.89], as expected for pressure broadening of absorption bands, and rotational relaxation.

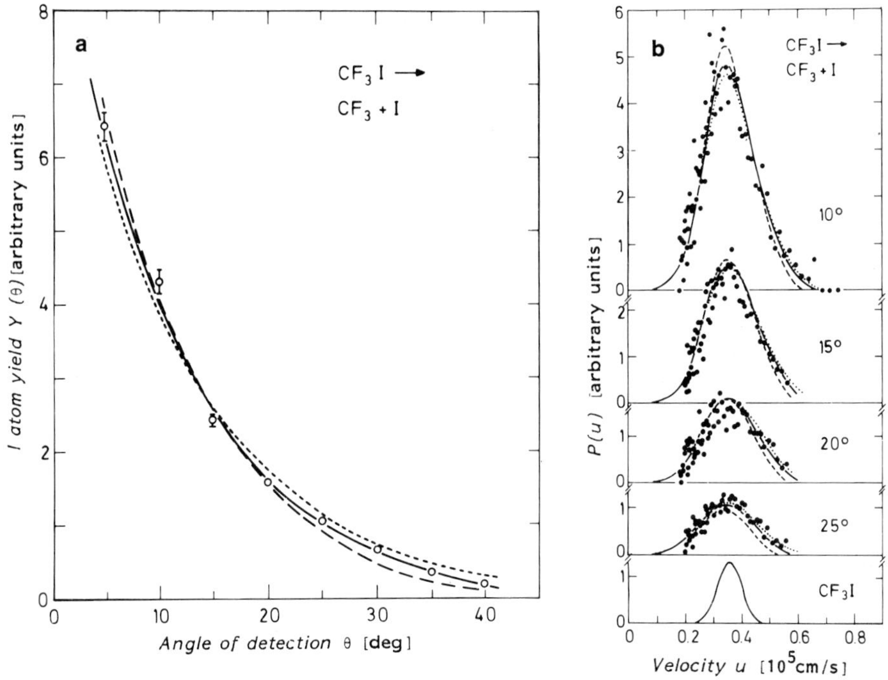

Fig.5.38. (a) Experimental (o) and calculated angular distribution of iodine atom from CF_3I multiphoton dissociation in a molecular beam experiment. Calculations used RRKM theory with the following excess energies beyond dissociation limit: (----) 3 kcal/mole, (——) 4.5 kcal/mole, (— — — —) 6 kcal/mole. (b) Velocity distribution of iodine atom from the same experiment at 10°, 15°, 20° and 25° from the molecular beam direction. Symbols as in (a). CF_3I beam velocity distribution is given in the bottom trace (adapted from [5.90])

Energy Disposal in Products

The most compelling evidence showing the unimolecular character of multiphoton dissociation comes from time-resolved experiments and from molecular beam studies. In the latter technique a laser beam is crossed with a molecular beam (effective pressure 10^{-5} Torr) and fragment yields are monitored as a function of the relative velocity and recoil angle from the direction of the center-of-mass trajectory prior to fragmentation. The results for the reaction $CF_3I \rightarrow CF_3 + I$ are compared with RRKM theory in Fig.5.38.

It is recalled that intensity (and not only fluence) was found to strongly affect the multiphoton dissociation yield of this reaction. The good agreement between the observed angular and velocity distributions and the predictions of RRKM theory (Fig.5.38) shows that even if the absorption process includes nonlinear contributions, all isoenergetic states are equiprobable.

Other reaction types that were also investigated in some detail are three- or four-center eliminations, such as

$$CF_2HCl \rightarrow CF_2 + HCl(v) \qquad (5.4.13)$$

$$H_2C{=}CHF \rightarrow HC{\equiv}CH + HF(v) \quad . \qquad (5.4.14)$$

Here, vibrational excitation of the products (e.g., HX, X=Cl,F) is possible, as there is a considerable barrier for the back reaction and a considerable change in H-X bond order is invoked (Sect.2.3.3). The extent of this excitation depends on the details of the potential energy surface. Indeed, strong infrared emission due to vibrationally excited HF was observed (Fig. 5.39).

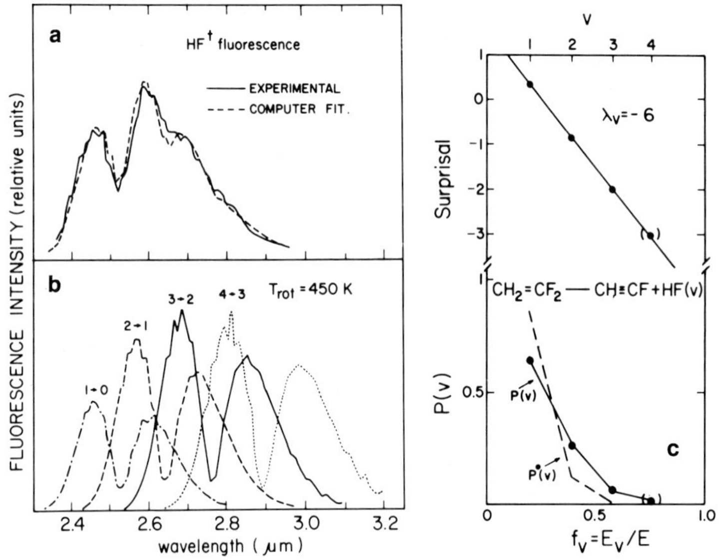

Fig.5.39a-c. Vibrational distribution of HF formed by the multiphoton dissociation of 1,1 difluoro ethylene.
(a) The observed fluorescence spectrum, and a computer fit from which the relative populations were derived. (b) The calculated emission spectrum from the first four excited vibrational states of HF, assuming equal populations. The computer fit in (a) was performed using this spectrum. (c) Surprisal analysis. The v=4 point is extrapolated; the experiment provided only an upper limit (after [5.91])

The translational energy distributions in (5.4.13) and (5.4.14) do not follow simple statistical behavior—one has to introduce constraints imposed by the nature of the potential. Figure 5.40 shows the comparison of experimental results with a prior distribution (dots) assuming a simple repulsive

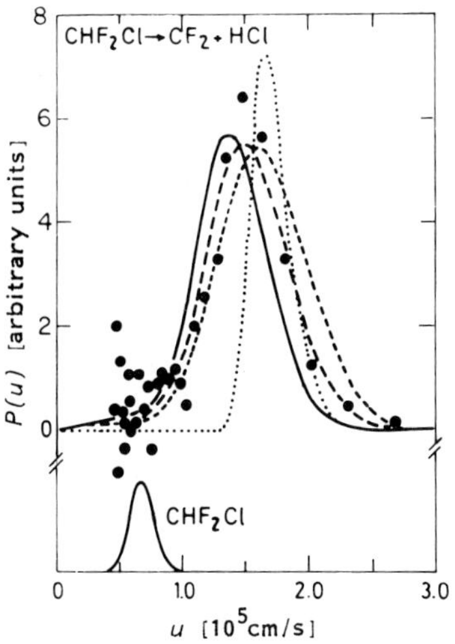

Fig.5.40. Velocity distribution in the $CF_2HCl \rightarrow CF_2 + HCl$ reaction. Experimental results (solid circles) are compared with fragment velocity distributions calculated by assuming a translational energy distribution of the form, $P(E_T) \propto E_T^2 \exp(-3E_T/\langle E_T\rangle)$, $[E_T=(1/2)\mu u^2,\ P(E_T)=P(u)du/dE_T \propto P(u)/u]$. Here $\langle E_T\rangle$ is the average translational energy reflecting a constraint needed to account for the results (Sect.2.4). The best fit is with $\langle E\rangle$=8 kcal/mole, and a FWHM of 8 kcal/mole. The dotted curve shows the fragment distribution expected from RRKM theory for a dissociation life-time of 1.5 ns (corresponding to 8 kcal/mole energy above the dissociation limit), assuming a back reaction barrier of 6 kcal/mole and that all excess energy appears as *translational* energy of the fragments. Note that the most probable velocity obtained by this method is near the experimental result, but the width of the distribution is too narrow. The result indicates vibrational excitation of the fragments (adapted from [5.92])

potential. As expected, agreement is poor. The introduction of a constraint on the mean translational energy of HF suffices, however, to account for the data.

Mode-Selective Chemistry

An interesting case may arise when two or more reaction channels are open. The rate equation model then predicts that one may control the branching ratio by varying laser fluence. Once the lowest dissociation threshold is reached, decomposition competes with further absorption. Increasing laser intensity tends to enhance the more endoergic reaction, much in the same way as increasing the temperature in thermal reactions. An example is the dissociation of CF_2Cl_2 according to

$$CF_2Cl_2 \rightarrow CF_2 + Cl_2 \quad \Delta H = 74\ \text{kcal/mole} \tag{5.4.15}$$

$$CF_2Cl_2 \rightarrow CF_2Cl + Cl \quad \Delta H = 78\ \text{kcal/mole}\ . \tag{5.4.16}$$

This reaction was studied in molecular beams [5.90] as well as by LIF (Sect. 5.5.2) of CF_2 [5.93]. Even though reaction (5.4.16) is more endoergic, it was the only one observed in the beam experiments, possibly due to a barrier to the four-center elimination.

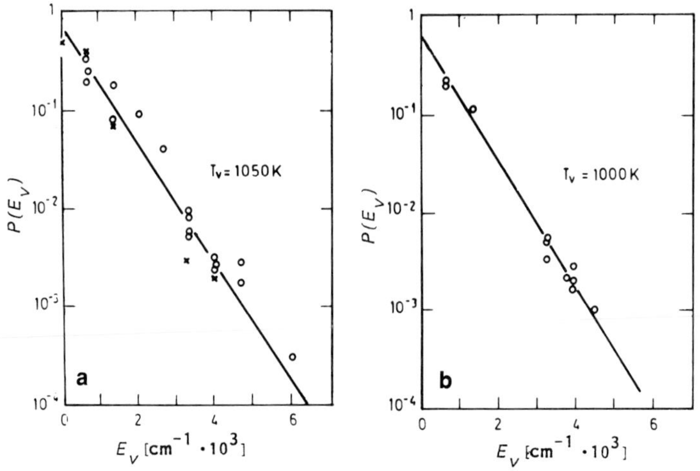

Fig.5.41. (a) The fraction $P(E_V)$ of CF_2 photofragments which were initially formed with vibrational energy E_V in the ν_2 bending mode when CF_2Cl_2 was dissociated by pulses from a CO_2 TEA laser. The CO_2 laser operated on the P(34) line of the 10.6 μm band. This transition excited the ν_8 rocking motion in CF_2Cl_2. The circles and crosses are data obtained for a CO_2 laser fluence of 5.7 J/cm^2 and 2.7 J/cm^2, respectively. The data yield a vibrational temperature T_V=1050 ± 100 K. (b) The fraction $P(E_V)$ of CF_2 photofragments which were initially formed with vibrational energy E_V when CF_2Cl_2 was dissociated by pulses from a CO_2 TEA laser. Data are given for the vibrational levels CF_2(0,n,0) for n=0,2,3,5, and 6, and for CF_2 (1,4,0) and CF_2 (1,5,0). The laser operated on the R(26) line of the 9.4 μm CO_2 band, with a fluence of 4 J/cm^2. This transition excited the CF_2Cl_2 ν_1 stretching motion. The data yield a vibrational temperature T_V=1000 ± 100 K (adapted from [5.93])

The LIF method monitors only the CF_2 fragment. It was estimated that the maximum probability for its collision-free formation is $P \sim 0.7 \pm 0.25$ at 29 J/cm^2. However, it is not clear whether CF_2 is formed by (5.4.15) or by a two-step mechanism in which CF_2Cl formed by (5.4.16) is dissociated by absorption of more infrared photons. Such sequential mechanisms were postulated for many MPD reactions [5.90]. LIF provides the rotational and vibrational distributions of the nascent CF_2 fragment. Figure 5.41 shows the distribution of vibrational energy in the ν_2 mode of CF_2. It is seen that the distribution may be described by a Boltzmann-like expression $P(E_V) \sim \exp(-E_V/kT_V)$ [cf. (2.2.21)] showing essentially the same temperature for two different laser lines, as well as for different fluence levels. The data of Fig.5.41, combined with the rotational distribution obtained from LIF measurements, allow the determination of the average kinetic energy of the products. This was found to be 1.5 ± 0.5 kcal/mole. For CF_2HCl and CF_2Br_2 parent molecules, the average kinetic energy was 1.7 ± 0.5 and 6.9 ± 2 kca/mole.

The CF_2HCl reaction can be directly compared with beam mass spectrometric measurements (Fig.5.40). Agreement between the two methods, as far as average energies are concerned, is remarkable. The results are consistent with an energy barrier of 6 kcal/mole for the back reaction. Our discussion so far failed to reveal evidence for mode selective chemistry in the sense of Sect. 2.3.5. The problem is whether energy in the excited molecules is randomized among all energetically available modes, or is perhaps localized in a specific mode or combination of modes. It appears that most dissociation data are compatible with the former: Energy is distributed in the molecule (at the time of dissociation) in the same way as if it were thermally excited. In other words, no extra constraints were introduced by this form of laser excitation. In principle, however, one can envision laser-induced mode-selective reactions, provided that energy acquisition from the electromagnetic field can compete with energy randomization. This may be the case, if, for instance, the vibrational modes of the molecule may be partitioned into subgroups, such that energy randomization is fast within each subgroup, but crosstalk between different subgroups is slow compared to either energy acquisition from the field or to the dissociation [5.94]. An experiment that may be interpreted along these lines is the irradiation of cyclopropane at two different frequencies [5.95]. It was found that multiphoton absorption of 3.2 μm photons leads primarily to isomerization, whereas excitation at 9.6 μm results in more pronounced dissociation. Inert gas addition increases isomerization yield in the case of 9.6 μm irradiation, as expected, since isomerization is the energetically more favorable route. On the other hand, addition of argon during irradiation at 3.2 μm leads to more extensive dissociation—a result explained by collision-induced vibrational redistribution of energy between otherwise noncommunicating subgroups.

A different interpretation of the experiment might be that at 9.6 μm isomerization is followed by fast dissociation, particularly in view of the fact that the 9.6 μm laser pulse was much longer than the 3.2 μm laser pulse (400 ns compared to 40 ns). Real time monitoring of the reaction products could be instrumental in elucidating this point.

Another process which is species selective and potentially "nonthermal" is unimolecular resonant multiphoton dissociative ionization of molecules such as benzene, discussed below (Sect.5.4.4).

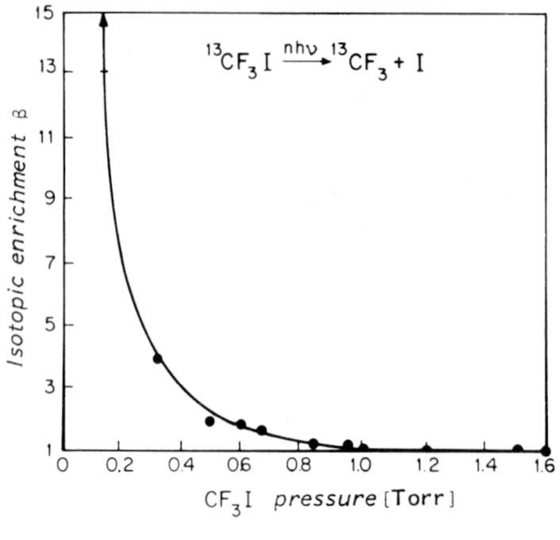

Fig.5.42. Isotopic selectivity of product formation as a function of pressure: a $^{12}CF_3I/^{13}CF_3I$ mixture is irradiated at the R(14) line of the 9.6 μm CO_2 laser transition. Power was 5.5 MW/cm^2. β is the isotope ratio in the products (P_{12}/P_{13}) compared to the ratio expected for a nonselective process (adapted from [5.96])

Collisional Effects

Isotope separation and other practical applications of MPD, call for maximum utilization of laser photons. Increasing the sample pressure is a simple way to maximize the probability of photon absorption, but carries the penalty of selectivity loss. Figure 5.42 shows that isotopic selectivity decreases rapidly with pressure in the case of CF_3I. A pressure-induced decrease of isotopic selectivity was observed in other systems, and appears to be a general phenomenon. It can be readily accounted for by collisional energy transfer from laser-excited molecules to nonabsorbing molecules. Complete collision-induced sharing of energy among all molecules would lead to a uniform temperature in the sample cell, and to total loss of species selectivity. This situation is sometimes referred to as thermal conditions or "laser heating", while a system for which energy sharing is incomplete is called "nonthermal". Obviously, nonthermal conditions may hold even under fairly high pressures, depending on reaction and energy transfer rates. The appearance of "different vibrational temperatures" as discussed in Sect.2.5.5 is an example of laser pumping with collisional energy transfer where energy sharing is incomplete.

A schematic representation of "low", "intermediate", and "high" pressure conditions is given in Fig.5.43. The figure shows that as long as τ_{V-V}, the vibrational relaxation time, is longer than the laser pulse, species-selective chemistry should be feasible. Thus "low" and "high" pressures are to be understood in terms of the associated V-V relaxation times.

Selective dissociation is easily monitored by irradiating a mixture of two different molecules, only one of which absorbs at the laser frequency.

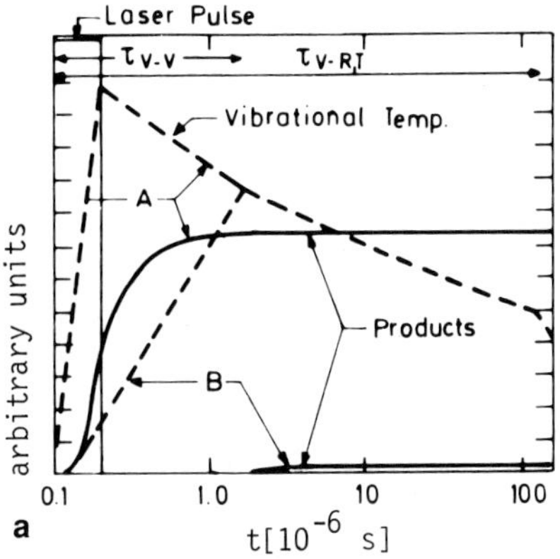

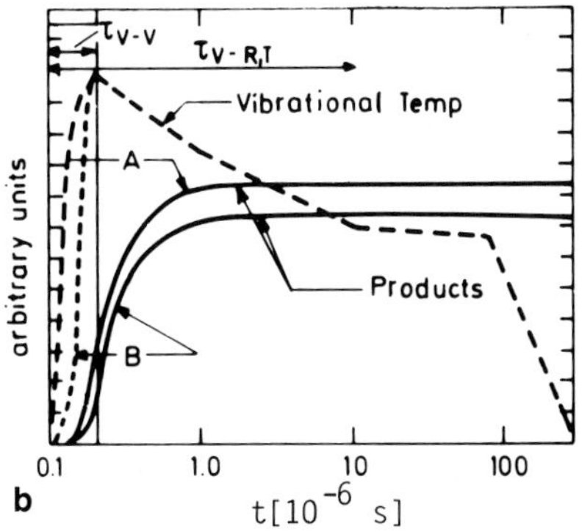

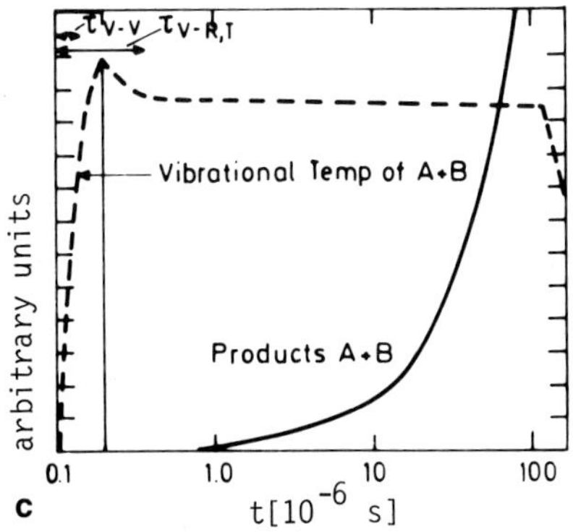

Fig.5.43a-c. Schematic representation of relaxation processes and chemical transformations occuring in a mixture of an absorbing molecule A and a non-absorbing molecule B.
(a) "Low-pressure" regime: The vibrational temperature of A is higher than that of B long after termination of the laser pulse. A is essentially the only reacting molecule. (b) Intermediate case. (c) "High-pressure" regimes: Vibrational temperature is uniform at all times. Both molecules react, as in a thermally activated system (adapted from [5.97])

Comparison can be made with shock tube (thermal) measurements in which comparable temperatures are nonselectively obtained by collisions. Complications due to subsequent radical chain reactions can be avoided by choosing reactions leading to stable products in one step. For example, the following reactions were studied [5.98], using a CO_2 TEA laser:

$$\text{i - }C_3H_7Br \rightarrow C_3H_6 + HBr \qquad k_{IBr} = 10^{13.70}\exp(-47.200/RT)s^{-1}$$

$$CH_3COOC_2H_5 \rightarrow C_2H_4 + CH_3COOH \qquad k_{EA} = 10^{12.59}\exp(-48.000/RT)s^{-1} \quad . \qquad (5.4.18)$$

The first-order rate constants are known and quite similar. Ethylacetate (EA) strongly absorbs the R(24) line of the 9.5 μm band, while isopropylbromide (IBr) is essentially transparent. Figure 5.44 shows the observed ratio of dissociation rate constants of these molecules as a function of total pressure. It is seen that selective excitation is possible even under comparatively high pressure conditions. This result may be accounted for in

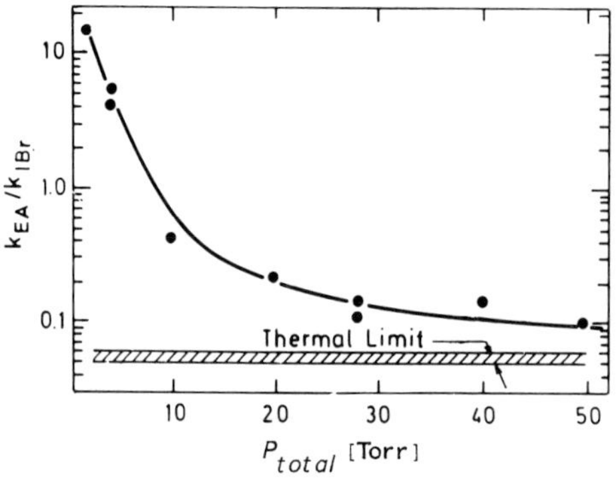

Fig.5.44. Plot of k_{EA}/k_{IBr} vs total pressure in the multiphoton dissociation of a mixture of ethylacetate and isopropylbromide. Concentration ratio was held constant at [EA]/[IBr] =3. Hatched stripe indicates k_{EA}/k_{IBr} under thermal conditions in the range 900-1300 K. The rate constant is defined by the equation $k=(1/jt)\sum_{n=1}^{j}\ln(C_{n-1}/C_n)$, where C_n is the concentration after n laser pulses, t is the heating time during and possibly after each pulse, and j is the total number of laser pulses employed in a single experiment. In view of the ambiguity in t, which can be as short as the laser pulse duration (100 ns) or as long as the heat pulse following it (200 μs) absolute rate constants cannot be determined. In the limit of low total conversions, where $C_n \simeq C_{n-1}$, the sum reduced to $k=(1/jt)\ln(C_0/C_j)$. For reactions with similar activation energies, the effective reaction times are the same, so that one can use $k(1)/k(2)=\ln(C_0^{(1)}/C_j^{(1)})\ln(C_0^{(2)}/C_j^{(2)})$ for the ratio of the rate constants (adapted from [5.99]

terms of competition between reaction and energy transfer. Note that the data do not imply nonrandomization of energy in the dissociating molecule. In fact, the products obtained from a given molecule in the mixture were independent of the total pressure.

The effect of inert gas addition has been mentioned before (P.418), as eliminating intensity effects. This can be accounted for by rotational relaxation and pressure broadening leading to removal of bottlenecking in region I. Evidence in favor of rotational relaxation is provided by Fig. 5.45. Small amounts of helium are seen to substantially increase the absorption cross section of SF_6 at 10^6 W/cm^2 power. Pressure broadening is about 10-100 MHz in this case, and cannot account for the effect (power broadening is about 1.6 GHz).

Foreign gas addition can serve to prevent scrambling and thus preserve species selectivity. Excited molecules that were not dissociated during the laser pulse may transfer energy to other species that were not directly activated by the laser. The collision-induced activation is avoided in the presence of a large excess of foreign gas. Overall yield may be lower, as only molecules that were directly activated by the laser can react, but selectivity is enhanced. This effect is shown in Figs.5.46 and 5.47. Figure 5.46 should be compared with Fig.5.44: The "thermalization" effect of increased total pressure is partially offset by addition of helium. Figure 5.47 displays a real time measurement of tetramethyldioxetane multiphoton

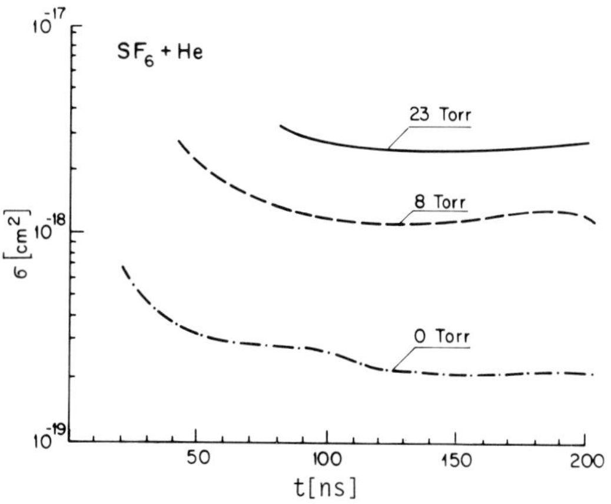

Fig.5.45. The absorption cross of SF_6 at the P(20) line of the 10.6 μm CO_2 laser band, with different pressures of added helium. The laser energy was 100 mJ in a 100 ns pulse (adapted from [5.100])

dissociation. In the absence of added nitrogen, dissociation persists for many microseconds after termination of the laser pulse due to collisions between hot molecules. Addition of nitrogen eliminates this process and only prompt dissociation (within the laser pulse duration) is observed.

The data of Figs.5.46 and 5.47 suggest that in some cases inert gas addition might help preserve isotopic selectivity. This possibility has not been extensively studied; in one case, chlorine isotope enrichment by CF_2Cl_2 dissociation, added helium appears to enhance selectivity slightly [5.101].

5.4.3 Nondissociative Reactions Induced by Multiphoton Absorption

Figure 5.23 shows that dissociation is only one channel open to vibrationally excited molecules. An important alternative is isomerization. Multiphoton activation has been used to drive isomerization reactions "uphill", namely to convert a thermodynamically favorable isomer to a less stable one. An example is provided by the pair hexafluorobutadiene (HFB) and hexafluorocyclobutene (HFCB) [5.102]. Figure 5.48 shows the energetics of the systems. The thermodynamically stable HFCB is found to absorb strongly at 10.54 μm, while absorption by HFB is negligible. At high fluence levels, HFCB is largely dissociated. Reducing the fluence makes isomerization the dominant pathway, with complete quantitative conversion to HFB obtained only on addition of helium in large excess. One way to account for this result along the lines of the previous section is the following: Complete conversion in

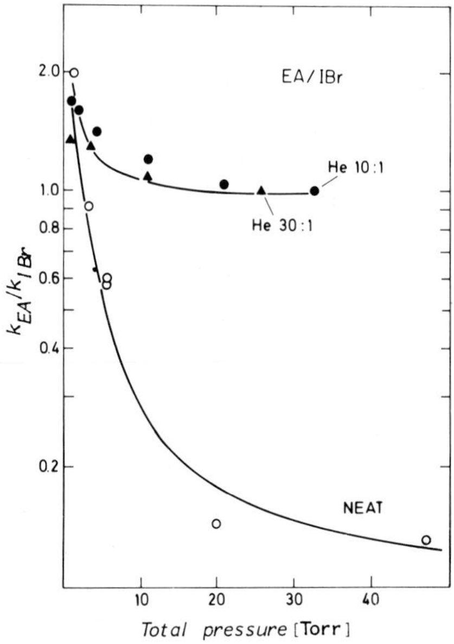

◄ Fig.5.46

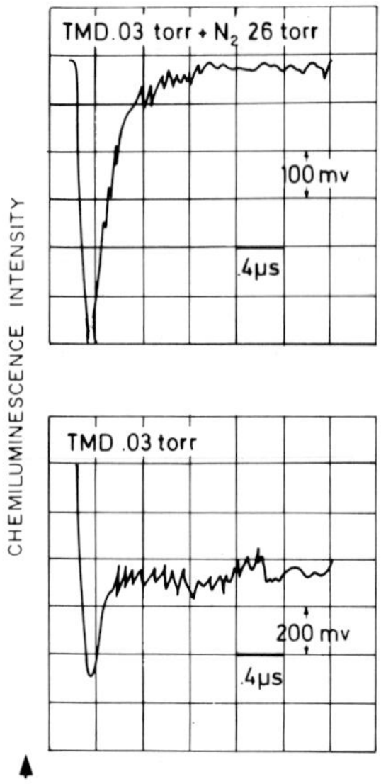

▲ Fig.5.47

Fig.5.46. Relative rates of decomposition of a 3:1 mixture of ethylacetate and isopropyl bromide. Compare with Fig.5.44 where no helium was added (a smaller laser fluence was used in this experiment). The solid line results from model calculations assuming collisional deactivation by helium (adapted from [5.98])

Fig.5.47. Effect of added N_2 on the multiphoton dissociation of tetramethyldioxetane. Laser pulse width was about 100 ns. In the absence of nitrogen, chemiluminescence persists long after the laser pulse. Adding nitrogen practically eliminates this delayed chemiluminescence, while barely affecting the instantaneous one. Luminescence intensity increases downwards in the figure (adapted from [5.87])

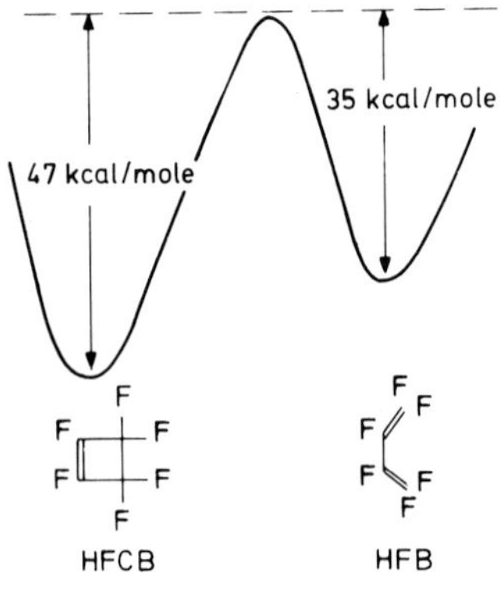

Fig.5.48. Energetics of the hexafluorocyclobutene reaction. The cyclic form is thermodynamically more stable by 12 kcal/mole (adapted from [5.102])

the pure sample is prevented by back reaction due to collisonally activated HFB molecules. An excess of helium eliminates this reaction by collisional cooling of the energy-rich HFB production. This example shows that the end composition of a mixture is kinetically and not thermodynamically controlled. Such cases are well known in nonlaser reaction schemes. The unique feature of laser chemistry in this respect is that one species is selectively heated due to better matching between ν_ℓ and a molecular vibration frequency (Sect. 1.3). In that sense "nonthermal" or "nonequilibrium" reactions may be brought about.

Another possible outlet for vibrational energy accumulated in the molecule is conversion to electronic energy. Referring again to Fig.5.23, it appears that the process is quite improbable in view of the smaller density of states of the accepting electronic level. Some possible examples of this process have been found in the multiphoton activation of chromylchloride (CrO_2Cl_2) [5.103], carbonyl fluoride (F_2CO) [5.104], and osmium tetroxide OsO_4 [5.105]. Emission from the parent molecule is reported under collision-free conditions. The process involved has been termed inverse electronic relaxation (IER) [5.106], and can be understood in terms of the general formalism of Sect.2.2.2 [cf. (2.2.12-15)]. In Sect.5.3.3 we considered the case of narrow-band excitation, which preferentially populates the S_1 component of the molecular eigenstate. The weak coupling case was assumed to be applicable, and (2.2.15) was used to derive rate constants for decay of the excited singlet. In the present case, we consider interaction between a dense manifold (high vibrational states of S_0, populated by multiphoton absorption) and an excited state (denoted here by S_1, though inverse intersystem crossing to a triplet state could be described by the same formalism). If the energy separation of different S_1 states is larger than their width, we obtain the black holes situation, while if they are dense, the overlap region is obtained (cf. Fig.5.19). In both cases, (2.2.15) leads to the following expression for the decay rate of S_1 [5.106]:

$$\gamma(E) = \Gamma(S_1)/D \quad , \tag{5.4.19}$$

where $\Gamma(S_1)$ is the radiative decay of zero order S_1 states, and D is a "dilution factor" given by

$$D(\text{black holes}) = \rho_{S_0}(E) \cdot R(S_1 \rightarrow S_0) \tag{5.4.20}$$

and

$$D(\text{overlap}) = \rho_{S_0}/\rho_{S_1} \tag{5.4.21}$$

for the black holes and overlap regions, respectively. $R(S_1 \rightarrow S_n)$ in (5.4.20) is given by (2.2.15). The IER rate is essentially determined by the diluted radiative decay rates (5.4.20) and (5.4.21). Experimental observation is feasible provided that the quantum yield of emission is not too small, i.e., that the dilution factor D is not too large. In the case of F_2CO only 10^{-7} of the molecules irradiated by the infrared laser are found to emit in the visible. The small yield is actually expected, since the effect should be observable mostly in cases where the lifetime of S_1 is dominated by radiationless processes (i.e., the quantum yield is small). Practical considerations would limit the dilution factor to 10^4-10^5 at most. Thus the process may be observable in some cases, but emission yields are expected to be very small.

5.4.4 Multiphoton Ionization (MPI)

One of the more striking demonstrations of laser power, familiar to anyone visiting a high-power ($\gtrsim$ 10 MW) laser laboratory, is the creation of a spark on focussing the laser pulse in air. The visible spark is due to plasma formation, initiated by electrons detached from molecules by a multiphoton absorption process [5.107]. This phenomenon, known as "dielectric breakdown", was mentioned above (Sect.5.4.1) as an undesirable side effect that should be avoided whenever selective excitation is attempted. It turns out that when carried out under appropriate conditions, to be detailed below, multiphoton ionization can be a highly selective process, and may be used to extract useful information on the molecule in question.

Ionization potentials for most molecules are of the order of 10 eV. Table 5.2 shows the minimum number of photons needed to provide the necessary energy for some common laser sources.

Table 5.2. Minimum number of photons (N_{min}) needed to ionize a molecule with ionization potential 10 eV

Laser	Emission wavelength [μm]	N_{min}
ArF	0.193	2
KrF	0.248	3
BBQ dye	0.390	4
R6G dye	0.590	5
HF	2.9	30
CO_2	10.6	90

In the absence of intermediate resonant states, simultaneous absorption of many photons is required. It turns out that when the laser frequency coincides with a molecular energy level, a sequential process, similar to that discussed above for multiphoton absorption of infrared radiation, takes place. Using UV or visible lasers, the total number of photons needed is rather small. Under these conditions it would seem that coherent phenomena, such as are expected in region I (Fig.5.23) might play a dominant role. In view of the high power densities involved, one may expect interference from effects such as self-induced transparency and Rabi oscillations between two intermediate levels. As in the case of infrared multiphoton excitation, absorption and stimulated emission take place with a net effect of up-pumping. If intramolecular relaxation (e.g., internal conversion), dissociation, or ionization occurs at the end, irreversibility is introduced and the coherent effects are damped. It has been shown that the rate equation approach (5.4.10) is applicable to a multistep ionization process, provided that each successive up-pumping process is faster than its predecessor [5.108]. In many cases, the rate-determining step is the one leading to the first resonance, usually a one- or two-photon process. This ensures species selectivity, in contrast with dielectric breakdown. At higher energies, molecular electronic energy levels are more densely spaced, allowing further excitation by sequential one-photon processes. Three possible cases are depicted in Fig.5.49. The number of photons needed to reach the first resonance m is usually singled out, and the process is often referred to as an m + n process. For example, the photoionization of benzene using a KrF laser [5.109] or a frequency-doubled dye laser at 260 nm [5.110] is a 1 + 1 process, whereas using a BBQ dye [5.111,112] is a 2 + 1 process.

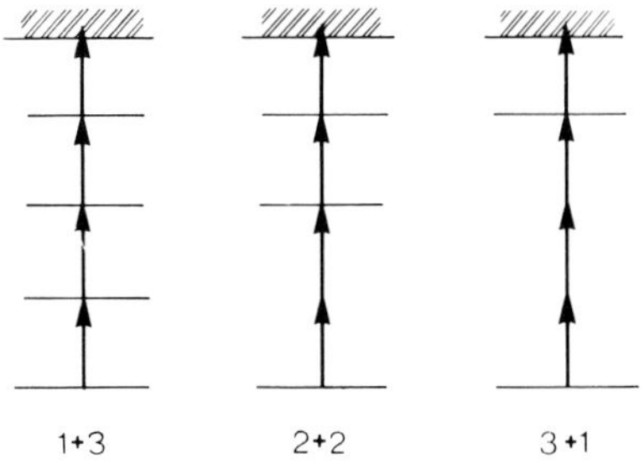

Fig.5.49. Some typical multiphoton ionization schemes

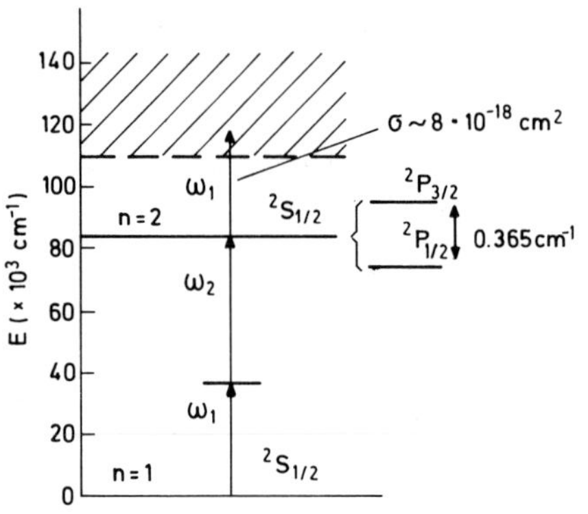

Fig.5.50. Resonantly enhanced three-photon ionization detection of hydrogen atoms. Two photons are absorbed at frequency ω_1 (266 nm), one at ω_2 ($\sim$224 nm). $\omega_1+\omega_2$ is resonant with the transition to the $2^2S_{1/2}$ state at 82259 cm^{-1} for H atoms, 82281 cm^{-1} for D atoms. These two species can easily be distinguished and concentration as low as 4×10^9 atoms/cm^3 were measured (adapted from [5.113])

The availability of a bound intermediate state allows selective excitation, particularly in conjunction with tunable lasers. This opens up possible applications, particularly ultrasensitive specific detection.[4] In one case H and D atoms were selectively monitored in the same sample, at concentrations of the order of 10^9 atoms/cm^3 in the presence of 10^{17} atoms/cm^3 of other components (mostly helium) [5.113] (Fig.5.50).

Electron impact ionization of polyatomic molecules often results in parent ion fragmentation, depending on the excess energy available. Multiphoton ionization can also lead to ion fragmentation whose character depends on the laser frequency and power. In the benzene MPI studies, cited in Sect.1.3.5 only the parent ion $C_6H_6^+$ was observed when a low-power, frequency-doubled dye laser was used [5.110]. With a higher power dye laser, utilizing a 2 + 1 process, extensive fragmentation could be observed. At the highest power levels, the most abundant ion was C^+ (cf. Fig.1.24), which was not observed by conventional electron impact ionization [5.112]. A similar effect was found with the 1 + 1 KrF laser-induced MPI (Fig.5.51).

Obviously, the appearance of C^+ indicates that a sequence of fragmentation and ionization steps has to be considered in which many photons are absorbed (at least nine in the case of the dye laser [5.112]). The details of the process are not yet clear. It is possible that autoionizing states play an important role in the up-pumping process. Another possibility that needs be considered is further absorption of light by fragments. Support for this mechanism arises from the different laser power laws pertaining to different

---

4 This book is not concerned with spectroscopic applications. MPI affords a highly sensitive probe of the intermediate state, particularly if it is reached by a simultaneous two-photon process. For examples of these applications the reader is referred to [5.114-116].

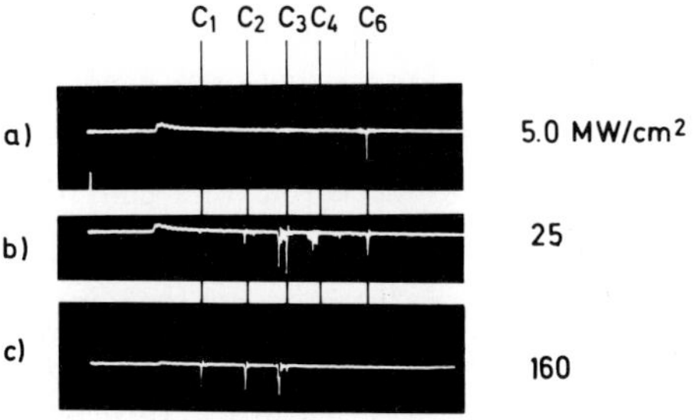

Fig.5.51a-c. Time-of-flight mass spectra of benzene under different conditions of irradiation by a KrF laser. (a) At low laser power the parent ions ($C_6H_6^+$, $C_6H_5^+$) are predominant. (c) At the highest laser power used only low mass fragments, predominantly C^+, are formed indicating a very high-order multiphoton process or alternatively excessive secondary fragmentation. By choosing the appropriate laser parameters, any fragmentation pattern in between the two extremes shown can be generated. (b) From [5.109]

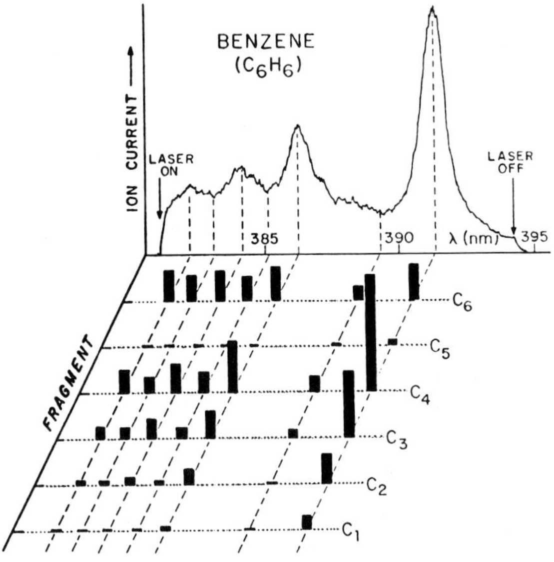

Fig.5.52. Multiphoton ionization of benzene using a tunable dye laser. Individual fragment yields as well as overall ion current are shown as a function of laser wavelength. It is evident that the fragmentation pattern varies with excitation energy (adapted from [5.112])

ions in the case of acetaldehyde [5.117]. Whatever the mechanism, the combined spectral selectivity and high sensitivity call for the implementation of a new mass spectrometric method, laser ionization mass spectrometry (Fig. 5.52), which will possibly provide better resolution and more information than conventional electron impact mass spectrometry. Advantages of laser-induced mass spectrometry, in addition to species selectivity, are good time resolution and intrinsic better collection efficiency due to the small volume occupied by the initially formed ions and absence of background, a considerable problem in electron impact mass spectrometry.

A further use of MPI is the creation and study of "hard-to-get-to" ions, such as C^+. The photoionization by vacuum UV sources has been one of the major tools in obtaining accurate ionization potentials. Lasers afford even better spectral resolution, but the use of MPI to determine ionization potentials must await elucidation of the mechanism, or at least unambiguous determination of the number of absorbed photons.

MPI turns out to be a very general phenomenon. In addition to the resonant process discussed above, high laser intensities appear to induce also nonresonant multiphoton ionization in many molecules and atoms. MPI may thus appear as an undesirable side effect in experiments requiring high-power lasers. On the other hand, researchers interested in ion chemistry, may welcome this possibility of replacing indiscriminate preparation methods such as electron bombardment or electric discharge.

5.5 The "Compleat" Laser Chemist

This section discusses some practical aspects of laser chemistry. Much of the experience gained in other, classical methods of reaction kinetics is utilized in laser experiments. These techniques include, for instance, flash photolysis and pulse radiolysis, shock tube and adiabatic compression, chemical relaxation studies, spectral line broadening, fluorescence quenching and flame spectroscopy. Here we provide an introduction to some of the newer, more common, and more versatile experimental schemes. We begin by comparing molecular beam experiments to bulk gas handling in flow systems and static gas cells, and proceed to describe the techniques of photofragmentation spectroscopy, intracavity absorption, laser induced fluorescence, double resonance and CARS.

5.5.1 Preparing the Sample

Three major types of experimental configurations are employed in laser chemistry experiments of gaseous molecules: Static gas (or bulk) cells, fast flows, and molecular beams. The beam arrangement provides the most detailed, state resolved, kinetic information and will be taken up first.

Molecular Beam Experiments

In a beam, effusive or supersonic, interaction between neighbouring molecules is minimized. Thus, collision free conditions are best approximated in this configuration. One can cross a molecular beam with a laser beam and activate a molecule to energies above a unimolecular reaction barrier (Sects.5.3.1 and 5.4) or even above the ionization threshold. Bimolecular reactions can also be studied, using either crossed molecular beam or a molecular beam incident into a bulk sample. In these cases a laser is used to excite one species downstream from the interaction zone. A schematic arrangement, in which a laser is also used to probe products, is shown in Fig. 5.53. The very low sample density (typically 10^{-5} Torr) precludes the use of many detection methods, including optical absorption. Laser-induced fluorescence and ion detection are two of the most widely employed diagnostic methods due to their high sensitivity.

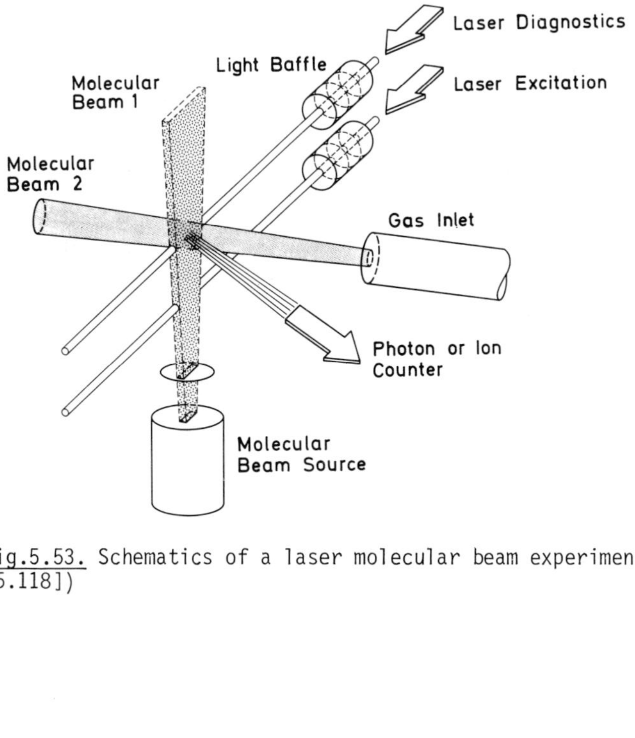

Fig.5.53. Schematics of a laser molecular beam experiment (adapted from [5.118])

The high monochromaticity of laser sources is probably best utilized on using molecular beams [5.119]. The characteristic velocity of a thermal beam is $v=(3kT/m)^{\frac{1}{2}} \approx 5\times 10^4$ cm/s. The resonant laser excitation now depends on the angle between the molecular beam and the light beam. The Doppler shift from line center is $\Delta\nu=\nu_0(v/c)\cos\theta$, where θ is the angle between the light and molecule directions. This effect can be used to advantage to compensate for any mismatch between laser frequency and molecular absorption frequency. Crossing the beams at a right angle minimizes the Doppler broadening of the transition and thereby provides higher spectral resolution. (If the frequency bandwidth of the laser is not wider than the absorption bandwidth this also improves the absorption). The effective transverse Doppler temperature of an effusive beam is $T_D \approx (3/2)T\sin^2\theta/2$ which for T=300 K and $\theta\approx 7^\circ$ yields $T_D\approx 1.6$ K. It is easily seen that this reduces the effective Doppler width by more than an order of magnitude.

Pulsed lasers are required for time-resolved experiments. They are often powerful enough to saturate the transition and promote a large fraction of the molecules irradiated to the excited state. The absolute number of excited molecules depends, however, on the "duty factor", defined as the product of the pulse duration (s) and the repetition rate (s^{-1}). In other words, the excitation is proportional to the average energy of the laser. Thus, in steady-state experiments, cw lasers are often more effective than pulsed lasers.

A special feature of molecular beam-laser interaction involves the conservation of linear momentum. In the absorption process the momentum of the photon, $h\nu/c$, is added to the original momentum of the molecule and causes recoil in the direction of the light. As the molecular momentum is typically three to four orders of magnitude larger than $h\nu/c$, the effect is minor for single-photon events (except at short wavelengths) but can become substantial in multiphoton excitation. This opens an interesting possibility to do

Fig.5.54. Experimental arrangement for studying reactions induced by lasers in low pressure gas flows. The reactant gases enter the cell by flowing through two concentric tubes in a "shower head" configuration. A detailed view of the cell cross section perpendicular to the gas flow is shown at the bottom right. The $^{14}N^{16}O$ molecules were pumped by radiation from a cw liquid nitrogen cooled CO laser delivering 3 W in the TEM_{00} mode at the $P_{9\to 8}(13)$ transition at 1884 cm^{-1}. Because of a slight frequency mismatch between this line and NO(v=0, J=3/2)→ NO(v=1, J=5/2) the molecular transition had to be tuned into coincidence by applying a magnetic field. The Helmholtz coils for this purpose are shown in the figure (adapted from [5.28])

beam-deflection spectroscopy and has also been proposed as a potential route for isotope separation [5.120].

So far the discussion referred to both effusive and supersonic beams. The latter are frequently used to cool molecules down to near the absolute zero. Molecules at a temperature T_0 are allowed to expand from a high pressure region (pressure P_0) through a small nozzle to an evacuated chamber (pressure P). The achievable temperature is given by $T=T_0(P/P_0)^{(\gamma-1)/\gamma}$; $\gamma=C_p/C_v$ where C_p and C_v are the heat capacities at constant pressure and volume, respectively. γ is usually determined by the carrier gas (noble gases are normally used).

Temperatures of very few K have been realized for many atomic and molecular species, e.g., NO_2, I_2, s-tetrazine, and CrO_2Cl_2 [5.121]. More recently, much larger molecules were studied including naphthalene [5.122] and pentacene [5.123] (Fig.1.22). The low temperature greatly simplifies the spectral features of the molecules as demonstrated by Fig.1.22. Consequently all the high resolution laser techniques previously used for atoms and, diatomics can now be used for polyatomics. In principle it is possible to cool all degrees of freedom of the molecule and reduce the velocity distribution to a very narrow range. Such supercooling is obtainable down to temperatures much below the boiling points of the compound. However, condensation and cluster formation can occur and are sometimes even desired, e.g., in the study of van der Waals molecules forming only under such conditions (Fig.2.31).

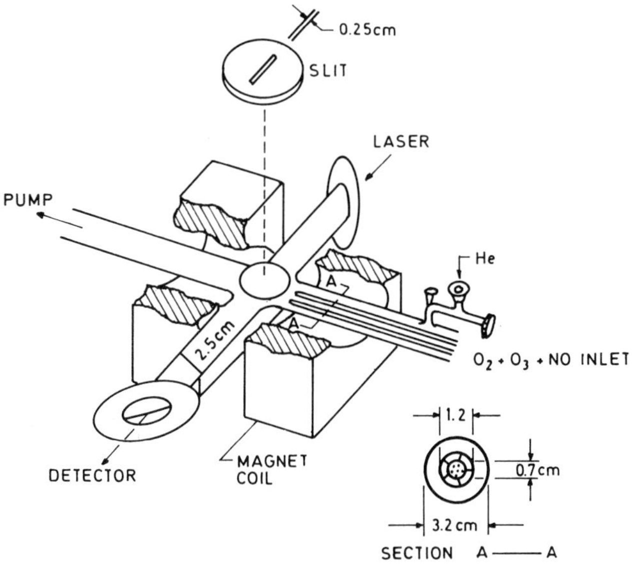

Fig.5.54. Figure caption see opposite page

Static Cells and Low-Pressure Flow Systems

Experimental arrangements of this type are often used in conventional kinetic studies and were discussed previously (Figs.5.6,29). Figure 5.54 shows one laser oriented version, used in the study of $NO(v) + O_3 \rightarrow NO_2 + O_2$ (Sect. 5.2.2.).

5.5.2 Excitation and Probing Techniques

Absorption of laser light by molecules can be used either to enhance their reactivity (laser-induced chemistry) or to facilitate their detection (laser diagnostics). Any excitation scheme can be used to achieve both ends, as shown below. Thus, laser chemistry is intimately related to laser probing of molecules, and very similar methods are used in both applications. Photofragmentation, laser induced fluorescence (LIF) and CARS (Coherent Anti-Stokes Raman Scattering) probe the fate of laser-excited molecules by monitoring scattered particles: photofragments in the first, photons in the two others. These are discussed in this section, while methods relying on the measurements of the attenuation of the light beam are taken up in Sect.5.5.3.

Photofragmentation Spectroscopy

Photofragmentation spectroscopy [5.124] is a technique used to study and characterize dissociative electronic transitions. Figure 5.55 shows the details of the photofragment spectrometer. A molecular beam is illuminated by a pulsed laser source, and the resulting photofragments are then detected, using a mass spectrometer. The distribution of photofragments is determined as a function of photon energy, photon flux, ion mass, time after light pulse and laboratory angle of recoil with respect to the electric vector of polarized light. A laser source is needed to provide the required large photon flux with a well defined frequency and polarization, at a reasonably short

Fig.5.55. (b) Side view of photofragment spectrometer. The laser and molecular beams intersect in an interaction chamber, and photodissociation fragments which recoil upward are detected by a quadrupole mass spectrometer in a separately pumped detection chamber. The Q-switched ruby or neodymium-glass laser assembly is shown with the following components: A—lens to adjust size of laser beam to match diameter of molecular beam, B—polarization rotator (half-wave plate or double Fresnel rhomb) adjustable under computer program control, C—second harmonic generator (ADP or KDP crystal), D—front reflector, E—laser rod and pumping flash lamp, F—polarizer (calcite Glan prism), G—adjustable aperture, H—Q switch (Pockels cell), I—rear reflector, and J—light monitor (photodiode) (adapted from [5.125])

a

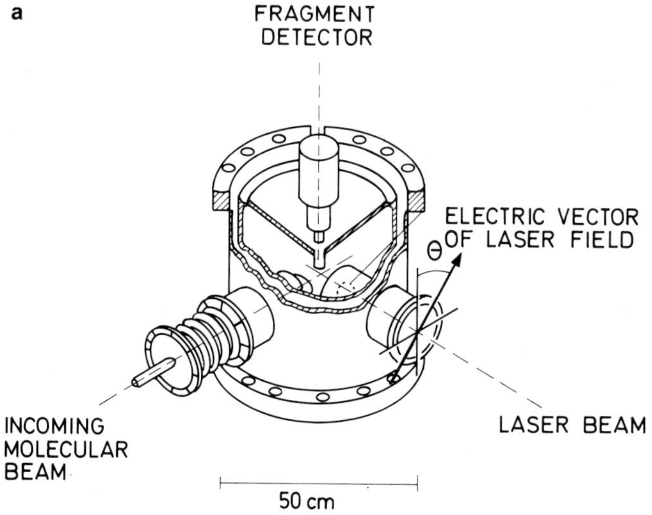

Fig.5.55. (a) Cutaway drawing of photofragment spectrometer. The beam of molecules to be photodissociated enters from the left and is crossed perpendicularly by pulses of polarized light, usually from a laser. The photodissociation fragments which recoil upward are detected by a mass spectrometer as a function of mass, of photon energy, of photon flux, of time after the laser pulse, and of angle of recoil θ measured from the electric vector of the light. (The θ shown in the drawing would be a negative angle of recoil.) The interaction region and the mass spectrometer are in separately pumped chambers connected by a small liquid nitrogen cooled tube, which collimates the fragments

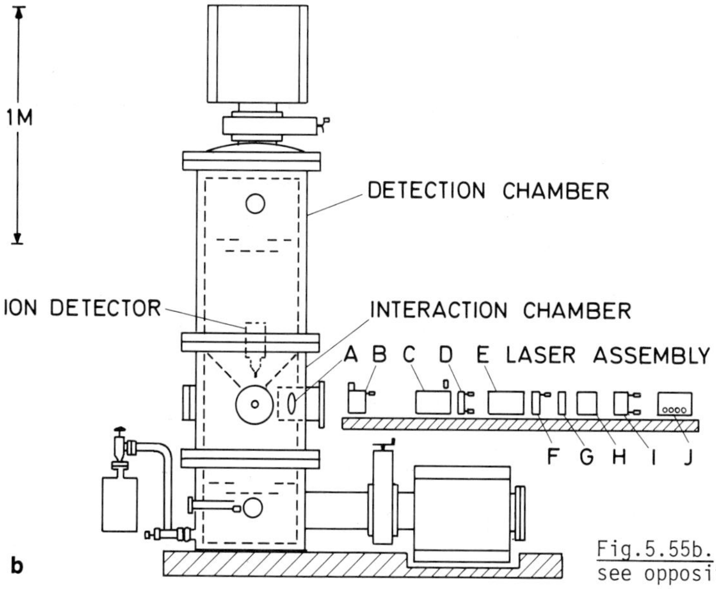

b

Fig.5.55b. Figure caption see opposite page

time. Appropriate sources are fixed frequency lasers (Nd:YAG, Ruby, excimer lasers) or dye lasers pumped by one of them.

In Sect.5.3.1 we discussed the photofragmentation of ICN. Figure 5.11 showed the center of mass translational energy and recoil angle (with respect to the electric vector of the light) distributions. Raw data are measured in the laboratory frame of reference. Such laboratory distributions of I and CN fragments are shown in Fig.5.56. Note that due to its heavier mass, the I atom distribution is broader.

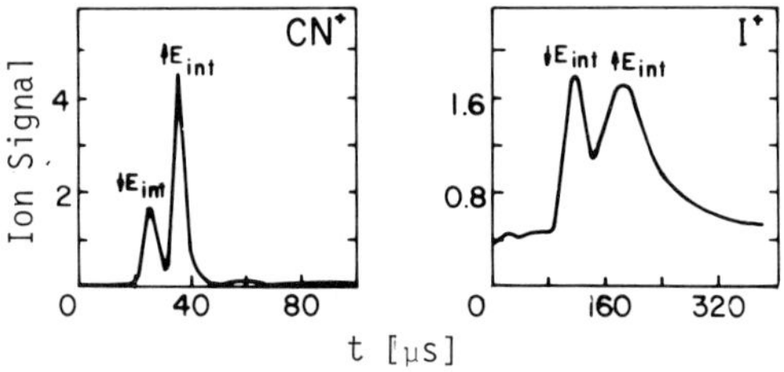

Fig.5.56a,b. Time-of-flight spectra resulting from the photodissociation of ICN by a quadrupled Nd:YAG laser. (a) and (b) show the signal intensity of the I atom and CN radical fragments, respectively. Transformation to the center-of-mass reference system leads to Fig.5.11 (adapted from [5.37])

Laser-Induced Fluorescence (LIF)

Optical excitation of molecules is often followed by radiative decay (fluorescence). Fluorescence can be induced by any light source, but the unique properties of lasers made LIF a most versatile and powerful tool [5.126]. The term LIF relates to many laser excited molecular fluorescence studies, but is most commonly used in one of the following specialized applications:

In the infrared region, lasers have been used to preferentially excite a given vibrational (sometimes vibrotational) level. Fluorescence can be observed either from the initially excited level or from levels coupled to it by intra- or intermolecular interactions. A major application of the method was the study of V-V and V-R/T energy transfer routes and rates (see, e.g., Sect.2.5.5).

In the visible/UV region lasers can be used to excite a particular vibronic transition (Fig.5.57). This variant of the method is often used sensitive detection of fragments formed by unimolecular and bimolecular reactions (see, e.g., Fig.5.4).

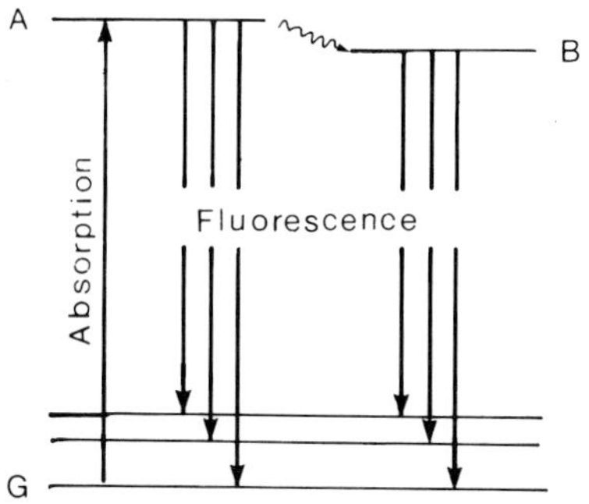

Fig.5.57. Schematics of an LIF experiment. A laser is used to populate level A. Re-emission of light follows, as well as collision-induced energy transfer to state B, which may also fluoresce. Note that emission from either energy level is spread over several spectral lines, and shifted to the red with respect to the exciting light

Three specific examples that illustrate the power of the method are shown in Fig.5.58.

LIF was extensively used to determine internal energy distributions (see, for instance, ICN studies, Sect.5.3.1). An interesting extension of the method is the measurement of translational energy release in reactions, e.g., photodissociation [5.127,128]. Figure 5.58 displays two alternative ways of obtaining the translational energy distribution; accurate measurement of the Doppler width of a single absorption line, and time of flight measurement of the fragments. In the case depicted, the multiphoton dissociation of C_2H_3CN, the results of these two methods are in relatively poor agreement. This may be due partly to different experimental conditions (e.g., laser fluence and power) and partly to inherent errors, which are possibly more pronounced in the time-of-flight method, which can be confidently used only in a molecular beam apparatus.

The widespread use of the LIF method is primarily due to the following properties [5.129]: 1) The method is selective for molecule and state. 2) It provides spatial resolution in a sample if so desired. 3) It is a very sensitive technique and applies even for weak transitions. 4) It is characterized by good time resolution, limited by the pulse width. LIF is, of course, applicable only to molecular transitions whose fluorescence quantum yield is sufficiently large ($\gtrsim 10^{-4}$). It was applied to product energy distribution mainly in the case of diatomic or triatomic molecules, where state assignment is relatively easy and where intra- and intermolecular energy transfer rates are slow.

Coherent Anti-Stokes Raman Spectroscopy (CARS)

Lasers can be used in a variety of probing techniques utilizing nonlinear effects. Some such effects (frequency doubling, summing, and subtracting)

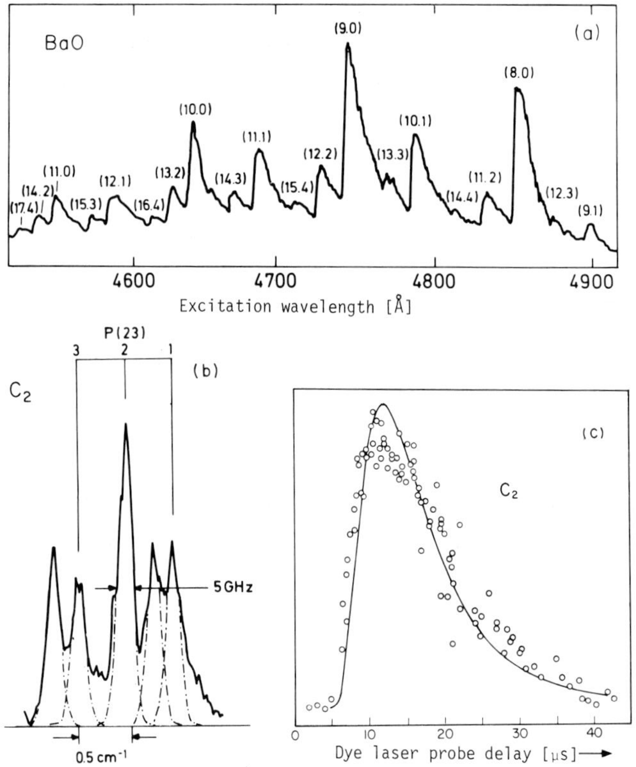

Fig.5.58a-c. Laser-induced fluorescence monitoring of reaction products. In each case the total emission from the laser-excited species is monitored. (a) Vibrational distribution of BaO formed by the reaction Ba + O_2→ B_aO + O. The numbers in parentheses (v',v") indicate the vibrational quantum numbers of the upper and lower states connected by the transition, respectively (adapted from [5.126]). The use of LIF to measure energy disposal in the multiphoton dissociation of C_2H_3CN is shown in (b) and (c). The C_2 fragment was monitored by exciting the $a^3\pi_u \rightarrow d^3\pi_g$ transition. (b) An experimental view of the P(23) lines of the 0-0 band. The Doppler width of the lines leads to a mean recoil energy of 5.9 kcal/mole (adapted from [5.127]). (c) Time of flight measurement. Visible fluorescence was induced by a dye laser whose beam direction is parallel to the CO_2 laser beam at a distance of 1 cm. The time of flight of the fragments is determined by varying the delay between the two laser pulses. The mean translational energy found by this method is only 1.1 kcal/mole (adapted from [5.128])

were shown in Sect.3.4 to expand the lasing range of Nd:YAG and dye lasers. Here we briefly introduce one nonlinear technique that found considerable use in systems where LIF or light-absorption techniques are restricted, for instance, in the study of flames or electric discharges.

In Sect.3.1 we have seen that a molecule can interact with the electromagnetic field associated with a light wave due to the existence of an electric dipole moment, μ. An electric field can also induce a dipole moment in a molecule, leading to the phenomenon of polarization. The polarization P induced by an electric field E in an ensemble of molecules or in crystal is the (vector) sum of individual induced dipole moments. At low radiation intensities it is linearly proportional to the field strength. Increasing the field leads to observation of contributions due to higher powers of the field, i.e.,

$$P = \alpha E + \beta E^2 + \gamma E^3 + \dots \quad . \tag{5.5.1}$$

Here α is the usual, linear susceptibility coefficient of the medium. It is actually a tensor, but for the present discussion only its magnitude is important. β and γ are the second- and third-order nonlinear coefficients of the susceptibility, respectively.

In the case of interaction between the sample and an oscillating electric field $E=E_0 \cos\omega_0 t$, it is easily verified that the nonlinear terms lead to a resonant response at new frequencies. For instance, the second term leads to a polarization $P^{(2)}$ which may be written as

$$P^{(2)} = \beta E^2 = \beta E_0^2 \cos^2\omega_0 t = (\beta/2)E_0^2(1 + \cos 2\omega_0 t) \quad . \tag{5.5.2}$$

Thus, energy at frequency $2\omega_0$ may be exchanged, leading to two photon absorption, frequency doubling and similar phenomena.

The third-order term can be used to triple the fundamental frequency, or to combine two or three different frequencies. Under suitable conditions these interactions can be resonantly enhanced, leading to the generation of coherent radiation at new frequencies [5.130]. A special case that found widespread applications is the combination

$$\omega_3 = 2\omega_1 - \omega_2 \tag{5.5.3}$$

where ω_1 and ω_2 are the frequencies of two independent laser sources. The effects is resonantly enhanced when ω_3 coincides with a molecular frequency that is Raman active. This is the basis for CARS, [5.131-133].

Ordinary Raman scattering is generally used for vibrational spectroscopy. Its main difference from infrared absorption spectroscopy is that it requires a change in the polarization rather than in the transition dipole moment. Therefore, in Raman spectroscopy one can often see transitions which are not dipole allowed, e.g., in homonuclear diatomics (Sect.3.3.1). The major disadvantage, however, rest with the fact that it is a weak phenomenon associated with small scattering cross sections. Typically, one photon is scattered back for every 10^8-10^{10} photons incident on the sample. In addition, collection efficiency is relatively poor as (for unpolarized light) scattering is into a 4π solid angle. These difficulties are partially compensated for by using lasers as a excitation sources. However, in fluorescent samples, LIF is often much more intense than the Raman signal and tends to obscure it. These shortcomings can, at least partically, be eliminated by the CARS technique.

Figure 5.59 shows some important scattering processes. Anti-Stokes scattering arises from interaction of the incident photon with a vibrationally excited molecule leading to a virtual state which emits at the sum frequency. The process is stimulated and enhanced in a CARS experiment, where two lasers (with frequencies ω_1, ω_2 and wave vectors $\underline{k}_1$ and $\underline{k}_2$) are employed. In addition to the resonance condition (5.5.3) phase matching of the wave vectors

$$2\underline{k}_1 - \underline{k}_2 = \underline{k}_3 \qquad (5.5.4)$$

is required to ensure coherence.

This is usually done by crossing the two laser beams at an appropriate angle, Figure 5.60. The term "scattering" in the acronym CARS is somewhat misleading; the emission is highly collimated and directional and, therefore, easily separable from the exciting laser beams as well as from nondirectional emission, such as fluorescence.

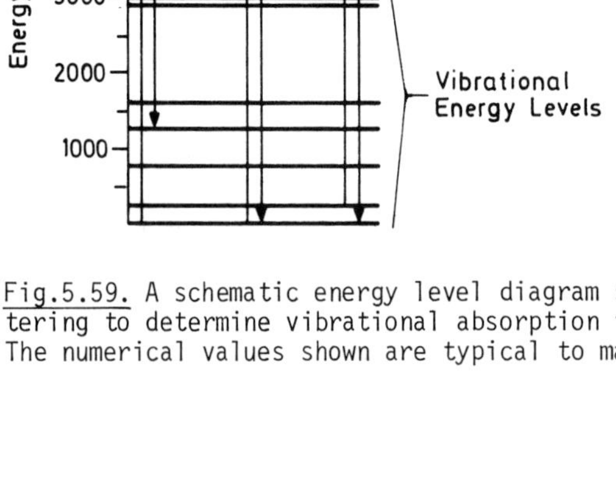

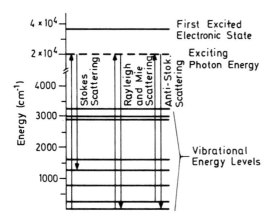

Fig.5.59. A schematic energy level diagram showing various types of scattering to determine vibrational absorption features (adapted from [5.141]). The numerical values shown are typical to many CARS experiments

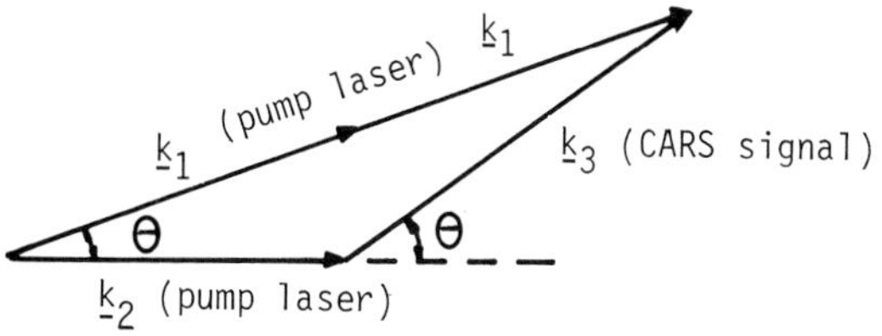

Fig.5.60. Phase matching condition for CARS. The anti-Stokes radiation is emitted at a direction different from the propagation direction of both pump lasers. This allows easy spatial separation of the CARS beam

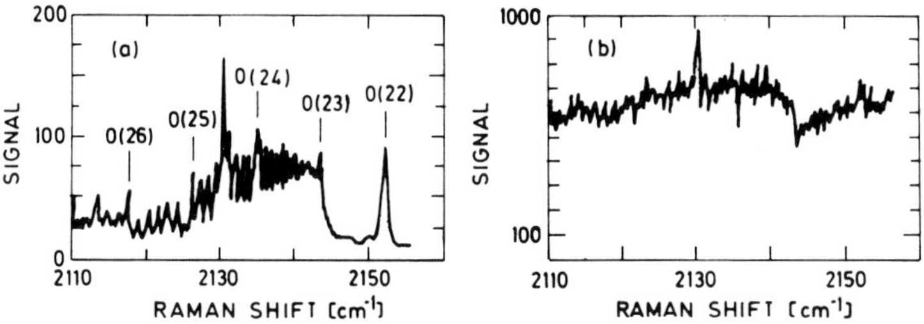

Fig.5.61a,b. CARS signal of CO from a rich, methane-air flame in the region of the Q branch. (a) Background-free spectrum (see text). The Q branch lines of ground state N_2 are labelled. (b) Conventional CARS spectrum with all polarizations parallel. Both vertical axes are scaled in the same arbitrary units (adapted from [5.133])

A prime problem in CARS is the ever-present background scattering caused by the nonresonant susceptibility of the medium. This interference is particularly severe in condensed phases. Currently it is the main cause for limiting the sensitivity to about 10^3 ppm for most diatomic gases [5.132]. Several methods were developed to remove this difficulty. One is by setting the two laser beam polarizations at predetermined angles and analyzing the resulting CARS signal with a polarizer, Fig.5.61. Other means are background subtraction or signal enhancement by tuning one of the lasers to near resonance with one photon absorption [5.134] (compare Fig.5.59 where the exciting laser is far from resonance with any molecular level). An example of resonance enhancement is provided by a mixture of iodine and air. On tuning the exciting laser to near coincidence with an I_2 absorption line, the signal due to 0.1% I_2 was twice as strong as that of O_2 whose volume percentage was 20% [5.135].

Figure 5.62 shows an example of the spectrum obtainable in an electric discharge. D_2 was monitored at 48 Torr, and excitation of v=1 could be easily observed. The results indicate a nonthermal distribution of vibrational states.

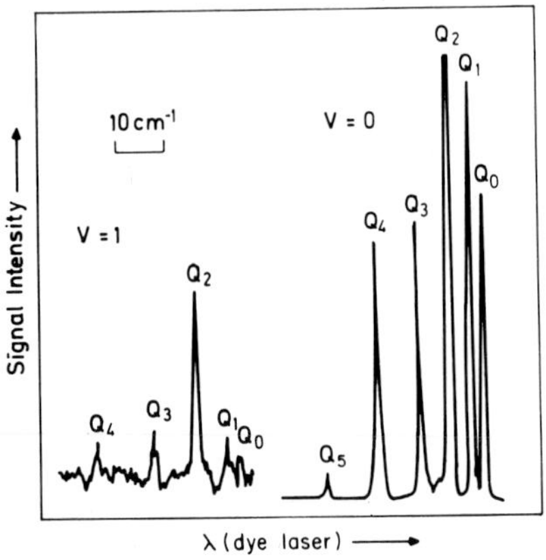

Fig.5.62. CARS spectrum of D_2 in an electric discharge. Population of both v=0 and v=1 are easily monitored without appreciable interference from the intense background radiation. The rotational and vibrational temperature can be estimated from the line strengths. T_{rot} is about ambient while T_{vib} is 1000 K (adapted from [5.135b])

5.5.3 Laser-Oriented Absorption Measurements

Methods based on light scattering and on fluorescence are usually superior to those relying on measuring the attenuation of a light beam passing through the sample. However, the latter are often easier to interpret and yield directly and quantitatively the amount of energy deposited in the sample. Lasers can be used to increase the sensitivity of absorption spectroscopy by various means, some of which are discussed in this section.

Fluorescence vs Absorption Detection

LIF is especially useful for low-pressure samples, where absorption techniques are impractical as they require excessively long absorption pass lengths. It is instructive to compare the sensitivity of LIF with absorption measurements, as these techniques are often complementary. The following analysis is based on [5.137].

Specifically, we would like to compare the sensitivity of fluorescence and absorption detection for probing excited level populations. In both methods the measured signal depends on the excited level population, N_1. Suppose that before the excitation practically all the molecules are in the ground state, i.e. $N_0 \gg N_1$, and that the excitation is due to a laser beam of constant cross sectional area passing through the sample. By Beer's law the number of molecules excited after the beam has traversed a length L is

$$N_1 = N_0[1 - \exp(-\sigma_0 I_0 L)] \simeq N_0 \sigma_0 I_0 L = n_1 L \qquad (5.5.5)$$

where the second equality applies to the practically important case of low optical density, $(\sigma_0 I_0 L \ll 1)$. σ_0 is the absorption cross section for the transition $0 \rightarrow 1$. I_0 is the laser beam intensity, expressed as the number of photons crossing unit area per unit time. (In other words, I_0 is the photon flux. More commonly the intensity is defined as the energy flux, see e.g., Sects.3.1 and 3.2). N_1 is the number of excited molecules per unit length.

The fluorescence signal is the power reaching the detector due to molecules in the observable length L_f,

$$S_f = \frac{N_1}{\tau_f}\frac{hc}{\lambda_f} F \cdot \phi = \frac{n_1 L_f}{\tau_f}\frac{hc}{\lambda_f} F \cdot \phi \quad . \tag{5.5.6}$$

Here λ_f and τ_f are the fluorescence wavelength and lifetime, respectively, F is the fraction of photons reaching the detector and ϕ is the quantum yield. The noise in fluorescence detection is mainly determined by NEP-"the noise equivalent power" of the detector which is, essentially, the minimum detectable power. (NEP $\sim 10^{-8}$ W for good infrared detectors). Thus, the signal to noise ratio for fluorescence detection is

$$(S/N)_{fluor.} = S_f/(NEP) \quad . \tag{5.5.7}$$

In absorption measurements N_1 is probed by the attenuation of a light beam with frequency corresponding to, say, the $1 \rightarrow 2$ transition, and intensity I_1. Using Beer's law again the absorption signal is proportional to the attenuation,

$$\Delta I_1 = I_1 \sigma_1 n_1 L_a \tag{5.5.8}$$

where σ_1 is the absorption cross section for the $1 \rightarrow 2$ transition and L_a is the absorption length. The noise in absorption measurements is usually determined by the fluctuations, δI_1, in the incident intensity, I_1. ($I_1/\delta I_1 \sim 100$ is a typical value). Hence

$$(S/N)_{abs.} = \frac{\Delta I_1}{\delta I_1} = \frac{I_1}{\delta I_1} \sigma_1 n_1 L_a \quad . \tag{5.5.9}$$

Comparing fluorescence and absorption detection we obtain

$$\frac{(S/N)_{abs.}}{(S/N)_{fluor.}} = \frac{I_1}{\delta I_1} \frac{\sigma_1 L_a}{F \cdot \phi \cdot L_f hc} (NEP) \tau_f \lambda_f \quad . \tag{5.5.10}$$

From (5.5.10) it is seen that absorption is favored in the case of long radiative lifetimes and long wavelengths. Thus, in the microwave region fluorescence detection is impractical. On the other hand, fluorescence detection is favored in the visible and UV range, particularly as the NEP of photomultipliers is much smaller than that of infrared detectors. In the infrared range absorption is preferred if L_a can be made considerably larger than L_f, which is usually of the order of 1 cm.

Double Resonance Techniques

These techniques are based on using two light sources: one prepares the excited state to be probed and is called the pump source, and the other, used to monitor the population of that state, is termed the probe source. Normally, the probe power is much smaller than the pump power. There are many varieties of the method as any combination of excitation and probing sources can be used. The reader is referred to [5.137] for an extensive review of the literature through 1976.

We have already seen several uses of the double resonance absorption technique, e.g., for spectroscopic studies in Sect.1.3.2, in connection with chemical laser applications in Sect.4.5, and infrared multiphoton excitation studies in Sect.5.4. A typical experimental arrangement for such measurements is shown in Fig.5.63.

Very detailed information on molecular energy distributions is available when, for instance, optical-microwave double-resonance techniques are used. The state analysis of BaO from the reaction $Ba + O_2 \rightarrow BaO + O$ may serve as one example to show this. By a cw dye laser excitation is achieved on selected vib-rotational transitions $X^1\Sigma(v'',J'') \rightarrow A^1\Sigma(v',J'')$. When the fluorescent interaction region is exposed to microwave radiation, transitions between neighbouring J levels are induced at certain microwave frequencies (cf. Fig.5.64). This in turn influences the absorption of the dye laser on the

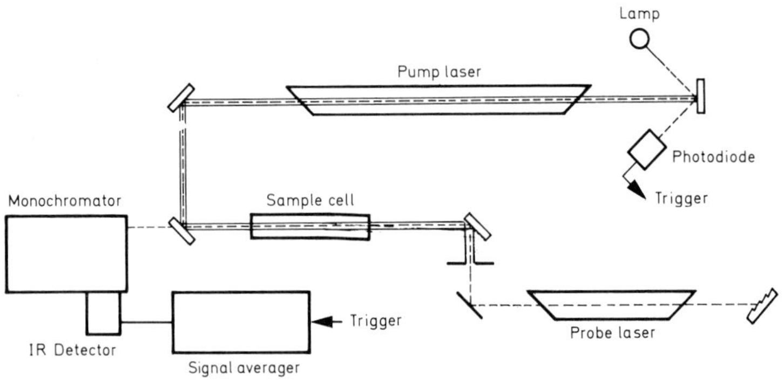

Fig.5.63. Schematic diagram of an infrared-infrared double-resonance apparatus (adapted from [5.137])

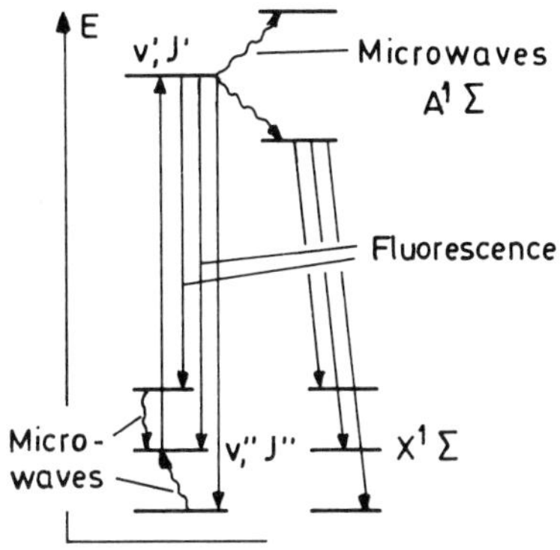

◄Fig.5.64. Microwave-optical double-resonance experiment. A dye laser is used to excite a well-defined rotational component of the v"=0 level of the $X^1\Sigma$ ground state of BaO, to another well-defined rotational level of the electronically excited $A^1\Sigma$ state. Visible fluorescence is monitored, providing high sensitivity. A microwave source is used to couple the rotational states involved in the optical transition to other states. Changes in the LIF signal reflect these population changes (after [5.138])

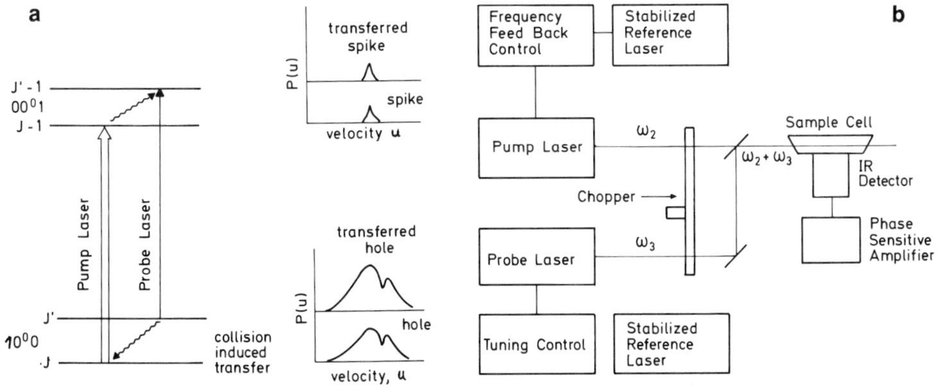

Fig.5.65. (a) The principle of the experiment measuring momentum transfer by inelastic collisions between molecules in well-defined rotational states. A powerful laser (pump) excites molecules in a given J state of the 10^00 level of CO_2. The laser interacts with only a small portion of the Doppler broadened profile, thus selecting a subgroup of molecules with a well defined translational energy. A hole is burned in the Doppler profile (cf. Fig.3.24). This hole can be transferred by collisions to an adjacent J' level. Concurrently, a spike (excess population) formed in the upper state of the pump transition (J-1 of the 00^01 vibrational level) can also be transferred by collisions. Population changes in J' and J'-1 are monitored by a second, probe laser. The width and amplitude of the hole and the spike are measured as a function of added gas pressure.
(b) Schematic of the experimental apparatus. The pump laser is stabilized by a feedback mechanism controlled by a stable low-power laser. The frequency of the probe laser is scanned slowly, and is constantly controlled by comparing it to another highly stabilized laser. The pump and probe beams are chopped with an off-center mechanical chopper, resulting in different chopping frequencies. They are then combined and sent into the sample cell. CO_2 molecules excited by the lasers emit in the infrared. The emission is recorded at the sum of the two chopping frequencies. Thus, only the signal due to excitation by both lasers is monitored (adapted from [5.139])

pumped transition and subsequently manifests itself in the laser induced fluorescence.

An intriguing extension of double-resonance techniques is the utilization of narrow-band lasers to select not only a given vibrational state but also a small velocity group within the Doppler profile (Sect.3.2.2) of an absorption line. Velocity selection has been traditionally associated only with molecular-beam experiments. Figure 5.65 demonstrates the feasibility of velocity selection in a bulk sample. The experiment deals with the effect of translational energy on the rate of rotational-energy transfer. It must be noted that this elegant experiment requires a fairly sophisticated (and expensive) setup; in the arrangement shown, four highly stabilized lasers were employed simultaneously (see Fig.5.65 for details).

Intracavity Absorption and Optoacoustic Detection

In Sect.3.2 we discussed the creation and amplification of a light pulse in a suitable cavity. Under appropriate conditions (small signal gain) the amplification was shown to be exponential. When a small amount of absorbing species is placed inside a laser cavity, it spoils the Q factor (Sect.3.2.3) at some particular frequencies. If the emission line is inhomogeneously broadened, the resulting spectrum will contain "holes" at these frequencies. The amplitude of these holes is related to the amount of absorbing molecules and may be used to monitor their concentration. This is the basis for the intracavity absorption technique which has been extensively used to observe trace amounts of material. It is most widely used in conjunction with dye lasers, where the broadening is essentially homogeneous. The effect is traced in this case to spatial inhomogeneities [5.140,141]. As an example the intracavity absorption spectrum of the radical HCO is shown in Fig.5.66.

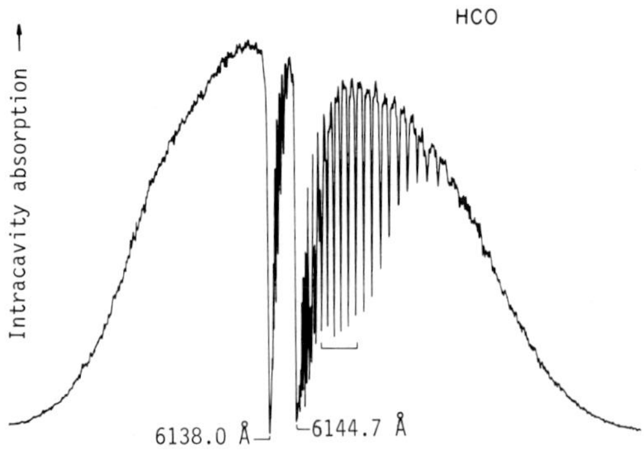

Fig.5.66. The absorption of HCO radicals formed in the photolysis of formaldehyde. The broad contour is that of the free-running flash laser pumped dye laser. No tuning elements were inserted to the cavity (cf. Fig.5.67). Several rotational lines are clearly observed (adapted from [5.142]

As compared to positioning the sample outside the laser cavity, the sensitivity increase due to an intracavity measurement depends on the quality factor of the cavity and can be quite large (~100 or more) since small changes in total absorption may cause large changes in the laser intensity, especially when operated near threshold. Ambiguities may arise from the presence of many laser frequencies, and, in the case of flash lamp pumped lasers, from stray light scattered into the absorption cell. These difficulties can be reduced by placing the probe in a coupled laser cavity and keeping it just below the threshold for self-oscillation. This allows one to measure fractional changes in the absorption coefficient down to 10^{-6} [5.143]. Single-mode tunable cw dye lasers which have now become available open the way to doing such measurements without the limitations of fixed-frequency lasers. Intracavity experiments with these sources have pushed this method to quite an extreme sensitivity (cf. Fig.1.18). If the dye laser oscillates on several modes close to threshold, the sensitivity to selective intracavity absorption on a single mode is proportional to the number of oscillating modes as a result of the coupling of different modes. An increase in absorption sensitivity by a factor of 10^5 as compared to single pass measurements can be achieved. The technique has been demonstrated for iodine absorption where a layer density of 5×10^{-11} g/cm^2 has been detectable [5.141]. The laser-induced fluorescence method has obvious advantages in cases where the sample cannot be put inside a laser cavity. Therefore the two methods may be considered complementary.

An alternative method of monitoring intracavity absorption is by direct measurement of the energy dissipated inside the cavity. This energy is eventually converted to heat and can be detected by sensitive calorimetric methods. One of these is the optoacoustic technique [5.144]. It has also been widely used with nonlaser sources, and with samples irradiated by lasers outside the cavity.

The principle of the method is as follows: Light is directed into a well-defined (small) volume inside the absorption cell. Care is taken to rapidly convert all the absorbed energy to heat. This is often accomplished by adding a large excess of inert gas to facilitate electronic and vibrational relaxation. The heat generated in the small volume causes a pressure increase, which in turn creates a radial acoustic perturbation that travels towards the vessel's walls. A microphone is placed in one of these walls and detects the acoustic signal. By properly calibrating the system, the amplitude of the microphone signal can be directly related to the amount of energy deposited in the sample.

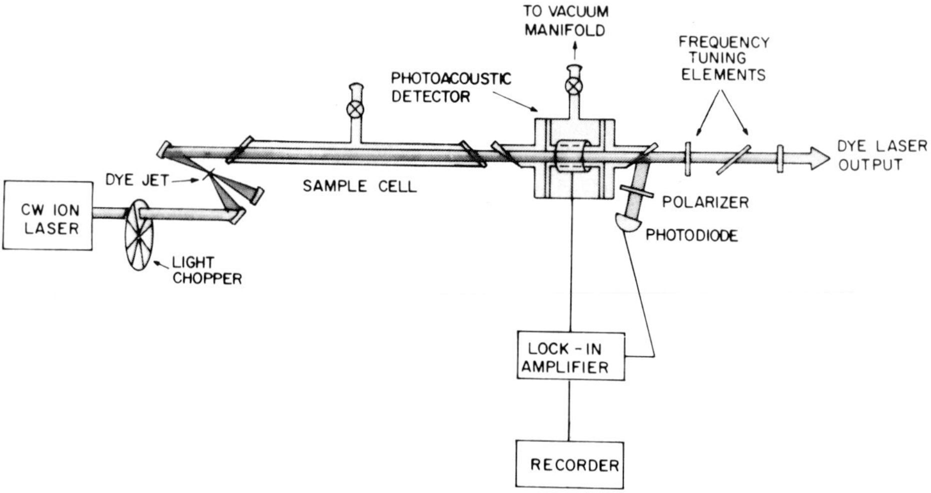

Fig.5.67. Intracavity cw dye laser apparatus for optoacoustic spectroscopy and excited-state chemistry involving weak transitions. A cw laser is used, with the optoacoustic (OA) cell placed inside the cavity. Laser energy absorbed by the gas in the OA cell is converted to heat, creating a perturbation that is detected by a sensitive microphone. Absorption cross sections as small as 10^{-25} cm^2 have been measured by this method (adapted from [5.145])

Figure 5.67 shows an apparatus that uses intracavity absorption to simultaneously monitor a very weak absorption and utilize it in a laser-induced chemistry experiment. An optoacoustic cell is used to tune the laser to the maximum of the absorption band (cf. Fig.1.18). This is one example of a situation where direct absorption would require prohibitively long cells and high pressures as a weak, forbidden overtone (Sect.3.3) is involved. Another example is that of multiphoton excitation by powerful infrared lasers (Sect.5.4) [5.71,146]. Here the fractional deposited energy ($\Delta E/E$) is often too small to be measured by direct attenuation of the beam. Optoacoustic detection involves measurement of ΔE — the absolute absorbed energy, rather than $\Delta E/E$.

5.6 From the Laboratory to Large-Scale Laser Chemistry

Potential industrial applications of laser chemistry [5.147,148] depend also on economic considerations. Optimistic estimates suggest that the cost per mole of laser photons can be brought to a few cents for infrared lasers and to about a dollar for ultraviolet lasers. Taking into consideration the quantum yields, transfer losses, etc., one mole of laser-induced change will cost above a dollar. Since energy costs in the chemical industry are often below 10% of the total, laser-induced processes are only of interest for fine chemicals, whose cost exceeds about ten dollars per mole. Indeed, present-day applications of conventional photochemistry are largely confined to such fields (e.g., synthesis of lactams, sulfones, vitamins, drugs, essences and fragrances, etc.). Another application area not so much different from photochemistry is material processing where lasers are indeed finding new applications by the day.

The economic picture does change, however, if lasers are used to initiate or trigger chain reactions. Here the laser is employed to generate species that act as radical starters or homogeneous gas phase catalysts. Chain lengths of roughly 10^3 have been reported [5.149] in the synthesis of vinyl chloride initiated by KrF laser photons.

$$H_2ClC - CClH_2 \xrightarrow[\lambda\sim 250\ \mathrm{nm}]{h\nu(\mathrm{KrF})} H_2\dot{C} - CClH_2 + Cl \qquad \text{chain initiation}$$

$$Cl + H_2ClC - CClH_2 \rightarrow H_2ClC - \dot{C}HCl + HCl \qquad \text{chain propagation}$$

$$\left.\begin{array}{l} H_2ClC - \dot{C}HCl \rightarrow H_2C = CHCl + Cl \\ Cl + H_2\dot{C} - ClCH_2 \rightarrow H_2C = CClH + HCl \end{array}\right\} \begin{array}{l} \text{product formation} \\ \text{chain termination} \end{array} \qquad (5.6.1)$$

(among many other reaction steps not shown here).

Vinyl chloride is produced in megaton quantities and so an important additional consideration is the technology required to ensure the stability and long-term reliability required for a large-scale operation.

Other examples of laser-initiated reactions include the isomerization of substituted ethylenes catalyzed by sulfur molecules in their ground triplet state [5.150]

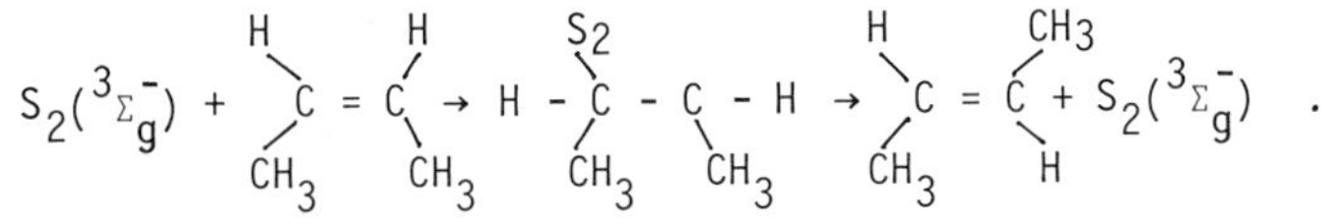

(5.6.2)

The $S_2(^3\Sigma_g^-)$ molecules can be produced from S atoms generated by the KrF laser photolysis of OCS , and the quantum yield of isomerization can be as high as 200. Trimerization of tetrachloroethylene to form hexachlorobenzene in the presence of borontrichloride irradiated by a cw CO_2 laser

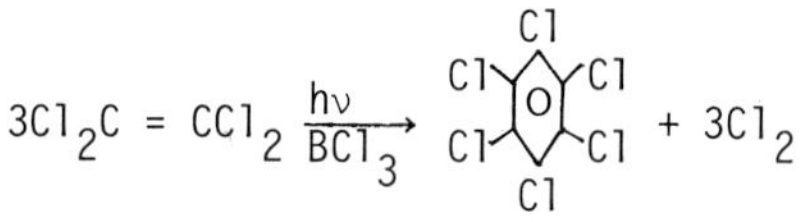

$$3Cl_2C = CCl_2 \xrightarrow[BCl_3]{h\nu} \underset{Cl}{\overset{Cl}{\underset{Cl}{\bigoplus}}} \, + \, 3Cl_2 \qquad (5.6.3)$$

has also been reported [5.151]. Another development in this general direction is laser-triggered explosions [5.152]. Since such reactions often exhibit very high rates hopes have been expressed that under these conditions bimolecular laser specific reactions may be able to effectively compete with collisional deactivation [5.153].

In addition to economic considerations, the discussion in this chapter has also noted some inherent, scientific, limitations on the potential applications. These refer both to the inherently nonstationary character of the laser prepared states in the larger molecules and to the collisional[5] scrambling of the molecular eigenstates (cf. Sects.2.3.7 and 5.4.2). The inter- and intramolecular relaxation processes tend to degrade the property valued most by laser chemists, that of selectivity. On the other hand, we have tried to emphasize that lasers are unique because they *combine* the well-defined narrow frequency bandwidth with other attributes such as high power, short pulses, and spatial coherence of the laser beam. These may not be essential for conventional photochemical applications, but may well point out the new and unique areas where the use of lasers is the sine qua non of photochemistry.

5.6.1 Practical Photoselective Chemistry

The most immediate applications of lasers are based on the selective absorption of one component in a mixture (e.g., Figs.1.16 or 1.19 or 5.44). At high pressures collisional effects (Fig.5.43) can play a dominant role and totally alter the product composition. A simple, but important, example is the conversion of an absorbing species to a nonabsorbing product. The backward reaction can be made much slower compared to the forward one by varying laser power and foreign gas pressure. Examples include the ring opening of hexafluorocyclobutene (Sect.5.4.3) double bond shifting [5.154] in open chain hexadienes

---

5 While collisional effects can be minimized under laboratory conditions, this is not necessarily the case for industrial processes.

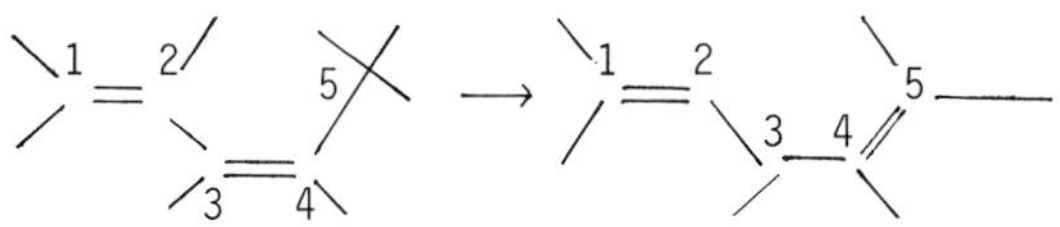 (5.6.4)

and the step-by-step substitution of $B(CH_3)_3$ in a mixture with HBr, using different lines of the CO_2 laser [5.155];

$$B(CH_3)_3 + HBr \xrightarrow{h\nu_1} B(CH_3)_2Br + CH_4$$
$$B(CH_3)_2Br + HBr \xrightarrow{h\nu_2} B(CH_3)Br_2 + CH_4$$
$$B(CH_3)Br_2 + HBr \xrightarrow{h\nu_3} BBr_3 + CH_4 \quad . \tag{5.6.5}$$

Here each methyl compound absorbs at a somewhat different frequency so that the reaction can be brought to a stop after any one of its three stages. Obviously such conversions should be possible in many other examples and, in particular, among pairs of isomers, and in the preparation of high purity materials.

For the removal of an impurity (or an isomer) by a selective chemical reaction of the pumped species it is often advantageous to use UV photons [5.156], since energy transfer among nonresonant electronic states is usually slow and allows operation at high pressures or even in condensed phases. Since electronic absorption bands are generally broad, a two-step photolysis is sometimes employed. A selective infrared mode is pumped first. This produces a shift in the UV absorption spectrum which is generally sufficient for selective absorption of the UV photon.

At higher pressures, the chemistry induced by high-power infrared lasers often resembles thermal reactions (pyrolysis or shock wave heating, cf. Fig.5.43). The high degree of collimation of laser light allows, however, homogeneous heating effects in regions which are not in contact with the vessel's walls. This is of importance where it is necessary to avoid wall-catalyzed side reactions or where the wall material shows insufficient chemical resistance. The high power available from infrared lasers allows heating to unusually high (thousands of degrees) temperatures. The high rate of heating means that molecules may well be pumped past the lowest threshold for reaction (Sects.1.3 and 5.4 and, e.g., Fig.1.21). Such energy-rich molecules tend to undergo unimolecular reactions before they have time to engage in the bimolecular processes which are typical of the intermediate temperature range. An example is ethylene which polymerizes when heated to about 1000 K. Infrared laser pumping of ethylene produces exclusively acetylene [5.157].

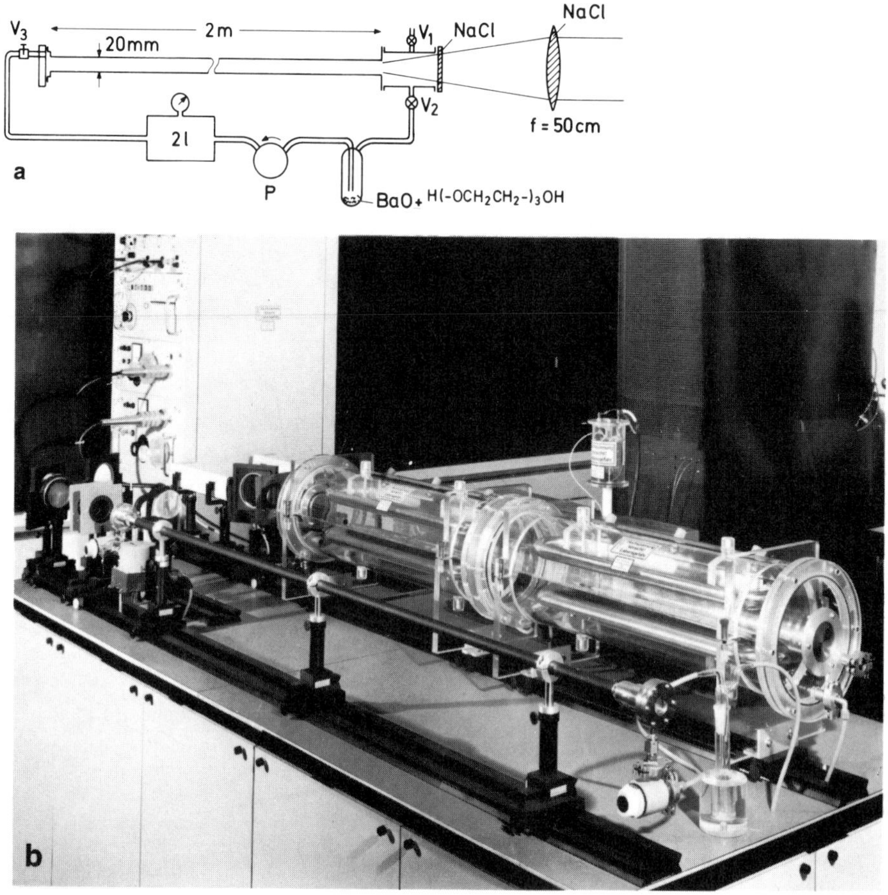

Fig.5.68. (a) Experimental arrangement for the separation of ^{32}S and ^{34}S (and other light) isotopes on a laboratory scale. The molecular compound irradiated by a pulsed, line-tunable CO_2 laser in a long waveguide-type absorption tube is SF_6 at $P \sim 0.1$ Torr. In order to have a larger materials throughput a cyclic gas flow is employed with a storage volume at atmospheric pressure and a compression stage/expansion valve included in the cycle. The dissociation product of interest is $^{32,34}SF_4$ (the isotopic labelling depending on the particular CO_2 laser line chosen). This is chemically trapped and obtained in the form of $Ba^{32,34}SO_3$ with >99% purity. A table top "separation factory" of these dimensions operated with a ~10 J CO_2 laser will produce up to gram amounts of light isotopes in $\sim 10^4$ laser shots. It is more or less representative for many other light element separation schemes (e.g., $^{12,13}C$ upon irradiation of $^{12,13}CF_3I$). For heavier elements with smaller vibrational isotope shifts, separation at room temperature is generally not efficient and the atomic or molecular samples may better be introduced in the form of cold (dynamically cooled) beams in order to eliminate line overlap and hot band absorption (adaped from [5.163]).
(b) Photograph of the apparatus

An application of laser high-temperature chemistry which is of potential industrial interest is the formation of high purity (99.999%) tetrafluoroethylene [5.88,89,98,158] required for teflon production,

$$2CHClF_2 \xrightarrow{h\nu} F_2C = CF_2 + 2HCl \quad , \qquad (5.6.6)$$

a reaction which was shown to be possible also by IR laser radiation.

Tunable dye lasers have been used to prepare electronically excited molecular oxygen [5.159]

$$O_2(^3\Sigma_g^-, v=0) \xrightarrow[\lambda = 761\ nm]{h\nu} O_2^*(^1\Sigma_g^+, v \geqq 0) \quad . \qquad (5.6.7)$$

While the chemistry of singlet oxygen molecules has been extensively studied [5.159] only laser means have been providing direct insight into its fate in, for example, reactions with unsaturated hydrocarbons yielding peroxy compounds as primary products,

$$O_2(^1\Sigma_g^+, v=0) + (CH_3)_2C = C(CH_3)_2 \rightarrow HO - O - C(CH_3)_2 - C(=CH_2)CH_3 \quad . \qquad (5.6.8)$$

Isotope selective excitation (Figs.1.16 and 5.32) is by far the most successful and best studied [5.160,161] practical application of laser chemistry. Infrared multiphoton dissociation (with many variations in detail) and two-step photolysis have been the two most commonly used techniques. Elements across the periodic table, from hydrogen to uranium (and, in particular, boron, carbon, nitrogen, oxygen, silicon, sulfur, titanium, selenium, tellurium, osmium, and molybdenum) have been separated in this fashion [5.162] (Fig.5.68).

5.6.2 State Selective Chemistry

In principle, laser pumping can be used to prepare diatomic molecules not only in a well-defined vibrotational level but even with a preferred spatial orientation (cf. Fig.1.14). In practice, the selective effect of the pumping is degraded by collisions. Alignment selectivity is very rapidly destroyed [5.164], rotational state soon thereafter, and even the often considerable enhancement of the reaction rate by vibrational excitation (Fig.2.1) need compete with rapid (e.g., V-R,T) energy loss mechanisms (Fig.5.7). In general, the practical chemist will limit considerations of state selected pumping to small molecules (or atoms) excited to a degree sufficient to make the reaction exothermic (cf. Sect.2.1.4) or where curve crossing among the dressed states (cf. Sect.2.4.8) is possible.

Dressing of atoms or molecules by very high power near infrared, visible, or even UV photons has been extensively discussed [5.165]. It is potentially both very sensitive (essentially all molecules in the irradiated volume can be clothed) and selective, both with respect to species and with respect to the distribution of the reaction products (cf. Figs.1.24 and 5.52). Since the absorption of energy occurs only during the desired bimolecular (cf. Fig.2.49) or unimolecular reaction event it avoids the degradation of the prepumped state by collisions.

For larger molecules, the interest in the near future will be limited to "group" or "mode" selective chemistry. Here the initial excitation is spread among many molecular states, the distribution being, however, different from a statistical one (where all isoenergetic states are equally populated). The selective effects of localizing the initial excitation has been demonstrated by chemical activation (Sects.2.2.7 and 2.3.6), and mode selective excitation using infrared lasers under collision-free conditions has been reported [5.95,166]. Mode selectivity at higher pressures has been claimed [5.78] but not unequivocally demonstrated.

As the total energy is increased, the equipartitioning of the energy among the accessible modes becomes more and more rapid. Even when this process is so fast that chemical rearrangements can hardly compete, the advantages of laser pumping are not lost. To begin with, intramolecular relaxation can, at worst, only thoroughly mix the isoenergetic states. As discussed in Sect.5.6.1 it should still be possible, using high-power lasers, to pump molecules to a nonthermallike distribution of the total energy (cf. Sect. 5.4.1). At higher pressures, collisional effects will first tend to mix nearly resonant states and only on much longer time scales will energy be dumped into the translation.

Finally, it should be remembered that energy very rich polyatomic molecules formed by chemical activation do dissociate in a nonrandom fashion [5.167]. Hence, if the energy deposition can be made fast enough [e.g., by a single or few photon absorption or, say, by a sideways internal conversion or inter-system crossing (Sect.2.4.9)] it should prove possible to beat nature's preference for equipartitioning.

5.7 Synergism

Synergism[6] has been the key to the past progress and also, we believe, to the future potential of our field. Lasing itself is made possible by the coherent emission from a large number of excited atoms or molecules. Imaginative laser chemistry reflects a combination of two or more of the properties of lasers, each property separately being readily found in more conventional sources, but the conjunction is unique. The rapid progress has been made possible by the blending of experiment and theory, each branch stimulating and sustaining the other. The very writing of this volume reflects this synergism.

Much of this book and indeed of current research is concerned with the elementary processes taking place in chemical lasers and in laser-induced chemistry. In the future we expect to see even more emphasis on the integration of these elementary processes into the description of complex molecular systems. The need to model chemical lasers has given a considerable impetus to such studies, and lately the same approach is being applied to laser-driven reactions.

Laser chemistry is synergistic not only with respect to the properties of the laser. We have tried to document and examine the matching of the capabilities of the laser to the energy levels, optical transition probabilities, and the possible (both inter- and intramolecular) dynamical processes of molecules. The judicious matching of source and substrate is itself made possible because of an earlier and ongoing synergism which gave birth to molecular reaction dynamics. Indeed, as we have seen in the last chapter, experiments using lasers not only require considerable input from chemical dynamics in their design but often yield much valuable information in return. Much of the current vitality of molecular reaction dynamics is due to the use of lasers. There is moreover another branch of chemistry where considerable progress in the elucidation of the detailed dynamics has recently been achieved: condensed phases and interfaces. Here too, lasers will soon be elevated from probing devices to full partners.

It is an opportune time to stop and take stock, not because the climb is over but because we can see what paths to follow. Much of the excitement

---

6 The combined activity of separate agencies such that the total effect is greater than the sum of the separate effects.

is still ahead but we have overcome the initial barrier and have climbed high enough to view and hence map the common grounds from which all furture expeditions should start. In other words, the base camp has been erected and is reasonably well stocked. The rest is up to you.

Other areas in chemistry, physics, biology, and indeed wherever molecules can interact with radiation are benefiting from the laser evolution. Revolutionary effects are, however, expected where the insight available on the system and the potential of the laser are coherently matched.

References

Chapter 1

1.1 D.O. Cowan, R.L. Drisko: *Elements of Organic Photochemistry* (Plenum, New York 1976)
1.2 J.G. Calvert, J.N. Pitts: *Photochemistry* (Wiley, New York 1966)
1.3 R.B. Woodward, R. Hoffmann: *The Conservation of Orbital Symmetry* (Verlag Chemie, Weinheim 1970)
1.4 R.G. Pearson: *Symmetry Rules for Chemical Reactions* (Wiley, New York 1976)
1.5 R.D. Levine, R.B. Bernstein: *Molecular Reaction Dynamics* (Clarendon Press, Oxford 1974)
1.6 P.R. Brooks, E.F. Hayes (eds.): *State-to-State Chemistry*, American Chemical Society Symposium Series, Vol. 56 (American Chemical Society, Washington 1977); see also the articles by A.H. Zewail, V.S. Letokhov; R.N. Zare, R.B. Bernstein, Y.T. Lee, Y.R. Chen: Phys. Today *33* (11), (1980)
1.7 M.J. Berry: J. Chem. Phys. *59*, 6229 (1973)
1.8 L.C. Lee, D.L. Judge: Can. J. Phys. *51*, 378 (1973)
1.9 S.C. Yang, A. Freedman, M. Kawasaki, R. Bersohn: J. Chem. Phys. *72*, 4058 (1980)
1.10 C.R. Jones, H.P. Broida: J. Chem. Phys. *60*, 4369 (1974)
1.11 A. Einstein: Phys. Z. *18*, 121 (1917)
1.12 W. Kauzmann: *Quantum Chemistry* (Academic, New York 1957)
1.13 A.E. Siegman: *An Introduction to Lasers and Masers* (McGraw-Hill, New York 1971)
1.14 J.V.V. Kasper, G.C. Pimentel: Phys. Rev. Lett. *14*, 352 (1965), (the first $H + Cl_2 \rightarrow HCl + Cl$ laser)
T.F. Deutsch: Appl. Phys. Lett. *10*, 234 (1967); K.L. Kompa, G.C. Pimentel: J. Chem. Phys. *47*, 857 (1967), ($F + H_2$)
M.A. Pollack: Appl. Phys. Lett. *8*, 237 (1966), (O + CS)
R.W.F. Gross: J. Chem. Phys. *50*, 1889 (1971); H.L. Chen, J.C. Stephenson, C.B. Moore: Chem. Phys. Lett. *2*, 593 (1968); T.A. Cool, R.R. Stephens, T.J. Falk: Int. J. Chem. Kinet. *1*, 495 (1969); Appl. Phys. Lett. *15*, 318 (1969), (HX-CO_2 transfer lasers)
C.B. Moore: IEEE J. QE-*4*, 52 (1968); J.R. Airey, S.F. McKay: Appl. Phys. Lett. *15*, 401 (1969), (Cl + HI)
1.15 J.V.V. Kasper, G.C. Pimentel: Appl. Phys. Lett. *5*, 231 (1964)
1.16 M.J. Berry, G.C. Pimentel: J. Chem. Phys. *51*, 2274 (1969)
For reviews, see *Handbook of Chemical Lasers*, ed. by R.W.F. Gross, J.F. Bott (Wiley, New York 1976)
1.17 M.J. Berry, G.C. Pimentel: J. Chem. Phys. *49*, 5190 (1968)
1.18 D.H. Maylotte, J.C. Polanyi, K.B. Woodall: J. Chem. Phys. *57*, 1547 (1972)

1.19 M.J. Berry: In *Molecular Energy Transfer*, ed. by R.D. Levine, J. Jortner (Wiley, New York 1976)
1.20 I.W.M. Smith: In *Molecular Energy Transfer*, ed. by R.D. Levine, J. Jortner (Wiley, New York 1976)
1.21 J.C. Polanyi, K.B. Woodall: J. Chem. Phys. *57*, 1574 (1972)
1.22 R.L. Wilkins: J. Chem. Phys. *59*, 698 (1973)
1.23 D.J. Douglas, J.C. Polanyi, J.J. Sloan: Chem. Phys. *13*, 15 (1976)
1.24 D. Arnoldi, J. Wolfrum: Ber. Bunsenges. Phys. Chem. *80*, 892 (1976)
1.25 S.R. Leone, R.G. McDonald, C.B. Moore: J. Chem. Phys. *63*, 4735 (1975)
1.26 A.M.G. Ding, J.C. Polanyi, J.L. Schreiber: Discuss.Faraday Soc. *55*, 252 (1973)
1.27 P.R. Brooks: Science *193*, 11 (1976)
1.28 Z. Karny, R.C. Estler, R.N. Zare: J. Chem. Phys. *69*, 5199 (1978); see also
Z. Karny, R.N. Zare: J. Chem. Phys. *68*, 3360 (1978)
1.29 D.M. Brenner, S. Datta, R.N. Zare: J. Am. Chem. Soc. *99*, 4554 (1977)
1.30 V.S. Letokhov, C.B. Moore: In *Chemical and Biochemical Applications of Lasers*, Vol. 3, ed. by C.B. Moore (Academic, New York 1977) p. 1
1.31 M. Stuke, E.E. Marinero: In *Laser-Induced Processes in Molecules*, Springer Series in Chemical Physics, Vol. 6, ed. by K.L. Kompa, S.D. Smith (Springer, Berlin, Heidelberg, New York 1978) p. 294; see also
M. Stuke, F.P. Schäfer: Chem. Phys. Lett. *48*, 271 (1977)
1.32 P.L. Houston, J.I. Steinfeld: In *Laser and Coherence Spectroscopy*, ed. by J.I. Steinfeld (Plenum, New York 1978)
1.33 K.V. Reddy, M.J. Berry: Chem. Phys. Lett. *66*, 223 (1979)
1.34 W. Fuss, T.P. Cotter: Appl. Phys. *12*, 265 (1977)
1.35 C.C. Jensen, T.G. Anderson, C. Reiser, J.I. Steinfeld: J. Chem. Phys. *71*, 3648 (1979)
1.36 Aa.S. Sudbø, D.J. Kranjovich, P.A. Schulz, Y.R. Chen, Y.T. Lee: In *Multiphoton Excitation and Dissociation of Polyatomic Molecules*, ed. by C.D. Cantrell: Topics in Current Physics (Springer, Berlin, Heidelberg, New York, in preparation)
1.37 A. Amirav, U. Even, J. Jortner: Opt. Common. *32*, 266 (1980)
1.38 S.J. Riley, K.R. Wilson: Discuss. Faraday Soc. *53*, 132 (1972)
1.39 L. Zandee, R.B. Bernstein: J. Chem. Phys. *71*, 1359 (1979)

Chapter 2

2.1 R.D. Levine, R.B. Bernstein: *Molecular Reaction Dynamics* (Clarendon, Oxford 1974)
2.2 A.M.G. Ding, L.J. Kirsch, D.S. Perry, J.C. Polanyi, J.L. Schreiber: Faraday Discuss. Chem. Soc. *55*, 252 (1973)
2.3 J.G. Pruett, R.N. Zare: J. Chem. Phys. *64*, 1774 (1976)
2.4 P.R. Brooks, E.F. Hayes(eds.): *State-to-State Chemistry* (ACS Symposium Series 56, Washington 1977)
2.5 S.H. Bauer: Chem. Rev. *78*, 147 (1978)
2.6 C.B. Moore, P.F. Zittel: Science *182*, 541 (1973)
2.7 R.N. Zare, P.J. Dagdigian: Science *185*, 739 (1974)
2.8 B.F. Gordietz, A.I. Osipov, E.V. Stupechenko, L.A. Shelepin: Sov. Phys. Usp. *15*, 759 (1973)
V.S. Letokhov, A.A. Makarov: J. Photochem. *3*, 249 (1974)
2.9 M.J. Berry: Ann. Rev. Phys. Chem. *26*, 259 (1975)
2.10 E. Pollak, R.D. Levine, Chem. Phys. Lett. *39*, 199 (1976)
2.11 R.D. Levine, J. Manz: J. Chem. Phys. *63*, 4280 (1975)
2.12 H. Kaplan, R.D. Levine, J. Manz: Chem. Phys. *12*, 447 (1976)
2.13 R.D. Levine, R.B. Bernstein: Accts. Chem. Res. *7*, 393 (1974)

2.14 L.J. Kirsch, J.C. Polanyi: J. Chem. Phys. *57*, 4498 (1972)
2.15 K.L. Kompa, J. Wanner: Chem. Phys. Lett. *12*, 560 (1972)
2.16 D.H. Maylotte, J.C. Polanyi, K.B. Woodall: J. Chem. Phys. *57*, 1547 (1972)
2.17 D.J. Douglas, J.C. Polanyi, J.J. Sloan: Chem. Phys. *13*, 15 (1976)
2.18 K. Bergmann, C.B. Moore: J. Chem. Phys. *63*, 643 (1975)
2.19 D.S. Perry, J.C. Polanyi: Chem. Phys. *12*, 419 (1976)
2.20 W.F. Calaway, G.E. Ewing: J. Chem. Phys. *63*, 2892 (1975)
S.R.J. Brueck, T.F. Deutsch, R.M. Osgood: J. Chem. Phys. *68*, 4941 (1978)
2.21 A. Laubereau, W. Kaiser: Ann. Rev. Phys. Chem. *26*, 83 (1975), and Chap. 3 in *Chemical and Biochemical Applications of Lasers*, Vol. 2, ed. by C.B. Moore (Academic, New York 1977)
2.22 S.R.J. Brueck, T.F. Deutsch, R.M. Osgood: Chem. Phys. Lett. *51*, 339 (1977); *60*, 242 (1979)
2.23 R.D. Levine: Chem. Phys. Lett. *39*, 205 (1976)
R.D. Levine, W.C. Gardiner, Jr.: J. Chem. Phys. *68*, 4524 (1978)
2.24 R.D. Levine: *Quantum Mechanics of Molecular Rate Processes* (Clarendon, Oxford 1969)
2.25 J.C. Light, J. Ross, K.E. Shuler: In *Kinetic Processes in Gases and Plasmas*, ed. by A.R. Hochstim (Academic, New York 1965)
2.26 K.J. Schmatjko, J. Wolfrum: Ber. Bunsenges. Phys. Chem. *82*, 419 (1978)
2.27 E.T. Jaynes: Phys. Rev. *106*, 620 (1957)
2.28 R.D. Levine, M. Tribus (eds.): *The Maximum Entropy Formalism* (MIT Press, Cambridge, MA 1979)
2.29 R.D. Levine, A. Ben-Shaul: *Chemical and Biochemical Applications of Lasers*, Vol. II, ed. by C.B. Moore (Academic, New York 1977) p. 145
2.30 M. Tribus: *Thermodynamics and Thermostatics* (Van Nostrand, New York 1961)
2.31 A. Katz: *Introduction to Statistical Mechanics* (Freeman, San Francisco 1967)
2.32 R. Bairelein: *Atoms and Information Theory* (Freeman, San Francisco 1971)
2.33 R.D. Levine: Ann. Rev. Phys. Chem. *29*, 59 (1978)
2.34 R.D. Levine, R.B. Bernstein: *Dynamics of Molecular Collisions*, ed. by W.H. Miller (Plenum, New York 1976) Part. B, p. 323
2.35 R.D. Levine, J.L. Kinsey: In *Atom-Molecule Collision Theory*, ed. by R.B. Bernstein (Plenum, New York 1979)
2.36 C.E. Shannon: Bell Syst. Tech. J. *27*, 397 (1948)
2.37 N. Wienner: *Cybernetics* (MIT Press, Cambridge, MA 1948)
2.38 S. Goldman: *Information Theory* (Dover, New York 1953)
R. Ash: *Information Theory* (Interscience, New York 1965)
2.39 Y. Alhassid, N. Agmon, R.D. Levine: Chem. Phys. Lett. *53*, 22 (1978); J. Comput. Phys. *30*, 250 (1979) and in Ref. 2.28
2.40 J.E. Mayer, M.G. Mayer: *Statistical Mechanics* (Wiley, New York 1940); see also J.C. Keck: In Ref. 2.28
2.41 Y. Alhassid, R.D. Levine: Phys. Rev. *A18*, 89 (1978); See also R.D. Levine: In Ref. 2.28
2.42 J.L. Kinsey, R.D. Levine: Chem. Phys. Lett. *65*, 413 (1979)
2.43 A. Ben-Shaul, R.D. Levine, R.B. Bernstein: J. Chem. Phys. *57*, 5427 (1972)
2.44 D.J. Bogan, D.W. Setser: J. Chem. Phys. *64*, 586 (1976)
2.45 M.J. Berry: J. Chem. Phys. *59*, 6229 (1973)
2.46 N. Jonathan, C.M. Melliar-Smith, S. Okuda, D.H. Slater, D. Timlin: Mol. Phys. *22*, 561 (1971)
2.47 G. Hancock, C. Morley, I.W.M. Smith: Chem. Phys. Lett. *12*, 193 (1973)
2.48 A. Ben-Shaul: Chem. Phys. *1*, 244 (1973)
2.49 D.M. Manos, J.M. Parson: J. Chem. Phys. *63*, 3575 (1975)
2.50 U. Dinur, R. Kosloff, R.D. Levine, M.J. Berry: Chem. Phys. Lett. *34*, 199 (1975)

2.51 R.G. Shortridge, M.C. Lin: J. Chem. Phys. *64*, 4076 (1976)
2.52 C.L. Chalek, J.L. Gole: J. Chem. Phys. *65*, 2845 (1976)
2.53 K. Liu, J.M. Parson: J. Chem. Phys. *68*, 1794 (1978)
2.54 F. Reif: *Statistical and Thermal Physics* (McGraw-Hill, New York 1965)
2.55 M.E. Umstead, R.G. Shortridge, M.C. Lin: J. Phys. Chem. *82*, 1455 (1978)
2.56 A. Ben-Shaul, K.L. Kompa, U. Schmailzl: J. Chem. Phys. *65*, 1711 (1976)
2.57 J.C. Polanyi, K.B. Woodall: J. Chem. Phys. *57*, 1547 (1972)
W.H. Chang, D.W. Setser: J. Chem. Phys. *55*, 2298 (1973)
J.G. Moehlmann, R.G. McDonald: J. Chem. Phys. *62*, 3061 (1975)
See also 2.19
2.58 R.D. Levine, B.R. Johnson, R.B. Bernstein: Chem. Phys. Lett. *19*, 1 (1973)
2.59 J.A. Silver, W.L. Dimpfl, J.H. Brophy, J.L. Kinsey: J. Chem. Phys. *65*, 1811 (1976)
R.P. Mariella, B. Lantzsch, V.T. Maxson, A.C. Luntz: J. Chem. Phys. *69*, 3436 (1978)
2.60 R.E. Wyatt: Chem. Phys. Lett. *34*, 167 (1975)
2.61 D.C. Clary, R.K. Nesbet: Chem. Phys. Lett. *59*, 437 (1978)
2.62 W.S. Struve, J.R. Krenos, D.L. McFadden, D.R. Herschbach: J. Chem. Phys. *62*, 404 (1975)
J.R. Krenos, J.C. Tully: J. Chem. Phys. *62*, 420 (1975)
2.63 R.D. Levine: In *Electronic Transitions Lasers*, ed. by J.I. Steinfeld (MIT Press, Cambridge, MA 1975)
2.64 M. Luria, D.J. Eckstrom, S.A. Edelstein, B.E. Perry, S.W. Benson: J. Chem. Phys. *64*, 2247 (1976)
2.65 M.B. Faist, R.D. Levine: Chem. Phys. Lett. *47*, 5 (1977)
2.66 C.L. Chalek, J.L. Gole: Chem. Phys. *19*, 59 (1977)
2.67 M.J. Berry: In Ref. 2.68
2.68 R.D. Levine, J. Jortner (eds.): *Molecular Energy Transfer* (Wiley, New York 1976)
2.69 G.A. West, R.E. Weston, Jr., G.W. Flynn: Chem. Phys. *35*, 275 (1978); see also
C.R. Quick, C. Wittig, Chem. Phys. *32*, 75 (1978)
2.70 M.J. Berry: J. Chem. Phys. *61*, 3114 (1974)
2.71 E. Zamir, R.D. Levine: Chem. Phys. Lett. *67*, 237 (1979)
2.72 Aa.S. Sudbø, P.A. Schulz, Y.R. Chenand, Y.T. Lee: J. Chem. Phys. *69*, 2312 (1978)
2.73 I. Kusunoki, Ch. Ottinger: J. Chem. Phys. *71*, 4227 (1979)
2.74 R.D. Levine: In Ref. 2.75
2.75 J. Jortner, R.D. Levine, S.A. Rice (eds.): *Photoselective Chemistry* (Wiley, New York 1980)
2.76 C. Lifshitz: Adv. Mass Spectrom. *7A*, 3 (1978)
2.77 W.H. Miller (ed.): *Dynamics of Molecular Collisions* (Plenum, New York 1976)
2.78 W. Jost: *Kinetics of Gas Reactions*, Vol. IV A in Physical Chemistry An Advanced Treatise, ed. by H. Eyring, D. Hendersen, W. Jost (Academic, New York 1974)
2.79 A.H. Zewail (ed.): *Advances in Laser Chemistry*, Springer Series in Chemical Physics, Vol. 3 (Springer, Berlin, Heidelberg, New York 1978)
2.80 R.B. Bernstein (ed.): *Atom-Molecule Collision Theory* (Plenum, New York 1979)
2.81 J.C. Polanyi, J.L. Schreiber: In Ref. 2.78
2.82 R.N. Porter, L.M. Raff: In Ref. 2.77
2.83 P.J. Kuntz: In Refs. 2.77 and 2.80
2.84 J.T. Muckerman, D.G. Truhlar: In Ref. 2.80
2.85 D.L. Thompson: Acc. Chem. Res. *9*, 338 (1976)
2.86 I.W.M. Smith: In Ref. 2.68

2.87 J. Wolfrum: Ber. Bunsenges. Phys. Chem. *81*, 114 (1977)
2.88 H.F. Schafer, III: In Ref. 2.80
2.89 G.G. Balint-Kurti: Adv. Chem. Phys. *30*, 137 (1975)
2.90 H.S. Johnston: *Gas Phase Reaction Rate Theory* (Ronald Press, New York 1966)
2.91 M.J. Berry, O. Kafri: Faraday Discuss. Chem. Soc. *62*, 127 (1977)
2.92 R.A. Marcus: J. Phys. Chem. *72*, 891 (1968)
2.93 G.S. Hammond: J. Am. Chem. Soc. *77*, 334 (1955)
2.94 N. Agmon, R.D. Levine: J. Chem. Phys. *71*, 3034 (1979)
2.95 M.J. Berry: Chem. Phys. Lett. *27*, 73 (1974); *29*, 323 (1974)
2.96 A. Kafri, E. Pollak, R. Kosloff, R.D. Levine: Chem. Phys. Lett. *33*, 202 (1975)
2.97 G.C. Schatz, J. Ross: J. Chem. Phys. *66*, 1037 (1977)
2.98 P.J. Robinson, K.A. Holbrook: *Unimolecular Reactions* (Wiley, New York 1972)
2.99 W.L. Hase: In Ref. 2.77
2.100 W. Forst: *Theory of Unimolecular Reactions* (Academic, New York 1973)
2.101 C.C. Jensen, J.I. Steinfeld, R.D. Levine: J. Chem. Phys. *69*, 1432 (1978)
2.102 J.D. Rynbrandt, B.S. Rabinovitch: J. Phys. Chem. *75*, 2164 (1971)
2.103 I. Oref, B.S. Rabinovitch: Acc. Chem. Res. *12*, 166 (1979)
2.104 I. Oref, N. Gordon: J. Phys. Chem. *82*, 2035 (1978)
2.105 G.W. Flynn: In *Chemical and Biochemical Applications of Lasers*, Vol. 1, ed. by C.B. Moore (Academic, New York 1974) p. 163
2.106 J.I. Steinfeld, R.G. Gordon: In Ref. 2.68
2.107 M. Quack, J. Troe: In *Gas Kinetics and Energy Transfer*, Vol. II (Chemical Society, London 1977) p. 175
2.108 G.E. Ewing: J. Chem. Phys. *71*, 3143 (1979); see also J.A. Beswick, J. Jortner: In Ref. 2.75
2.109 J. Ford: Adv. Chem. Phys. *24*, 155 (1973); in *Lectures in Statistical Physics*, Lecture Notes in Physics, Vol. 28, ed. by W.C. Schieve, J.S. Turner (Springer, Berlin, Heidelberg, New York 1974)
2.110 S.A. Rice: In Ref. 2.79
2.111 J.D. McDonald, R.A. Marcus: J. Chem. Phys. *65*, 2180 (1976)
2.112 P. Brumer: In Refs. 2.75 and 2.79
2.113 I.C. Percival: Adv. Chem. Phys. *36*, 1 (1977)
2.114 B.R. Henry: Acc. Chem. Res. *10*, 207 (1977)
2.115 M.L. Elert, P. Stannard, W.M. Gelbart: J. Chem. Phys. *67*, 5395 (1977)
2.116 A. Albrect: In Ref. 2.79
2.117 K.V. Reddy, R.G. Bray, M.J. Berry: In Ref. 2.79
2.118 D.F. Heller, S. Mukamel: J. Chem. Phys. *70*, 463 (1979)
2.119 D.D. Smith, A.H. Zewail: J. Chem. Phys. *71*, 540 (1979)
2.120 M. Rubinson, J.I. Steinfeld: Chem. Phys. *4*, 467 (1974)
2.121 I. Procaccia, R.D. Levine: J. Chem. Phys. *63*, 4261 (1975)
2.122 C. Ennen, Ch. Ottinger: Chem. Phys. *3*, 404 (1974)
2.123 J.F. Bott, N. Cohen: J. Chem. Phys. *58*, 4539 (1973)
2.124 I. Procaccia, R.D. Levine: J. Chem. Phys. *65*, 495 (1976); Phys. Rev. *14A*, 1569 (1976)
2.125 C.B. Moore: J. Chem. Phys. *43*, 2979 (1965)
2.126 E. Sirkin, Ph.D. Thesis: University of California Berkeley (1980); see also
J.H. Smith, D.W. Robinson: J. Chem. Phys. *68*, 5474 (1978); *71*, 271 (1979)
E. Cuellar, G.C. Pimentel: J. Chem. Phys. *71*, 1385 (1979)
2.127 D.C. Tardy, B.S. Rabinovitch: Chem. Rev. *77*, 369 (1977)
2.128 K.F. Hertzfeld, T.A. Litovitz: *Absorption and Dispersion of Ultrasonic Waves* (Academic, New York 1959)
2.129 D. Rapp, T. Kassal: Chem. Rev. *61*, 61 (1968)

2.130 R.D. Levine, R.B. Bernstein: J. Chem. Phys. *56*, 228 (1972)
2.131 J.F. Bott: J. Chem. Phys. *63*, 2253 (1975)
2.132 J.E. Dove, H. Teitelbaum: Chem. Phys. *6*, 431 (1974)
2.133 R.V. Steele, C.B. Moore: J. Chem. Phys. *60*, 2794 (1974)
2.134 D.J. Seery: J. Chem. Phys. *56*, 1796 (1973)
2.135 P.F. Zittel, C.B. Moore: J. Chem. Phys. *59*, 6636 (1973)
2.136 R.A. Lucht, T.A. Cool: J. Chem. Phys. *63*, 3962 (1975)
2.137 J.F. Bott, N. Cohen: J. Chem. Phys. *58*, 934 (1973)
2.138 R.D. Levine, B.R. Johnson: Chem. Phys. Lett. *14*, 132 (1972)
2.139 R. Bersohn: In Ref. 2.68
2.140 A.P.M. Baede: Adv. Chem. Phys. *30*, 463 (1975)
2.141 E. Bauer, E.R. Fisher, F.R. Gilmore: J. Chem. Phys. *51*, 4173 (1969)
2.142 R. Olsen, F.T. Smith, E. Bauer: Appl. Opt. *10*, 1848 (1971)
R. Grice, D.R. Herschbach: Mol. Phys. *27*, 159 (1974)
2.143 M.B. Faist, R.D. Levine: J. Chem. Phys. *64*, 2953 (1976)
2.144 A.M.C. Moutinho, J.A. Aten, J. Los: Physica *53*, 471 (1971)
2.145 J.C. Tully: In Ref. 2.77
2.146 A.B. Petersen, C. Wittig, S.R. Leone: Appl. Phys. Lett. *27*, 305 (1975); J. Appl. Phys. *47*, 1051 (1976)
2.147 W.R. Green, J. Lukasik, J.R. Wilson, M.D. Wright, J.F. Young, S.E. Harris: Phys. Rev. Lett. *42*, 970 (1979)
2.148 T.F. George, I.H. Zimmermann, J.M. Yuan, J.R. Laing, P.L. De Vries: Acc. Chem. Res. *10*, 449 (1977)
J.M. Yuan, T.F. George: J. Chem. Phys. *70*, 990 (1979)
2.149 A.M.F. Lau, C.K. Rhodes: Phys. Rev. *15*, 1570 (1977)
2.150 S. Mukamel, J. Jortner: In Ref. 2.68
2.151 K.F. Freed: Top. Appl. Phys. *15*, 1 (1976)
2.152 W. Siebrand: In Ref. 2.77
2.153 P. Avouris, W.M. Gelbart, M.A. El-Sayed: Chem. Rev. *77*, 793 (1977)
2.154 I. Oppenheim, K.E. Shuler, G.H. Weiss: *Stochastic Processes in Chemical Physics: The Master Equation* (MIT Press, Cambridge, MA 1977)
2.155 E.W. Montroll, K.E. Shuler: J. Chem. Phys. *26*, 454 (1957)
2.156 J.C. Polanyi, K.B. Woodall: J. Chem. Phys. *56*, 1563 (1972)
A.M.G. Ding, J.C. Polanyi: Chem. Phys. *10*, 39 (1975)
2.157 I. Procaccia, Y. Shimoni, R.D. Levine: J. Chem. Phys. *63*, 3181 (1975)
2.158 N.C. Lang, J.C. Polanyi, J. Wanner: Chem. Phys. *24*, 219 (1977)
2.159 J.J. Hinchen, R.M. Hobbs: J. Chem. Phys. *65*, 2732 (1975); See also L.H. Sentman: J. Chem. Phys. *67*, 966 (1977)
2.160 T. Feldmann, A. Ben-Shaul: Chem. Phys. Lett. *64*, 286 (1979)
2.161 S.D. Rockwood, E.S. Brau, W.A. Proctor, G.H. Canavan: IEEE, J. QE-*9*, 120 (1973) and references cited therein; See also
W.Q. Jeffers, J.D. Kelley: Chem. Phys. *55*, 4433 (1971)
2.162 H.T. Powell: J. Chem. Phys. *63*, 2635 (1975)
2.163 M.R. Verter, H. Rabitz: J.Chem. Phys. *64*, 2939 (1976)
2.164 M. Tabor, R.D. Levine, A. Ben-Shaul, J.I. Steinfeld: Mol. Phys. *37*, 141 (1979)
For the use of (2.4.17); see also H. Pummer, D. Proch, U. Schmailzl, K.L. Kompa: Opt. Commun. *19*, 273 (1976)
2.165 C.E. Treanor, J.W. Rich, R.G. Rehm: J. Chem. Phys. *48*, 1798 (1968)
2.166 S. Tsuchiya, N. Nielsen, S.H. Bauer: J. Phys. Chem. *77*, 2455 (1973); see also
Y.S. Liu, R.A. McFarlane, G.J. Wolga: J. Chem. Phys. *63*, 235 (1975)
2.167 K.E. Shuler: J. Chem. Phys. *32*, 1692 (1960)
2.168 E. Weitz, G.W. Flynn: In Ref. 2.75
R.E. McNair, S.F. Fulghum, G.W. Flynn, M.S. Feld: Chem. Phys. Lett. *48*, 24 (1977)
I. Shamah, G.W. Flynn: J. Am. Chem. Soc. *99*, 319 (1977)

2.169 A. Ben-Shaul, K.L. Kompa: Chem. Phys. Lett. *55*, 560 (1978); see also R.K. Huddleston, E. Weitz: J. Chem. Phys. *66*, 1740 (1977)
2.170 R.M. Osgood, P.B. Sacket, Jr., A. Javan: J. Chem. Phys. *60*, 1464 (1974)
2.171 I. Procaccia, Y. Shimoni, R.D. Levine: J. Chem. Phys. *65*, 3284 (1976)
2.172 R.D. Levine, M. Berrondo: Chem. Phys. Lett. *47*, 399 (1977)
2.173 I. Procaccia, R.D. Levine: J. Chem. Phys. *65*, 3357 (1976)
2.174 R.D. Levine: J. Chem. Phys. *65*, 3302 (1976)
2.175 R.D. Levine, O. Kafri: Chem. Phys. Lett. *27*, 175 (1974); Chem. Phys. *8*, 426 (1975)
O. Kafri, R.D. Levine: Isr. J. Chem. *16*, 342 (1977)
2.176 A. Ben-Shaul, O. Kafri, R.D. Levine: Chem. Phys. *16*, 342 (1977); *36*, 307 (1979)
For reviews see A. Ben-Shaul: In Ref. 2.75; R.D. Levine: In *Proceedings of the 12th Symposium on Shock Tubes and Waves, Jerusalem, 1979*, ed. by A. Lifshitz (Magnes Press, Jerusalem 1979)
2.177 R.A. Marcus, O.K. Rice: J. Phys. Colloid Chem. *55*, 894 (1951)
2.178 G.Z. Whitten, B.S. Rabinovitch: J. Chem. Phys. *38*, 2466 (1963)
2.179 R.D. Levine: J. Phys. Chem. *83*, 159 (1979)
2.180 P. Pechukas: In Ref. 2.77
2.181 R.B. Bernstein, A. Dalgarno, H.S.W. Massey, I.C. Percival: Proc. R. Soc. London *A 274*, 427 (1963)
P. Pechukas, J.C. Light: J. Chem. Phys. *42*, 328 (1965)
A. Ben-Shaul: Chem. Phys. *22*, 341 (1977)
M. Quack, J. Troe: Ber. Bunsenges. Phys. Chem. *80*, 1140 (1976)
2.182 R.A. Marcus: J. Chem. Phys. *62*, 1372 (1975)
2.183 J.C. Keck: Adv. Chem. Phys. *13*, 85 (1967)

Chapter 3

3.1 A. Yariv: *Quantum Electronics* (Wiley, New York 1967, 1975)
3.2 A. Yariv: *Introduction to Optical Electronics* (Holt, Rinehart and Winston, New York 1971)
3.3 A.E. Siegman: *An Introduction to Lasers and Masers* (McGraw-Hill, New York 1968)
3.4 O. Svelto: *Principles of Lasers* (Plenum, New York 1976)
3.5 B.A. Lengyel: *Lasers*, 2nd ed. (Wiley, New York 1971)
3.6 M. Sargent, III, M.O. Scully, W.E. Lamb, Jr.: *Laser Physics* (Addison-Wesley, London 1974)
3.7 A. Maitland, M.H. Dunn: *Laser Physics* (North-Holland, Amsterdam 1969)
3.8a G. Herzberg: *Atomic Spectra and Atomic Structure* (Dover, New York 1944)
3.8b G. Herzberg: *Diatomic Molecules*, *Molecular Spectra and Molecular Structure*, Vol. I (Van Nostrand, Princeton 1950)
3.8c G. Herzberg: *Infrared and Raman Spectra of Polyatomic Molecules*, *Molecular Spectra and Molecular Structure*, Vol. II (Van Nostrand, Princeton 1945)
3.8d G. Herzberg: *Electronic Spectra and Electronic Structure of Polyatomic Molecules*, *Molecular Spectra and Molecular Structure*, Vol. III (Van Nostrand, Princeton 1966)
3.9 I.N. Levine: *Quantum Chemistry*, Vols. I and II (Allyn and Bacon, Boston 1970)
3.10 G.W. King: *Spectroscopy and Molecular Structure* (Holt, Rinehart and Winston, New York 1964)
3.11 G.M. Barrow: *Introduction to Molecular Spectroscopy* (McGraw-Hill, New York 1962)
3.12 M.D. Harmony: *Introduction to Molecular Energies and Spectra* (Holt, Rinehart and Winston, New York 1972)

3.13 J.I. Steinfeld: *Molecules and Radiation; An Introduction to Modern Molecular Spectroscopy* (Harper and Row, New York 1974)
3.14 E.B. Wilson, Jr., J.C. Decius, P.C. Cross: *Molecular Vibrations* (McGraw-Hill, New York 1955)
3.15 N. Bloembergen: *Non Linear Optics* (Benjamin, New York 1965)
3.16 V.S. Letokhov, V.P. Chebotayev: *Nonlinear Laser Spectroscopy*, Springer Series in Optical Sciences, Vol. 4 (Springer, Berlin, Heidelberg, New York 1977)
3.17 G.C. Baldwin: *An Introduction to Nonlinear Optics* (Plenum, New York 1969)
3.18 P.G. Harper, B.S. Wherret (eds.): *Nonlinear Optics* (Academic, New York 1977)
3.19 G. Emanuel: In *Handbook of Chemical Lasers*, ed. by R.W.F. Gross, J.F. Bott (Wiley, New York 1976) Chap. 8
3.20 V.S. Zuev, V.A. Katulin, V.Y. Nosach, O.Y. Nosach: Sov. Phys. JETP *35*, 870 (1972)
3.21 J.E. Geusic, H.M. Marcos, L.G. Van Vitert: Appl. Phys. Lett. *4*, 182 (1964)
3.22 F.R. Gilmore: J. Quant. Spectrosc. Radiat. Transfer *5*, 369 (1965)
3.23 C.K.N. Patel: Sci. Am. *219*, 22 (Aug. 1968)
3.24 T. Shimanouchi: "Tables of Molecular Vibrational Frequencies", USRSD, NBS 39 (1972)
3.25 E. Weitz, G.W. Flynn: J. Chem. Phys. *58*, 2679 (1973)
3.26 J.C. Slater: *Quantum Theory of Atomic Structure*, Vol. I (McGraw-Hill, New York 1960)
3.27 W.H. Bennet, C.F. Meyer: Phys. Rev. *32*, 888 (1928)
3.28 J.D. Louck, H.W. Galbraith: Rev. Mod. Phys. *48*, 69 (1976)
3.29 W.G. Harter, C.W. Patterson, F.J. da Paixao: Rev. Mod. Phys. *50*, 37 (1978)
3.30 J.P. Aldridge, H. Filip, H. Flicker, R.F. Holland, R.S. McDowell, N.G. Nereson: J. Mol. Spectrosc. *58*, 165 (1975)
3.31 J. Moret Bailly: Cah. Phys. *15*, 237 (1961); J. Mol. Spectrosc. *15*, 344 (1965)
3.32 For a review, see D.C. Moule, A.D. Walsh: Chem. Rev. *75*, 67 (1975)
3.33 V.A. Job, V. Sethuraman, K.K. Innes: J. Mol. Spectrosc. *30*, 365 (1969)
3.34 P.L. Houston, C.B. Moore: J. Chem. Phys. *65*, 757 (1976)
3.35 F.P. Schäfer (ed.): *Dye Lasers*, Topics in Appl. Phys., Vol. 1 (2nd ed.) (Springer, Berlin, Heidelberg, New York 1977)
3.36 H. Kressel: In *Lasers*, Vol. 3, ed. by A.K. Levine, A.J. DeMaria (Dekker, New York 1971) p. 1
3.37 Y.R. Shen (ed.): *Nonlinear Infrared Generation*, Topics in Appl. Phys., Vol. 16 (Springer, Berlin, Heidelberg, New York 1977)
3.38 V.T. Nguyen, T.J. Bridges: In Ref. 3.37, p. 139
S.R.J. Brueck, A. Mooradian: Appl. Phys. Lett. *18*, 229 (1971)
3.39 C.K. Rhodes (ed.): *Excimer Lasers*, Topics in Appl. Phys., Vol. 30 (Springer, Berlin, Heidelberg, New York 1979)
3.40 J.J. Ewing, C.A. Brau: Appl. Phys. Lett. *27*, 350 (1977)
3.41 J.H. Jacob, M. Rokni, J.A. Mangano, R. Brochu: Appl. Phys. Lett. *32*, 109 (1978)
3.42 H. Pummer, K. Hohla, M. Diegelmann, J.P. Reilly: Opt. Commun. *28*, 104 (1979)
M. Diegelmann, K. Hohla, K.L. Kompa: Opt. Commun. *29*, 334 (1979)
For detailed kinetic modeling see, e.g., H. Pummer, K. Hohla, F. Rebentrost: Appl. Phys. *20*, 129 (1979)
W. Chow, K. Hohla, F. Rebentrost: Report PLF 8, Projektgruppe für Laserforschung der MPG Garching (1978)

3.43 J.R. McDonald, A.P. Baronavski, V.M. Donnelly: Chem. Phys. *33*, 161 (1978)
3.44 P.W. Smith: Opt. Acta *23*, 901 (1976)
3.45 G. Marowsky, R. Condray, F.K. Tittel, W.L. Wilson, C.B. Collins: Appl. Phys. Lett. *33*, 59 (1978)
3.46 P.K. Cheo: In *Lasers*, Vol. 3, ed. by A.K. Levine, A.J. DeMaria (Dekker, New York 1971) p. 111; See also
C.K.N. Patel: In *Lasers*, Vol. 2, ed. by A.K. Levine, A.J. DeMaria (Dekker, New York 1971) p. 1
3.47 S.D. Smith, W.E. Schmid: Report PLF 7, Projektgruppe für Laserforschung der MPG Garching (1978); See also
S.D. Smith, W.E. Schmid, F.M.G. Tablas, K.L. Kompa: In *Laser Induced Processes in Molecules*, ed. by K.L. Kompa, S.D. Smith (Springer, Berlin, Heidelberg, New York 1978)

Chapter 4

4.1 J.C. Polanyi: J. Chem. Phys. *34*, 347 (1961); Appl. Opt. Suppl. 2 (Chemical Lasers) 106 (1965), and other articles in this volume
4.2 J.V.V. Kasper, G.C. Pimentel: Appl. Phys. Lett. *5*, 231 (1964)
4.3 J.V.V. Kasper, G.C. Pimentel: Phys. Rev. Lett. *14*, 352 (1965)
4.4 K.L. Kompa: *Chemical Lasers*, Topics in Current Chemistry, Vol. 37 (Springer, Berlin, Heidelberg, New York 1973)
4.5 M.J. Berry: In *Molecular Energy Transfer*, ed. by R.D. Levine, J. Jortner (Wiley, New York; Israel University Press, Jerusalem 1976) p. 114
4.6 R.W.F. Gross, J.F. Bott (eds.): *Handbook of Chemical Lasers* (Wiley, New York 1976)
4.7 M.J. Berry: Ann. Rev. Phys. Chem. *26*, 259 (1975)
4.8 C.K. Rhodes (ed.): *Excimer Lasers*, Topics in Applied Physics, Vol. 30 (Springer, Berlin, Heidelberg, New York 1979)
4.9 A.N. Oraevskii: Sov. Phys. JETP *32*, 856 (1971)
4.10 L.E. Wilson, S.N. Suchard, J.I. Steinfeld (eds.): *Electronic Transition Lasers II* (MIT Press, Cambridge, MA 1977)
J.I. Steinfeld (ed.): *Electronic Transition Lasers* (MIT Press, Cambridge, MA 1976)
4.11 R.J. Jensen: In Ref. 4.6, p. 703
4.12 W.E. McDermott, N.R. Pchelkin, D.J. Benard, R.R. Bousek: Appl. Phys. Lett. *32*, 469 (1978); *34*, 40 (1979)
4.13 T.F. Deutsch: Appl. Phys. Lett. *11*, 18 (1967)
4.14 E. Cuellar, J.H. Parker, G.C. Pimentel: J. Chem. Phys. *61*, 422 (1974)
E. Sirkin, G.C. Pimentel (to be published)
4.15 J.P. Sung, D.W. Setser: J. Chem. Phys. *69*, 3868 (1978)
4.16 N. Skribanovitch, I.P. Herman, R.M. Osgood, M.S. Feld, A. Javan: Appl. Phys. Lett. *20*, 428 (1972)
4.17 G.D. Downey, D.W. Robinson, J.H. Smith: J. Chem. Phys. *66*, 1685 (1977)
J.H. Smith, D.W. Robinson: J. Chem. Phys. *68*, 5474 (1978)
4.18 J.H. Smith, D.W. Robinson: J. Chem. Phys. *71*, 271 (1979)
4.19 M.J. Berry: J. Chem. Phys. *59*, 6229 (1973)
4.20 A. Ben-Shaul, O. Kafri, S. Feliks: Chem. Phys. *36*, 291 (1979)
4.21 M.J. Berry, G.C. Pimentel: J. Chem. Phys. *53*, 3453 (1970); See also
M.J. Berry: J. Chem. Phys. *61*, 3114 (1974)
4.22 R. Herman, R.F. Wallis: J. Chem. Phys. *23*, 637 (1955)
R. Herman, R.W. Rotheny, R.J. Rubin: J. Mol. Spectrosc. *2*, 369 (1958)
4.23 R.D. Meredith, F.G. Smith: J. Quant. Spectrosc. Radiat. Transfer *13*, 89 (1973); J.M. Herbelin, G. Emanuel: J. Chem. Phys. *60*, 689 (1974)
4.24 A. Ben-Shaul, G.L. Hofacker, K.L. Kompa: J. Chem. Phys. *59*, 4664 (1973)

4.25 C.K.N. Patel: Phys. Rev. *136A*, 1187 (1964); See also
C.K.N. Patel: In *Lasers*, Vol. 2, ed. by A.K. Levine (Dekker, New York 1967) p. 79
4.26 R.S. Mulliken: Phys. Rev. *51*, 310 (1937)
4.27 C. Turner, N.L. Papagani: "Laser Fusion Program - Semiannual Report"; UCRL-50021-73-1, Lawrence Livermore Laboratory, California (Jan.-June, 1973), UCID-16935
4.28 K. Hohla, K.L. Kompa: In Ref. 4.6, p. 667
4.29 V.S. Zuev, V.A. Katulin, O.Y. Nosach: Sov. Phys. JETP *35*, 870 (1972); for a recent more general discussion see
V.I. Babkin, S.V. Kuzuetsova, A.I. Maslov: Sov. J. Quantum Electron. *8*, 285 (1978)
4.30 W. Fuss, K. Hohla: Z. Naturforsch. *31a*, 569 (1976)
4.31 E.A. Yukov: Sov. J. Quantum Electron. *3*, 117 (1973)
4.32 G. Brederlow, K.J. Witte: Laboratory Report PLF 19, MPG Projektgruppe für Laserforschung, Garching (1979)
4.33 N.G. Basov, V.S. Zuev: Nuovo Cimento 3113, 129 (1976)
M.E. Riely, R.E. Palmer: Laboratory Report SAND 77-0775, Sandia Laboratories, Albuquerque, NM (1977)
G. Brederlow, K.J. Witte, E. Fill, K. Hohla, R. Volk: IEEE J. QE-*12*, 152 (1976); for a detailed description see also
G. Brederlow, R. Brodermann, K. Eidmann, H. Krause, M. Nippus, R. Petsch, R. Volk, S. Witkowski, K.J. Witte: Laboratory Report PLF 5, MPG Projektgruppe für Laserforschung, Garching (1979)
4.34 S.N. Suchard, J.R. Airey: In Ref. 4.6, p. 389
4.35 H. Pummer, K.L. Kompa: Appl. Phys. Lett. *9*, 356 (1972)
4.36 N.R. Greiner, L.S. Blair, E.L. Patterson, R.A. Gerber: Report given at the 8th Int. Quant. Electr. Conf., San Francisco, CA (June 1974)
4.37 G.G. Dolgor-Savel'yev, V.F. Zharov, Y.S. Neganov, G.M. Chumak: Sov. Phys. JETP *34*, 34 (1972)
N.G. Basov, V.T. Golochkin, V.I. Igoshin, L.V. Kulakov, E.P. Markin, A.I. Nikitin, A.N. Oraevskii: Appl. Opt. *10*, 1814 (1971)
4.38 N.G. Basov, L.V. Kulakov, E.P. Markin, A.I. Nikitin, A.N. Oraevskii: JETP Lett. *9*, 375 (1969); for a general discussion see Ref. 4.34
4.39 N. Cohen, J.F. Bott: In Ref. 4.6, p. 33
4.40 R.W.F. Gross, D.J. Spencer: In Ref. 4.6, p. 205
4.41 D.J. Spencer, H. Mirels, T.A. Jacobs: Appl. Phys. Lett. *16*, 384 (1970)
4.42 D. Proch, J. Wanner, H. Pummer: Report IPP IV/73, Institut für Plasmaphysik, Garching, Germany (1974)
4.43 T.A. Cool: In Ref. 4.6, p. 431
4.44 J.A. Shirley, R.N. Sileo, T.A. Cool: "Purely Chemical Laser Operation in the HF, DF, HF-CO_2 and DF-CO_2 Systems", presented at the AIAA 9th Aerospace Science Meeting, New York (Jan. 1971)
4.45 B.R. Bronfin, W.Q. Jeffers: In Ref. 4.6, p. 619
4.46 L.R. Boedeker, J.A. Shirley, B.R. Bronfin: Appl. Phys. Lett. *21*, 247 (1972)
4.47 G. Emanuel: In Ref. 4.6, p. 469
4.48 R.I. Cukier, C.M. Fortuin, K.E. Shuler, A.G. Petschek, J.M. Schaibly: J. Chem. Phys. *59*, 3873 (1973)
J.H. Schaibly, K.E. Shuler: J. Chem. Phys. *59*, 3879 (1973)
J.T. Hwang, E.P. Dougherty, S. Rabitz, H. Rabitz: J. Chem. Phys. *69*, 5180 (1978)
4.49 A. Ben-Shaul, K.L. Kompa, U. Schmailzl: J. Chem. Phys. *65*, 1711 (1976)
4.50 J.R. Airey: J. Chem. Phys. *52*, 156 (1970)

4.51 R.L. Kerber, G. Emanuel, J.S. Whittier: Appl. Opt. *11*, 1112 (1972)
4.52 E. Keren, R.B. Gerber, A. Ben-Shaul: Chem. Phys. *21*, 1 (1976)
4.53 T.D. Padrick, M.A. Gusinow: Chem. Phys. Lett. *24*, 270 (1974)
H.L. Chen, R.L. Taylor, J. Wilson, P. Lewis, W. Fyfe: J. Chem. Phys. *61*, 306 (1974)
L.H. Sentman: J. Chem. Phys. *62*, 3523 (1975); Appl. Opt. *15*, 744 (1976)
J.J.T. Hough, R.L. Kerber: Appl. Opt. *14*, 2960 (1975)
G.K. Vasil'ev, E.F. Makarov, A.G. Ryabenko, V.L. Tal'roze: Sov. Phys. JETP *44*, 690 (1976)
M. Baer, Z.H. Top, Z.B. Alfassi: Chem. Phys. *22*, 485 (1977); R.L. Kerber, J.T. Hough: Appl. Opt. *17*, 2369 (1978); see also Refs. 4.20, 4.24, 4.49, 4.59
4.54 H.V. Lillenfeld, W.Q. Jeffers: J. Appl. Phys. *47*, 2520 (1976)
4.55 W.Q. Jeffers, C.E. Wiswall: IEEE J. QE-*10*, 860 (1974)
4.56 J.E. Broadwell: Appl. Opt. *13*, 962 (1974)
4.57 H. Mirels, R. Hofland, W.S. King: AIAA J. *11*, 156 (1973)
4.58 G. Emanuel, J.S. Whittier: Appl. Opt. *11*, 2049 (1972)
4.59 R.J. Hall: IEEE J. QE-*12*, 453 (1976)
4.60 K.L. Kompa, J. Wanner: Chem. Phys. Lett. *12*, 560 (1972)
V.I. Igoshin, L.V. Kulakov, A.I. Nikitin: Sov. J. Quantum Electron. *3*, 306 (1974); for a background discussion see
N. Cohen, J.F. Bott: Report of the Aerospace Corporation El Segundo, CA, AD/A 024 332 (April 1976)
4.61 A. Persky: J. Chem. Phys. *59*, 3612, 5578 (1973)
4.62 E.R. Grant, J.W. Root: Chem. Phys. Lett. *27*, 484 (1974)
4.63 R.L. Williams, F.S. Rowland: J. Chem. Phys. *77*, 301 (1973)
4.64 M.J. Molina, G.C. Pimentel: IEEE J. QE-*9*, 64 (1973)
4.65 K. Hohla, K.L. Kompa: Chem. Phys. Lett. *14*, 445 (1972); Z. Naturforsch. *27a*, 938 (1972)
4.66 L. Henry: 3rd Conf. on Chemical and Molecular Lasers, St. Louis, MO (May 1972)
4.67 P. Genzel, K.L. Kompa, J.R. MacDonald: 3rd Conf. on Chemical and Molecular Lasers, St. Louis, MO (May 1972)
4.68 G. Hancock, C. Morley, I.W.M. Smith: Chem. Phys. Lett. *12*, 193 (1971)
4.69 Y. Nachshon, P.D. Coleman: J. Chem. Phys. *61*, 2520 (1974)
4.70 A. Tönnissen, J. Wanner, K.W. Rothe, H. Walther: Appl. Phys. *18*, 297 (1979)
4.71 E.D. Hinkley (ed.): *Laser Monitoring of the Atmosphere*, Topics in Applied Physics, Vol. 14 (Springer, Berlin, Heidelberg, New York 1976)
K.W. Rothe, H. Walther: In *Tunable Lasers and Applications*, Springer Series in Optical Sciences, Vol. 3, ed. by A. Mooradian, T. Jaeger, P. Stokseth (Springer, Berlin, Heidelberg, New York 1976)

Chapter 5

5.1 T.J. Odiorne, P.R. Brooks, J.V.V. Kasper: J. Chem. Phys. *55*, 1980 (1971)
5.2 H.H. Dispert, R.W. Geis, P.R. Brooks: J. Chem. Phys. *70*, 5317 (1979)
5.3 B.A. Blackwell, J.C. Polanyi, J.J. Sloan: Chem. Phys. *30*, 299 (1978)
5.4 S. Stolte, A.E. Proctor, W.M. Pope, R.B. Bernstein: J. Chem. Phys. *66*, 3468 (1977)
5.5 J.G. Pruett, R.N. Zare: J. Chem. Phys. *64*, 1774 (1976)
5.6 Z. Karny, R.N. Zare: J. Chem. Phys. *68*, 3360 (1978)
5.7 J.C. Polanyi: Acc. Chem. Res. *5*, 161 (1972)
5.8 R.W. Solarz, S.A. Johnson, R.K. Preston: Chem. Phys. Lett. *57*, 514 (1978)
5.9 Y. Zeiri, M. Shapiro: Chem. Phys. *31*, 217 (1978)

5.10 R.C. Estler, R.N. Zare: J. Am. Chem. Soc. *100*, 1323 (1978), after completion of this text a re-interpretation of reaction 5.2.9 is reported describing the observed emission as laser-induced fluorescence of the ground-state reaction product rather than as laser-induced chemiluminescence. C.T. Rettner, L. Woeste, R.N. Zare: To be published
5.11 P. Siegbahn, B. Liu: J. Chem. Phys. *68*, 2457 (1978)
5.12 M. Kneba, U. Wellhausen, J. Wolfrum: Ber. Bunsenges. Phys. Chem. *83*, 940 (1979)
5.13 Z. Karny, B. Katz, A. Szöke: Chem. Phys. Lett. *35*, 100 (1975)
5.14 M. Kneba, J. Wolfrum: J. Phys. Chem. *83*, 69 (1979)
5.15 S.R. Leone, R.G. McDonald, C.B. Moore: J. Chem. Phys. *63*, 4735 (1975)
5.16 D. Arnoldi, K. Kaufmann, J. Wolfrum: Phys. Rev. Lett. *34*, 1597 (1975)
5.17 S.H. Bauer, D.M. Lederman, E.L. Resler, Jr., E.R. Fisher: Int. J. Chem. Kinet. *5*, 93 (1973)
5.18 S.H. Bauer, E. Ossa: J. Chem. Phys. *45*, 434 (1966)
R.D. Kern, G.G. Niki: J. Phys. Chem. *35*, 1615, 2541 (1971)
5.19 H. Conroy, G. Mali: J. Chem. Phys. *50*, 5049 (1969)
C.W. Wilson, Jr., W.A. Goddard: J. Chem. Phys. *51*, 716 (1969)
M. Rubenstein, I. Shavitt: J. Chem. Phys. *51*, 2014 (1969)
5.20 B.M. Gimarc: J. Chem. Phys. *53*, 1623 (1970)
5.21 D.A. Dixon, R.M. Stevens, D.R. Herschbach: Faraday Discuss. Chem. Soc. *62*, 110 (1977)
5.22 R.J. Gordon, M.C. Lin: Chem. Phys. Lett. *22*, 262 (1973); J. Chem. Phys. *64*, 1058 (1976)
5.23a M.J. Kurylo, W. Braun, A. Kaldor, S.M. Freund, R.P. Wayne: J. Photochem. *3*, 71 (1974/5)
5.23b M.J. Kurylo, W. Braun, C.N. Xuan, A. Kaldor: J. Chem. Phys. *62*, 2065 (1975)
5.24 P.N. Clough, B.A. Thrush: Trans. Faraday Soc. *63*, 915 (1967)
5.25 A.E. Redpath, M. Menzinger: Can. J. Chem. *49*, 3063 (1971); J. Chem. Phys. *62*, 1987 (1975)
5.26 D.I. Rosen, T.A. Cool: J. Chem. Phys. *62*, 466 (1975)
5.27 E. Bar-Ziv, J. Moy, R.J. Gordon: J. Chem. Phys. *68*, 1013 (1978)
5.28 K.K. Hui, T.A. Cool: J. Chem. Phys. *68*, 1022 (1978)
5.29 J.C. Stephenson, S.M. Freund: J. Chem. Phys. *65*, 4303 (1976)
5.30 J.G. Calvert, J.N. Pitts: *Photochemistry* (Wiley, New York 1966)
5.31 D.O. Cowan, R.L. Drisko: *Elements of Organic Photochemistry* (Plenum, New York 1976)
5.32 M.J. Berry: Chem. Phys. Lett. *56*, 423 (1978)
5.33 G.W. King, A.W. Richardson: J. Mol. Spectrosc. *21*, 339 (1966)
5.34 N. Basco, J.E. Nicholas, R.G.W. Norrish, W.H.F. Wickers: Proc. R. Soc. *A272*, 147 (1963)
5.35 R.J. Donovan, J. Konstantatos: J. Photochem. *1*, 75 (1972/3)
5.36 K.E. Holdy, L.C. Klotz, K.R. Wilson: J. Chem. Phys. *52*, 4588 (1970)
5.37 J.H. Ling, K.R. Wilson: J. Chem. Phys. *63*, 101 (1975)
5.38 J.W. Rabelais, J.M. McDonald, V. Schurr, S.P. McGlynn: Chem. Rev. *71*, 73 (1971)
5.39 M.J. Sabety-Dzvonik, R.J. Cody: J. Chem. Phys. *66*, 125 (1977)
5.40 A.P. Baronavski, J.R. McDonald: Chem. Phys. Lett. *45*, 172 (1977)
5.41 A.P. Baronavski, J.R. McDonald: Quoted by A.P. Baronavski, M.E. Umstead, M.C. Lin in Ref. 2.75
5.42 R.D. McQuigg, J.G. Calvert: J. Am. Chem. Soc. *91*, 1590 (1969)
5.43 R.G. Miller, E.K.C. Lee: Chem. Phys. Lett. *33*, 109 (1972)
5.44 E.S. Yeung, C.B. Moore: Appl. Phys. Lett. *21*, 109 (1972)
5.45 J.H. Clark, Y. Haas, P.L. Houston, C.B. Moore: Chem. Phys. Lett. *35*, 82 (1975)

5.46 J.B. Marling: J. Chem. Phys. *66*, 4200 (1977)
5.47 J.C. Weisshar, A.P. Baronavski, A. Cabello, C.B. Moore: J. Chem. Phys. *69*, 4720 (1978)
5.48 J.C. Weisshar, C.B. Moore: J. Chem. Phys. *70*, 5135 (1979)
5.49 A.C. Luntz: J. Chem. Phys. *69*, 3436 (1978)
5.50 Y. Haas, C.B. Moore: Unpublished
5.51 P.L. Houston, C.B. Moore: J. Chem. Phys. *65*, 757 (1976)
5.52 D.F. Heller, M.L. Elert, W.M. Gelbart: J. Chem. Phys. *69*, 4061 (1978)
5.53 J.R. Sodeau, E.K.C. Lee: Chem. Phys. Lett. *57*, 71 (1978)
5.54 W.H. Miller, C.W. McCurdy: J. Chem. Phys. *69*, 5163 (1978)
5.55 *The Triplet State* 1, 2, 3, Topics in Current Chemistry, Vols. 54, 55, and 56 (Springer, Berlin, Heidelberg, New York 1975/76)
5.56 H. Okabe, W.A. Noyes, Jr.: J. Am. Chem. Soc. *79*, 801 (1957)
5.57 C.M. Almy, P.R. Gillette: J. Chem. Phys. *11*, 188 (1943)
5.58 R. van der Werf, J. Kommandeur: Chem. Phys. *6*, 125 (1976)
5.59 R. van der Werf, E. Schulten, J. Kommandeur: Chem. Phys. *11*, 281 (1975); Chem. Phys. *16*, 151 (1976)
5.60 R. Naaman, D.M. Lubman, R.N. Zare: J. Chem. Phys. *71*, 4192 (1979)
5.61 R.A. Beyer, W.C. Lineberger: J. Chem. Phys. *62*, 4024 (1975)
5.62 H.B. Lin, M.R. Topp: Chem. Phys. Lett. *67*, 273 (1979)
5.63 W.M. McClain, R.A. Harris: In *Excited States*, Vol. 3, ed. by E.C. Lim (Academic, New York 1978)
5.64 J.E. Bjorkholm, P.F. Liao: Phys. Rev. Lett. *32*, 643 (1974)
5.65 J.B. Marling, I.P. Herman: Appl. Phys. Lett. *34*, 439 (1979)
5.66 S.M. Freund, C.D. Cantrell, J.L. Lyman: In *Laser Handbook*, Vol. IIIb, ed. by M.L. Stitch (North Holland, Amsterdam 1979)
5.67 R.V. Ambartzumian, V.S. Letokhov: In *Chemical and Biochemical Applications of Lasers*, Vol. III, ed. by C.B. Moore (Academic, New York 1977)
5.68 Aa.S. Sudbø, D.J. Krajnovich, P.A. Schulz, Y.R. Shen, Y.T. Lee: In Ref. 5.69
5.69 C.D. Cantrell (ed.): *Multiple Photon Excitation and Dissociation of Polyatomic Molecules*, Topics in Current Physics (Springer, Berlin, Heidelberg, New York, in preparation)
5.70 J.R. Ackerhardt, H.W. Galbraith: J. Chem. Phys. *69*, 1200 (1978)
5.71 V.N. Bagratashvili, I.N. Knyazev, V.S. Letokhov, V.V. Lobko: Opt. Commun. *18*, 525 (1976)
5.72 M. Quack: J. Chem. Phys. *69*, 1282 (1978)
5.73 J.J. Tiee, C. Wittig: Opt. Commun. *27*, 379 (1978)
5.74 W. Fuss, J. Hartmann: J. Chem. Phys. *70*, 5468 (1980)
5.75 J.G. Black, E. Yablonovitch, N. Bloembergen, S. Mukamel: Phys. Rev. Lett. *38*, 1131 (1977); See also
N. Bloembergen, E. Yablonovitch: Phys. Today *31*, 23 (1978)
5.76 E.R. Grant, P.A. Schulz, A. Sudbø, Y.R. Shen, Y.T. Lee: Phys. Rev. Lett. *40*, 115 (1978)
N.C. Petersen, R.G. Manning, W. Braun: J. Res. Nat. Bur. Stand. *83*, 117 (1978)
5.77 W. Fuss: Chem. Phys. *36*, 135 (1979)
5.78 G.A. Hill, E. Grunwald, P.M. Keehn: J. Am. Chem. Soc. *99*, 6521 (1977) See also
E. Grunwald, D.F. Dever, P.M. Keehn: *Megawatt Infrared Laser Chemistry* (Wiley, New York 1978)
5.79 J. Stone, M.F. Goodman: J. Chem. Phys. *71*, 408 (1979)
5.80 R.V. Ambartzumian, V.S. Letokhov: Acc. Chem. Res. *10*, 61 (1977)
5.81 R.V. Ambartzumian, Y.A. Gorokhov, V.S. Letokhov, G.N. Makarov, A.A. Puretskii: JETP Lett. *23*, 22 (1976)
5.82 J.C. Stephenson, D.S. King: J. Am. Chem. Soc. *100*, 7151 (1978)
5.83 P. Kolodner, C. Winterfeld, E. Yablonovitch: Opt. Commun. *20*, 119 (1977)

5.84 M. Rossi, J.R. Barker, D.M. Golden: Chem. Phys. Lett. *65*, 523 (1979)
5.85 R. Naaman, M. Rossi, J.R. Barker, D.M. Golden, R.N. Zare: To be published
5.86 R.L. Woodin, D.S. Bomse, J.L. Beauchamp: In *Chemical and Biochemical Applications of Lasers*, Vol. 4, ed. by C.B. Moore (Academic, New York 1979)
5.87 Y. Haas: In Ref. 2.75
5.88 J.C. Stephenson, D.S. King, M.F. Goodman, J. Stone: J. Chem. Phys. *70*, 4996 (1979)
5.89 D.S. King, J.C. Stephenson: Chem. Phys. Lett. *66*, 33 (1979)
5.90 Aa.S. Sudbø, P.A. Schulz, E.R. Grant, Y.R. Shen, Y.T. Lee: J. Chem. Phys. *70*, 912 (1979)
5.91 Experimental data from C.R. Quick, C. Wittig: J. Chem. Phys. *72*, 1694 (1980); surprisal analysis by E. Zamir
5.92 Aa.S. Sudbø, P.A. Schulz, Y.R. Shen, Y.T. Lee: J. Chem. Phys. *68*, 2312 (1978)
5.93 D.S. King, J.C. Stephenson: Chem. Phys. Lett. *51*, 49 (1977)
5.94 E. Thiele, M.F. Goodman, J. Stone: Opt. Eng. *19*, 10 (1980)
5.95 R.B. Hall, A. Kaldor: J. Chem. Phys. *70*, 4027 (1979)
5.96 S. Bittenson, P.L. Houston: J. Chem. Phys. *67*, 4819 (1977)
5.97 W. Braun, W. Tsang: J. Chem. Phys. *67*, 4291 (1977)
5.98 W. Tsang, J.A. Walker, W. Braun, J.P. Herron: Chem. Phys. Lett. *59*, 487 (1978)
5.99 D. Gutman, W. Tsang, W. Braun: J. Chem. Phys. *67*, 4291 (1977)
5.100 S.D. Smith, W.E. Schmid, F.M.G. Tablas, K.L. Kompa: In *Laser Induced Processes in Molecules*, Springer Series in Chemical Physics, Vol. 6, ed. by S.D. Smith, K.L. Kompa (Springer, Berlin, Heidelberg, New York 1979) p. 121
5.101 R.E. Huie, J.T. Herron, W. Braun, W. Tsang: Chem. Phys. Lett. *56*, 193 (1978)
5.102 A. Yogev, R.M.J. Ben Mair: Chem. Phys. Lett. *46*, 290 (1977)
5.103 Z. Karny, A. Gupta, R.N. Zare, S.T. Lin, J. Nieman, A.M. Ronn: Chem. Phys. *37*, 15 (1979)
5.104 J.W. Hudgens, J.L. Durant, D.J. Bogan, R.A. Coveleskie: J. Chem. Phys. *70*, 5906 (1979)
5.105 R.V. Ambartzumian, G.N. Makarov, A.A. Puretzkii: Presented at 6th International Conference on Molecular Energy Transfer, Rodez, France, July 16-20 (1979)
5.106 A. Nitzan, J. Jortner: Chem. Phys. Lett. *60*, 1 (1978); J. Chem. Phys. *71*, 3524 (1979)
5.107 Y.P. Raizer: *Laser Induced Discharge Phenomena* (Consultant Bureau, New York 1977)
5.108 J.R. Ackerhaldt, J.H. Eberly: Phys. Rev. *A14*, 1705 (1976)
5.109 J.P. Reilly, K.L. Kompa: J. Chem. Phys. *73*, 5468 (1980)
5.110 U. Boesl, H.J. Neusser, E.W. Schlag: Z. Naturforsch. A*33*, 1546 (1978)
5.111 P.M. Johnson: J. Chem. Phys. *62*, 4563 (1975)
5.112 L. Zandee, R.B. Bernstein: J. Chem. Phys. *71*, 1359 (1979)
5.113 G.C. Bjorklund, C.P. Ausschnitt, R.R. Freeman, R.H. Storz: Appl. Phys. Lett. *33*, 54 (1978)
5.114 G.C. Nieman, S.D. Colson: J. Chem. Phys. *68*, 5656 (1978)
5.115 M.B. Robin, N.A. Kuebler: J. Chem. Phys. *69*, 806 (1978)
5.116 D.H. Parker, J.O. Berg, M.A. El-Sayed: In *Advances in Laser Chemistry*, Springer Series in Chemical Physics, Vol. 3, ed. by A.H. Zewail (Springer, Berlin, Heidelberg, New York 1978) p. 320
5.117 G.J. Fisanick, T.S. Eichelberger IV, B.A. Heath, M.B. Robin: J. Chem. Phys. *72*, 5571 (1980)

5.118 F. Engelke, J.C. Whitehead, R.N. Zare: Faraday Discuss. Chem. Soc. *62*, 222 (1977)
5.119 P.R. Brooks: In *Chemical and Biochemical Applications of Lasers*, Vol. I, ed. by C.B. Moore (Academic, New York 1974) p. 139. For a general discussion of molecular beam techniques and applications, see, e.g., M.A.D. Fluendy, K.P. Laweley: *Chemical Applications of Molecular Beam Scattering* (Chapman and Hall, London 1973)
5.120 A. Herrmann, E. Schumacher, L. Woeste: J. Chem. Phys. *68*, 2327 (1978)
5.121 D.H. Levy, L. Wharton, R.E. Smalley: In *Chemical and Biochemical Applications of Lasers*, Vol. II, ed. by C.B. Moore (Academic, New York 1977) p. 1
5.122 S.M. Beck, D.L. Monts, M.G. Liverman, R.E. Smalley: J. Chem. Phys. *70*, 1062 (1979)
5.123 A. Amirav, U. Even, J. Jortner: Opt. Comm. *32*, 266 (1980)
5.124 See. e.g., G. Hancock, K.R. Wilson: In *Fundamental and Applied Laser Physics*, ed. by M.S. Feld, N. Kurnit (Wiley, New York 1973) p. 257
5.125 G.E. Busch, J.F. Cornelius, R.T. Mahoney, R.I. Morse, D.W. Schlosser, K.R. Wilson: Rev. Sci. Instrum. *41*, 1066 (1970)
5.126 A. Schultz, H.W. Cruse, R.N. Zare: J. Chem. Phys. *57*, 1354 (1972)
5.127 R. Schmiedl, R. Böttner, H. Zacharias, U. Meier, D. Feldman, K.H. Welge: In *Laser Induced Processes in Molecules*, Springer Series in Chemical Physics, Vol. 6, ed. by K.L. Kompa, S.D. Smith (Springer, Berlin, Heidelberg, New York 1979) p. 186
For a discussion of using Doppler width measurements for translational energy effects, see
J.L. Kinsey: J. Chem. Phys. *66*, 2560 (1977)
5.128 J.D. Campbell, M.H. Yu, M. Mangir, C. Wittig: J. Chem. Phys. *69*, 3854 (1978)
5.129 W. Demtröder: *Laser Spectroscopy*, Topics in Current Chemistry, Vol. 17 (Springer, Berlin, Heidelberg, New York 1971)
5.130 For a discussion of nonlinear techniques see, e.g., A. Yariv: *Quantum Electronics*, 2nd ed. (Wiley, New York 1975) Chaps. 16-18
5.131 W.M. Tolles, J.W. Nibler, J.R. McDonald, A.B. Harvey: Appl. Spectrosc. *31*, 253 (1977)
5.132 A.B. Harvey, J.R. McDonald, W.M. Tolles: Prog. Anal. Chem. *8*, 211 (1976)
5.133 N. Omenetto, P. Benetti, L.P. Hart, J.D. Winefordner, C.Th. Alkemade: Spectrochim. Acta *28B*, 289 (1973)
5.134 L.A. Rahn, L.J. Zych, P.L. Mattern: Opt. Commun. *30*, 249 (1979)
5.135 S.A. Druet, B. Attal, T.K. Gustafson, J.P. Taran: Phys. Rev. *A18*, 1529 (1978)
B. Attal, O.O. Schnepp, J.-P. Taran: Opt. Commun. *24*, 77 (1978)
5.136 J.W. Nibler, J.R. McDonald, A.B. Harvey: Opt. Commun. *18*, 134 (1976)
5.137 J.I. Steinfeld, P.L. Houston: In *Laser and Coherence Spectroscopy*, ed. by J.I. Steinfeld (Plenum, New York 1978) p. 1
5.138 R.W. Field, A.D. English, T. Tanaka, D.O. Harris, P.A. Jennings: J. Chem. Phys. *59*, 2191 (1973)
5.139 W.K. Bischel, C.K. Rhodes: Phys. Rev. *A14*, 176 (1976)
5.140 W. Brunner, H. Paul: Opt. Commun. *12*, 252 (1974)
5.141 For a general discussion see *Laser Probes for Combustion Chemistry*, ed. by D.R. Crosley, ACS Symp. Ser. 134 (Amer. Chem. Soc., Washington, D.C. 1980)
5.142 J.P. Reilly, J.H. Clark, C.B. Moore, G.C. Pimentel: J. Chem. Phys. *69*, 4381 (1978)
5.143 W. Demtröder: *Principles and Techniques of Laser Spectroscopy* (Springer, Berlin, Heidelberg, New York 1977) cf. Chap. 8
5.144 Y.-H. Pao (ed.): *Optoacoustic Spectroscopy and Detection* (Academic, New York 1977)

5.145 K.V. Reddy, R.G. Bray, M.J. Berry: In Ref. 5.147, p.48

5.146 J.G. Black, P. Kolodner, M.J. Schultz, E. Yablonovitch, N. Bloembergen: Phys. Rev. A-*19*, 704 (1979)

5.147 A.H. Zewail (ed.): *Advances in Laser Chemistry*, Springer Series in Chemical Physics, Vol. 3 (Springer, Berlin, Heidelberg, New York 1978)

5.148 M.J. Berry: In *Future Sources of Organic Raw Materials*, ed. by L.E. St. Pierre (Pergamon Press, Elmsford, NY 1980)

5.149 J. Wolfrum: In: Proceedings of the 1st SFB Kolloquium "Photochemie und Lasern", Göttingen, October 1979

5.150 J.H. Clark, K.M. Leary, T.R. Loree, L.B. Harding: In Ref. 5.147, p. 74

5.151 H.R. Bachmann, R. Rinck, H. Noth, K.L. Kompa: Chem. Phys. Lett. *45*, 169 (1977)

5.152 S.H. Bauer, E. Bar-Ziv, J.A. Haberman: IEEE J. QE-*14*, 237 (1978)

5.153 S.H. Bauer, J.A. Haberman: IEEE J. QE-*14*, 233 (1978); See also S.H. Bauer: Chem. Rev. *78*, 147 (1978)

5.154 J.L. Buechele, E. Weitz, F.D. Lewis: J. Am. Chem. Soc. *101*, 3700 (1979)

5.155 H.R. Bachmann, H. Noth, H. Rinck, K.L. Kompa: Chem. Phys. Lett. *33*, 261 (1975)

5.156 T.H. Clark, R.G. Anderson: Appl. Phys. Lett. *32*, 46 (1978)

5.157 J.T. de Maleissye, F. Lempereur, C. Marsal: Chem. Phys. Lett. *42*, 472 (1976)

5.158 R.C. Slater, J.H. Parks: Chem. Phys. Lett. *60*, 275 (1979)

5.159 H.H. Wasserman, R.W. Murray (eds.): *Singlet Oxygen* (Academic, New York 1979)

5.160 V.S. Letokhov, C.B. Moore: In *Chemical and Biochemical Applications of Lasers*, Vol. 3, ed. by C.B. Moore (Academic, New York 1978)

5.161 V.S. Letokhov: Ann. Rev. Phys. Chem. *28*, 133 (1977)

5.162 J.P. Aldridge, J.H. Birely, C.D. Cantrell, D.C. Cartwright: In *Laser Photochem., Tunable Lasers and Other Topics*, Physics of Quantum Electronics, Vol. 4, ed. by S.F. Jacobs, M. Sargent, III, M.O. Scully, C.T. Walker (Addison-Wesley, Mass., 1976) p. 57

5.163 W. Fuss, W.E. Schmid: Ber. Bunsenges. Phys. Chem. *83*, 1148 (1979)

5.164 R. Bersohn: Acc. Chem. Res. *5*, 200 (1972)

5.165 T.F. George, I.H. Zimmermann, P.L. DeVries, J.M. Yuan, K.S. Lam, J.C. Bellum, H.W. Lee, M.S. Slutzky, J. Lin: In *Chemical and Biochemical Applications of Lasers*, Vol. 4, ed. by C.B. Moore (Academic, New York 1979)
For an introduction to the literature see, for instance, T.F. George, I.H. Zimmermann, J.M. Yuan, J.R. Laing, P.L. DeVries: Acc. Chem. Res. *10*, 449 (1979)

5.166 A.H. Kung, C.B. Moore: In *Laser Spectroscopy IV*, Springer Series in Optical Sciences, Vol. 21, ed. by H. Walther, K.W. Rothe (Springer, Berlin, Heidelberg, New York 1979)

5.167 J.D. McDonald: Ann. Rev. Phys. Chem. *30*, 29 (1979)

Author Index

Ackerhardt, J.R. 404,431
Agmon, N. 63,65,93,94, 186
Airey, J.R. 11,315,319, 333,336
Albrect, A. 111,112
Aldridge, J.P. 261,457
Alfassi, Z.B. 339
Alhassid, Y. 63,65,86, 108,186
Almy, C.M. 393
Ambartzumian, R.V. 401, 414,415,429
Amirav, A. 39,437
Anderson, R.G. 455
Anderson, T.G. 34
Arnoldi, D. 23,24,375
Ash, R. 58
Aten, J.A. 138
Attal, B. 445
Ausschnitt, C.P. 432
Avouris, P. 141

Babkin, V.I. 308
Bachmann, H.R. 453,454
Baede, A.P.M. 135
Baer, M. 339
Bagratashvilli, V.N. 403,452
Bairelein, R. 58,63
Baldwin, G.C. 201
Balint-Kurti, G.G. 92
Barker, J.R. 415,416
Baronavski, A.P. 277, 383,384,386,387
Barrow, G.M. 192,235, 238
Bar-Ziv, E. 378,453
Basco, N. 381
Basov, N.G. 312,318,319
Bauer, E. 136,137
Bauer, S.H. 46,156,376, 453,454
Beauchamp, J.L. 416,417
Beck, S.M. 437
Bellum, J.C. 457
Benard, D.J. 294
Ben Mair, R.M.J. 427, 428
Bennet, W.H. 259,260
Ben-Shaul, A. 62,71,72, 76,150,156,160,171, 174,189,296,336,338, 340
Benson, S.W. 79
Berg, J.O. 432
Bergmann, K. 47
Bernstein, R.B. 2,43,46, 58,62,63,65,71,76,77, 83,89,92,93,96,115, 131,132,133,134,185, 186,187,188,189,366, 431,432,443
Berrondo, M. 168
Berry, M.J. 3,11,14,31, 46,70,71,79,81,82, 93,95,112,290,291, 292,294,295,296,297, 331,350,351,352,354, 380,381,452,455
Bersohn, R. 4,135,457
Beswick, J.A. 108
Beyer, R.A. 397
Birely, J.H. 457
Bischel, W.K. 449
Bittenson, S. 424
Bjorkholm, J.E. 400
Bjorklund, G.C. 432
Black, J.G. 408,452
Blackwell, B.A. 366
Blair, L.S. 318,319
Bloembergen, N. 201,408, 452
Boedeker, L.R. 326
Boesl, U. 431,432
Bogan, D.J. 71,80,81, 429
Bomse, D.S. 416,417
Bott, J.F. 11,121,132, 134,290,291,292,293, 294,305,308,315,319, 320,321,324,325,327, 330,333,336,341,343, 345,348,349
Böttner, R. 441,442
Bousek, R.R. 294
Bran, J.S. 155
Brau,C.A. 273
Braun, W. 376,378,411, 425,426,427,428,455
Bray, R.G. 112,452,455
Brederlow, G. 310,312, 313,314
Brenner, D.M. 27,28

Bridges, T.J. 268
Broadwell, J.E. 343,344, 345
Brochu, R. 277
Brodermann, R. 312
Broida, H.P. 6
Bronfin, B.R. 324,325, 326
Brooks, P.R. 2,26,46,87, 365,366,436
Brophy, J.H. 78
Brueck, S.R.J. 48,268
Brumer, P. 111
Brunner, W. 450
Buechele, J.L. 454
Busch, G.E. 438

Cabello, A. 386,387
Calaway, W.F. 48
Calvert, J.G. 1,380,385, 392,457
Campbell, J.D. 441,442
Canavan, G.H. 155
Cantrell, C.D. 36,401, 457
Cartwright, D.C. 457
Chalek, C.L. 70,79
Chang, W.H. 76,77
Chebotayev, V.P. 201
Chen, H.L. 11,339
Chen, Y.R. 2,36
Chenand, Y.R. 83
Cheo, P.K. 284,286
Chow, W. 278
Chumak, G.M. 318
Clark, J.H. 386,450,453
Clark, T.H. 455
Clary, D.C. 78
Clough, P.N. 376
Cody, R.J. 383
Cohen, N. 121,134,320, 348,349
Coleman, P.D. 357,358
Collins, C.B. 281
Colson, S.D. 432
Condray, R. 281
Conroy, H. 376
Cool, T.A. 11,134,320, 324,377,378,436
Cornelius, J.F. 438
Cotter, T.P. 32
Coveleskie, R.A. 429
Cowan, D.O. 1,380
Cross, P.C. 192,235,238, 248
Cruse, H.W. 440,442
Cuellar, E. 127,294
Cukier, R.I. 327

Dagdigian, P.J. 46,86
Dalgarno, A. 189
da Paixao, F.J. 260
Datta, S. 27,28
Decius, J.C. 192,235, 238,248
DeMaria, A.J. 268,284, 286
Demtröder, W. 441,451
Deutsch, T.F. 11,48,294
Dever, D.F. 412,458
DeVries, P.L. 139,141, 457
Diegelmann, M. 278
Dimpfl, W.L. 78
Ding, A.M.G. 25,46,122, 150,151
Dinur, U. 70,79
Dispert, H.H. 365,366
Dixon, D.A. 376
Dolgor-Savel'yev, G.G. 318
Donnelly, V.M. 277
Donovan, R.J. 381,384
Dougherty, E.P. 327
Douglas, D.J. 23,24,47, 49,54,122
Dove, J.E. 132
Downey, G.D. 294
Drisko, R.L. 1,380
Druet, S.A. 445
Dunn, M.H. 192,222
Durant, J.L. 429

Eberly, J.H. 431
Eckstrom, D.J. 79
Edelstein, S.A. 79
Eichelberger, T.S.IV, 433
Eidmann, K. 312
Einstein, A. 7,9
Elert, M.L. 111,389
El-Sayed, M.A. 141,432
Emanuel, G. 205,300,327, 330,333,336,337,341, 343,344,345
English, A.D. 448
Ennen, C. 119
Estler, R.C. 26,370
Even, U. 39,437
Ewing, G.E. 48,108
Ewing, J.J. 273
Eyring, H. 89,93,96

Faist, M.B. 79,138
Falk, T.J. 11
Feld, M.S. 160,161,162, 294,438
Feldman, D. 441,442
Feldmann, T. 150
Feliks, S. 296,297,340
Field, R.W. 448
Filip, H. 261
Fill, E. 312
Fisanick, G.J. 433
Fisher, E.R. 136,376
Flicker, H. 261
Fluendy, M.A.D. 436
Flynn, G.W. 82,106,107, 160,161,162,252
Ford, J. 110,111
Forst, W. 99,100,105, 107,181,182,183,184
Fortuin, C.M. 327

Freed, K.F. 141
Freedman, A. 4
Freeman, R.R. 432
Freund, S.M. 376,379, 401
Fulghum, S.F. 160,161, 162
Fuss, W. 32,309,409,412, 456
Fyfe, W. 339

Galbraith, H.W. 260,404
Gardiner, W.C.Jr. 53
Geis, R.W. 365,366
Gelbart, W.M. 111,141, 389
Genzel, P. 355
George, T.F. 139,141, 457
Gerber, R.A. 318,319
Gerber, R.B. 336,338, 339
Geusic, J.E. 235
Gillette, P.R. 393
Gilmore, F.R. 136,240
Gimarc, B.M. 376
Goddard, W.A. 376
Golden, D.M. 415,416
Goldman, S. 58
Gole, J.L. 70,79
Golochkin, V.T. 318
Goodman, M.F. 413,416, 417,423,455
Gordietz, B.F. 46
Gordon, N. 104,105
Gordon, R.G. 106
Gordon, R.J. 376,378
Gorokhov, Y.A. 414,415
Grant, E.R. 351,411,419, 421,422
Green, W.R. 139,140
Greiner, N.R. 318,319
Grice, R. 137
Gross, R.W.F. 11,205, 290,291,292,293,294, 308,315,319,320,321, 324,325,327,330,333, 336,341,343,345
Grunwald, E. 412,458
Gupta, A. 429
Gusinow, M.A. 339
Gustafson, T.K. 445
Gutman, D. 426

Haas, Y. 386,387,418, 428
Haberman, J.A. 453,454
Hall, R.B. 405,423,458
Hall, R.J. 339,346,347
Hammond, G.S. 93,94
Hancock, G. 71,356,438
Hänsch, T.W. 444,450,451
Harding, L.B. 453
Harmony, M.D. 192,235, 238
Harper, P.G. 201
Harris, D.O. 448
Harris, R.A. 400,401, 403
Harris, S.E. 139,140
Harter, W.G. 260
Hartmann, J. 409
Harvey, A.B. 441,445, 446
Hase, W.L. 99,100,107
Hayes, E.F. 2,46,87
Heath, B.A. 433
Heller, D.F. 112,389
Hendersen, D. 89,93,96
Henry, B.R. 111
Henry, L. 355
Herbelin, J.M. 300
Herman, I.P. 294,401
Herman, R. 300
Herrmann, A. 437
Herron, J.P. 425,427, 428,455
Herschbach, D.R. 79,137, 376
Hertzfeld, K.F. 131,132, 155
Herzberg, G. 192,235, 237,238,250,254,258, 260,264
Hill, G.A. 412,458
Hinchen, J.J. 150,151
Hinkley, E.D. 359
Hobbs, R.M. 150,151
Hochstim, A.R. 53,69, 189
Hofacker, G.L. 301,339
Hoffmann, R. 1
Hofland, R. 343
Hohla, K. 278,308,309, 312,354,355,356
Holbrook, K.A. 99,100, 107
Holdy, K.E. 381
Holland, R.F. 261
Hough, J.J.T. 339
Houston, P.L. 30,265, 386,388,390,424,446, 448
Huddleston, R.K. 160, 161
Hudgens, J.W. 429
Hui, K.K. 378,436
Huie, R.E. 427
Hwang, J.T. 327

Igoshin, V.I. 318,348, 349
Innes, K.K. 264

Jacob, J.H. 277
Jacobs, S.F. 457
Jacobs, T.A. 321
Jaeger, T. 359,441,445
Javan, A. 167,294
Jaynes, E.T. 58,63
Jeffers, W.Q. 155,324, 325,342

Jennings, P.A. 448

Jensen, C.C. 34,103,129, 131

Jensen, R.J. 290,293, 455

Job, V.A. 264

Johnson, B.R. 76,134

Johnson, P.M. 431

Johnson, S.A. 369

Johnston, H.S. 93,94

Jonathan, N. 71

Jones, C.R. 6

Jortner, J. 14,18,39,81, 82,86,89,96,106,108, 111,135,141,160,161, 162,174,290,291,292, 294,350,352,354,384, 418,428,429,437

Jost, W. 89,93,96

Judge, D.L. 4

Kafri, A. 99

Kafri, O. 93,172,174, 296,297,340

Kaiser, W. 48

Kaldor, A. 376,378,405,423, 458

Kaplan, H. 46,48,55,125

Karny, Z. 26,372,376,429

Kasper, J.V.V. 11,289,365

Kassal, T. 131,132,155

Katulin, V.A. 234,308

Katz, A. 58,63

Katz, B. 372

Kaufmann, K. 375

Kauzmann, W. 9

Kawasaki, M. 4

Keck, J.C. 64,189

Keehn, P.M. 412,458

Kelley, J.D. 155

Kerber, R.L. 336,337,339

Keren, E. 336,338,339

Kern, R.D. 376

King, D.S. 414,415,416, 417,418,422,455

King, G.W. 192,235,238, 262,380,381,383

King, W.S. 343

Kinsey, J.L. 58,63,65, 73,78,115,185,186, 188,441,442

Kirsch, L.J. 46,47,122

Klotz, L.C. 381

Kneba, M. 371,373

Knyazev, I.N. 403,452

Kolodner, P. 415,452

Kommandeur, J. 394,395, 396,397

Kompa, K.L. 11,28,47, 76,155,160,278,287, 290,291,292,294,301, 308,331,339,348,349, 353,354,355,356,427, 431,433,441,442,453, 454

Konstantatos, J. 381, 384

Kosloff, R. 70,79,99

Kranjovich, D.J. 36,401

Krause, H. 312

Krenos, J.R. 79

Kressel, H. 268

Kuebler, N.A. 432

Kulakov, L.V. 318,319, 348,349

Kung, A.H. 458

Kuntz, P.J. 89,92,93,96

Kurnit, N. 438

Kurylo, M.J. 376,378

Kusunoki, I. 84

Kuzuetsova, S.V. 308

Laing, J.R. 139,141,457

Lam, K.S. 457

Lamb, W.E.Jr. 192,199, 206, 222

Lang, N.C. 150,151

Lantzsch, B. 78

Lau, A.M.F. 139

Laubereau, A. 48

Laweley, K.P. 436

Leary, K.M. 453

Lederman, D.M. 376

Lee, E.K.C. 385,386,389

Lee, H.W. 457

Lee, L.C. 4

Lee, Y.T. 2,36,83,401, 411,419,421,422

Lengyel, B.A. 192,222, 231,271

Leone, S.R. 23,24,139, 374,375

Letokhov, V.S. 2,29, 46,201,401,403,414, 415,452,457

Levine, A.K. 268,284, 286,301

Levine, I.N. 192,235, 238

Levine, R.D. 2,14,18,46, 48,53-189,290-294, 350,352,354,384,418, 428

Levy, D.H. 437

Lewis, F.D. 454

Lewis, P. 339

Liao, P.F. 400

Lifŝhitz, A. 174

Lifshitz, C. 89

Light, J.C. 53,69,189

Lillenfeld, H.V. 342

Lin, E.C. 400,401

Lin, H.B. 398

Lin, J. 457

Lin, M.C. 70,74,79,376, 384

Lin, S.T. 429

Lineberger, W.C. 397

Ling, J.H. 382,383,440

Litovitz, T.A. 131,132, 155

Liu, B. 371

Liu, K. 70,79

Liu, Y.S. 156

Liverman, M.G. 437

Lobko, V.V. 403,452
Loree, T.R. 453
Los, J. 138
Louck, J.D. 260
Lubman, D.M. 396,398
Lucht, R.A. 134
Lukasik, J. 139,140
Luntz, A.C. 78,388
Luria, M. 79
Lyman, J.L. 401

MacDonald, J.R. 355
Mahoney, R.T. 438
Maitland, A. 192,222
Makarov, A.A. 46
Makarov, E.F. 339
Makarov, G.N. 414,415, 429
Mali, G. 376
Mangano, J.A. 277
Mangir, M. 441,442
Manning, R.G. 411
Manos, D.M. 72,79
Manz, J. 46,48,55,87, 115,125,129,185
Marcos, H.M. 235
Marcus, R.A. 93,111,184, 189
Mariella, R.P. 78
Marinero, E.E. 28
Markin, E.P. 318,319
Marling, J.B. 386,401
Marowsky, G. 281
Maslov, A.I. 308
Massey, H.S.W. 189
Mattern, P.L. 445
Maxson, V.T. 78
Mayer, J.E. 64
Mayer, M.G. 64
Maylotte, D.H. 11,12,23, 47,49,71,77
McClain, W.M. 400,401, 403
McCurdy, C.W. 390
McDermott, W.E. 294
McDonald, J.D. 111,458
McDonald, J.M. 383
McDonald, J.R. 277,383, 384,441,445,446
McDonald, R.G. 23,24, 76,77,374,375
McDowell, R.S. 261
McFadden, D.L. 79
McFarlane, R.A. 156
McGlynn, S.P. 383
McKay, S.F. 11
McNair, R.E. 160,161, 162
McQuigg, R.D. 385
Meier, U. 441,442
Melliar-Smith, C.M. 71
Menzinger, M. 377
Meredith, R.D. 300
Meyer, C.F. 259,260
Miller, R.G. 385,386
Miller, W.H. 58,62,63, 89,92,93,96,99,100, 107,138,141,143,188, 189,390
Mirels, H. 343
Moehlmann, J.G. 76,77
Molina, M.J. 353
Montroll, E.W. 146,148
Monts, D.L. 437
Mooradian, A. 268,359, 441,445
Moore, C.B. 11,23,24, 29,46,47,48,58,62,63, 70,72,79,106,127,133, 134,151,156,185,374, 375,385,386,387,388, 390,401,416,417,436, 437,450,457,458
Moret-Bailly, J. 261,281
Morley, C. 71,356
Morse, R.I. 438
Moule, D.C. 262
Moutinho, A.M.C. 138
Moy, J. 378
Muckerman, J.T. 89,96
Mukamel, S. 112,141,408
Mulliken, R.S. 304
Murray, R.W. 457

Naaman, R. 396,398,416
Nachshon, Y. 357,358
Neganov, Y.S. 318
Nereson, N.G. 261
Nesbet, R.K. 78
Neusser, H.J. 431,432
Nguyen, V.T. 268
Nibler, J.W. 441,446
Nicholas, J.E. 381
Nielsen, N. 156
Nieman, G.C. 432
Nieman, J. 429
Niki, G.G. 376
Nikitin, A.I. 318,319, 348,349
Nippus, M. 312
Nitzan, A. 429
Norrish, R.G.W. 381
Nosach, O.Y. 234,308
Nosach, V.Y. 234
Noth, H. 453,454
Noyes, W.A.Jr. 393,457

Odiorne, T.J. 365
Okabe, H. 393
Okuda, S. 71
Olsen, R. 137
Oppenheim, I. 144,145, 148,149,164,168
Oraevskii, A.N. 290,318, 319
Oref, I. 104,105,129
Osgood, R.M. 48,167, 294
Osipov, A.I. 46
Ossa, E. 376
Ottinger, Ch. 84,119

Padrick, T.D. 339
Palmer, R.E. 312
Pao, Y.-H. 451
Papagani, N.L. 306,307
Parker, D.H. 432
Parker, J.H. 294
Parks, J.H. 455
Parson, J.M. 70,72,79
Patel, C.K.N. 249,284, 286,301
Patterson, C.W. 260
Patterson, E.L. 318,319
Paul, H. 450
Pchelkin, N.R. 294
Pearson, R.G. 1
Pechukas, P. 189
Percival, I.C. 111,189
Perry, B.E. 79
Perry, D.S. 46,47,55, 76,77,122
Persky, A. 351
Petersen, N.C. 411
Peterson, A.B. 139
Petsch, R. 312
Petschek, A.G. 327
Pimentel, G.C. 11,127, 289,294,297,353,450
Pitts, J.N. 1,380,392, 457
Polanyi, J.C. 11,12,18, 23,24,25,46,47,49, 54,55,71,76,77,89, 93,96,122,150,151, 289,366,369
Pollack, M.A. 11
Pollak, E. 46,99,122,129
Pope, W.M. 366
Porter, R.N. 89,96
Powell, H.T. 155
Preston, R.K. 369
Procaccia, I. 115,119, 121,150,165,167,169
Proch, D. 155,323
Proctor, A.E. 366
Proctor, W.A. 155
Pruett, J.G. 46,367,369
Pummer, H. 155,278,317, 323
Puretskii, A.A. 414,415, 429

Quack, M. 107,189,406
Quick, C.R. 82,107,420

Rabelais, J.W. 383
Rabinovitch, B.S. 104, 105,129,184
Rabitz, H. 155,327
Rabitz, S. 327
Raff, L.M. 89,96
Rahn, L.A. 445
Raizer, Y.P. 430
Rapp, D. 131,132,155
Rebentrost, F. 278
Reddy, K.V. 31,112,452, 455
Redpath, A.E. 377
Rehm, R.G. 156,160,191
Reif, F. 74,110,173, 178,217
Reilly, J.P. 278,431, 433,450
Reiser, C. 34
Resler, E.L.Jr. 376
Rhodes, C.K. 139,272, 274,290,291,292,293, 294,449
Rice, O.K. 184
Rice, S.A. 86,89,108, 110,111,160,161,162, 174,384,418,428
Rich, J.W. 156,160,191
Richardson, A.W. 380, 381,383
Riely, M.E. 312
Riley, S.J. 41
Rinck, H. 454
Robin, M.B. 432,433
Robinson, C.P. 455
Robinson, D.W. 127,294
Robinson, P.J. 99,100, 107
Rockwood, S.D. 155
Rokni, M. 277
Ronn, A.M. 429
Root, J.W. 351
Rosen, D.I. 377
Ross, J. 53,69,99,189
Rossi, M. 415,416
Rothe, K.W. 358,359, 458
Rotheny, R.W. 300
Rowland, F.S. 351
Rubenstein, M. 376
Rubin, R.J. 300
Rubinson, M. 115
Ryabenko, A.G. 339
Rynbrandt, J.D. 104,105

Sabety-Dzvonik, M.J. 383
Sacket, P.B.Jr. 167
Sargent, M.III,192,199, 206,222,457
Schäfer, F.P. 28,268, 279,280
Schafer, H.F.III,92
Schaibly, J.M. 327
Schatz, G.C. 99
Schawlow, A.L. 444,450, 451
Schieve, W.C. 110,111
Schlag, E.W. 431,432
Schlosser, D.W. 438
Schmailzl, U. 76,155, 331,339
Schmatjko, K.J. 55,123
Schmid, W.E. 287,427, 456
Schmiedl, R. 441,442
Schnepp, O.O. 445
Schreiber, J.L. 25,46, 89,93,96,122
Schulten, E. 396,397

Schultz, A. 440, 442
Schultz, M.J. 452
Schulz, P.A. 36,83, 401,411,419,421, 422
Schumacher, E. 437
Schurr, V. 383
Scully, M.O. 192,199, 206,222,457
Seery, D.J. 134
Sentman, L.H. 150,151, 339
Sethuraman, V. 264
Setser, D.W. 71,76,77, 80,81,294
Shamah, I. 160,161,162
Shannon, C.E. 58
Shapiro, M. 370
Shavitt, I. 376
Shelepin, L.A. 46
Shen, Y.R. 268,401, 411,419,421,422
Shimanouchi, T. 251
Shimoni, Y. 150,151, 166,167
Shirley, J.A. 324,326
Shortridge, R.G. 70,74, 79
Shuler, K.E. 53,69,144, 145,146,148,149,158, 160,164,168,170,189, 327
Siebrand, W. 141,143
Siegbahn, P. 371
Siegman, A.E. 9,192, 203,218,222,271
Sileo, R.N. 324
Silver, J.A. 78
Sirkin, E. 127,294
Skribanovitch, N. 294
Slater, D.H. 71
Slater, J.C. 262
Slater, R.C. 455
Sloan, J.J. 23,24,47, 49,54,122,366
Slutzky, M.S. 457
Smalley, R.E. 437
Smith, D.D. 112
Smith, F.G. 300
Smith, F.T. 137
Smith, I.W.M. 18,71,89, 96,356
Smith, J.H. 127,294
Smith, P.W. 281
Smith, S.D. 28,287, 427,441,442
Sodeau, J.R. 389
Solarz, R.W. 369
Spencer, D.J. 320,321
Stannard, P. 111
Steele, R.V. 134
Stein, L. 435
Steinfeld, J.I. 30,34, 79,103,106,115,129, 131,155,171,192,235, 238,290,446,448
Stephens, R.R. 11
Stephenson, J.C. 11, 379,414,415,416,417, 418,422,455
Stevens, R.M. 376
Stitch, M.L. 401
Stokseth, P. 359,441, 445
Stolte, S. 366
Stone, J. 413,416,417, 423,455
Storz, R.H. 432
St. Pierre, L.E. 452
Struve, W.S. 79
Stuke, M. 28
Stupechenko, E.V. 46
Suchard, S.N. 290,315, 319
Sudbø, Aa.S. 36,83,401, 411,419,421,422
Sung, J.P. 294
Svelto, O. 192,222
Szöke, A. 372
Tablas, F.M.G. 287,427
Tabor, M. 155,171
Tal'rose, V.L. 339
Tanaka, T. 448
Taran, J.-P. 441,445
Tardy, D.C. 129
Taylor, R.L. 339
Teitelbaum, H. 132
Thiele, E. 423
Thompson, D.L. 89,96
Thrush, B.A. 376
Tiee, J.J. 408
Timlin, D. 71
Tittel, F.K. 281
Tolles, W.M. 441,445
Tönnissen, A. 358
Top, Z.H. 339
Topp, M.R. 398
Toschek, P.E. 444,450, 451
Treanor, C.E. 156,160, 191
Tribus, M. 58,62,63,64, 65,86,108,186
Troe, J. 107,189
Truhlar, D.G. 89,96
Tsang, W. 425,426,427, 428,455
Tsuchiya, S. 156
Tully, J.C. 79,138
Turner, C. 306,307
Turner, J.S. 110,111

Umstead, M.E. 74,384

van der Werf, R. 394, 395,396,397
Van Vitert, L.G. 235
Vasil'ev, G.K. 339
Verter, M.R. 155
Villani, S. 455
Volk, R. 312

Walker, C.T. 457
Walker, J.A. 425,428, 455
Wallis, R.F. 300
Walsh, A.D. 262
Walther, H. 358,359, 435,458
Wanner, J. 47, 150, 151,323,348,349, 358,435
Wasserman, H.H. 457
Wayne, R.P. 376
Weiss, G.H. 144,145, 148,149,164,168
Weisshar, J.C. 386, 387,388
Weitz, E. 160,161,162, 252,454
Welge, K.H. 441,442
Wellhausen, U. 371
West, G.A. 82,107
Weston, R.E.Jr. 82, 107
Wharton, L. 437
Wherret, B.S. 201
Whitten, G.Z. 184
Whittier, J.S. 336,337, 344
Wickers, W.H.F. 381
Wiener, N. 58
Wilkins, R.L. 21
Williams, R.L. 351
Wilson, C.W.Jr. 376
Wilson, E.B.Jr. 192, 235,238,248
Wilson, J. 339
Wilson, J.R. 139,140
Wilson, K.R. 41,381, 382,383,438,440
Wilson, L.E. 290
Wilson, W.L. 281
Winterfield, C. 415
Wiswall, C.E. 342
Witkowski, S. 312
Witte, K.J. 310,312, 313,314
Wittig, C. 82,107,139, 408,420,441,442
Woeste, L. 437
Wolfrum, J. 23,24,55, 91,123,371,372,373, 375,453
Wolga, G.J. 156
Woodall, K.B. 11,12,18, 23,47,49,71,76,77, 150,151
Woodin, R.L. 416,417
Woodward, R.B. 1
Wright, M.D. 139,140
Wyatt, R.E. 78

Xuan, C.N. 376,378

Yablonovitch, E. 408, 415,452
Yahav, G. 418,428
Yang, S.C. 4
Yariv, A. 192,199,203, 206,218,221,222,443
Yeung, E.S. 385
Yogev, A. 427,428
Young, J.F. 139,140
Yu, M.H. 441,442
Yuan, J.M. 139,141,457
Yukov, E.A. 311

Zacharias, H. 441,442
Zamir, E. 81,420
Zandee, L. 43,431,432, 433
Zare, R.N. 2,26,27,28, 46,86,367,369,370, 396,398,416,429,440, 442
Zeiri, Y. 370
Zewail, A.H. 2,89,110, 111,112,432,452,453, 455
Zharov, V.F. 318
Zimmermann, I.H. 139, 141,457
Zittel, P.F. 46,106,134
Zuev, V.S. 234,308, 312
Zych, L.J. 445

Subject Index

Absorption cross section 198

Activation energy 21,47

Adiabatic limit for V-T transfer 132

Adiabaticity parameter 124ff.

Alignment of reagents 26,366

Allowed transitions 9,233

Alkyl iodides, photodissociation of 41

Allyl isocyanide, photoisomerization of 31,32

Allyl isocyanide spectrum 30

Amplification condition 8,214,295

Anharmonicity 17,25,113,154,242, 259,401

  correction term 17,154,242

  in symmetrical tops 259

Anti-Stokes scattering 444

Applications of chemical lasers 347

Arrhenius plot 21,47

Atomic spectra 232ff.

Attractive energy release 96

Autoionization 42

BaF, from Ba+HF 367ff.

BaO, formation from $Ba+N_2O$ 6

BaO, LiF detection of 442

Background absorption 260

BEBO method 93

Benzene, multiphoton ionization 43,433

Benzene, CH overtone 112

Biacetyl absorption spectrum 391

Biacetyl energy level diagram 391

Biacetyl fluorescence and phosphorescence 393

Biexponential decay 396

Black holes in radiationless transitions theory 395,430

Bond index 93

Bond order 93

Bond selective chemistry 32

Bond-tightening model 94,107

Born-Oppenheimer approximation 37, 235ff.

Bose-Einstein statistics 217

Bottleneck effect in multiphoton excitation 33,404

Br, reaction rate with HCl 23,46

Branching ratios 79,87

Brightness 223,224

Broadening mechanism 201

CARS (Coherent Anti-Stokes Raman Spectroscopy) 441

  of CO in a flame 445

  phase matching condition 445

  spectrum of D_2 446

Canon 51,54,56,57,154,188

Canonical invariance 148

Carbonyl compounds, ultraviolet spectrum of 261

Carbonyl group, molecular orbitals of 262

Cavity decay lifetime 214
Cavity modes 216,217,218
CD^+ from $C^+ + D_2$ 84
Chapman-Enskog expansion 191
Chemical activation 11,82,104
Chemical laser 11,289ff.
  apparatus 13,304,317,322,341, 350
  application 347
  CO 324ff.
  cw 340
  efficiency 318
  HF 323
  HX 314ff.
  kinetics 326ff.
  lasing conditions 295,303
  mixing model 343
  operation 304
  output characteristics 332
  pumping 289,330
  rotational equilibrium 334ff.
  rotational nonequilibrium 337ff.
  , temperature rise in 333
  triatomic 293
  types 11,292
Chemical transfer laser 293,294, 320
Chemi-ionization collisions 135
Chemiluminescence 6,417ff.
CH_3F, energy level diagram of 252
CH_3F normal modes 251
CH_3F vibrational relaxation 161
Chromophore 255
CH stretching vibrations 252
CH overtone 30,112
Classical trajectories 90ff.
ClF laser 278
Clusters, in the infrared spectrum of spherical tops 260
C_2 in multiphoton dissociation 442
CN, rotational excitation 384
CO, appearance time in H_2CO photolysis 388
CO chemical laser 324
CO laser 342
Coherent absorption 400
Coherent anti-Stokes Raman spectroscopy 441ff.
Coherent interaction 205
Collision dynamics, kinematic models 97
Collision frequency 15
  number 17,19
  rate constant 15
Collisional deactivation 25,106
Combination principle 232
Combination transitions 258
Computational studies of reaction dynamics 89ff.
Concerted reaction 93
Condon approximation 239
Cooling by supersonic beams 29
Constraints 63ff.,85,102,108,124, 157,164,166,185
Constraints for chemical equilibrium 64
Continuous wave HX lasers 320
Coriolis splitting 30,260,404
CO vibrational distribution
  from acetaldehyde photodissociation 74
  from $(CH_3CHCO)^+$ 74
  from CO_2 photodissociation 4
  from O+CN(v) 55,91
  from O+CS 71
  in flame 156
CO, vibrational relaxation 155
CO_2, energy level diagram of 250
CO_2 infrared spectrum 258
CO_2 photodissociation 4,5
CO_2 laser 284ff.
  applications 287
  properties 286

pumping 285
rate constants 286
spectroscopy and kinetics 284
temporal evolution 287
tuning range 284

Cross section for stimulated emission 198ff.,300

Crossed molecular beams 365

CS, vibrational distribution from the photodissociation of CS_2 4,83

CS_2 photodissociation 4

Curve crossing 41,79,92,136,142, 237
in excimer laser 275
laser assisted 141

cw chemical laser 323,340ff.
efficiency 345
model 341

cw lasers 269,320,341

cw operation 320

Cyclopropane absorption spectrum 405

Cyclopropane, multiphoton excitation of 37,423

Degenerate vibrations 248

Density matrix 110

Density of translational states 178

Density of states 66,69,182

Density of states in RRK theory 99

Dephasing 110,143,203
broadening 208
collisions 203
time 203

Detailed balance 8,16,23,24,51,55, 114

Detailed rate constant 20,49ff,57, 114,363

Detailed rate constant, prior 113ff.

Detuning frequency 194

Diatomic molecules, electronic spectra of 238ff.

Diatomic molecules, infrared spectrum of 242ff.

Dielectric breakdown 405

Diffraction angle 222

Dilution factor 429

Dipole derivative 242

Dipole moment 9

Direct reactions 90

Direct reactive collisions, structure discrimination relations in 96

Dissociation rate constant 409
RRK theory 100,104,189
,RRKM theory 100

Dissociative continuum 40

Distribution of maximal entropy 62ff.,185ff.

Doorway states 30,407

Doppler
broadening 205
effect 202
profile 449
sub-Doppler spectroscopy 449
temperature 436
width 204

Double resonance 357
experiment 35,448ff.
glyoxal 398

Dressed states 141

Duty factor 436

Dye laser 278,281
applications 283
pumping 281,282
spectroscopy and kinetics 279
tuning range 282

Dye molecule, energy level diagram 279

Dynamical factor 69

Effective cross section 15
Effective number of oscillators 100
Einstein coefficients 7,196
Electronic energy transfer 134ff.
by curve crossing 137
intramolecular 37ff.,141ff.
Electronic excitation 4,6,27,41, 79,91,134,139,369
prior distribution 79,180
polyatomic molecules 380
Endoergic reaction 46
Endoergic reaction, role of reagent vibrational excitation 48
Energy consumption 53ff.
Energy disposal 4,53ff.
Energy equipartitioning 100
Energy localization, by V-V processes 163
Energy redistribution and multiphoton absorption 413
Energy release - attractive 96
mixed 369
repulsive 96
Energy-transfer rates 17ff.
Entropy 59ff.,62ff.,84,169ff.,173, 177
Entropy deficiency 61,64,169,186
Entropy and laser radiation 173
Equilibrium constant 52
E → V processes 139
E-V transfer 37,139,141
Excimer 272
Exciplex 272
Exciplex laser 272ff.
kinetics 276
spectroscopy 274
Exoergicity 12
Exponential gap 24,108,119ff.,122, 124,128,139,142,150,152
bridge using lasers 139
Exponential gap law in vibrational and rotational relaxation 119, 127,152
Exponential gap law in predissociation 108
Far field diffraction angle 222
F atoms, reaction with D_2 3,77
$F+H_2 \rightarrow HF+H$ reaction
lasing action 14,296
rotational state distribution 18,76
$F+H_2$ reaction, detailed rate constants 71,76
F, reaction
with HCl 25,46,71
with CH_3X 71
with polyatomics 73,80,81
Fermi resonance 249
Fermi's golden rule 69,193
Flux density 68
Forbidden transition 9,194,233, 261,263
Formaldehyde
absorption spectra 387
electronic energy levels 264
excited states lifetime 386
matrix isolation photochemistry 389
normal modes of 263,264
photochemistry 385
predissociation 388
spectroscopy 265
zero pressure lifetimes 388
Franck-Condon factor 239,369
Franck-Condon, principle 17,37,70, 95,239
Free energy 170
Funnelling in chemical lasers 303
Gain coefficient 209,213,214,219, 301,351
Gain-equal-loss approximation 228, 335

Gain profile 219,220

Gain probing 354

Gain saturation 219,220

Gain, total 211

Gas kinetic cross section 15,20

Gaussian beam 223

Gaussian line shape 204

Golden rule 70,193

Good quantum numbers 233

Grating selection method 352,354

Glyoxal : fluorescence decay 397

Glyoxal : zero pressure lifetime 396

Half collision 40,381

Hammond's principle 94

Harpoon mechanism 136

HBr, infrared spectrum 246

HCl+Br → Cl+HBr laser 336

HCl+Cl reaction, isotopic substitution 373

HCl+O → OH+Cl reaction 371ff.

HCl laser 339,373,374

HCl laser pulse pattern 297

HCl, reaction rate with halogen atoms 23,25,46

HCl, vibrational distribution from photoelimination of chlorethylene 81

HCl, vibrational levels of 22

HCl(v), reaction of with halogen atoms 46,374

HCl(v), reaction with K atoms 365

HCO absorption spectrum 450

H_2+D_2 → 2HD reaction 375ff.

Hexafluorocyclobutene multiphoton isomerization 428

H_2(v)+D reaction 371

HF laser 358,371

high power 319

kinetics 329

laser action on rotational transitions 127

pulsed 318

reactions 315

HF, rotational state distribution in F+H_2 reaction 18

HF+Sr, electronic excitation of Sr 369

HF, vibrational distribution

from multiphoton dissociation of CH_2CF_2 420

from the dissociation of $(CH_3CF_3)^+$ 82

from F+CH_2O 80

from F+polyatomics 73,81

HF, vibrational relaxation 155

HF(v), alignment of 26

HF(v), reaction with Br and Cl 54

HF(v), reaction with Ca,Sr,Ba atoms 26,367

HF(v), temperature dependence of reaction rate with H atoms 47

H+H_2, rotational distributions and surprisal 78

H+HBr(v), branching ratio 87

High-power lasers 311,319

iodine 311

HF 321

Highest gain transiton 303

H+NO_2 → OH+NO, rotational distribution 78

Hole burning 33,220,338

Homogeneous broadening 201,220

Hot phosphorescence 394

Hydrogen halide chemical lasers 314

Hyperfine splitting 234

I, energy level diagram 234

I_2^*, reaction with In and Tl 370

ICl, isotopic shift in electronic spectrum 27

ICl, photoselective reaction with acetylene 28

ICN
- absorption spectrum 381
- energy level diagram 380
- photodissociation 380ff.
- photofragment spectrum 382,440

Image hole 221
Induced transparency 33
Inelastic collisions 14ff.
Information theory 58
Infrared active vibrations 257
Infrared-infrared double resonance 34,448
Inhomogeneous broadening 201,220
Internal conversion 139,142,279ff.
Intersystem crossing 142,280
Intersystem crossing, biacetyl 392ff.
Intracavity absorption 450
Intramolecular coupling 403
Intramolecular energy transfer 31, 35,37,82,101,106,109,141,412,421
Intramolecular E-V transfer 38,141
Intramolecular relaxation 31,37, 38,101,105,109ff.,141,159,426
Inverse electronic relaxation 40, 429
Inversion density 214
- dye laser 215
- HF laser 215

Inversion level 209
Inversion, minimal 214
Iodine laser 304ff.
- chemical 294
- design 310
- efficiency 313
- Einstein coefficients 308
- high power 313
- modelling 306
- transitions 309

Iodine photochemical laser 305
Iodine photodissociation laser 289
Ionization, benzene 432
Ionization by multiphoton absorption 42
Irregular region of the spectrum 111
Isotope separation 28,283,374,401, 413
- of chlorine 28,375
- in formaldehyde 385
- preparative scale 456

Isotopic shift 27,28,414

J shifting 303
Johnson-Parr relation 93

Kinetic isotope effects 348
KrF laser 276ff.
KrF laser, energy levels 275

Lagrange parameters 63,185
Lamb dip 221
Landau-Teller model 145
Landau-Teller plots 132
Landau-Zener model 138
LaPorte's rule 233
Laser-assisted energy transfer 140
Laser
- activation of reagents 365ff.
- activation of polyatomic molecules 405
- amplifier 209
- average power 268
- beam divergence 223
- beam, hot spots 271
- beam quality 223,271
- beam waist 223,271
- cavity 13,26
- chemical 289ff.
- ClF 278
- CO 342
- CO_2 284ff.

conversion efficiency 271
cw 269
diagnostics 283
diode 268
dye 215,278,281
excimer 272ff.
exciplex 272ff.
fluence 411
halogen 273
high power 311,319
HCl 289,339,372
iodine 41,233,241,305
KrF 273
loss mechanism 212
Nd-YAG 235,270,381,383
OH 294
oscillators 211
output patterns 296
parasitic light absorption 275
pulse duration, Q switched laser 232
pulsed 269,318
radiation as a form of work 172
rate gas halide 273
rate equations 225
rhodamine 6G 280
sources 266,267
supersonic 321
thermodynamics 172
wavelength range 267,270

Laser chemistry industrial applications 452

Laser-assisted energy transfer 140

Laser-induced fluorescence 5,27, 368,384,421ff.,440ff.
experiment 441
of ICN photolysis 383

Laser-induced transfer 139

Laser-initiated explosion 453

Laser-initiated isomerization 453

Laser-initiated polymerization 453

Laser-molecular beam experiment 435

Lasing conditions 176,211ff.

LEPS method 93

Level degeneracy 10

Level density 109

Lifetime broadening mechanisms 202

Lifetime for photon emission 9

Light amplification 3,209

Line shape function 196,197,201

Local modes 31,111,142,251

Localized excitation 29

Localized orbitals, of a C=O bond 262

Loose oscillator 94

Lorentzian line shape 203

Loss mechanisms 212

Loss per round trip 213

Luminescence condition 8

Mass ratio 98

Master equation 144

Maximal work 171

Mean free path 15

Measures of specificity and selectivity 58ff.

Microcanonical distribution 62

Microcanonical ensemble 188

Microscopic and macroscopic descriptions 165

Microscopic reversibility 53
See also Detailed balance

Microwave-optical double-resonance 448

Mixed energy release 369

Mixing of vibrational states 111

Mode-selective chemistry 31,37, 399,421,423

Mode selectivity 399

Mode temperature 161

Modes in a resonant cavity 216
Modes, active 219
Modes density 217
Molecular beams 365ff.,435ff.
Moments of distribution 164
Momentum transfer measurement 449
Multi-mode operation 221
Multiphoton absorption
- bottleneck effect 404
- in cyclopropane 405
- mechanism of 33,414
  - rotational compensation 404

Multiphoton activation 398,427
Multiphoton dissociation 284,398
- CCl_2F_2 412
- C_2F_5Cl 36
- CF_2Cl_2 415,421,427
- CF_2HCl 418,421
- CF_3I 419,424
- C_2H_3CH 442
- $[(C_2H_5)_2O]_2H^+$ dimer 417
- collisional effects 424
- cw laser 417
- cyclopropanes 37,423
- energy disposal in products 419
- ethylacetate 425
- 1 difluoro ethylene 420
- incubation period 418
- inert gas addition 420,423
- ions 416
- intensity effects 418
- isotope separation 457
- isotopic selectivity 401,413, 424
- mode selective 421ff.
- molecular beam studies 419
- nonthermal 412,425
- power and fluence dependence 414
- rate equation 411
- saturation 414
- scambling 426
- selectivity 101,401
- SF_6 32,412ff.
- tetramethyldioxetane 418,428
- thermal 101,410
- three-region model 408
- threshold 414
- translational energy distribution 421
- UF_6 408
- ultrashort pulses 415

Multiphoton excitation 82,401
- bottleneck effect 404
- of a mixture 425

Multiphoton ionization 43,430
- acetaldehyde 433
- benzene 43,433
- nonresonant 434
- resonance enhanced 431

Multiphoton isomerization 427

Natural lifetime 199
Natural linewidth 202
Nd : YAG laser 381
Nd^{3+}, energy level diagram 235
Negative temperatures 173
Noise equivalent power 447
$NO + O_3$ reaction 376ff.
Normal coordinates 247
Normal mode analysis 248
Normal modes 249
$NO(v)+O_3$ reaction 376ff.
$n \rightarrow \pi^*$ transition 262,263
$n \rightarrow \sigma^*$ transition 263
Number density 68
Number of product states 66

O + CN reaction 55,91
O_2, energy level diagram 240

O_3(v)+NO reaction 376

OH laser 294

Optical cavity 212

Optoacoustic detection 31,450

Optoacoustic spectroscopy 452

Orientation effect in photochemical reactions 26,370

Oscillator strength 9,262

Oscillator-amplifier experiment 355

Output characteristics, chemical lasers 332

Output coupling 212

Output patterns 302

Overtone absorption 9,30,112,244, 245,258,374

Overtone spectroscopy 112

Oxygen spectroscopy 240

Ozone, vibrational frequencies 377

π pulse 207

Pt energy transfer rate 15,19

P and R branch inversion 176,299

Parallel bands, spectrum of 259

Parasitic oscillations 312

Partial population inversion 176, 300

Partition function 18,63,246

Patel gain equation 301

Passive optical cavity 213

Pauling relation 93

Pentacene, supersonic jet spectrum of 38

Perpendicular bands, symmetric top 260

Photodissociation 4,11,40,74,380, 438

- chemical lasers 293
- ICN 380

Photoelimination 11,83

Photoelimination lasers 291

Photofragment spectrometer 438ff.

Photofragmentation 382ff.

Photofragmentation spectroscopy 438ff.

Photoionization 42

Photon lifetime, in laser cavity 214

Photopredissociation : formaldehyde 385ff.

Photoselective chemistry 21

Planck equilibrium density of radiation 8,218

Point of no return 94

Polarized light 26

Polyatomic molecule

- electronic energy level diagram 256
- energy distribution by collisions 129
- infrared spectrum of 257
- vibrational and rotational energies of 247

Population inversion 3,10,157,172, 209ff.,298

Population inversion, partial 176, 300

Potential energy surface 92ff.,236

Power broadening 33,34,206,402ff.

Practical photoselective chemistry 454

Practical surprisal analysis 185

Predissociation 37,40,241,263,385

Pressure broadening 309,426

Prior detailed rate constant 113, 185

Prior distribution 60,64,65,69, 113,177ff.,187

- electronic 180
- molecule-molecule collision 181ff.
- number density 69
- polyatomics 73,130,183ff.
- rotational 68,179
- vibrational 67,76,116,118,178, 180

Probability matrix 55ff.

Pulse patterns 339

Pulse patterns, F+H_2 laser 14,296, 302

Pulsed lasers 269,318

Pulsed lasers, HF 315

Pump and probe experiment 33,398

Pumping-dumping model 405,413

Pumping rate 226

Q branch 244

Q switching 229

Quasicontinuum 29,30,31,32,35,39, 40,111,402,406

Quasicontinuum onset 402

Quasiequilibrium 153

Rabi frequency 34,194,200,206ff.

Radiationless transitions 37,70, 110,141,143,280

- biacetyl 393
- glyoxal 396
- pentacene 39

Rare gas halides, energy level diagram of 274

Rate constant

- detailed 20,364
- for stimulated emission 7,226
- total 20,364

Rayleigh range 224,271

Reactive reactants 118

Reagent translation 87

Recurrence time 106,109

Red shift in multiphoton dissociation 414

Reduced energy variables, f_X 67ff.

Reduced energy variables, g_R 76

Reduced mass 15

Relaxation terms, in laser rate equations 226

Repulsive energy release 96

Repulsive potentials 237

Resonance enhanced transition 400, 431

Resonance function 139

Rigid rotor approximation 18,180, 243

Role of reagent excitation 46ff., 58,96,367

- electronic 27,369,370
- prior 113ff.
- rotation 366
- translation 87ff.
- vibration 23ff.,46ff.,54,55,87, 91,96ff.,117,119,122,136,365ff., 371ff.,377

Rotational band contours 31

Rotational compensation in multiphoton absorption 403

Rotational degrees of freedom 29

Rotational disequilibrium 337,346

Rotational effects on fluorescence efficiency 387

Rotational energy transfer 14,19, 35,120,150

Rotational equilibrium in chemical lasers 18,174

Rotational lasing 244

Rotational relaxation 150

Rotational relaxation time 19,151, 153

Rotational state distribution 18, 74ff.,121,127

- H_2 79
- HCl 77
- HD 121
- HF 18,76, 127
- NO 78

Rotational surprisal plots 78,121

R,P branches 244

RRHO approximation 67,180,243,248

RRHO level scheme 16,243

RRK approximation 101

RRK theory 99,189

RRKM theory 100

R-T relaxation 19,150

Russel-Saunders coupling case 233

Saturable absorber 230

Scattering matrix 189

Schawlow-Townes threshold condition 214

Scrambling 29

Secondary dissociation processes 36

Selection rules 233

for electronic transitions 241

parallel bands 258

perpendicular bands 259

Selective energy consumption 49,58

Selective reagent excitation 48, 58,96

Selective vibrational excitation 23,46,55,87,91,96,117,119

Self-induced transparency 207

Sequential absorption 400

SF_6 absorption spectra 30,409

SF_6 absorption spectra, in the presence of He 427

SF_6 high resolution spectrum of ν_3 mode 30,261

infrared double resonance spectrum of 34

multiphoton induced dissociation 32

normal modes 252

Signal-to-noise ratio in fluorescence and in absorption 446

Signal-to-noise ratio in LIF experiments 367

Single-line operation 303,335

Single-mode operation 221

Singlet states 256,266,279,385ff., 391ff.

Slater model 107

Small signal gain coefficient 210

Snarled trajectories 90

Species selectivity 33,399

Specificity of energy disposal 12, 23,49,58,73

Spherical top 255,260

Spin exchange 357

Spin-orbit splitting 234

Spontaneous emission 7,9,199

SSH model 155

Stability diagram 222

State selective chemistry 457

State-to-state rate constants 55ff.,113,365

State-to-state rate constants, prior 113ff.

State-to-state reaction probabilities 55ff.,369

Stationary wave patterns 218

Steric effect 26,366,370

Stimulated emission 7,9,196

Stimulated radiation as a bimolecular collision 198

Strong collisions 106

Sudden collision 124

Sum rule 164

Supersonic expansion 38

Supersonic HX lasers 321

Supersonic molecular beams 38,437

Surface crossing 41,79,92,137,143, 237

Surface hopping 92

Surprisal 62ff.,68ff.,83ff.,185, 187

Surprisal analysis 65ff.,83ff., 119ff.,185ff.

dynamical significance 69,83ff.

rotational energy disposal 78

rotational energy transfer 121, 151

vibrational energy consumption 87,122

vibrational energy disposal 71,72,74,81-84,90,420

vibrational energy transfer 119,121,125,155

Surprisal synthesis 119,125,155, 165ff.,185

Symmetric top 258

Symmetric top, energy levels of 254

TEA HF laser 317

TEA lasers 286,316,418,422,425

See also High power lasers

TEM transverse mode 219

Term numbers 232

Term symbols 240

Thermal reaction rate, k(T) 50

Three-photon ionization-hydrogen atoms 432

Three-region model of multiphoton dissociation 407

Threshold

condition 214,298,332

inversion density 228

pumping rate 228

time 316,350,352

Time-of-flight spectra in photo-fragmentation of ICN 440

Time-resolved gain spectroscopy 355

Time to threshold 316

Total collision rate 20

Total rate constant 364

Trajectory computations 108

Transition dipole 9,194

Transition linewidth 218

Transition state theory 100,189

Translational energy defect 17

Translational energy release, in multiphoton dissociation 422ff.

Translational-rotational density of states 183

Transport phenomena 15

Transverse electromagnetic mode 212,222

Treanor distribution 156,190,326

Treanor distribution, V-V pumping 326

Triplet states 256,266,279ff., 385ff.,391ff.

Triplet-triplet annihilation 391

Tunability 267

of CO_2 lasers 284,285

of dye lasers 282,283

of molecular infrared gas lasers 270

Two-level system 193

Two-photon absorption 399

applications 401

cross section 400

UF_6 multiphoton dissociation 407

Ultrahigh purification 455

Unimolecular dissociation rate 99, 189,409,419,425

Unimolecular process 31,99,106

Unimolecular process, selectivity and specificity 106,109

Useful output coupling 228

van der Waals molecules 108

Velocity group selection, by sub-Doppler spectroscopy 449

Vertical transition 239

Vibrational degrees of freedom 29

Vibrational density of states 184

Vibrational distribution

of BaO from $Ba+N_2O$ 6

of BaO from $Ba+O_2$ 442

of CD^+ from $C^+ + D_2$ 84

of CF_2 produced by multiphoton dissociation 422

of CO from H_2CO photolysis 390

of CO from O+CN 54

of CO from photodissociation of CO_2 4

of CO from $(CH_3CHCO)^+$ 74

of CS from photodissociation of CS_2 4,5,83

of DF from $F+D_2$ 3
of HCl from Cl+HX 12,49,71
of HCl from photoelimination 81
in HF+Br reactions 368
of HF from $(CH_3CF_3)^{\dagger}$ 82
of HF from $F+CH_3X$ 71
of HF from F+HX 25,71
of HF from F + polyatomics 80, 81
of HF from multiphoton dissociation of CF_2CH_2 420
of SrF from Sr+HF 26
Vibrational energy transfer 126
Vibrational excitation 4
in endoergic reactions 54,96, 117,119,122
in exoergic reactions 54,96, 117,119,122
by the stimulated Raman effect 375
Vibrational quasicontinuum 38,407
Vibrational relaxation 17,48, 129ff.,131ff.,144,156,163
matrix 147
temperature dependence of 131ff.
time 19,48,132,146,153,167
Vibrational surprisal 70ff.
Vibrational temperature 149,160, 422,424
Vibrotational distribution for Cl+HI 77
Vibrotational distribution for $F+D_2$ 77
Vib-rotational levels of HCl 22
Voigt profile 205
VR approximation 67,180
V-R, T transfer 20,25,128,134
V-T relaxation 48,156,167
of anharmonic oscillators 154ff.,163ff.,167
of harmonic oscillators 144ff.
V-T relaxation time 19,48
V-T relaxation time, temperature dependence 131
V-T transfer table 133
V-V transfer 17,120,154,156,326, 373,374,424
intermode 159
intramolecular 31,35,37,82,101 105,109
pathways 159
in CH_3F 161
in CO 18,121,155
in HF 121,127,155
V-V up-pumping 154,326
V-V up-pumping, Treanor distribution 156,190,326

XeF laser kinetics 276
XeI laser 276

Zero-gain temperature 353

High-Power Lasers and Applications

Proceedings of the Fourth Colloquium on Electronic Transition Lasers in Munich, June 20–22, 1977
Editors: K.-L. Kompa, H. Walther

2nd printing. 1979. 142 figures, 27 tables. IX, 228 pages
(Springer Series in Optical Sciences, Volume 9)
ISBN 3-540-08641-2

Contents:
Excimer Lasers. – Chemical Lasers. – Other Laser Systems. – Frequency Conversion. – Applications.

V. S. Letokhov, V. P. Chebotayev

Nonlinear Laser Spectroscopy

1977. 193 figures, 22 tables. XVI, 466 pages
(Springer Series in Optical Sciences, Volume 4)
ISBN 3-540-08044-9

Contents:
Introduction. – Elements of the Theory of Resonant Interaction of a Laser Field and Gas. – Narrow Saturation Resonances On Doppler-Broadened Transition. – Narrow Resonances of Two-Photon Transitions Without Doppler Broadening. – Nonlinear Resonances on Coupled Doppler-Broadened Transitions. – Narrow Nonlinear Resonances in Spectroscopy. – Nonlinear Atomic Laser Spectroscopy. – Nonlinear Molecular Laser Spectroscopy. – Nonlinear Narrow Resonances in Quantum Electronics. – Narrow Nonlinear Resonances in Experimental Physics.

Raman Spectroscopy

of Gases and Liquids

Editor: A. Weber
1979. 103 figures, 25 tables. XI, 318 pages
(Topics in Current Physics, Volume 11)
ISBN 3-540-09036-3

Contents:
A. Weber: Introduction. – *S. Brodersen:* High-Resolution Rotation-Vibrational Raman Spectroscopy. – *A. Weber:* High-Resolution Rotational Raman Spectra of Gases. – *H. W. Schrötter, H. W. Klöckner:* Raman Scattering Cross Sections in Gases and Liquids. – *R. P. Srivastava, H. R. Zaidi:* Intermolecular Forces Revealed by Raman Scattering. – *D. L. Rousseau, J. M. Friedman, P. F. Williams:* The Resonance Raman Effect. – *J. W. Nibler, G. V. Knighten:* Coherent Anti-Stokes Raman Spectroscopy.

Springer-Verlag
Berlin
Heidelberg
New York